EMO

MARTIN MARIETT
U. S

AF683894

Nonlinear Dynamics and Turbulence

Interaction of Mechanics and Mathematics Series

EDITORIAL BOARD

A Jeffrey (Secretary), University of Newcastle-upon-Tyne
G I Barenblatt, Institute of Oceanology, Moscow
G Fichera, Università di Roma
Fritz John, Courant Institute of Mathematics and Sciences
D D Joseph, University of Minnesota
W T Koiter, Delft University of Technology
R J Knops, Heriot-Watt University
I Müller, Technische Universität Berlin
O A Oleinik, University of Moscow
H Zorski, Institute of Fundamental Technical Research, Warsaw

Nonlinear Dynamics and Turbulence

Edited by

G I Barenblatt
P P Shirsov Institute of Oceanology
USSR Academy of Sciences

G Iooss
Institut de Mathématiques et Sciences Physiques
Université de Nice

D D Joseph
Department of Aerospace Engineering and Mechanics
University of Minnesota

Pitman Advanced Publishing Program

BOSTON · LONDON · MELBOURNE

PITMAN PUBLISHING LIMITED
128 Long Acre, London WC2E 9AN

PITMAN PUBLISHING INC.
1020 Plain Street, Marshfield, Massachusetts 02050

Associated Companies
Pitman Publishing Pty Ltd, Melbourne
Pitman Publishing New Zealand Ltd, Wellington
Copp Clark Pitman, Toronto

First published 1983

Copyright © by G I Barenblatt, G Iooss and D D Joseph 1983

Library of Congress Cataloging in Publication Data
Main entry under title:

Nonlinear dynamics and turbulence.

Includes bibliographical references.
1. Dynamics. 2. Nonlinear theories. 3. Turbulence.
I. Barenblatt, G. I. II. Iooss, Gerard. III. Joseph, Daniel D.
QA845.N65 1983 530.1′5 83-8078
ISBN 0-273-08560-3

British Library Cataloguing in Publication Data

Nonlinear dynamics and turbulence.—(Interaction of mechanics and mathematics)
1. Turbulence 2. Mathematics
I. Barenblatt II. Iooss, G
III. Joseph, D. D. IV. Series
532′.0527′0151510 QA911

ISBN 0-273-08560-3

All rights reserved. No part of this publication may be reproduced, stored in a retrieval system, or transmitted in any form or by any means, electronic, mechanical, photocopying, recording and/or otherwise, without the prior written permission of the publishers. This book may not be lent, resold or hired out or otherwise disposed of by way of trade in any form of binding or cover other than that in which it is published, without the prior consent of the publishers.

Printed in Northern Ireland at The Universities Press (Belfast) Ltd.

Foreword

The International Society for the Interaction of Mechanics and Mathematics (ISIMM) was founded in 1977. Its purpose is to promote cooperative research involving the fields of mechanics and pure mathematics.

Its Executive Committee decided that, from time to time, scholarly works relevant to the Society's interests should, by invitation, be published under its auspices. The present volume is one in this series which, it is hoped, will help to advance the objective of the Society.

The Editorial Board

Edward Oblow
2132 Berrywood Dr
Knoxville TN 37932

Contents

List of Contributors

V S Afraimovich Research Institute of Applied Mathematics and Cybernetics, N I Lobachevsky Gorky State University
D G Aronson School of Mathematics, University of Minnesota
G I Barenblatt P P Shirsov Institute of Oceanology, USSR Academy of Sciences
Yu N Belayev Institute of Mechanics, Moscow State University
L A Bunimovich P P Shirsov Institute of Oceanology, USSR Academy of Sciences
F H Busse Department of Planetary and Space Science, University of California, Los Angeles
A I Chernykh Institute of Automation and Electrometry, Siberian Branch, USSR Academy of Sciences
M A Chory The Analytic Sciences Corporation, Reading, Massachusetts
M J Feigenbaum Physics Department, Cornell University
C Foias Department of Mathematics, Indiana University
J P Gollub Haverford College and the University of Pennsylvania
B D Greenspan Department of Mathematics, Vanderbilt University
G R Hall School of Mathematics, University of Minnesota
P J Holmes Department of Theoretical and Applied Mechanics and Center for Applied Mathematics, Cornell University
W F Langford Department of Mathematics and Statistics, University of Guelph, Ontario
V S L'vov Institute of Automation and Electrometry, Siberian Branch, USSR Academy of Sciences
R P McGehee School of Mathematics, University of Minnesota
R V Plykin Moscow Institute of Physical Engineering
Y Pomeau Service de Physique Théorique, Commissariat à l'Énergie Atomique, France
A A Predtechensky Institute of Automation and Electrometry, Siberian Branch, USSR Academy of Sciences
George R Sell School of Mathematics, University of Minnesota
L P Shil'nikov Research Institute of Applied Mathematics and Cybernetics, N I Lobachevsky Gorky State University
F Takens Mathematisch Instituut, Rijksuniversiteit Groningen
R Temam Mathématique Université de Paris-Sud, Orsay
Y M Treve La Jolla Institute, California
I M Yavorskaya Space Research Institute, USSR Academy of Sciences
G M Zaslavsky L V Kirensky Institute of Physics, Siberian Branch, USSI Academy of Sciences

Preface

Turbulence, narrowly viewed, is a mode of motion of fluids. There are *at least* two modes of motion: a laminar one whose dynamics mimics the symmetries of imposed external conditions, and a turbulent one which is unsteady and chaotic even when the external conditions are perfectly steady and regular. Osborne Reynolds was the first to write down a systematic study of laminar and turbulent flow in pipes. He showed that the transition between laminar and turbulent flows depends on a single dimensionless number, now called the Reynolds number. For Reynolds numbers inferior to a critical one the (laminar) flow is as steady and regular as the imposed data (the pressure drop and boundary conditions), whereas in the other case the laminar flow loses stability to turbulent flow, which is unsteady and irregular even though the imposed data are steady and regular. Since all motions of fluids are governed by equations of motion, the existence of a connection between dynamics and turbulence is apparent. However, we cannot solve the equations in the turbulent case and are unable even to establish some well-known properties of the observed motions in a mathematically satisfying way.

Dynamics is a larger and more general subject than turbulence. It applies to all motions of all material bodies. The mathematical theory of dynamics seeks properties of solutions of the equations of motion. It has been formulated and studied in abstract form as a topic in the branch of mathematics known as dynamical systems. As is frequently the case, the subject was simplified in some sense by looking at it in a more general context. In this context it was learned that deterministic problems, governed only by some few benign nonlinear ordinary differential equations or by iterations of maps in finite-dimensional spaces, could give rise to chaotic turbulent-like behaviour, highly sensitive to initial conditions and unpredictable after long times or many iterations.

The fact that dynamical systems with a few degrees of freedom can possess stochasticity due to strong (exponential repulsivity) sensitivity to initial conditions has been known for many years. E. N. Lorenz, in 1963, was the first to show that a deterministic problem of fluid mechanics could lead to turbulent-like behaviour with chaotic dynamics. Lorenz considered the problem of thermal convection governed by Navier–Stokes-type partial differential (Oberbeck–Boussinesq) equations. He supposed that the solutions of these equations could be expressed by a Galerkin approximation which he truncated drastically. In this way he came to three ordinary differential equations with bilinear nonlinearities.

Lorenz studied the solution trajectories numerically and interpreted the results in a mathematically precise way, using the qualitative theory of differential equations. He showed numerically that solutions were attracted to a certain special set in the three-dimensional phase space. This subset is more complicated than a two-dimensional surface and the trajectories are sensitive to changes in initial conditions: they have a very complicated structure and they give rise to continuous spectra. Sensitivity to initial conditions means that very slight changes in initial conditions give rise eventually to entirely different trajectories. The system has the property of mixing. This means that typical trajectories essentially 'forget' their initial conditions so that autocorrelations computed on functionals of the solutions decay eventually to zero: that is, two events, widely separated, are uncorrelated so that it is impossible to predict one event from another if the intervening time is large. Turbulence in fluids has this property. A continuous frequency spectrum is associated with mixing flows, because a discrete spectrum implies some kind of (almost) periodicity for which the autocorrelation between the same quantity at different times need not decrease to zero as the time lapse between events approaches infinity.

Lorenz's paper attracted attention some years after the appearance in 1972 of a paper by Ruelle and Takens which used the theory of topological dynamics to introduce the idea that turbulence in fluids would appear after a few bifurcations. This notion ran against the then popular theory which had been introduced earlier by Landau (1944) and Hopf (1948). The ideas of Landau and Hopf had a natural evolution. In the earlier days of the study of hydrodynamic problems it was thought that the conditions for the development of turbulence could be ascertained from the theory of stability of laminar flows. This study of stability showed that laminar flows are indeed unstable but that the flow which replaces them is also a laminar one, with a different symmetry pattern in either space or time; for example, a time-periodic motion may replace a steady motion after the loss of stability. It was easy to imagine that the loss of the stability of the periodic motion could lead to a flow with two frequencies, and so on, to n frequencies, as the Reynolds number was increased. So their theory was that, when transition is progressive, at any stage there exists in the phase space of solutions corresponding to the n frequencies an n-dimensional torus which is invariant and attracting.

The ideas of Landau and Hopf do not give a good description of transition to turbulence because quasiperiodic solutions are not mixing and do not exhibit a continuous spectrum, decaying autocorrelation or high sensitivity to initial conditions. Ruelle and Takens noted that it was possible for attracting sets having the stochasticity (mixing, etc.) required for turbulence to arise after a few bifurcations of the Landau type. A small perturbation of a quasiperiodic flow on a three-dimensional torus can lead to a strange attractor. In fact, the structure of bifurcations changes drastically after a few elementary bifurcations of the Landau–Hopf type. For instance, we have already noted that quasiperiodic flows can undergo transition to chaotic behaviour. Another possible transition is by means of a cascade of period-doubling bifurcations of the Feigenbaum type.

Mathematical studies require good definitions on which theories can be built. Turbulence in fluids has never had so precise a mathematical definition, though the general notion of turbulence in dynamical systems can be made precise. It remains, of course, to show that this general notion is actually useful in making theories to explain what we observe as turbulence in fluid flows. Many of the papers in this volume have as their aim the elucidation of the implications of the generalized concept of turbulence and the study of its utility in applications to observed turbulence in fluids. To achieve such an aim it is necessary to combine analytical, computer and experimental studies. The editors have tried here to represent all such approaches.

It will be clear to readers of this volume that the new results coming from the theory of nonlinear dynamics fit very well some simple systems of small dimension—for example, convection in small boxes, flow between rotating cylinders and rotating spheres. The dynamics of such problems are governed by partial differential equations of the Navier–Stokes type but are actually controlled by finite-dimensional ordinary differential equations of a type that can be obtained by appropriate Galerkin approximations. So the dynamics is actually set in a finite-dimensional phase space even though the governing problem is an infinite-dimensional one. When the configuration of the experiments and the Reynolds number are appropriately chosen, it appears that the number of relevant dimensions is small and there is encouraging agreement between the new theories and experiments. The study of the number of finite dimensions necessary to represent the qualitative features of the true dynamics is a new and important one, which is less well understood when the number of finite dimensions is large. The classical cases of turbulence in jets and in pipe and boundary layer flows are among those which presumably require a large (if finite) number of dimensions. It remains to be seen if these classical cases, and other practical problems of turbulence in fluids, can be usefully treated by these new ideas from the theory of nonlinear dynamics.

The editors and authors hope that this collection of papers will help readers to understand what has been achieved, and to achieve more. We also believe that this volume reflects the best efforts of our time, by scientists of various countries, to understand the nonlinear dynamics governing turbulence.

G. I. B.
G. I.
D. D. J.

1 Strange Attractors and Quasiattractors

V S Afraimovich and L P Shil'nikov

1.1 Introduction

The attention of specialists in various fields has frequently been drawn in recent years to certain special dynamical models with a complicated behaviour of trajectories. This relates primarily to an attempt to answer the following fundamental question: can stochastic oscillations arise in systems of markedly pronounced dynamical character? In this case the investigation of hydrodynamic models becomes especially important in order to explain the turbulence phenomenon.

The problem of the onset of turbulence and the description of developed turbulence are basic problems in hydrodynamics. Whereas the approach to the second problem is based on ideas and methods of the probability theory (see Monin and Yaglom [1]), the modern approach to the problem of turbulence onset relies upon the idea that a dynamical system, i.e., the actual Navier–Stokes equations or a finite-dimensional system which is their Galerkin approximation, can serve as a model for fluid motion. Hence the methods of quantitative theory of dynamical systems can be applied to this system. And if this is the case, a question arises: what is the mathematical prototype of turbulent motion? (Let us note that in a similar situation the answer to the question about the mathematical character of self-oscillations was given in the 1930s by A. A. Andronov on the basis of the Poincaré–Lyapunoff theory.)

Attempts to answer this question have been made for decades. For instance, in the 1940s Landau and Hopf suggested that an asymptotically stable torus with a quasiperiodic trajectory can be an adequate mathematical prototype for turbulent motion (at that time limiting sets consisting of almost periodic motions seemed to be the most 'complicated' ones). As for stochastic motions, objects that are the mathematical prototypes for them are more complicated than periodic motions or the motions on a torus with a quasiperiodic trajectory. These are the so-called strange attractors (see, e.g., [2]), in which the stability of the set as a whole is combined with instability of the individual trajectories. Mathematicians are well familiar with hyperbolic strange attractors such as submanifolds with Y-structure [3], Smale–Williams's solenoids [4, 5], R. V. Plykin's solenoids [6], etc. Since the methods of ergodic theory, as applied to hyperbolic sets, have been developed in some depth (see Chapter 5 by Bunimovich), it would seem that an apparatus now exists for the study of systems with a complicated behaviour of trajectories. It is this approach which is used by Ruelle and Takens [7], who propose Smale–Williams's solenoids as a mathematical prototype for turbulent flow.

One has to admit, however, that the possibility of the appearance of hyperbolic attractors in simple models still remains problematic. Among nonstructurally stable strange attractors, those are of primary importance in which the following characteristic properties of hyperbolic sets are present: transitivity, an everywhere density of saddle periodic motions, the existence of a homoclinic curve for each periodic motion (for definitions see below). A strange attractor will be called quasihyperbolic if it possesses the above-mentioned properties and preserves them under small perturbations and if, at the same time, it can be nonstructurally stable. The existence of quasihyperbolic attractors has been proved by Afraimovich *et al.* [8]. Numerical analyses performed by Lorenz [9], Afraimovich *et al.* [8] and others (see, e.g., [10, 11]) show that quasihyperbolic attractors may exist in simple three-dimensional models, and above all in Lorenz's triplet:

$$\dot{x} = -\sigma(x-y), \qquad \dot{y} = -xz + rx - y, \qquad \dot{z} = xy - bz. \tag{1.1}$$

It should be noted that Afraimovich *et al.* have formulated analytical conditions for Poincaré mapping; if these conditions are satisfied, a quasihyperbolic attractor does exist (see below). An appropriate 'transversal', namely the plane $z = r-1$, does exist for system (1.1). However, there is almost no information concerning the analytical properties of the Poincaré mapping of this plane along the trajectories of system (1.1). That is why Afraimovich *et al.* verify the theoretical conditions for Lorenz's model numerically.

Let us also present a model for the excitation of oscillations in a laser [12]:

$$\dot{x} = -\gamma_3(x-y), \qquad \dot{y} = \gamma_2(xz-y), \qquad \dot{z} = -\gamma_1(z-z_0) + \gamma_1(z_0-1)xy \tag{1.2}$$

and a model for a simple 'dynamo' [13]:

$$\frac{dw}{dt} = R - zy - \nu w, \qquad \frac{dz}{dt} = wy - z, \qquad \frac{dy}{dt} = \gamma(z-y). \tag{1.3}$$

The specific feature of these models is that their mathematical descriptions are similar in that they can be reduced to Eq. (1.1) by a linear change of variables. The reason for this 'similarity' appears to lie in the important fact that there is usually present a stable limiting set that causes stochastic behaviour of trajectories in three-dimensional systems (a quasihyperbolic attractor without lacunae—see below for an exact definition), which we shall call a Lorenz attractor; this therefore requires detailed investigation. The major part of this chapter (Sections 1.2–1.4) is devoted to quasihyperbolic attractors.

In special cases, however, it is frequently difficult to state whether a system is hyperbolic or quasihyperbolic on an attracting set, though certain easily verified consequences of hyperbolicity are valid: strong dependence on the initial conditions; irregular, chaotic behaviour of the trajectory; etc. Moreover, the following phenomenon has recently been discovered: in certain cases, where the existence of stable periodic motions together with a nontrivial hyperbolic subset (in an invariant region), to which the trajectory appears to be attracted, is proved theoretically, these motions cannot be 'picked up' either by computer or by other simulation methods. However, the observer is dealing with a stochastic motion. We propose the term 'quasiattractor' to denote such situations. We shall say, to be more exact, that an attracting set is an ε-quasiattractor if (i) it contains a nontrivial hyperbolic subset and (ii) it possesses no equilibrium state or stable periodic motion whose period would be less than $1/\varepsilon$. In cases where the value of ε is not specified, such a limiting set will be called a quasiattractor.

Numerical calculations show that in the Lorenz model (Eq. (1.1)) a quasiattractor exists for values of r from 28 to $\simeq 100$, though, as was proved by Afraimovich *et al.* [14], stable periodic motions may also exist. The reason for their appearance may be twofold:

(1) destruction of the saddle-focus separatrix loop whose existence, in turn, follows from the existence of a contour formed by the saddle, saddle-focus, and trajectories which connect them [15, 16];
(2) stable periodic motions. Stable periodic motions arise as a result of bifurcations taking place in systems close to systems with a nonstructurally stable homoclinic curve, the saddle value of periodic motion having a homoclinic curve being less than 1, which is determined by the fact that the divergence of the vector field (1.1) is negative (see [17, 18]).

Other cases where quasiattractors arise were considered by Afraimovich and Shil'nikov ([19–21]); they are connected with the action of small periodic perturbations on a self-oscillating system with weak interaction between two self-oscillating systems, and with the disappearance of nonstructurally stable motion of the saddle-node type. Mathematical effects that arise in such problems have partially been considered in [22–24]. In the above-mentioned cases a quasiattractor arises as a result of the destruction of a stable two-dimensional torus, and from the mathematical standpoint the problem is reduced to the mapping of a ring into itself. In Section 1.5 a mechanism is considered for the appearance of an attractor for the simplest mapping of a plane domain. These mappings, the simplest diffeomorphisms of the plane, are smooth analogs of Poincaré mappings with 'hooks' in situations similar to that found in the Lorenz system when the parameter values lie in some region of the plane (τ, σ) [25].

The main part of this chapter deals with extensive mathematical formulation of properties of quasihyperbolic attractors (following the results of Afraimovich *et al.* [8, 26] and Shil'nikov [25]) in terms of the qualitative theory of dynamical systems. A short review of the main concepts and facts of this theory is offered at the beginning for those who are not specialists in the qualitative theory. It is natural to divide these concepts and facts into three groups, in accordance with three aspects of the qualitative theory: topological dynamics, differential dynamics and symbolic dynamics. In topological dynamics, by a dynamical system we mean a one-parameter continuous group of homeomorphisms (mappings that are single-valued and continuous in both directions), which act on a metric space M. Differential dynamics considers smooth dynamical systems, i.e., one-parameter groups of diffeomorphisms (smooth together with inverse homeomorphisms) of smooth manifolds. If the parameter belongs to the group of real numbers, the dynamical system is called a flow; if to the group of integers, it is called a cascade. The simplest example of a flow is flow determined by a vector field:

$$\dot{x} = X(x), \quad x \in \mathbb{R}^n \tag{1.4}$$

where the vector field X obeys the conditions that ensure the existence and uniqueness of the solution to the Cauchy problem for system (1.4) (e.g., the field is differentiable or satisfies the Lipschitz condition) and is such that solutions of system (1.4) can be continued for all $t \in \mathbb{R}^1$ (otherwise, this can be achieved by change of time, i.e., by multiplication of the field X by a function $f(x)$, which does not vanish on $\mathbb{R}^n$). Then the mapping $T^t : x_0 \to x(t, x_0)$, which relates the solution of the Cauchy problem for system (1.4) to the origin x_0, will, after time t, $x = x(t, x_0)$, (a) be defined for all $t \in \mathbb{R}^1$ and (b) form a group, i.e., will satisfy the identity

$$x[t_1, x(t_2, x_0)] \equiv x(t_1 + t_2, x_0)$$

or

$$T_0^{t_1} T^{t_2} x_0 \equiv T^{t_1 + t_2} x_0$$

for all $t_1, t_2 \in \mathbb{R}^1$, $x_0 \in \mathbb{R}^n$. Here t is the parameter and $M = \mathbb{R}^n$.

The simplest example of a cascade is a group $\{T^k\}_{k \in \mathbb{Z}}$ of all powers of the homeomorphism T of a certain metric space M into itself.

The trajectory or orbit of a flow (a cascade) is a set

$$L(x_0) = \bigcup_{t \in \mathbb{R}^1} T^t x_0 \quad \left(L(x_0) = \bigcup_{n \in \mathbb{Z}} T^n x_0 \right)$$

i.e., the trajectory of a flow (a cascade) is a continuous image of $\mathbb{R}^1(\mathbb{Z})$ in

M. The point $T^t x_0$ ($T^k x_0$) is called an image point on the trajectory $L(x_0)$. Further definitions will be given for flow; similar definitions of a cascade are obtained if $k \in \mathbb{Z}$ is substituted for $t \in \mathbb{R}^1$. The trajectory $L(y_0)$ is said to be ω-limiting (α-limiting) for the trajectory $L(x_0)$ if

$$L(y_0) \subset CL\left(\bigcup_{t \geq 0} T^t x_0\right) \quad \left(L(y_0) \subset CL\left(\bigcup_{t \leq 0} T^t x_0\right)\right).$$

The trajectory $L(x_0)$ is said to be periodic if there exists $\tau \in \mathbb{R}^1$ such that $T^\tau x_0 = x_0$. The trajectory $L(x_0)$ is said to be positively (negatively) stable, in the Poisson sense, if for any point $x \in L(x_0)$, any $\varepsilon > 0$ and any $k > 0$, there exists $t > k$ ($t < -k$) such that $T^t x$ belongs to an ε-neighbourhood of the point x. In other words, a trajectory that is stable in the Poisson sense preserves some properties of periodicity, yet the trajectory does not return exactly into the same position, but only into its neighbourhood, and no *a priori* relationship is assumed concerning the time of this return.

The main property of trajectories stable in the Poisson sense can be expressed by the following Birkhoff theorem: *the closure of a nonperiodic trajectory stable in the Poisson sense contains a continual set of non-periodic trajectories, stable in the Poisson sense, which are everywhere dense in it.*

Thus, nonperiodic trajectories stable in the Poisson sense are continual collectives. It is these trajectories that perform ergodicity, mixing, etc., i.e., that determine the metric properties of a system.

A weaker property of the return of trajectories can be formulated on the basis of the concept of nonwandering. A point $x \in M$ is called nonwandering if for any of its neighbourhood U and any $K > 0$ there exists $t > k$ such that $T^t U \cap U \neq \varnothing$. It is a simple matter to show that, if a trajectory contains even one nonwandering point, all other points of the trajectory are also nonwandering—the set Ω of all nonwandering points consists of whole trajectories, i.e., it is invariant. If the space M is compact, the set of nonwandering points is not an empty one, and the trajectory spends a finite time outside any small neighbourhood of the nonwandering set. Thus, all the points are attracted to the nonwandering set and 'turn' about it.

Since Ω is invariant, a dynamical system can be considered on Ω and another nonwandering set Ω_1 can be distinguished in it, and so on. Birkhoff has shown that this process will cease in no more than a countable number of steps, and, for a dynamical system considered on the set Ω_B thus obtained, all the points will be nonwandering. Ω_B is called the centre of a dynamical system. Trajectories that are stable in the Poisson sense appear to be everywhere dense in Ω_B, that is, on the set of points at the centre a dynamical system behaves, to some extent, as a conservative one. For dynamical systems with quasihyperbolic attractors, which are considered in this chapter, the centre is reached in one step.

Some further definitions

An invariant set I is called:

limiting, if it consists of α-limiting or ω-limiting trajectories;
locally maximal, if it possesses a neighbourhood where no invariant set exists other than subsets of I;
attracting, if it has an invariant neighbourhood whose each point lies on a trajectory, for which a subset of I is an ω-limiting set:
transitive, if a trajectory exists which is everywhere dense in I;
an *attractor*, if I is simultaneously attracting and transitive.

The flow T^t on an invariant set I is called transitive if I is transitive, and it is called topological mixing if, for any open subsets $U, V \in I$, such t_0 exists that $T^t U \cap V \neq \varnothing$ for all $t \geq t_0$.

Let us now recall some concepts of differential dynamics. The equilibrium state $O: x = 0$ of an n-dimensional system of differential equations in $\mathbb{R}^n$ (for simplicity)

$$\dot{x} = X(x), \qquad X(0) = 0, \qquad X \in C^\tau, \quad \tau \geq 1 \tag{1.5}$$

is called structurally stable (or hyperbolic) if the roots $\lambda_1, \ldots, \lambda_n$ of the characteristic equations (which are called characteristic values of the equilibrium state),

$$\det\left(\frac{\partial X(0)}{\partial x} - \lambda E\right) = 0,$$

do not lie on the imaginary axis. A structurally stable equilibrium state will be assigned a type (m, p) where m is the number of roots in the left-hand side half-plane, and p is the number of roots in the right-hand side half-plane. If $m = n$ $(m = 0)$, the equilibrium state is called a stable (unstable) node. For $m \neq n$ and $m \neq 0$, O is called a saddle. If $n = 3$, $m = 2$ and the roots with negative real parts are complex, O is sometimes called a saddle-focus. A set of all points of a phase space such that trajectories passing through these points tend to O for $t \to +\infty$ $(t \to -\infty)$ is called a stable (unstable) manifold W^s_O (W^u_O) of the equilibrium state O. It is known that if O is of (m, p)-type, the manifolds W^s_O and W^u_O have dimensions m, p, respectively, and are each submanifolds of $\mathbb{R}^n$, homeomorphic to $\mathbb{R}^m$ and $\mathbb{R}^p$, in the neighbourhood of their point.

Assume now that system (1.5) has a periodic motion $L: x = \alpha(t)$, $\alpha(t+\tau) \equiv \alpha(t)$. Let us write for this motion the equation in variations:

$$\dot{\zeta} = \frac{\partial X[\alpha(t)]}{\partial x} \zeta \equiv B(t)\zeta. \tag{1.6}$$

It follows from Floquet's theory that the fundamental matrix of system (1.6) is of the form $Z(t) = C(t)\exp(Rt)$, where R is a constant matrix and $C(t)$ is a periodic one. The eigenvalues $\gamma_1, \ldots, \gamma_n$ of the matrix $\exp(Rt)$ are called multipliers of the periodic motion L. One of them (which can be taken as γ_n) is always equal to 1. A periodic motion is called structurally stable (hyperbolic) if neither of the multipliers $\gamma_1, \ldots, \gamma_{n-1}$ is equal to 1, in modulus. A periodic motion is called stable (unstable) if $|\gamma_i|<1$, $(|\gamma_i|>1)$, $i=1,\ldots,n-1$. If not all of values $\gamma_1, \ldots, \gamma_{n-1}$ lie inside (outside) a unit circle of the complex plane, such a periodic motion is called a saddle motion. A periodic motion will be assigned a type (m, p) if $m-1$ multipliers lie inside the unit circle and $p-1$ outside it. For a structurally stable periodic motion, $m+p=n+1$.

A set of all points of the phase space, such that trajectories passing through these points tend to L for $t\to+\infty$ $(t\to-\infty)$, is called a stable (unstable) manifold W_L^s (W_L^u) of the periodic motion L. It is clear that L belongs simultaneously both to W_L^s and W_L^u. W_L^s and W_L^u are each smooth submanifolds of $\mathbb{R}^m$ in the neighbourhood of their point, and have the dimensions m and p, respectively. For $n=3$ and $m=p=2$, W_L^s (W_L^u) is homeomorphic to a Möbius band, if the multiplier which is less (greater) than 1, in modulus, is negative, and to a cylinder if it is positive. In the latter case L divides W_L^s (W_L^u) into two unconnected pieces.

To study the behaviour of trajectories in the neighbourhood of L, it is frequently more convenient to study their traces on a transversal. Let S be a smooth, $(n-1)$-dimensional disc orthogonal to L at the point of intersection $S\cap L$. Let us introduce on S Euclidean coordinates $s(s_1,\ldots,s_{n-1})$ such that point O of the intersection $L\cap S$ has the coordinates $O=(0,0,\ldots,0)$. The mapping $T:S\to S$ along trajectories passing in the neighbourhood of L, which relates the point $s_0\in S$ to the point of the first intersection (for $t>0$) of the trajectory passing through s_0, with the transversal S, is called the Poincaré mapping. It can be written in the form $\bar{S}=Ds+\cdots$, where D $(n-1)\times(n-1)$ is a matrix whose eigenvalues are multipliers of the periodic motion L.

Let us now define the transversal intersection of stable and unstable manifolds. Let W_L^s and W_L^u be stable and unstable manifolds of periodic motions or of equilibrium states, and $W^s\cap W^u\neq\varnothing$. W^s and W^u are said to intersect each other transversally if

$$\dim \tilde{W}_x^s+\dim \tilde{W}_x^u-n=\dim(\tilde{W}_x^s\cap \tilde{W}_x^u)$$

where $\tilde{W}_x^s$ $(\tilde{W}_x^u)$ denotes a plane tangent to W^s (W^u) at point x. The property of intersection transversality does not change under small perturbations of a system and is, in this sense, a structurally stable one. Let us note that a saddle periodic motion L belongs to the transversal intersection of its stable and unstable manifolds. All other trajectories belonging to $W_L^s\cap W_L^u$ are called homoclinic curves (or trajectories). Those of the curves along which W^s and W^u intersect transversally are called structurally stable homoclinic curves. If a system has a structurally stable homoclinic curve Γ, this means [27, 28] that a countable set of saddle periodic motions (trajectories stable in the Poisson sense, etc.) exists in a certain neighbourhood of Γ. However, this cannot be, in general, a criterion for the existence of a stochastic regime, since this invariant set of trajectories in the neighbourhood of Γ is not an attracting one.

Trajectories from $W^s_{L_1} \cap W^u_{L_2}$, where L_1 and L_2 are different periodic motions, are called heteroclinic trajectories; those along which $W^s_{L_1}$ and $W^u_{L_2}$ intersect transversally are called structurally stable heteroclinic trajectories.

The basic bifurcations of dynamical systems are connected precisely with the violation of the structural stability of equilibrium states and periodic motions and with the violation of the transversality of intersection of their stable and unstable manifolds. By bifurcation we mean rearrangement of the topological structure of a set of phase trajectories or, as it is frequently expressed, a change in the topological structure of the partition of a phase space into trajectories under continuous transition from one system to another in the space of dynamical systems. In this case the space of dynamical systems can be identified with the Banach space of C^τ-smooth vector fields with one C^τ-norm, which generate these systems. The simplest bifurcations are characterized by transitions through films of codimension 1 on which nonstructurally stable systems are lying. These films can be represented as nondegenerate levels of smooth functionals.

Of special interest are transitions through bifurcational films when a countable set of saddle periodic motions arises or disappears. It is convenient to describe a nontrivial limiting set that arises under such bifurcations in terms of symbolic dynamics. In general, the concepts of symbolic dynamics are a convenient tool for studying systems with a complicated behaviour of trajectories. That is why we present first some principles of this theory. Here we shall confine ourselves to concepts and definitions concerning the so-called topological Markov chains (TMC).

Let us define a TMC. Let G be an orientable multigraph such that each vertex is the beginning of a certain edge and, at the same time, the end of this same edge or some other one. Let us numerate the edges of G by symbols $1, 2, \ldots, m$ and construct an $(m \times m)$-matrix $A(a_{ij})$ according to the following rule: $a_{ij} = 1$ if the end of edge i is the beginning of edge j; $a_{ij} = 0$ otherwise. For instance, for G in the form

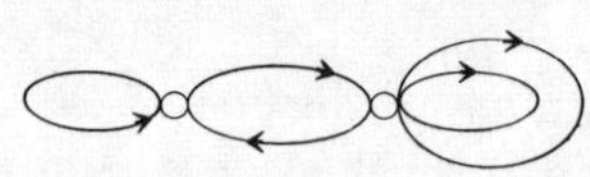

the matrix A is

$$A = \begin{pmatrix} 1 & 1 & 1 & 0 & 0 \\ 1 & 1 & 1 & 0 & 0 \\ 0 & 0 & 0 & 1 & 1 \\ 1 & 1 & 1 & 0 & 0 \\ 0 & 0 & 0 & 1 & 1 \end{pmatrix}.$$

Further, consider a set Ω of sequences, infinite in both directions, composed of symbols $1, \ldots, m$,

$$\Omega = \{(\ldots \omega_{-1}\underset{\uparrow}{\omega_0}\omega_1 \ldots \omega_k \ldots)\},$$

with a fixed position of the zero coordinate such that the symbol ω_{k+1} may follow the symbol ω_k if and only if $a_{\omega_k\omega_{k+1}} = 1$. In other words, Ω can be identified by a set of paths, infinite in both directions, along the graph G. Let us define a distance d on Ω:

$$d(\alpha, \beta) = \sum_{i=-\infty}^{\infty} \frac{|\alpha_i - \beta_i|}{2^{|i|}}$$

for $\alpha = (\alpha_{-1}, \alpha_0 \ldots)$, $\beta = (\ldots \beta_{-1}\beta_0 \ldots)$. It is a simple matter to verify that Ω with the distance d is a compact metric space. It can also be shown that if Ω has the cardinality of continuum, it is homeomorphic to a reference Cantor's set on the segment $[0, 1]$. Let us define a transformation σ on $\Omega: \sigma\omega = \omega'$ if

$$\omega = (\ldots \omega_{-k} \ldots \omega_{-1}\omega_0 \ldots),$$
$$\omega' = (\ldots \omega'_{-1}\omega'_0 \ldots),$$
$$\omega'_k = \omega_{k+1}, \quad k \in \mathbb{Z},$$

i.e., in the sequence ω the position of the zero coordinate is translated (shifted) by one symbol to the right: $(\ldots \omega_{-1}\omega_0\underset{\uparrow}{\omega_1} \ldots)$.

The transformation σ is called a shift. It can easily be verified that σ is single-valued and continuous together with the inverse mapping σ^{-1}. It is the cascade (σ^k, Ω) which is called TMC, and this will be denoted hereafter as (G, Ω, σ). If there exists a path from any vertex of the graph G to any other one, $\sigma \mid \Omega$ is transitive, and if moreover $h \equiv \ln \lambda_1 > 0$, σ has a countable set of periodic points and they are everywhere dense in Ω; here λ_1 is the maximal eigenvalue of the matrix A and h is called the topological entropy of the mapping σ.

TMCs have become a convenient tool for the study of hyperbolic sets of cascades. It is sufficient, for instance, to mention the following result [29, 30]: the limitation of a cascade onto a null-dimensional hyperbolic locally maximal set is topologically conjugated with a TMC. TMCs not only represent adequately null-dimensional sets, but can be used as an apparatus for the description of one-dimensional (etc.) invariant sets. The concept of Markov partition makes it possible to relate a TMC to a cascade invariant set.

Let us give a simplified definition of this concept for the case of one-dimensional hyperbolic attractors on an n-dimensional manifold M^n. Here, as is known [5], the attractor A of the diffeomorphism $T: M^n \to M^n$ consists of unstable manifolds of their points, each being homeomorphic to a straight line $\mathbb{R}^1$. Moreover, a stable manifold locally homeomorphic to an $(n-1)$-dimensional disc passes through each point of A, and each unstable manifold intersects transversally with each stable one. A Markov partition of an attractor A is its representation as a finite union of sets $A_1 \ldots A_k$ such that:

(1) each A_i consists of such segments of unstable manifolds that their boundary points lie on the same two $(n-1)$-dimensional 'discs', B_{i_1} and B_{i_2}, which are pieces of stable manifolds of certain points of A;
(2) the set $B = \bigcup_{i=1}^{k} (B_{i_1} \cup B_{i_2})$ is T-invariant in the sense that the image of each disc $B_{i_1j_1}$ from B belongs to one disc of the set

$$B_{i_2j_2}, \quad i_1, i_2 \in \{1, \ldots, k\}, j_1, j_2 \in \{1, 2\}.$$

Using Markov partition, one can construct a TMC as follows. A graph G has k vertices numerated by the numbers $1, \ldots, k$; a_{ij} edges lead from vertex 'i' to vertex 'j' if the image of a segment from A_i, obtained by mapping T and intersected with A_j, consists of a_{ij} segments. The TMC (G, Ω, σ) constructed from the graph G describes a cascade on an attractor in the sense that the trajectory of each is in one-to-one relation to the cascade trajectory. To be more exact, there exists a continuous mapping $\beta : \Omega \to A$ such that the diagram

$$\begin{array}{ccc} \Omega & \xrightarrow{\sigma} & \Omega \\ \downarrow{\scriptstyle\beta} & & \downarrow{\scriptstyle\beta} \\ A & \xrightarrow{T} & A \end{array}$$

is commutative. Under mapping β the points of Ω are matched by entire trajectories in such a way that a one-dimensional set A is formed from the null-dimensional set Ω (this is precisely the matching that makes us consider the numbers 0.99... and 1.00... to be the same). The mapping β is one-to-one on those trajectories whose points do not intersect with B (here we concentrate only on the topological aspect of the concept of Markov partition and do not consider problems concerning the ergodic theory (see [31])). For further details see Yakobson [32].

Thus, TMCs are standard discrete dynamical systems which are convenient for the description of invariant sets of cascades. Such reference systems also exist for flows; these are suspensions over TMCs. Let us formulate the definition of this object. In the direct product $\Omega \times I$, where I is the segment $[0, 1]$, we identify the points $(\omega, 1)$ and $(\sigma\omega, 0)$ for all $\omega \in \Omega$. The topological space thus obtained is the phase space of such a system and is denoted by Σ. Σ can be provided with a metric [33]. Let us define a flow T^t on Σ by the following relation:

$$T^t(\omega, s) = \begin{cases} (\omega, t+s), & \text{if } \; -s < t \leq 1-s \\ (\sigma\omega, t+s-1), & \text{if } \; 1-s < t \leq 1 \end{cases}$$

and for the rest, $t \in \mathbb{R}^1$, the flow is defined by the condition that $\{T^t\}$ forms a group (e.g., for $-1 < t \leq -s$, $T^t(\omega, s) = (\sigma^{-1}\omega, 1+t+s)$, etc.). It can be seen that the mapping T^1 preserves the set of points $\{(\omega, 0)\} \equiv \Omega$ invariant and acts on it exactly as σ does. In other words, σ is the Poincaré mapping of the transversal Ω of the phase space Σ for the flow T^t.

For a more detailed presentation of the above concepts and facts, see [4, 34–38].

1.2 Statement of the problem

We shall first distinguish a certain class of three-dimensional systems which have a structurally stable equilibrium state of the saddle type with a one-dimensional unstable manifold consisting of a saddle and two trajectories, called separatrices. This class is assumed to contain systems whose separatrices return to the saddle, being tangent to each other, i.e., the situation is exactly similar to that described by Eq. (1.1) for $b = 8/3$, $\sigma = 10$, $\tau \simeq 13.92$ [8]. Here we shall not confine ourselves to the case where Eq. (1.1) is invariant with respect to the change $(x, y, z) \rightarrow (-x, -y, z)$, but consider a general two-parameter family. Let us note that such a generalization is also directly relevant to applications. Moreover, we consider here two more cases related to the generation of periodic motions with negative multipliers from separatrix loops.

The fact that we are considering only three-dimensional systems is not important, in principle, since the general case where only one characteristic value is positive for the saddle, while the others have negative real parts, the value least in modulus being real, is completely similar to the three-dimensional case. Let B denote the Banach space of C^τ-smooth dynamical systems $(\tau \geqslant 1)$ with the C^τ-topology, which are specified on a smooth three-dimensional manifold M. Suppose that in a domain $U \subset B$ each system X has an equilibrium state O of the saddle type. In this case the inequalities $\lambda_1 < \lambda_2 < 0 < \lambda_3$ hold for the roots $\lambda_i = \lambda_i(X)$, $i = 1, 2, 3$, of the characteristic equation in O, and the saddle value $\sigma(X) \equiv \lambda_2 + \lambda_3 > 0$. A stable two-dimensional manifold of the saddle O will be denoted by $W^s = W^s(X)$ and the unstable one, consisting of O and two trajectories $\Gamma_1 = \Gamma_1(X)$ and $\Gamma_2 = \Gamma_2(X)$ proceeding from it, by $W^u(X)$. It is known [34] that $W^s(X)$ and $W^u(X)$ depend smoothly on X on each compact subset. Here it is assumed that in a certain local map $V = \{(x_1, x_2, x_3)\}$, containing O, X can be written in the form

$$\dot{x}_i = \lambda_i x_i + P_i(x_1, x_2, x_3), \quad i = 1, 2, 3. \tag{1.7}$$

Suppose that the following conditions are satisfied for the system $X_0 \subset U$ (see Fig. 1.1):

(1) $\Gamma_i(X_0) \subset W^s(X_0)$, $i = 1, 2$, i.e., $\Gamma_i(X_0)$, is doubly asymptotic to O;
(2) $\Gamma_1(X_0)$ and $\Gamma_2(X_0)$ come in to O, being tangent to each other.

Further considerations will require some concepts and facts presented

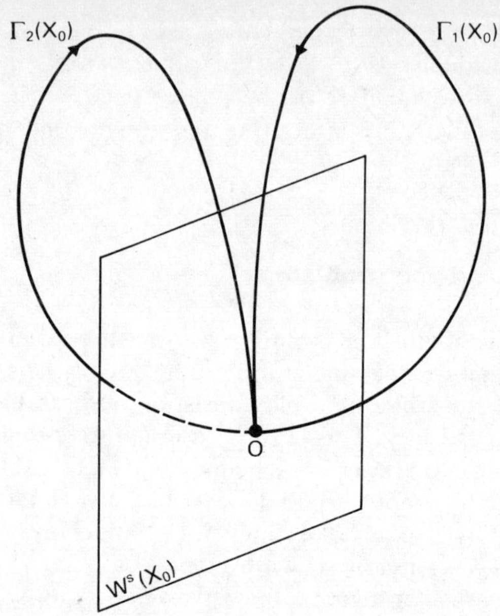

Fig. 1.1

in [39, 40]. The condition $\lambda_1 < \lambda_2 < 0$ implies that the nonleading submanifold $W^s_0(X)$ of the manifold $W^s(X)$, consisting of O and two trajectories tangent to the axis x_1 at point O, divides $W^s(X)$ into two open domains: $W^s_+(X)$ and $W^s_-(X)$. Without loss of generality one may assume that $\Gamma_i(X_0) \subset W^s_+(X_0)$, and hence $\Gamma_i(X_0)$ is tangent to the positive semiaxis x_2. Let v_1 and v_2 be sufficiently small neighbourhoods of the 'separatrix loops' $\overline{\Gamma_1(X_0)} = \Gamma_1(X_0) \cup O$ and $\overline{\Gamma_2(X_0)} = \Gamma_2(X_0) \cup O$. Let $\mathcal{M}_i$ stand for the connection component of the intersection of $\overline{W^s_+(X_0)}$ with v_i, which contains $\Gamma_i(X_0)$. It follows from [40] that in the general case $\mathcal{M}_i$ is a two-dimensional C^0-submanifold homeomorphic either to a cylinder or to a Möbius band. The generality condition lies in the fact that certain values $A_1(X_0)$ and $A_2(X_0)$, called the separatrix values, should not be equal to zero.

It follows from the above assumptions that X_0 belongs to the bifurcational set B^2_1 of codimension two, and B^2_1 is the intersection of two bifurcational 'films' B^1_1 and B^1_2 each of codimension 1, where B^1_1 corresponds to one separatrix loop of the saddle and B^1_2 to the other loop. In such situations it is natural to consider a two-parameter family of dynamical systems $X(\mu)$, $\mu = (\mu_1, \mu_2)$, $|\mu| < \mu_0$, $X(0) = X_0$, such that $X(\mu)$ intersects with B^2_1 only along X_0 and only for $\mu = 0$. It is also convenient to assume the family $X(\mu)$ to be transversal to B^2_1, the transversality meaning that for the system $X(\mu)$ the loop $\Gamma_1(X(\mu))$ 'deviates' from $W^s(X(\mu))$ by a value of the order of μ_1, and $\Gamma_2(X(\mu))$ by a value of the order of μ_2.

It is known [40] that the above assumptions imply that in the transition to a system close to X_0 the separatrix loop can generate only one periodic motion, which is of the saddle type. Let us assume, for certainty, that the loop $\Gamma_1(X_0) \cup O$ generates the periodic motion $L_1 = L_1(X)$ for $\mu_1 > 0$ and that $\Gamma_2(X_0) \cup O$ generates $L_2(X)$ for $\mu_2 > 0$.

The corresponding domain in U, which is the intersection of stability regions for L_1 and L_2, i.e., the domain in which periodic motions $L_1(X)$ and $L_2(X)$ of each system are structurally stable, will be denoted by U_0. A stable manifold of $L_i(X)$ for the system $X \subset U_0$ will be denoted by $W_i^s = W_i^s(X)$, and an unstable one by $W_i^u = W_i^u(X)$. If the separatrix value $A_i(X_0) > 0$, $W_i^u(X)$ is a cylinder; if $A_i(X_0) < 0$, $W_i^u(X)$ is a Möbius band. Let us note that, in the case where M is an orientable manifold, $W_i^s(X)$ will also be a cylinder if $A_i(X_0) > 0$, otherwise it will be a Möbius band. However, in the forthcoming analysis the signs of the separatrix values will play an important role. Therefore it is natural to distinguish between three main cases:

Case A (orientable)	$A_1(X_0) > 0,$	$A_2(X_0) > 0,$
Case B (semiorientable)	$A_1(X_0) > 0,$	$A_2(X_0) < 0,$
Case C (nonorientable)	$A_1(X_0) < 0,$	$A_2(X_0) < 0.$

In each of the above three cases the domain U_0 also contains two bifurcational films B_3^1 and B_4^1:

(1) in Case A, B_3^1 corresponds to the inclusion $\Gamma_1(X) \subset W_2^s(X)$ and B_4^1 to the inclusion $\Gamma_2(X) \subset W_1^s(X)$;
(2) in Case B, B_3^1 corresponds to the inclusion $\Gamma_1(X) \subset W_1^s(X)$ and B_4^1 to $\Gamma_2(X) \subset W_1^s(X)$;
(3) in Case C, along with the above-mentioned generated motions L_1 and L_2, there also arises a saddle periodic motion L_3 which makes one revolution 'along' $\Gamma_1(X_0)$ and $\Gamma_2(X_0)$, and if $W_i^u(X_0)$ are Möbius bands, $i = 1, 2$, the unstable manifold $W_3^u(X)$ of the periodic motion $L_3(X)$ is a cylinder. In this case the inclusions $\Gamma_1(X) \subset W_3^s(X)$, $\Gamma_2(X) \subset W_3^s(X)$ correspond to the films B_3^1 and B_4^1, respectively.

Suppose that B_3^1 and B_4^1 intersect transversally over the bifurcational set B_2^2 of codimension 2 (Fig. 1.2). In a two-parameter family $X(\mu)$ this means that the curves B_3^1 and B_4^1 intersect at some point $\mu^1 = (\mu_{11}, \mu_{12})$. In a small neighbourhood of the origin of the parameter plane equations of B_3^1 will be of the form: $\mu_1 = a_1 \mu_2^{1/\alpha}(1 + \ldots)$, $\mu_1 = b_1 \mu_2^{1/\alpha}(1 + \ldots)$ in Cases A and B, and $\mu_1 = c_1 \mu_2^{1/\alpha^2}(1 + \ldots)$ in Case C. Equations of B_4^1 will be $\mu_2 = a_2 \mu_1^{1/\alpha}(1 + \ldots)$ in Case A, and $\mu_2 = b_2 \mu^{1/\alpha^2}(1 + \ldots)$, $\mu_2 = c_2 \mu_1^{1/\alpha^2}(1 + \ldots)$ in Cases B and C. Here $\alpha = -\lambda_2/\lambda_3$, $a_1, \ldots, c_2$ are positive, and the dots stand for terms which tend to zero when the argument tends to zero. Let us denote a domain lying between B_3^1 and B_4^1 by U_1. It can be shown (this can also be deduced from the results of Shil'nikov [39]) that there exists in U a sufficiently small ε-neighbourhood U_ε of the system X_0 such that a one-dimensional limiting

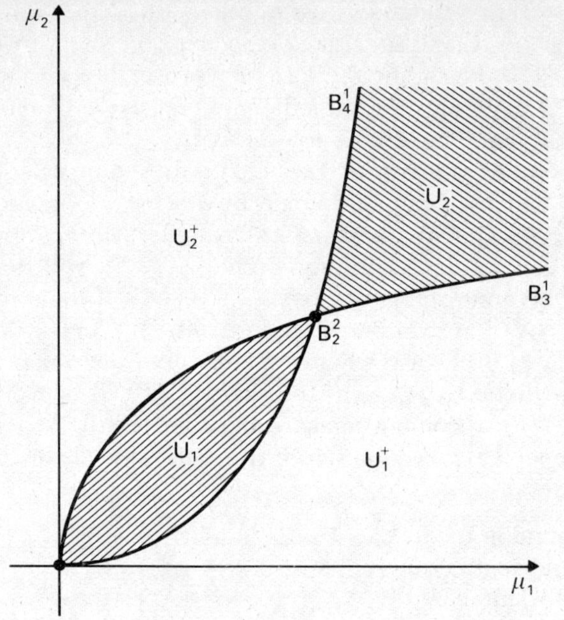

Fig. 1.2

set, homeomorphic to the suspension over the Bernoulli scheme with two symbols, will exist for each $X \in U_\varepsilon \cap U_1$. This statement may be proved if one considers the mapping T of a certain transversal to $\Gamma_1(X_0)$ and $\Gamma_2(X_0)$ along the trajectories of the system X (see also [41]).

The generalization of this situation requires the existence of a global transversal. Let us assume, therefore, that for each $X \in U$ there exists a transversal with the following properties:

(1) Euclidean coordinates (x, y) can be introduced on D such that

$$D = \{(x, y) \mid |x| \leq 1, |y| < 2\}.$$

(2) The equation $y = 0$ describes a connection component S of the intersection $W^s(X) \cap D$ such that no ω-semitrajectory that begins on S possesses any point of intersection with D for $t > 0$.

(3) Mappings $T_1(X): D_1 \to D$ and $T_2(X): D_2 \to D$ are defined along the trajectories of the system X, where

$$D_1 = \{(x, y) \mid |x| \leq 1, 0 < y \leq 1\},$$

$$D_2 = \{(x, y) \mid |x| \leq 1, -1 \leq y < 0\},$$

and $T_i(X)$ is written in the form

$$\bar{x} = f_i(x, y), \qquad \bar{y} = g_i(x, y); \quad f_i, g_i \in C^r, i = 1, 2. \tag{1.8}$$

(4) f_i and g_i admit continuous extension on S, and

$$\lim_{y\to 0} f_i(x, y) = x_i^{**}, \qquad \lim_{y\to 0} g_i(x, y) = y_i^{**}, \quad i = 1, 2.$$

(5) $T_1D_1 \subset \Pi_1 = \{(x, y) \mid \frac{1}{2} \leqslant x \leqslant 1, |y| < 2\}$,

$$T_2D_2 \subset \Pi_2 = \{(x, y) \mid -1 \leqslant x \leqslant -\tfrac{1}{2}, |y| < 2\}.$$

Let $T(X) \equiv T_i(X) \mid D_i$, $(f, g) \equiv (f_i, g_i)$ on D_i, $i = 1, 2$.
(6) Let us impose the following restrictions on $T(X)$:

$$\left.\begin{array}{l} \text{(a) } \|f_x\| < 1, \\ \text{(b) } 1 - \|g_y^{-1}\| \cdot \|f_x\| > 2\sqrt{(\|g_y^{-1}\| \cdot \|g_x\| \cdot \|g_y^{-1} \cdot f_y\|)}, \\ \text{(c) } \|g_y^{-1}\| < 1, \\ \text{(d) } \|g_y^{-1} \cdot f_y\| \cdot \|g_x\| < (1 - \|f_x\|)(1 - \|g_y^{-1}\|). \end{array}\right\} \tag{1.9}$$

Hereinafter, $\|\cdot\| = \sup|\cdot|$, $(x, y) \in D \backslash S$.

It follows from the analysis of the behaviour of trajectories near $W^s(X)$ [39, 40] that in a small neighbourhood of S the following representations are valid:

$$\begin{array}{ll} f_1 = x_1^{**} + \varphi_1(x, y)y^\alpha, & g_1 = y_1^{**} + \psi_1(x, y)y^\alpha, \\ f_2 = x_2^{**} + \varphi_2(x, y)(-y)^\alpha, & g_2 = y_2^{**} + \psi_2(x, y)(-y)^\alpha \end{array} \tag{1.10}$$

where $\varphi_1, \ldots, \psi_2$ are smooth with respect to x, y for $y \neq 0$, and $T_i(X)$ satisfies estimates (1.9) for sufficiently small y. Moreover, the limits of $\varphi_1, \ldots, \psi_2$ for $y \to 0$ do exist and are independent of x. The limit of ψ_1 will be denoted by $A_1(X)$ and of ψ_2 by $A_2(X)$. The functionals $A_1(X)$ and $A_2(X)$ will also be called separatrix values, in analogy with $A_1(X_0)$ and $A_2(X_0)$ which were considered above. Let us note that for a system lying in a small neighbourhood of the system X_0 all the conditions (1–6) are satisfied near S. Moreover, the concepts of orientable, semiorientable and nonorientable cases can be extended to any system $X \in U$. It is convenient to assume, for simplicity, that $A_i(X)$, $i = 1, 2$, do not vanish. It should also be noted that the point P_i with the coordinates $x_i^{**} = x_i^{**}(X)$, $y_i^{**} = y_i^{**}(X)$ is the first point of intersection of $\Gamma_i(X)$ with D, $i = 1, 2$.

Let us consider a constant

$$q = \frac{1 + \|f_x\| \cdot \|g_y^{-1}\| + \sqrt{[(1 - \|f_x\| \cdot \|g_y^{-1}\|^2) - 4\|g_y^{-1}\| \cdot \|g_x\| \cdot \|g_y^{-1} \cdot f_y\|]}}{2\|g_y^{-1}\|} \tag{1.11}$$

Conditions (1.9) imply that $q > 1$ and this (together with conditions (1.9)) means that all periodic points are of the saddle type.

In terms of mappings, the conditions for the existence of periodic motions $L_1(X)$ and $L_2(X)$ can be formulated quite simply: it is required that $P_1 \in D_2$, $P_2 \in D_1$ in Case A; $P_1 \in D_2$, $P_2 \in D_2$ in Case B; $P_1 \in D_1$, $P_2 \in D_2$, in Case C.

The point of intersection of $L_i(X)$ with D will be denoted by M_i, and its

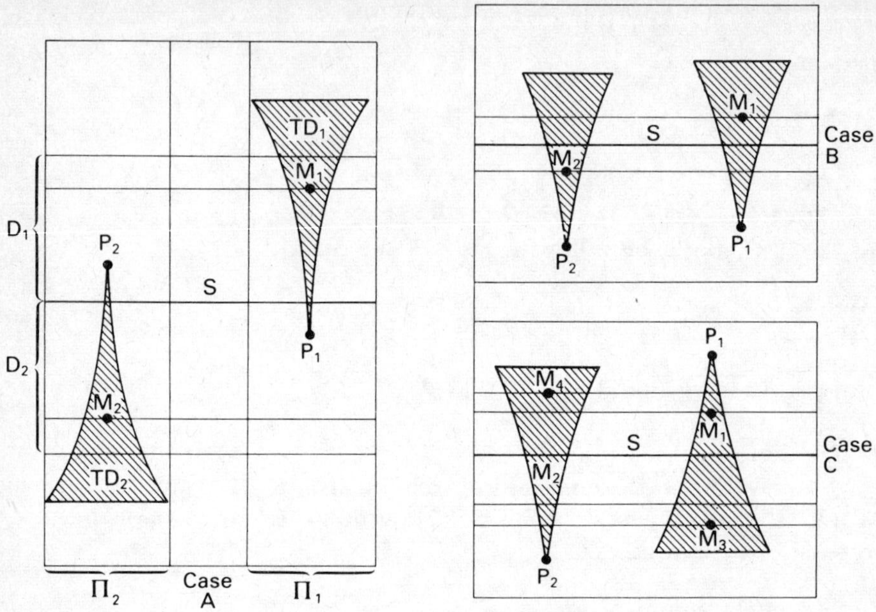

Fig. 1.3

coordinates by $x_i^* = x_i^*(X)$, $y_i^* = y_i^*(X)$. As already mentioned, the periodic motion $L_3(X)$ will exist in Case C together with $L_1(X)$ and $L_2(X)$. It intersects D at two points $M_3(x_3^*, y_3^*)$ and $M_4(x_4^*, y_4^*)$, where $TM_3 = M_4$, $TM_4 = M_3$ (Fig. 1.3). The periodic motions L_1, L_2, L_3 will be called basic. The conditions imposed above imply that equations for the intersections of connection components of stable manifolds $W_i^s(X)$ of periodic motions $L_i(X)$, $i = 1, 2$, with D, containing fixed points M_i, can be represented in the form $y = y_i(x)$, $|x| \leqslant 1$. In Case C the equation for the connection component $D \cap W_3^s(X)$, containing point M_j, can also be written in the form $y = y_j(x)$, $j = 3, 4$.

Let us define the functionals $R_1(X)$ and $R_2(X)$ as follows:

in Case A $\quad R_1(X) = -[y_1^{**} - y_2(x_2^{**})], \qquad R_2(X) = y_2^{**} - y_1(x_2^{**});$

in Case B $\quad R_1(X) = v_2^{**} - y_1(u_2^{**}), \qquad R_2(X) = y_2^{**} - y_1(x_2^{**}),$
where (u_i^{**}, v_i^{**}) are the coordinates of TP_i;

in Case C $\quad R_1(X) = [y_1^{**} - y_4(x_1^{**})], \qquad R_2(X) = -[y_2^{**} - y_3(x_2^{**})].$

It can easily be seen now that $R_1(X) = 0$ for the system $X \in B_3^1$ and $R_2(X) = 0$ if $X \in B_4^1$, and the domain U consists of systems X for which $R_1(X) > 0$ and $R_2(X) > 0$. The condition $R_1(X) > 0$ means that:

(1) in Case A, $L_1(X)$ and $L_2(X)$ have a heteroclinic trajectory (i.e., a trajectory which is asymptotic to $L_1(X)$ for $t \to -\infty$ and to $L_2(X)$ for $t \to +\infty$;

(2) in Case B, $L_1(X)$ has a structurally stable homoclinic curve;
(3) in Case C, $L_3(X)$ also has a structurally stable homoclinic curve.

If $R_2(X)>0$, it is only $L_3(X)$, which has a homoclinic curve in Case C. In Cases A and B, $L_1(X)$ and $L_2(X)$ will have a heteroclinic curve for $R_2(X)>0$.

The domain $\{X\in U_0 \mid R_1(X)<0,\ R_2(X)>0\}$ will be denoted by U_2^+, and $\{X\in U_0 \mid R_1(X)<0,\ R_2(X)<0\}$ by U_1^+. The domain $\{X\in U_0 \mid R_1(X)<0,\ R_2(X)<0\}$ will be denoted by U_2^+.

1.3 Structure of a nonwandering set—basic results

In this section we consider limiting sets of this class of dynamical systems. Special attention will be paid to the case where a domain can be specified in the phase space into which all the trajectories come and which does not contain stable periodic motions. If this domain contains only one limiting set, it appears to be a quasihyperbolic attractor. However, this domain may contain several limiting sets, among which all except one are unstable and this one is ω-limiting and, in the general case, where it is locally maximal, is an attractor.

Let $\Sigma(X)$ denote the closure of the set of points of all the trajectories of the mapping $T(X)$, which are contained entirely in D. Due to the above considerations, $\Sigma(X)$ is nontrivial and contains a countable set of periodic trajectories in $U_1\cup U_1^+\cup U_2^+$ for Case C and in $U_1\cup U_1^+$ for Case B. For Case A in $U_1^+\cup U_2^+$ and for Case B in U_2^+ the above-mentioned situation is possible and also a trivial one where $\Sigma(X)$ consists of M_1, M_2 and heteroclinic points. The corresponding bifurcation surfaces that divide these sets of systems will satisfy the following equations:

in Case A: $R_3(X)=0$ and $R_4(X)=0$

where

$$R_3=-v_1^{**}+y_1(u_1^{**}),\qquad R_4=v_2^{**}-y_2(u_2^{**})$$

where $[u_i^{**}(X),\ v_i^{**}(X)]$ are coordinates of points $T(X)P_i$, $i=1,2$;

in Case B: $R_3(X)=0$

where $R_3(X)=-v_3^{**}+y_1(u_3^{**})$ and u_3^{**}, v_3^{**} are coordinates of the point $T^2(X)P_1$.

It is obvious that in Case A the surface $R_4=0$ lies in U_2^+, and $R_3=0$ in U_1^+ (the same is true for Case B).

$\Sigma(X)$ is described most simply in the domain U_1. Here the following theorem holds.

Theorem 1.1. *If $X\in U_1$, $T(X)\mid\Sigma(X)$ is topologically conjugated with the Bernoulli scheme (σ,Ω_2) with two symbols.*

If $X \in \bar{U}_2 \backslash U_2$, $\Sigma(X)$ will be a one-dimensional set. We shall not concentrate on this case, let us only note that $\Sigma(X)$ is unstable and its structure is similar, in many respects, to the structure of $\Sigma(X)$ for $X \in \text{int } U_2$. We confine ourselves to the following theorem.

Theorem 1.2. *If $X \in B_2^2$ (i.e., $R_i(X) = 0$, $i = 1, 2$), $\Sigma(X)$ is one-dimensional, consists of two connection components and is the image of the space Ω_2 of the Bernoulli scheme with two symbols under continuous mapping $\beta : \Omega_2 \to \Sigma$, which is a homeomorphism on the residual set of points (of Ω_2) and is such that the diagram*

$$\begin{array}{ccc} \Omega_2 \backslash \eta^{-1}(\Sigma \cap S) & \xrightarrow{\sigma} & \Omega_2 \\ & \searrow^{\beta} & \downarrow{\scriptstyle \beta} \\ & \Sigma \backslash S & \xrightarrow{T} \Sigma \end{array}$$

is commutative.

It follows, in particular, from this theorem that $\Sigma(X)$ is a nonwandering set and periodic points are everywhere dense in it. Generally, the conditions $R_i(X) \leq 0$ imply that an unstable invariant manifold of any periodic point, except M_1 and M_2 in Case A, M_1 in Case B, and M_3, M_4 in Case C, lies in D and is a union of a countable set of intervals. Points $T^k P_i$ are the boundaries of these intervals. Of course, for a dynamical system generated by a vector field X, invariant manifolds of the corresponding periodic motions are either embedded cylinders or embedded Möbius bands. However, such an unusual behaviour of these motions on the transversal D is due to the fact that they intersect stable manifolds of the saddle O (and hence, $O \cup \Gamma_1(X) \cup \Gamma_2(X)$ is contained in their closure). Therefore, in a neighbourhood of any of its points not containing points $T^k P_i$, $i = 1, 2$, $k \in \mathbb{Z}_+$, $\Sigma(X)$ can be represented as a direct product of a Cantor set by a segment. For $X \in B_2^2$, $\Sigma(X)$ will be transformed into a Cantor set of segments only if points $T^k P_i$ and M_i, $k \in \mathbb{Z}_+$, are removed from $\Sigma(X)$. Each point $T^k P_i$ will be a boundary one for the continuum of segments of unstable invariant manifolds of points from $\Sigma(X)$.

For systems from U_2 we first formulate the following general theorem.

Theorem 1.3. *The system $X \in U_2$ has a two-dimensional limiting set $\Omega(X)$, which satisfies the following conditions:*

(1) *$\Omega(X)$ is nonstructurally stable.*
(2) *$[\Gamma_1(X) \cup \Gamma_2(X) \cup O] \subset \Omega(X)$.*
(3) *Structurally stable periodic motions are everywhere dense in $\Omega(X)$.*
(4) *Under perturbation of X periodic motions in $\Omega(X)$ disappear as a result of matching to the saddle separatrix loops $\overline{\Gamma_1(X)}$ and $\overline{\Gamma_2(X)}$.*

Note that in this case the basic periodic motions will not belong to $\Omega(X)$. In terms of mappings, the properties of $\Omega(X)$ can be formulated in more detail. Let us first single out a domain $\tilde{D}$ on D as follows: we assume that in Case A

$$\tilde{D}=\{(x, y)\in D_1\cup D_2 \mid y_2(x)<y<y_1(x)\}$$

and in Case B

$$\tilde{D}=\{(x, y)\in D_1\cup D_2 \mid y_{12}(x)<y<y_1(x)\}$$

where $y=y_{12}(x)$, $|x|\leq 1$, denotes a curve in D whose image lies on the curve $y=y_1(x)$; and finally, in Case C

$$\tilde{D}=\{(x, y)\in D_1\cup D_2 \mid y_3(x)<y<y_4(x)\}.$$

The closure of points of all the trajectories of the mapping $T(X)$, which are entirely contained in $\tilde{D}$, will be denoted by $\tilde{\Sigma}(X)$.

Theorem 1.4. *Let $X\in U_2$. Then:*

I. $\tilde{\Sigma}(X)$ is compact, one-dimensional and consists of two connection components in Cases A and C, and of a finite number of connection components in Case B.
II. $\tilde{D}$ is foliated by a continuous stable foliation $H^+(X)$ into leaves, satisfying the Lipschitz conditions, along which a point is attracted to $\tilde{\Sigma}(X)$; inverse images of the discontinuity line $S: y=0$ (with respect to the mapping T^k, $k=1, 2, \ldots$) are everywhere dense in $\tilde{D}$.
III. There exists a sequence of $T(X)$-invariant null-dimensional sets $\Delta_k(X)$, $k\in\mathbb{Z}_+$, such that $T(X)\mid\Delta_k(X)$ is topologically conjugated with a finite topological Markov chain (TMC) with a nonzero entropy, the condition $\Delta_k(X)\subset\Delta_{k+1}(X)$ being satisfied, and $\Delta_k(X)\to\tilde{\Sigma}(X)$, $k\to\infty$.
IV. The nonwandering set $\Sigma_1(X)\subset\tilde{\Sigma}(X)$ is a closure of saddle periodic points $T(X)$ and either $\Sigma_1(X)=\tilde{\Sigma}(X)$ or $\Sigma_1(X)=\Sigma^+(X)\cup\Sigma^-(X)$, where:
(1) $\Sigma^-(X)$ is null-dimensional and is an image of the space Ω^- of a certain TMC (G^-, Ω^-, σ) under the homeomorphism $\beta:\Omega^-\to\Sigma^-(X)$, which conjugates $\sigma\mid\Omega^-$ and $T(X)\mid\Sigma^-(X)$;

$$\Sigma^-(X)=\bigcup_{m=1}^{l(X)}\Sigma_m^-(X),\ l(X)<\infty,$$

where

$$T(X)\Sigma_m^-(X)=\Sigma_m^-(X),\qquad \Sigma_{m_1}^-(X)\cap\Sigma_{m_2}^-(X)=\varnothing$$

for $m_1\neq m_2$ and $T(X)\mid\Sigma_m^-(X)$ is transitive;

(2) $\Sigma^+(X)$ is compact, one-dimensional, and
(3) if $\Sigma^+(X)\cap\Sigma^-(X)=\varnothing$, $\Sigma^+(X)$ is an attracting set in a certain neighbourhood;
(4) if $\Sigma^+(X)\cap\Sigma^-(X)\neq\varnothing$, then $\Sigma^+(X)\cap\Sigma^-(X)=\Sigma^+(X)\cap\Sigma_m^-(X)$ for a certain m, and this intersection consists of periodic points of no more than

two periodic orbits, and
(a) *if $\Sigma_m^-(X)$ is finite, $\Sigma^+(X)$ is ω-limiting for all the trajectories in a certain neighbourhood;*
(b) *if $\Sigma_m^-(X)$ is infinite, $\Sigma^+(X)$ is not locally maximal, but is ω-limiting for all the trajectories in $\tilde{D}$, excluding those asymptotic to $\Sigma^-(X)\backslash\Sigma^+(X)$.*

Let U_2' denote a set of systems from U_2, for which the separatrices $\Gamma_1(X)$ and $\Gamma_2(X)$ tend either to 0 or to periodic motions, and let U_2'' stand for a set of such systems from U_2', for which $\Gamma_1(X)$ and $\Gamma_2(X)$ tend to 0 as $t\to\infty$.

Theorem 1.5.
I. *U_2'' is dense in U_2.*
II. *For $X\in U_2'$:*
(1) *the mapping $T(X)$ admits a finite Markov partition of $\tilde{\Sigma}(X)$ whose boundary belongs to a finite set of leaves from $H^+(X)$;*
(2) *for a TMC (G, Ω, σ), constructed according to this partition, there exists a continuous mapping $\beta:\Omega\to\tilde{\Sigma}(X)$, which is one-to-one on the residual set of Ω and is such that the diagram*

$$\begin{array}{ccc} \Omega\backslash\beta^{-1}(\tilde{\Sigma}(X)\cap S) & \xrightarrow{\ \sigma\ } & \Omega \\ \downarrow{\scriptstyle\beta} & \swarrow{\scriptstyle\beta} & \\ \tilde{\Sigma}(X)\backslash S & \xrightarrow{\ T\ } & \tilde{\Sigma}(X) \end{array}$$

is commutative;
(3) *$\tilde{\Sigma}(X)$ is locally homeomorphic to the direct product of a Cantor set by a segment at each point, which does not belong to the Markov partition boundary.*
III. *$\Sigma^+(X)$ is an attractor for $X\in U_2''$.*
IV. *$T(X)$ does not admit a finite Markov partition of $\tilde{\Sigma}(X)$ for $X\in U_2\backslash U_2'$.*

We shall say that the rational case takes place if $X\in U_2'$, and the irrational one if $X\in U_2\backslash U_2'$. It follows from the definition of the rational case that a bifurcation set B_X^2 of codimension 2, which is an intersection of two bifurcation surfaces B_{1X}^1 and B_{2X}^1 of codimension 1, is connected with each system $X\in U_2'$. Let us call the surface B_{iX}^1 a rational one, if $\Gamma_i(X)$ tends either to 0 or to a periodic motion; if $\Gamma_i(X)$ tends to 0 then B_{iX}^1 will be called a bifurcation surface of the first type, and if $\Gamma_i(X)$ tends to the periodic motion, a bifurcation surface of the second type. Surfaces of the first type are of interest in that the bifurcation of periodic motions is connected only with them—a periodic motion may disappear only via matching either to a separatrix loop or to a contour. Transitions through surfaces of the second type are responsible, in general, for local rearrangements of an attractor. However, there may be exceptional rearrangements, namely those related to global rearrangements of $\Sigma(X)$ that result in the appearance of lacunae (see below).

It follows from Theorem 1.5 that two families of rational surfaces exist in the neighbourhood of any system, the rational surfaces of the first type being everywhere dense. In general, the neighbourhood of X is foliated into a family of bifurcational surfaces, including those that can naturally be found as limits of the rational surfaces of the first type. However, this problem will not be considered here because its consistent presentation requires the use of topological invariants of the 'kneading' type, which were introduced by Milnor and Thurston [42] in connection with the study of continuous piecewise-monotonous mappings of an interval. This concept can easily be extended to the case where the mapping of a segment is discontinuous (see, e.g. [43]). Since a stable foliation exists, we can reduce our mapping, by means of a change of variables, to the form

$$\bar{x} = F(x, y), \qquad \bar{y} = G(y) \tag{1.12}$$

where the right-hand sides only are, in general, continuous (everywhere except the segment $y = 0$), and G is, at the same time, piecewise-monotonous. It is natural, therefore, to relate to (1.12) the one-dimensional mapping

$$\bar{y} = G(y). \tag{1.13}$$

A question naturally arises: will 'kneadings' be a complete topological invariant for (1.12) in those cases where they are such an invariant for (1.13)? Williams [44, 45] and Guckenheimer [46] deal with this question: note that Williams uses, along with 'kneadings', an equivalent concept of η-functions. As for the technical aspect, both these authors develop the method of limits of inverse spectra which permits one to construct a two-dimensional mapping on the basis of a one-dimensional mapping, and Williams considers straightforward flows and semiflows. It has been shown by these authors that kneadings are complete topological invariants also for a two-dimensional mapping provided that $\inf |G_y(y)| > \sqrt{2}$ (the latter condition is, obviously, possible only if a stable foliation is smooth). This condition is not used in [45], though it is difficult to understand without it why the theorem on the everywhere dense sets of periodic motions in the 'Lorenz attractor' is valid.

It is quite natural to expect that there exists a two-parameter family $X(\mu)$ where $X(1, 1) \in B_2^2$ and that the parameters should appear in it in such a way that μ_1 controls the separatrix Γ_1 and μ_2 controls Γ_2, in the sense that the path $X(\mu)$ intersects monotonously the bifurcational films $\{B_{1X}^1\}$ if μ_1 changes monotonously and μ_2 is fixed, and the films $\{B_{2X}^1\}$ if μ_1 is fixed and μ_2 changes monotonously.

Note. All the statements formulated above have been proved by us earlier under conditions (1.9), which imply that $q > 1$. Examples show, however, that these conditions are rather restrictive and we have therefore made an attempt to improve them. It was found that if conditions (a) and (b) are assumed to be satisfied, conditions (c) and (d) can be replaced by exactly similar ones, but for some arbitrary power $T^n(X)$, rather than for $T(X)$. The mapping $T^n(X)$ will be smooth at all the points of D, except a finite set of curves of the form $y=f_j(x)$, which are inverse images (for the mapping T^k, $k \leq n$) of the discontinuity line. A new q_n, as calculated from the equation similar to (1.11) for the mapping T^n, will now be greater than unity.

1.4 Lorenz attractors—lacunae

In this section we concentrate on 'actual' bifurcations of a Lorenz attractor which lead to the formation of unstable, one-dimensional limiting sets that are structurally stable.

Let us first consider in a greater detail the case where $\tilde{\Sigma}(X)$ is a nonwandering set. In this case $\tilde{\Sigma}(X)$ will be called the Lorenz attractor. To prove the fact of nonwandering, let us introduce the concept of complete extension.

As was noted above, the domain $\tilde{D}$ containing $\tilde{\Sigma}(X)$ admits a continuous foliation $H^+(X)$. Let us select in $H^+(X)$ two extreme leaves $h^-(X)$ and $h^+(X)$ such that $\tilde{\Sigma}(X) \cap h^+(X) \neq \emptyset$, $\tilde{\Sigma}(X) \cap h^-(X) \neq \emptyset$, but there are no points of $\tilde{\Sigma}(X)$ above $h^+(X)$ and below $h^-(X)$. Such leaves do exist and, for instance, in Case A pass through the points P_1 and P_2. A set of leaves located between $h^+(X)$ and $h^-(X)$ will be denoted by $H_1^+(X)$. Let us choose two different leaves $h_1, h_2 \subset H_1^+(X)$. The domain between these leaves will be denoted by $V_{h_1h_2}$, and let $V^n_{h_1h_2}$ stand for a maximal set in $V_{h_1h_2}$, on which $T^n(X)$ is defined. Obviously, $V^n_{h_1h_2}$ is an open set consisting of a finite number of domains whose adjacent boundaries are inverse images of the discontinuity line S of the mapping T^k, $k<n$. We shall say that complete extension takes place if, for arbitrary leaves h, h_1 and h_2 of the foliation H_1^+, a number n can be found such that $T^nV^n_{h_1h_2} \cap h \neq \emptyset$.

Statement 1.1. *If complete extension takes place, $\tilde{\Sigma}(X)$ is a Lorenz attractor.*

The inverse statement does not hold, but the following statement is valid.

Statement 1.2. *In order that complete extension may take place, it is necessary and sufficient that $T(X) \mid \tilde{\Sigma}(X)$ should be the limit of transitive $T(X) \mid \Delta_k(X)$, $k \in \mathbb{Z}_+$.*

Let us recall that, according to Theorem 1.4, $\tilde{\Sigma}(X)$ is the limit of null-dimensional $\Delta_k(X)$, where $\Delta_k(X) \subset \Delta_{k+1}(X)$, and such that $T(X) \mid \Delta_k(X)$ is homeomorphic to a finite TMC.

Statement 1.3. *If the condition of complete extension is not satisfied, the nonwandering set $\Sigma_1(X)$ can be represented in the form $\Sigma_1(X) = \Sigma^+(X) \cup \Sigma^-(X)$, and:*

(1) $\Sigma^+(X)$ is the limit of null-dimensional $\Delta_k^+(X)$ such that $T(X) \mid \Delta_k^+(X)$ is homeomorphic to a finite transitive TMC;
(2) $\Sigma^-(X) = \bigcup_{m=1}^{l(X)} \Sigma_m^-(X)$, where $1 \leqslant l(X) < \infty$, $\Sigma_i^-(X) \cap \Sigma_j^-(X) = \varnothing$ for $i \neq j$ and $T(X) \mid \Sigma_j^-(X)$ is transitive and homeomorphic to a finite TMC.

If $l(X) = 1$, $\Sigma_1^-(X)$ is a periodic orbit, and if moreover $\Sigma_1(X) = \tilde{\Sigma}(X)$, we obtain precisely the case where the condition of complete extension is not satisfied on a Lorenz attractor.

Let us now introduce the very useful concept of lacunae. A leaf $h \in H_1^+(X)$ will be called exceptional if leaves h_1 and h_2 can be found such that, for any $n \geqslant 0$, $T^n V_{h_1 h_2}^n \cap h = \varnothing$. A set K consisting of points lying on exceptional leaves will contain a finite number of components. It can be shown that their number does not exceed $-2(\ln q)^{-1} \cdot \ln(q-1)$. The union of components $K_e^1, \ldots, K_e^{s_e} < K$ will be called a lacuna if $\Sigma_e^- \subset \bigcup_{i=1}^{s_e} K_e^i$, but $\Sigma_e^- \not\subset (\bigcup_{i=1}^{s_e} K_e^i) \backslash K_e^j$ for any $1 \leqslant j \leqslant s_e$. It is clear that K is the union of all the lacunae. The set $H_1^+ \backslash K$ will be denoted by $H_1(X)$, and complete extension will take place on it.

The main question that will be considered below is: do systems with a Lorenz attractor exist and how do lacunae arise?

Sufficient conditions for the existence of complete extension are formulated below.

Let us define natural numbers corresponding approximately to the number of iterations up to which $T^k(X)$ can be assumed continuous in the neighbourhoods of points P_1 and P_2. More specifically, let Case A hold: the number $N_1 = N_1(X)$ is defined such that $T^{N_1}(X)P_1 \notin D_2$ and $T^k(X)P_1 \in D_2$ for $k = 1, 2, \ldots, N_1 - 1$; the number $N_2 = N_2(X)$ is defined such that $T^{N_2}(X)P_2 \notin D_1$ and $T^k(X)P_2 \in D_1$ for $k = 1, 2, \ldots, N_2 - 1$. In Case C, N_1 and N_2 are defined in a similar way. In Case B, $N_i = N_i(X)$ are defined such that $T^{N_i}(X)P_i \notin D_2$ and $T^k(X)P_i \in D_2$ for $k = 1, 2, \ldots, N_i - 1$.

Let us recall that the conditions imposed on $T(X)$ imply that, from Eq. (1.11), $q > 1$.

Theorem 1.6. *Let $X \in U_2$. If the condition*

$$q^{N_1+N_2} > q^{N_1} + q^{N_2}, \quad N_i > 1, i = 1, 2$$

is satisfied, the complete extension takes place.

Since on approaching the boundary of U_2 at least one of the numbers $N_1(X)$ and $N_2(X)$ tends to infinity, we find that a domain exists, in which $\tilde{\Sigma}(X)$ is the Lorenz attractor.

Statement 1.4. *The property of complete extension is structurally stable.*

It follows from Theorem 1.6 that if $\Sigma^-(X) \neq \varnothing$, two cases are possible:

$$\Sigma^+(X) \cap \Sigma^-(X) = \varnothing \qquad \text{or} \qquad \Sigma^+(X) \cap \Sigma^-(X) \neq \varnothing;$$

then for a certain m

$$\Sigma^+(X) \cap \Sigma^-(X) = \Sigma^+(X) \cap \Sigma_m^-(X).$$

Statement 1.5. *If $\Sigma^+(X^*) \cap \Sigma^-(X^*) = \varnothing$, then $\Sigma^+(X) \cap \Sigma^-(X) = \varnothing$ for all the systems of a certain small neighbourhood of X^*.* (Note in this connection that the proof of Theorem 1 of [47] is incorrect.)

This implies, in particular, that systems with only one lacuna are present at the boundary of the domain containing systems with complete extension.

It follows from the property of complete extension that all periodic motions have homoclinic orbits. Moreover, each component $W_L^u \backslash L$ of an unstable manifold of a periodic motion L intersects a stable manifold W_L^s. However, systems located at the boundary of the domain of systems with the Lorenz attractor do not possess this property. In the orientable case these systems can be divided into four classes, namely:

(1) $\Gamma_1(X)$ and $\Gamma_2(X)$ tend to a periodic motion $L(X)$, which has no homoclinic curves. In this case $\tilde{\Sigma}(X)$ is the Lorenz attractor and $\Sigma_1^-(X) = L(X)$.

(2) Only one separatrix, say $\Gamma_1(X)$, tends to a periodic motion L, which possesses no homoclinic curves. In this case each lacuna component has a finite height. Therefore $\tilde{\Sigma}(X) \neq \Sigma_1(X)$ and $\Sigma_1(X)$ is unstable, though it is ω-stable in $\Sigma(X)$.

(3) The separatrices $\Gamma_1(X)$ and $\Gamma_2(X)$ tend to periodic motions $L_1(X)$ and $L_2(X)$, respectively, which possess the following property: one of the components $W_{L_i}^u \backslash L_i$ does not intersect $W_{L_i}^s$. For this case

$$\Sigma^+(X) \cap \Sigma^-(X) = L_1(X) \cup L_2(X)$$

(4) Only one separatrix $\Gamma_i(X)$ tends to a periodic motion $L(X)$, for which only one component $W_L^u \backslash L$ has homoclinic curves, and $\Sigma^+(X) \cap \Sigma^-(X) = L(X)$.

In Case (2) and Case (4) the systems will belong to the rational surfaces of the second type. The set of such systems will be denoted by B_{iL}^1. In Case (1) and Case (3) the systems will form bifurcational sets of codimension two. They will be denoted by B_L^2 and $B_{L_1L_2}^2$, respectively. B_L^2 and $B_{L_1L_2}^2$ also belong to the rational surfaces of the second type and divide them into two components. In Case (1) one of the components, which will be denoted by B_{iL}^2, $i = 1, 2$, is characterized by the fact that it consists of systems corresponding to Case (2). In Case (3) we may also single out in

B_L^2 those components that are filled with systems corresponding to Case (4). These components will be denoted by $B_{1L_1}^1$ and $B_{2L_2}^1$.

Let $X^* \in B_L^2(B_{L_1L_2}^2)$ and let v_1 be its neighbourhood such that $B_{1L}^1 \cup B_L^2 \cup B_{2L}^1$ $(B_{1L_1}^1 \cup B_{L_1L_2}^2 \cup B_{2L_2}^1)$ divides v_1 into two domains: v^+ and v^-. In Case (1), by v^+ we mean a domain in which L has no homoclinic curves. In Case (2), by v^+ we mean a domain in which one of the periodic motions (and, hence, also the other one) has homoclinic curves belonging to both components of an unstable manifold. Let us recall that $\Gamma_1(X)$ and $\Gamma_2(X)$ are boundaries for each component of an unstable manifold of any periodic motion. Similarly, in the neighbourhood of $X^* \in B_{iL}^1(B_{iL_i}^1)$, B_{iL}^1 $(B_{iL_i}^1)$ divides v into two domains v^+ and v^-.

Statement 1.6. *Let $\Sigma(X^*)$ have only one lacuna. Then there exists a sufficiently small neighbourhood v such that for $X \in v^+$ $\tilde{\Sigma}(X)$ is the Lorenz attractor and for $X \in v^-$ $\Sigma_1(X) = \Sigma^+(X) \cup \Sigma_1^-(X)$, where $\Sigma^+(X) \cap \Sigma_1^-(X) = \varnothing$, and also the property of complete extension is satisfied on $\Sigma^+(X)$.*

Situations connected with the transition through B_L^2 and $B_{L_1L_2}^2$ correspond to the appearance of lacunae or systems with a symmetry. In Case A the corresponding mapping $T(X)$ will be written in the form (1.8) where $f_2(-x, -y) = -f_1(x, y)$, $g_2(-x, -y) = -g_1(x, y)$. Therefore in this case the separatrices behave similarly, i.e., they tend either to one and the same limiting set or to different, but naturally symmetrical, ones. This means that if one of the separatrices tends to a periodic motion such that $L(X) \cap D$ consists of an odd number of points, then under the above conditions a lacuna with a nontrivial hyperbolic set arises. For instance, if the number of points is five, $\tilde{\Sigma}(X)$ will be described by a TMC shown in Fig. 1.4. It can be shown that in going over from $\tilde{\Sigma}$ to $\tilde{\Sigma}^+$ the mapping topological entropy suffers a discontinuity, and $\text{ent}(T \mid \Sigma_1^-) > \text{ent}(T \mid \Sigma^+)$.

All the above considerations correspond to the orientable case. Similar bifurcations may, in general, take place also in Cases B and C. However, another type of bifurcations may arise here. It is connected with the fact that the separatrix Γ_i may tend to a periodic motion whose unstable manifold is a Möbius band and this, in turn, may lead to the formation of a trivial lacuna. In the symmetrical Case C the separatrices may tend to different periodic motions with simultaneous formation of trivial lacunae.

Let $U_A \subset U_2$ stand for a domain in which systems have the Lorenz attractor. A countable set of domains, consisting of systems with lacunae, is adjacent to this domain. The latter domains will be denoted by $U_{\Sigma_1^-}$ where the symbol Σ_1^- indicates that besides the attractor Σ^+ the system $X \in U_{\Sigma_1^-}$ contains basic sets Σ_1^-. The properties of Σ^+ are rather similar to those of the Lorenz attractor; in particular, the property of complete extension is satisfied for it. Therefore all the above considerations can be applied to Σ^+, hence the domains $U_{\Sigma_1^-\Sigma_2,\dots}$ can be constructed by successive selection of lacunae.

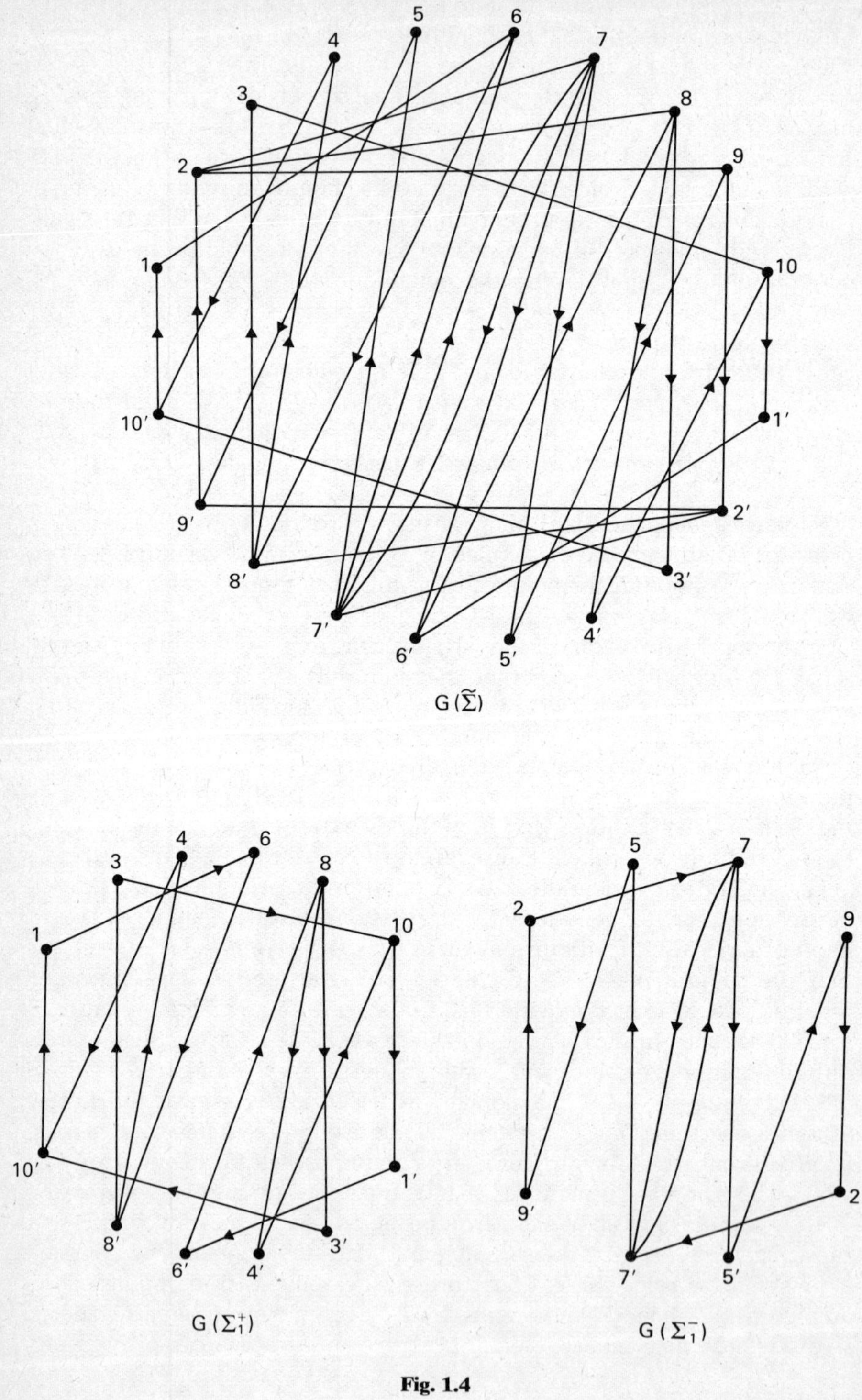

Fig. 1.4

1.5 Quasiattractors

To conclude this chapter, we shall discuss problems concerning an analogy between discontinuous and smooth mappings. When the separatrix values change their sign, the Poincaré mapping T in the Lorenz system will not satisfy conditions (1.9) but will act as shown in Fig. 1.5. The images of rectangles lying above and below the discontinuity line resemble 'hooks'. In such a situation an attempt can be made to simulate the mapping by a diffeomorphism [48–50]. It is clear that the analogy cannot be complete, but one may hope that many of the effects of the model take place for the original mapping too. (It seems more natural to relate mapping (1.14)—rather than (1.15) for $b<0$, as in [48, 49]—to mapping T in the Lorenz system.)

Let us consider examples of the simplest diffeomorphisms, which are similar, to a certain extent, to discontinuous mappings in Cases A, B and C.

Case A. A discontinuous mapping of a segment whose plot has two intervals of monotonous increase is a one-dimensional prototype for the mapping T in Case A. Proceeding from this fact, it is natural to seek 'basic' one-dimensional mappings among polynomials with not less than two intervals of increase. The simplest is y^3-ay, $a>0$. Let us extend this mapping to the diffeomorphism of a plane:

$$\bar{x}=y, \quad \bar{y}=y^3-ay-bx. \tag{1.14}$$

Mapping (1.14) is a diffeomorphism for all $b\neq 0$. If $b=0$, this mapping has an invariant cubic parabola $y=x^3-ax$, onto which any point of the plane is mapped for one iteration.

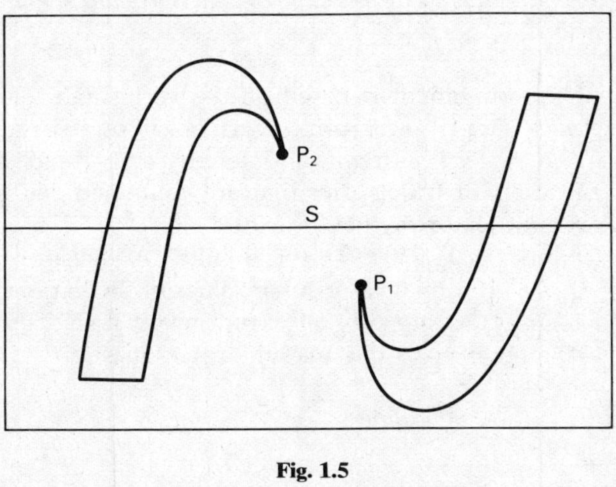

Fig. 1.5

Case B. For the semiorientable case let us take the Hénon diffeomorphism

$$\bar{x} = y, \quad \bar{y} = 1 - ay^2 - bx. \tag{1.15}$$

(This diffeomorphism is usually written in the form $\bar{x} = by$, $\bar{y} = 1 - ay^2 - x$. Obviously, it can be reduced to the form of (1.15) by the change $x \to x/b$, $y \to y$. Note that mapping (1.15) was considered in [17] in connection with the study of systems similar to that with a nonstructurally stable homoclinic curve.)

Case C. It is natural to consider the mapping

$$\bar{x} = y, \quad \bar{y} = ay - y^3 - bx \tag{1.16}$$

as an analog of the nonorientable case.

Like systems from domain U_1, these diffeomorphisms possess a nontrivial hyperbolic set: namely, the following statement is valid.

Statement 1.7.
(a) *If the condition*

$$a > \frac{2 + 3\sqrt{2}}{2}(1 + |b|) \tag{1.17}$$

holds for mappings (1.14) *and* (1.16), *each of them has an invariant hyperbolic set, on which both the mappings are conjugated with the topological Bernoulli scheme with three symbols;*
(b) *if the condition*

$$a > \frac{5 + 2\sqrt{5}}{4}(1 + |b|)^2 \tag{1.18}$$

is satisfied for mapping (1.15), *it has a nontrivial hyperbolic set, on which this mapping is conjugated with the topological Bernoulli scheme with two symbols* (*see also* [51]).

The proof of this statement is based on the verification of conditions (1.9) for mappings (1.14)–(1.16) on a certain subset of the rectangle $D = \{(x, y) \mid |x| \leq \beta, |y| \leq \alpha\}$ with appropriately chosen α and β.

The set Σ of points of trajectories that are contained entirely in D is unstable. All semitrajectories that do not belong to Σ and to stable manifolds of points of Σ leave D for infinity. A question arises: do mappings (1.14) and (1.16) have, for some values of the parameters a and b, domains that the trajectory cannot leave, once it finds itself in them? The following statement gives the answer to this question.

Statement 1.8. *If the condition*

$$a < 3(1 - |b|), \qquad |b| < 1 \tag{1.19}$$

holds for mappings (1.14), (1.16) *and the condition*

$$a<2(1-|b|)^2, \quad |b|<1 \tag{1.20}$$

for mapping (1.15), *then for each of these mappings a rectangle exists on the plane* $\{(x, y)\}$ *which is mapped into itself.*

To prove this statement, one has to find a rectangle $D=\{(x, y) \mid |x|\leqslant\beta, |y|\leqslant\alpha\}$, which is mapped into itself. If the inequalities (in α and β)

$$\alpha^3-a\alpha+|b|\,\beta\leqslant\alpha, \qquad \tfrac{2}{3}a\sqrt{\left(\frac{a}{3}\right)}+|b|\,\beta\leqslant\beta, \quad \alpha<\beta$$

are satisfied, the images of the boundaries of D do belong to D; the solution of these inequalities exists if (1.19) holds. For the mapping (1.15) the inequalities

$$1-a\alpha^2-|b|\,\beta\geqslant-\alpha, \qquad 1-|b|\,\beta\leqslant\alpha, \quad \alpha<\beta$$

should be satisfied; their solution exists if (1.20) holds.

Let us now find out how quasiattractors arise when the parameter a varies from the values satisfying inequalities (1.17), (1.18) to those satisfying (1.19) and (1.20), respectively. In other words, let us determine bifurcations responsible for their generation. To do this, we have to know which among stable and unstable manifolds of points Σ are 'extreme' ones. The following statement is valid.

Statement 1.9. *The set* Σ *is located in the cross-hatched domains* (*Fig.* 1.6) *whose boundaries are*:

(1) *in Case A, segments of stable and unstable manifolds of fixed points* M_1 *and* M_2 *for* $b>0$, *segments of stable manifolds of points* M_1 *and* M_2 *and of unstable manifolds of periodic points* Q_1 *and* Q_2 ($TQ_1=Q_2$, $TQ_2=Q_1$) *for* $b<0$ (*hereinafter T is one of the mappings* (1.14)–(1.16));
(2) *in Case B, segments of stable and unstable manifolds of the fixed point* M_1 *for* $b>0$, *segments of a stable manifold of the point* M_1 *and of an unstable manifold of the point* M_2 *for* $b<0$;
(3) *in Case C, segments of stable and unstable manifolds of periodic points* Q_1 *and* Q_2 ($TQ_1=Q_2$, $TQ_2=Q_1$) *for* $b>0$, *segments of stable manifolds of periodic points* Q_1 *and* Q_2, *and of unstable manifolds of fixed points* M_1 *and* M_2 *for* $b<0$.

It can easily be verified directly that for $a=3(1-|b|)$ for mappings (1.14) and (1.16), and for $a=2(1-|b|)^2$ for mapping (1.15) (and hence for values that differ little from the above ones) the points M_1, M_2 and Q_1, Q_2 remain saddles. Therefore, it follows from Statement 1.3 that with decreasing a the first bifurcation connected with the violation of the transversality of intersection of $W^s(\Sigma)$ and $W^u(\Sigma)$ occurs at the moment when stable and unstable manifolds, specified in Statement 1.3, become tangent. The moment of formation of a quasiattractor is also determined

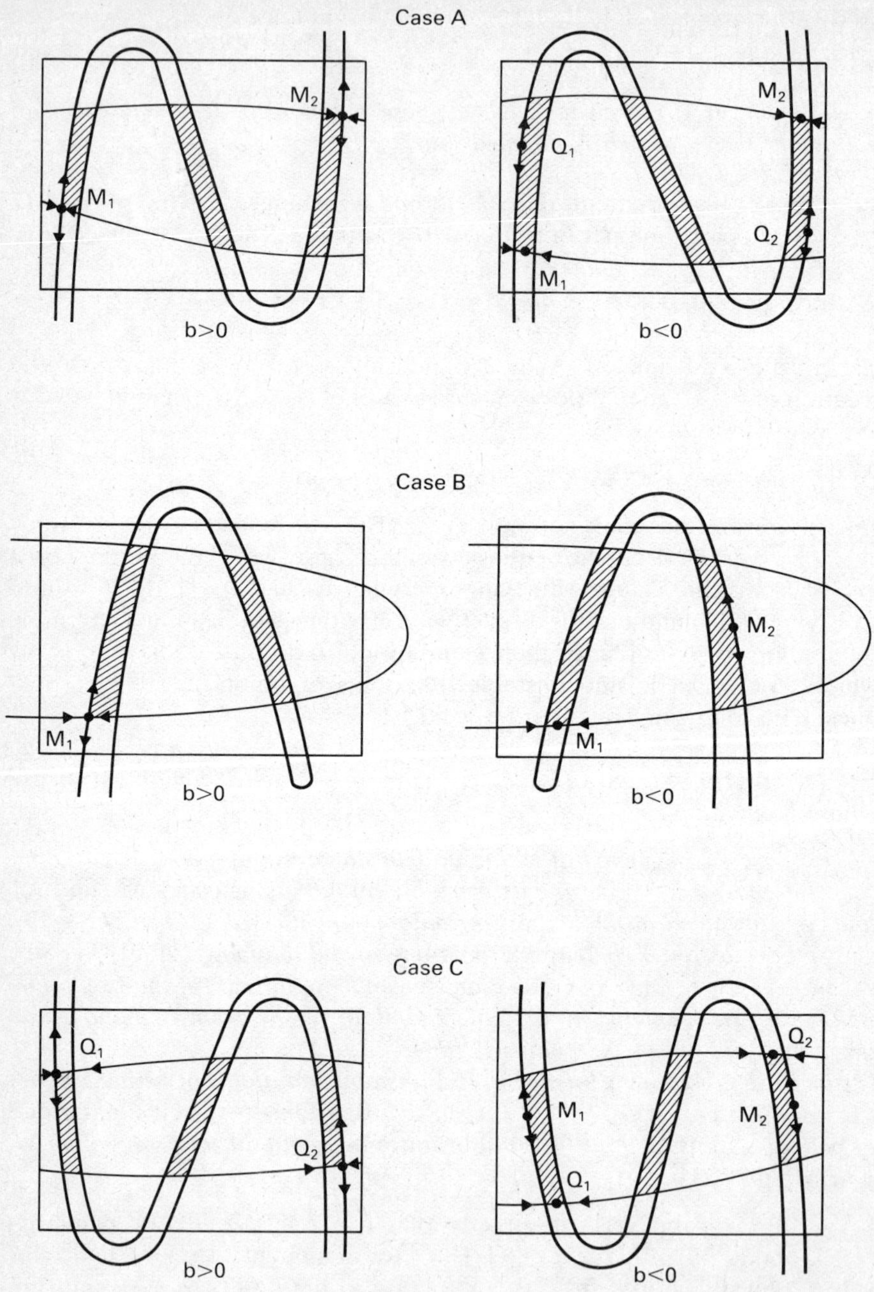
Case A
M$_2$
M$_1$
b>0
M$_2$
Q$_1$
Q$_2$
M$_1$
b<0
Case B
M$_1$
b>0
M$_2$
M$_1$
b<0
Case C
Q$_1$
Q$_2$
b>0
Q$_2$
M$_1$
M$_2$
Q$_1$
b<0

Fig. 1.6

by the tangency of just those manifolds. To be exact, another statement holds.

Statement 1.10. *A quasiattractor for the mapping T arises for such values of the parameters a and b when:*

(1a) *in Case A for $b>0$, an unstable manifold of the point $M_1(M_2)$ becomes tangent to a stable manifold of the point $M_2(M_1)$;*
(1b) *in Case A for $b<0$, an unstable manifold of the point $Q_1(Q_2)$ becomes tangent to a stable manifold of one of the points M_1, M_2 and does not contain structurally stable heteroclinic trajectories ω-asymptotic to M_1 or M_2;*
(2a) *in Case B for $b>0$, there exists a nonstructurally stable homoclinic trajectory of the point M_1 and there exist no structurally stable homoclinic trajectories;*
(2b) *in Case B for $b<0$, a stable manifold of the point M_1 is tangent to an unstable manifold of the point M_2 (see [49, 50], which report computation results for stable and unstable manifolds and bifurcational values corresponding to the formation of quasiattractors);*
(3a) *in Case C for $b>0$, an unstable manifold of the point $Q_1(Q_2)$ is tangent to a stable manifold of the point $Q_2(Q_1)$;*
(3b) *in Case C for $b<0$, an unstable manifold of the point $M_1(M_2)$ is tangent to a stable manifold of one of the points Q_1, Q_2.*

Since the model mappings (1.14)–(1.16) perform nonstructurally stable tangency of stable and unstable manifolds, these mappings are also characterized by the whole spectrum of bifurcations typical of systems with a nonstructurally stable homoclinic curve of a periodic motion whose saddle value is less than unity (see [17, 18]). These are, in particular, generation of stable periodic motions, alternation of stability and instability, etc.

References

1. Monin, A. and Yaglom, A., *Statistical Hydromechanics*: Moscow, Nauka, 1965, 1967, Parts 1 and 2 (in Russian).
2. Ruelle, D., Strange attractors, *Mathematical Intelligencer,* **2,** 126–137, 1980.
3. Anosov, D., Geodesic flows on compact Riemannian manifolds of negative curvature, *Trudy Matematicheskogo Instituta im. V. A. Steklova,* **90,** 1967 (in Russian).
4. Smale, S., Differentiable dynamical systems, *Bulletin of the American Mathematical Society,* **73,** 747–817, 1967.
5. Williams, R., Classification of one-dimensional attractors, *Proceedings of Symposium in Pure Mathematics,* No. 14: Providence, American Mathematical Society, 1970, 361–393.

6. Plykin, R., Sources and sinks of A-diffeomorphisms of surfaces, *Matematichesky Sbornik*, **94,** 243–264, 1974.
7. Ruelle, D. and Takens, F., On the nature of turbulence, *Communications in Mathematical Physics*, **20,** 167–192, 1971.
8. Afraimovich, V., Bykov, V. and Shil'nikov, L., On the appearance and structure of Lorenz attractor, *Doklady Akademii Nauk USSR*, **234**(2), 336–339, 1977.
9. Lorenz, E., Deterministic nonperiodic flow, *Journal of the Atmospheric Sciences*, **20,** 134–141, 1963.
10. Ruelle, D., The Lorenz attractor and the problem of turbulence, *Lecture Notes in Mathematics*, **565,** 146–158, 1976.
11. Kaplan, J. and York, J., A metastable regime in the system of Lorenz (preprint), 1977.
12. Haken, H., Analogy between higher instabilities in fluid and lasers, *Physics Letters*, **53A,** 77–79, 1975.
13. Robbins, K., A new approach to subcritical instability and turbulent transitions in a simple dynamo, *Mathematical Proceedings of the Cambridge Philosophical Society*, **82,** 309–325, 1977.
14. Afraimovich, V., Bykov, V. and Shil'nikov, L., On the existence of stable periodic motions in the Lorenz model, *Uspekhi Matematicheskikh Nauk*, **35,** 164–165, 1980.
15. Bykov, V., On the structure of the neighbourhood of a separatrix contour with a saddle-focus, in *Methods of Qualitative Theory of Differential Equations*: Gorky State University, 1978, 3–32 (in Russian).
16. Bykov, V., On bifurcations of dynamical systems similar to those with a separatrix contour containing a saddle-focus, in *Methods of Qualitative Theory of Differential Equations*: Gorky State University, 1980, 44–72 (in Russian).
17. Gavrilov, N. and Shil'nikov, L., On three-dimensional dynamical systems similar to those with a nonstructurally stable homoclinic curve, *Matematichesky Sbornik*, **88,** 475–492, 1972; **90,** 139–156, 1973.
18. Newhouse, S., Diffeomorphisms with infinitely many sinks, *Topology*, **12,** 9–18, 1974.
19. Afraimovich, V. and Shil'nikov, L., On small periodic perturbations of autonomous systems, *Doklady Akademii Nauk USSR*, **214,** 739–742, 1974.
20. Afraimovich, V. and Shil'nikov, L., On certain global bifurcations connected with the disappearance of a fixed point of the saddle-node type, *Doklady Akademii Nauk USSR*, **219,** 1281–1285, 1974.
21. Afraimovich, V. and Shil'nikov, L., The ring principle and the problem of the interaction of two self-oscillating systems, *Prikladnaya Matematika i Mekhanika*, **41,** 618–627, 1977.
22. Shil'nikov, L., Theory of bifurcations of dynamical systems and dangerous boundaries, *Doklady Akademii Nauk USSR*, **234,** 1046–1051, 1975.

23. Lukyanov, V. and Shil'nikov, L., On certain bifurcations of dynamical systems with homoclinic structures, *Doklady Akademii Nauk USSR*, **243,** 26–29, 1978.
24. Shil'nikov, L., Theory of bifurcations of dynamical systems with homoclinic Poincaré curves, *VII Internationale Konferenz über nichtlineare Schwingungen*, Vols 1, 2: Berlin, Akademie-Verlag, 1977, 279–293.
25. Shil'nikov, L., Theory of bifurcations and the Lorenz model, Supplement II to the Russian translation of Marsden, J. and McCracken, J., *The Hopf Bifurcation and Applications*: Moscow, Mir, 1980, 317–335 (in Russian).
26. Afraimovich, V., Bykov, V. and Shil'nikov, L., On the attracting nonstructurally stable limiting sets of the Lorenz attractor type, *Trudy Moskovskogo Matematicheskogo Obshchestva*, **44,** 150–212, 1982.
27. Smale, S., Diffeomorphisms with many periodic points, *Differential and Combinatorial Topology*, No. 9: Princeton University Press, 1965, 63–80.
28. Shil'nikov, L., On a certain Poincaré–Birkhoff problem, *Matematichesky Sbornik*, **74,** 378–397, 1967.
29. Alexeev, V., Perron sets and topological Markov chains, *Uspekhi Matematicheskikh Nauk*, **24,** 227–228, 1969.
30. Bowen, R., Markov partitions for Axiom A diffeomorphisms, *American Journal of Mathematics*, **92,** 907–918, 1970.
31. Katok, A. *et al.*, Theory of dynamical systems and general groups and transformations with an invariant measure, in *Matematichesky Analiz*, Moscow, Viniti, **13,** 129–262, 1975 (in Russian).
32. Yakobson, M., On certain properties of Markov partitions, *Doklady Akademii Nauk USSR*, **226,** 1021–1024, 1976.
33. Bowen, R. and Walters, P., Expansive one-parameter flows, *Journal of Differential Equations*, **12,** 180–193, 1972.
34. Nemytsky, V. and Stepanov, V., *Qualitative Theory of Differential Equations*: Moscow–Leningrad, Gostekhizdat, 1949 (in Russian).
35. Nitecki, Z., *Differentiable Dynamics*: Cambridge, Massachusetts and London, MIT Press, 1971.
36. Alexeev, V., *Symbolic Dynamics*: Kiev, XI Matematicheskaya Shkola, Institute of Mathematics of the USSR Academy of Sciences, 1976 (in Russian).
37. Sinai, Ya., Markov partitions and Y-diffeomorphisms, *Funktsionalny Analiz i yego Prilogeniya* (*Functional Analysis and its Applications*), **2,** 64–89, 1968 (in Russian).
38. Alexeev, V. and Yakobson, M., Symbolic dynamics and hyperbolic dynamical systems, Supplement to the Russian translation of Bowen, R., *Methods of Symbolic Dynamics*: Moscow, Mir, 1979 196–240.
39. Shil'nikov, L., On certain cases of generation of periodic motions from singular trajectories, *Matematichesky Sbornik*, **61,** 433–466, 1963.

40. Shil'nikov, L., On generation of periodic motion from a trajectory doubly asymptotic to the saddle-type equilibrium state, *Matematichesky Sbornik*, **77,** 461–472, 1968.
41. Simonov, A., The study of bifurcations in certain dynamical systems by symbolic dynamics methods, *Doklady Akademii Nauk USSR*, **240,** 1297–1300, 1978.
42. Milnor, J. and Thurston, W., On iterated maps of the interval (preprint), 1977.
43. Rand, D., The topological classification of Lorenz attractor, *Mathematical Proceedings of the Cambridge Philosophical Society*, **83,** 451–460, 1978.
44. Williams, R., The structure of Lorenz attractors (preprint), 1977.
45. Williams, R., The structure of Lorenz attractors, *Lecture Notes in Mathematics*, **615,** 94–112, 1977.
46. Guckenheimer, J., Structural stability of Lorenz attractor (preprint), 1977.
47. Guckenheimer, J., A strange, strange attractor, in Marsden J. and McCracken, J., *The Hopf Bifurcation and Applications*, Applied Mathematical Sciences No. 19: Berlin, Springer-Verlag, 1976.
48. Hénon, M., A two-dimensional mapping with a strange attractor, *Communications in Mathematical Physics*, **50,** 69–77, 1976.
49. Simó, C., On the Hénon-Pomeau attractor (preprint), Barcelona, 1978.
50. Gumowski, I. and Mira, C., *Dynamique Chaotique, Transformations Ponctuelles, Transition Ordre–Désordre*: Toulouse, Editions Cepadues, 1980.
51. Devaney, R. and Nitecki, Z., Shift automorphisms in the Hénon mapping, *Communications in Mathematical Physics*, **67,** 137–146, 1979.

2 Resonance Phenomena for Two-parameter Families of Maps of the Plane: Uniqueness and Nonuniqueness of Rotation Numbers

D G Aronson, M A Chory, G R Hall and R P McGehee

In their 1945 paper on the periodically forced van der Pol equation [1], Cartwright and Littlewood observed that for certain parameter values the equation possesses stable periodic solutions with distinct periods. These costable periodic orbits were found also by Levinson [2] for a piecewise linear version of the van der Pol equation and have been studied numerically by Flaherty and Hoppensteadt [3]. A geometric interpretation of these results has been given by Levi [4]. In this chapter we describe and interpret similar phenomena that we have observed in numerical studies of certain maps of the plane.

2.1 Observations on a one-parameter family

We begin by considering the so-called delayed logistic model in population dynamics:

$$N_{n+1} = aN_n(1 - N_{n-1}). \tag{2.1}$$

Here N_n represents the density of some population in its nth generation and a is a positive parameter. Since Eq. (2.1) is a second-order difference equation, we can rewrite it as a system of two first-order equations. Thus, if we set

$$x_n = N_{n-1} \qquad \text{and} \qquad y_n = N_n,$$

Eq. (2.1) becomes

$$\mathbf{x}_{n+1} = F_a(\mathbf{x}_n), \tag{2.2}$$

where $\mathbf{x} = (x, y)$ and $F_a(\mathbf{x}) \equiv [y, ay(1-x)]$. We shall regard $\{F_a\}$ as a one-parameter family of maps of R^2 into itself.

The map F_a has fixed points at $(x, y) = (0, 0)$ and at

$$(x, y) = \mathbf{p}_a \equiv \frac{a-1}{a}(1, 1).$$

The origin is stable for $0<a<1$, while $\mathbf{p}_a$ is stable for $1<a<2$. When the value of the parameter a is near 2, the Jacobian of F_a at $\mathbf{p}_a$ has a complex conjugate pair of eigenvalues which lie inside the unit circle. As a passes through the value 2, these eigenvalues pass to the outside of the unit circle through the points $\exp(\pm i\pi/3)$. Thus the fixed point loses stability and spawns an attracting invariant circle A_a via a phenomenon known as Hopf bifurcation. For sufficiently small $a-2>0$, A_a is a smooth circle, as ensured by the Hopf bifurcation theorem. As a increases, the invariant set A_a can be continued as an attractor, although it may lose its integrity as a circle. Indeed, computer simulations indicate that the attractor A_a passes through a complicated sequence of bifurcations, eventually becoming a so-called strange attractor.

The set A_a is illustrated in Fig. 2.1 for various values of a. Each of the sets shown in Fig. 2.1 consists of iterates 1000 through 2500 of the point $(0.1, 0.2)$ under the map F_a. Note that, as a increases, A_a grows and

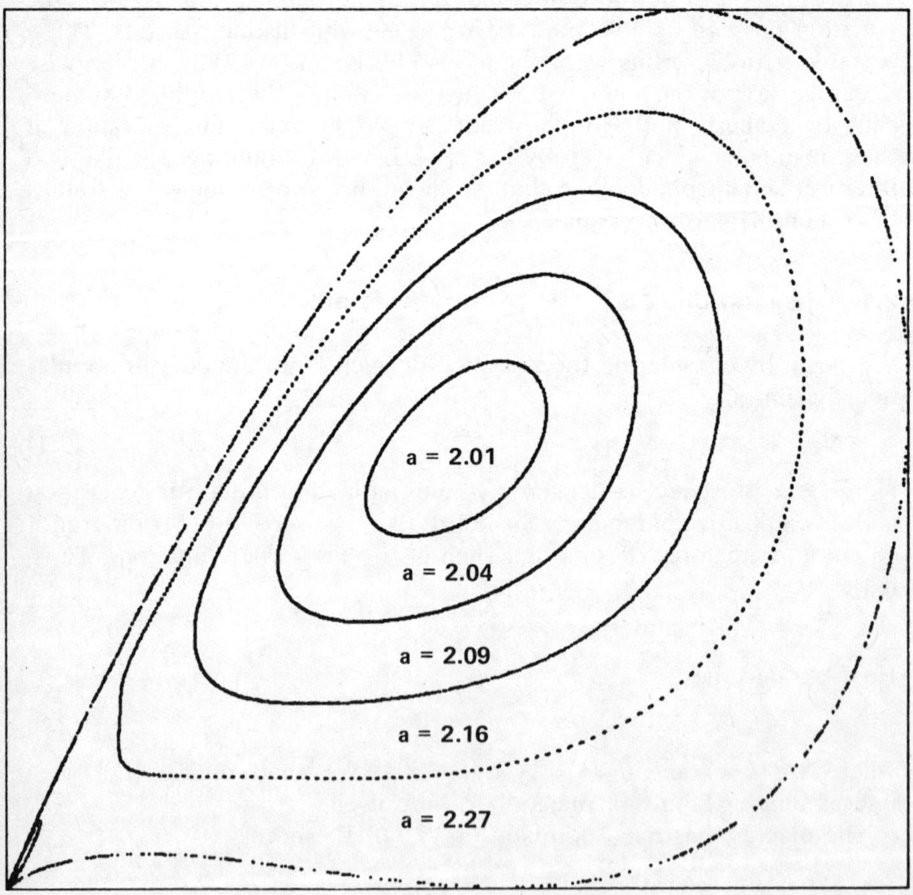

Fig. 2.1 The attractor A_a for various values of a

changes shape, becoming noticeably warped by the time a reaches a value of 2.16. For a larger than a certain critical value $a^* \approx 2.27$, there is no attractor in the first quadrant. For values of a just below a^*, the set A_a is what is usually called a strange attractor, illustrated in various magnifications in Fig. 2.2. For more detail concerning the structure of the map F_{a^*} the reader is referred to [5] and [6].

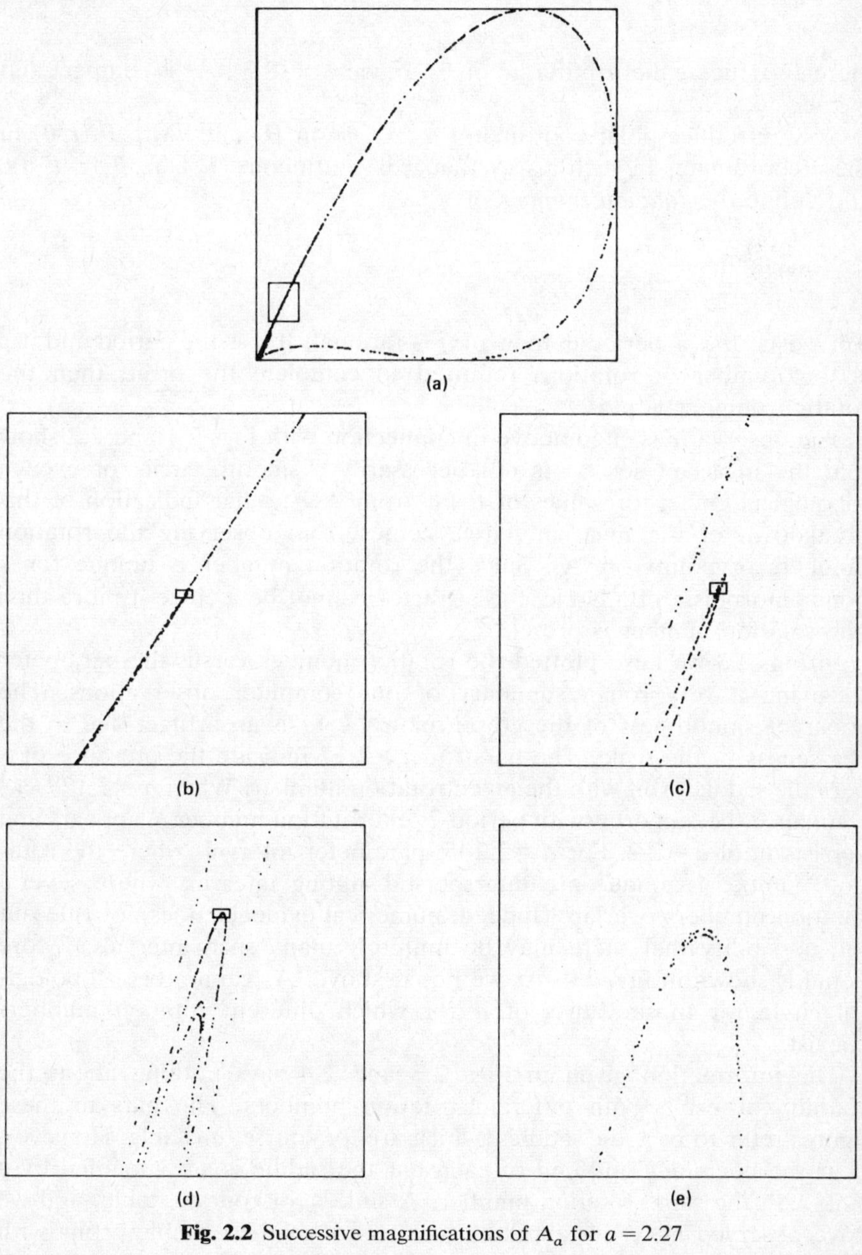

Fig. 2.2 Successive magnifications of A_a for $a = 2.27$

For $2<a<a^*$, the attractor A_a is contained in an annular attractor block B_a: that is, there exists a closed annulus B_a such that

$$F_a(B_a) \subset \mathrm{int}(B_a)$$

and

$$A_a = \bigcap_{n>0} F_a^n(B_a). \tag{2.3}$$

Here F_a^n denotes the nth iterate of F_a. In view of (2.3), A_a is compact and connected.

Now introduce polar coordinates $\mathbf{x}=(r, \theta)$ on B_a and write $F_a(r, \theta)$ in these coordinates in such a way that it is continuous. Let $(r_n, \theta_n)=F_a(\mathbf{x})$ and define the *rotation number* of $\mathbf{x}$ as

$$\rho(\mathbf{x}) \equiv \lim_{n\to\infty} \frac{\theta_n}{2\pi n},$$

if it exists. If $\mathbf{x}$ is periodic, then $\rho(\mathbf{x})$ is rational. If q is the period and if p is the number of rotations required to complete the orbit, then the rotation number is p/q.

The observations cited above in connection with Figs 2.1 and 2.2 show that the invariant set A_a is not necessarily a smooth circle, or even a topological circle, for values of a far from 2. Another indication of this breakdown of the invariant circle comes from observing the rotation numbers of points on A_a. Since the rotation number is unique for a homeomorphism of a circle, the attractor cannot be a circle if more than one rotation number is present.

In Fig. 2.3 we have plotted the rotation number versus the parameter a, giving a very rough summary of our computer observations. The apparent smoothness of the graph for $a<2.17$ is an artifact due to the coarseness of the scale. The bars for $a>2.17$ indicate the presence of a periodic saddle orbit with the given rotation number. When $a\approx 2.177$, a $\frac{1}{7}$ resonance (i.e., an orbit with period 7 and rotation number $\frac{1}{7}$) appears and persists until $a\approx 2.2$. For $a>2.205$, parameter intervals where the rotation number is unique are interspersed among intervals where several rotation numbers overlap. (Indeed, numerical evidence does not rule out the possibility that there may be infinitely many such intervals.) More detail is shown in Fig. 2.4. As we noted above, A_a cannot be a topological circle for those values of a for which different rotation numbers coexist.

The information given in Figs 2.3 and 2.4 says nothing about the stability of orbits with different rotation numbers. The bars in these figures refer to periodic saddles, which are, of course, unstable. However, near the beginning and end of each bar the saddle is accompanied by a sink with the same rotation number. A sink is, of course, stable, and we have observed many instances of coexistence of these stable orbits with

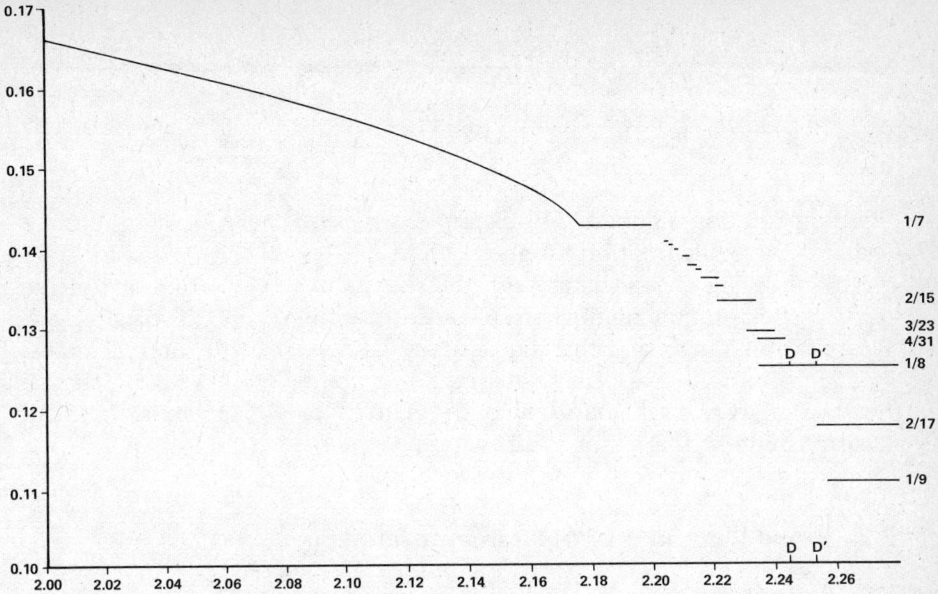

Fig. 2.3 Rotation number versus parameter a

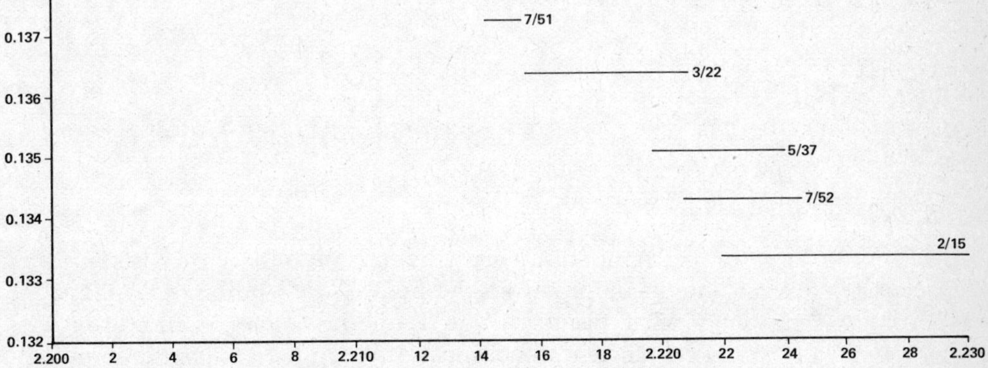

Fig. 2.4 Enlargement of part of Fig. 2.3

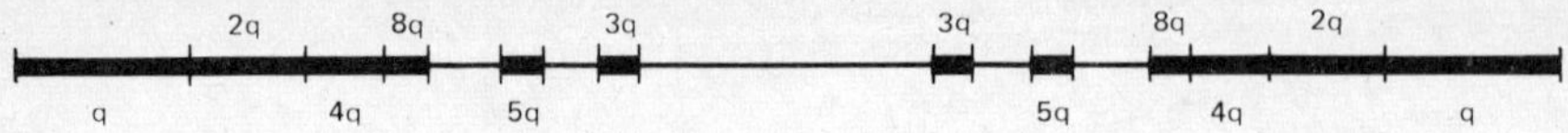

Fig. 2.5 Some of the stable periodic orbits observed along a typical resonance line

different rotation numbers. For example, when a is near 2.2344 there are stable orbits with rotation numbers $\frac{1}{8}$ and $\frac{2}{15}$.

The bifurcations associated with the sink as a moves across one of the bars in the rotation number graph are quite interesting. Typically, one sees the full range of behaviour usually associated with interval maps: cascading bifurcations $\{q2^n\}$ and $\{kq2^n\}$ for odd k followed by 'chaos'. Figure 2.5 gives a schematic summary of the typical behaviour. We shall return briefly to this point later on.

2.2 Embedding in a two-parameter family

We found it difficult to understand the transitions we observed until we were inspired by the work of Arnol'd [7] to embed $\{F_a\}$ in a two-parameter family of maps. For computational and conceptual simplicity we chose the family

$$F_{(a,b)}(\mathbf{x}) \equiv [y+bx,\ ay(1-x)]$$

for $0 \leq b < 1$. Note that $F_a = F_{(a,0)}$. The map $F_{(a,b)}$ has a fixed point

$$\mathbf{x} = \mathbf{p}_{(a,b)} \equiv \frac{a-1}{a}(1, 1-b),$$

and the eigenvalues of the Jacobian $DF_{(a,b)}(\mathbf{p}_{(a,b)})$ are complex for $\frac{5}{3} < a < 4$. For each $b \in [0, 1)$ a Hopf bifurcation to an attracting invariant circle occurs as a crosses the line $a = 2$ in the (a, b)-plane. The eigenvalues of $DF_{(a,b)}(\mathbf{p}_{(a,b)})$ are $\exp(\pm i\alpha)$ where

$$\cos \alpha = \frac{1+b}{2}.$$

Therefore the p/q resonance occurs for $b = b^*$, where b^* satisfies

$$\cos \frac{2\pi p}{q} = \frac{1+b^*}{2}.$$

According to the Arnol'd theory, the p/q resonance persists in a horn-shaped region emanating from the point $(2, b^*)$ in the (a, b)-plane. Figure 2.6 shows what occurs along a one-parameter section which traverses the p/q resonance horn near its tip. When the parameter point is outside the horn, no periodic point with rotation number p/q is present. When the parameter point encounters the boundary of the horn, q saddle-nodes appear on the invariant circle. These saddle-nodes are the iterates of a periodic point of period q, and three of them are labelled ab, $a'b'$ and $a''b''$ in Fig. 2.6. As the parameter point passes into the interior of the horn, the saddle-nodes bifurcate to form saddle–sink pairs. Figure 2.6 shows three of the sinks, labelled a, a' and a'', and three of the saddles, labelled b, b' and b''. There are now two distinct periodic orbits on the invariant circle, one consisting of saddles and the other consisting of sinks. The qth iterate of the map has $2q$ fixed points, alternating between saddles and sinks around the invariant circle. As the parameter point continues to move across the horn, the saddles and sinks move apart. As the point approaches the other boundary of the horn, the periodic points form different saddle–sink pairs. The saddle labelled b in Fig. 2.6 pairs with the sink labelled a', the saddle labelled b' pairs with the sink labelled a'', etc. Finally, when the parameter point reaches the other boundary of the horn, these new saddle–sink pairs combine to form saddle-nodes. Two of these saddle-nodes are shown in Fig. 2.6, labelled $a'b$ and $a''b'$.

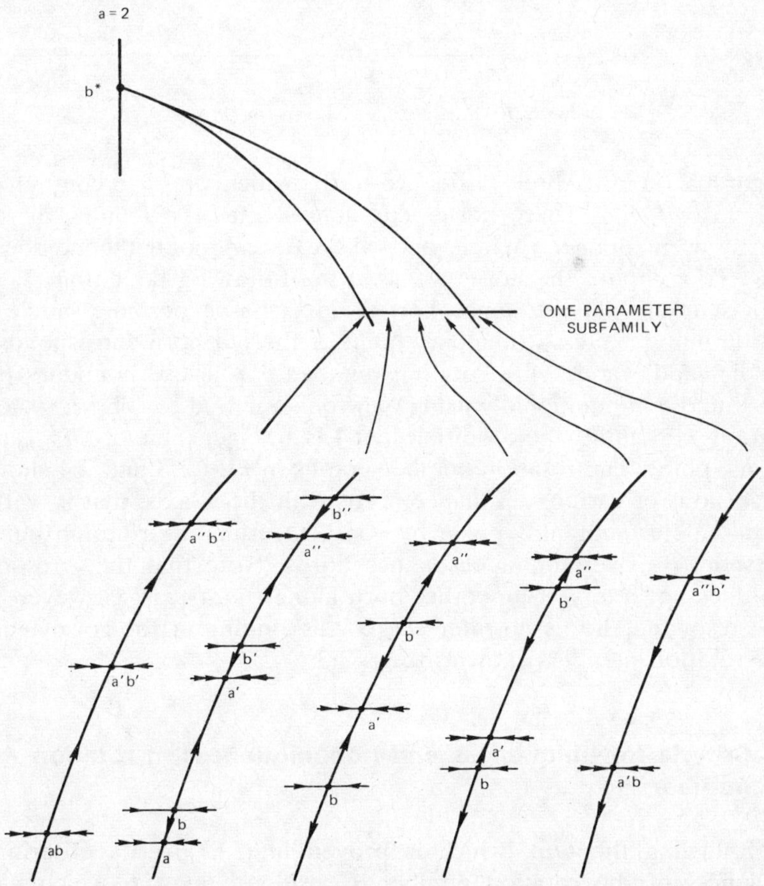

Fig. 2.6 Relative positions of saddles and sinks across a resonance horn

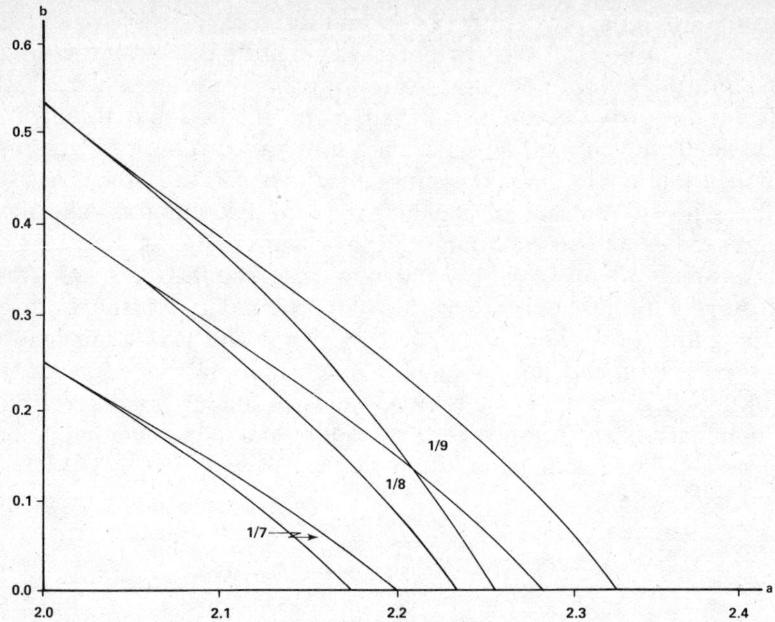

Fig. 2.7 Three resonance horns for $\{F_{(a,b)}\}$

Figure 2.7 shows three resonance horns which we have computed for the family $\{F_{(a,b)}\}$. These horns extend well into the region where the attractor is no longer a circle and where the Arnol'd theory does not apply. To compute the horns we used the following definition. A point (a, b) is in the p/q resonance horn if $F_{(a,b)}$ has a periodic saddle with rotation number p/q. A boundary point of the p/q horn corresponds to a periodic saddle-node with rotation number p/q. These boundary points were found computationally using Newton's method to solve a system of equations containing the condition that 1 is an eigenvalue of $DF_{(a,b)}$ at the periodic point. The rotation number graphs in Figs 2.3 and 2.4 show the intersections of various resonance horns with the a-axis, that is, with the one parameter subfamily given by $b=0$. Overlapping rotation numbers correspond to overlapping resonance horns. Note that the $\frac{1}{7}$ resonance horn does not intersect any other horn along the a-axis. However, the $\frac{1}{8}$ and $\frac{1}{9}$ resonance horns do intersect, corresponding to the coexistence of these rotation numbers shown in Fig. 2.3.

2.3 Criteria for uniqueness and nonuniqueness of rotation number

The following theorem, which is proved in [5], gives a criterion for uniqueness of the rotation number. Consider a map F which has an annular attractor block B with corresponding attractor A. The attractor separates the block into two components: B_i, containing the inside boundary of B, and B_0, containing the outside boundary of B. If $\mathbf{y}$ is a fixed point of F^q, then we denote by $W^s(\mathbf{y}; F^q)$ the stable manifold and by $W^u(\mathbf{y}; F^q)$ the unstable manifold of the point $\mathbf{y}$.

Theorem 2.1. *Suppose that F satisfies the following three hypotheses:*

(a) *There exists a periodic saddle point* $\mathbf{y} \in A$ *with positive eigenvalues.*
(b) $W^s(\mathbf{y}; F^q) \cap A = \{\mathbf{y}\}$, *where q is the period of* $\mathbf{y}$.
(c) *One branch of* $W^s(\mathbf{y}, F^q)$ *intersects only* B_i, *while the other branch intersects only* B_0.

Then $\rho(\mathbf{z}) = \rho(\mathbf{y})$ *for all* $\mathbf{z} \in B$.

For the maps $\{F_{(a,b)}\}$ our computer observations indicate that the hypotheses of Theorem 2.1 are satisfied for points (a, b) which are inside a resonance horn and sufficiently close to its tip. Inside the p/q resonance horn and near its tip, the attractor $A_{(a,b)}$ is a smooth circle which consists of q saddles, q sinks, and the unstable manifolds of the saddles, as shown in Fig 2.8(a). However, as the parameter point moves away from the tip of the horn, the invariant circle becomes distorted until ultimately there are transverse homoclinic points. This process is shown in Figs 2.8(b), (c) and (d). A point $\mathbf{x}$ not contained in the saddle orbit $\{\mathbf{y}, F(\mathbf{y}), \ldots, F^{q-1}(\mathbf{y})\}$ is said to be *homoclinic* to that periodic orbit if

$$\mathbf{x} \in \left[\bigcup_{k=0}^{q-1} W^u(F^k(\mathbf{y}); F^q)\right] \cap \left[\bigcup_{k=0}^{q-1} W^s(F^k(\mathbf{y}); F^q)\right],$$

and it is a *transverse homoclinic point* if the intersection of the stable and unstable manifolds is nontangential at $\mathbf{x}$. Note that in Fig. 2.8 the homoclinic point occurs as the intersection of the stable and unstable manifolds of distinct iterates of the periodic point. The occurrence of such homoclinic points for the map $F_{(a,b)}$ means that the hypotheses of the following theorem are satisfied and that there is no unique rotation number for points on $A_{(a,b)}$.

Theorem 2.2. *Let* $\mathbf{y} \in A$ *be a periodic point of period q and rotation number p/q, where p and q are relatively prime. Suppose that* $W^u(\mathbf{y}; F^q)$ *intersects* $W^s(F^k(\mathbf{y}); F^q)$ *transversely for some integer k with* $0<k<q$. *Then there exists a nontrivial interval I containing p/q such that, for every* $\alpha \in I$, *there is a* $\mathbf{z} \in A$ *with* $\rho(\mathbf{z}) = \alpha$. *Furthermore, there are points in A for which the rotation number fails to exist.*

Note that the existence of a nondegenerate homoclinic point, which we assume in Theorem 2.2, also implies the existence of a Smale horseshoe and hence of an invariant Cantor set on which the map is equivalent to a Bernoulli shift [8]. From this one can deduce the existence of orbits with infinitely many different periods, but, without further construction, one

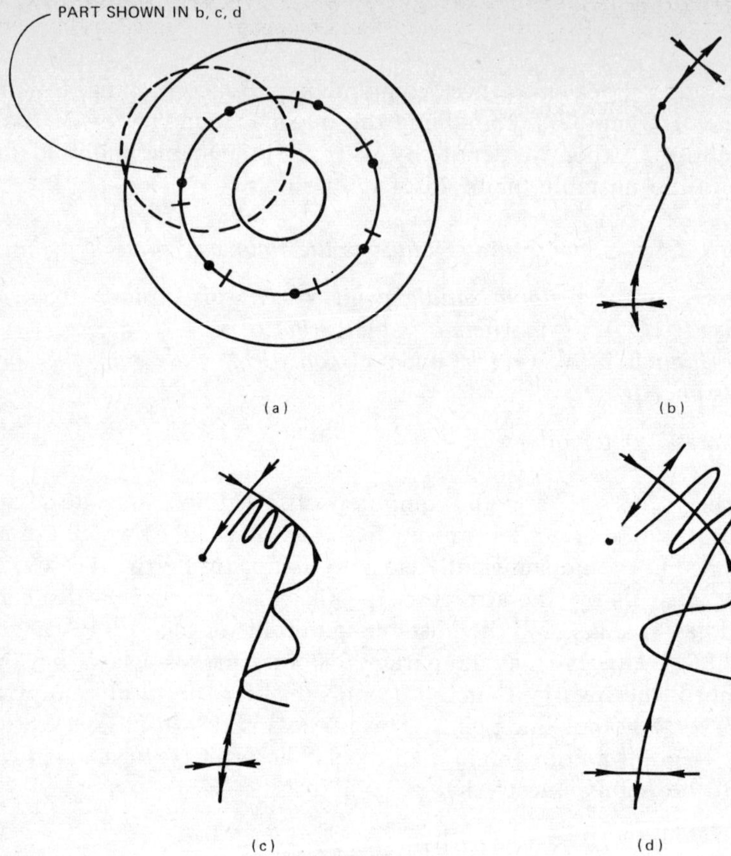

Fig. 2.8 The development of homoclinic points

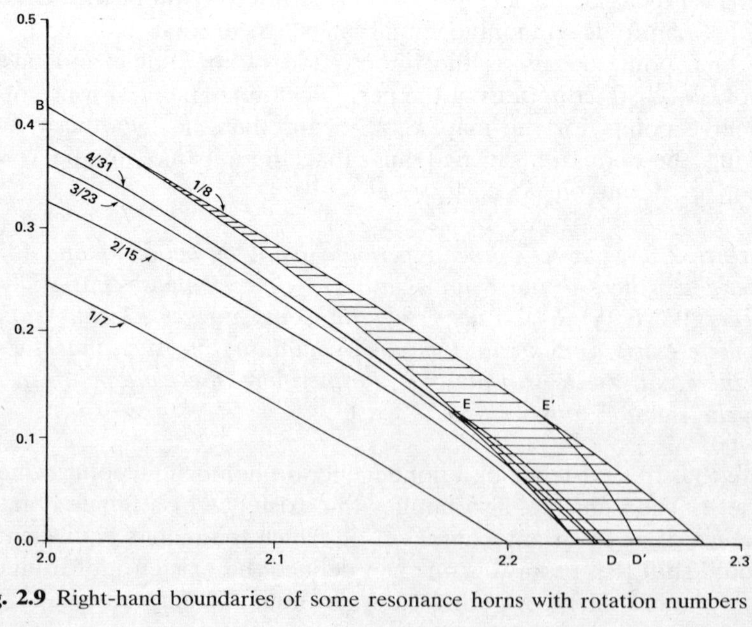

Fig. 2.9 Right-hand boundaries of some resonance horns with rotation numbers $>\frac{1}{8}$

cannot conclude the existence of an interval of rotation numbers. Our proof of Theorem 2.2, given in [5], uses the theory of pseudo-orbits and shadows ([9], [10]) instead of the horseshoe construction.

Computer observations show that the homoclinic tangencies which lead to nonunique rotation numbers (*cf.* Fig. 2.8(*c*)) occur along arcs in the resonance horns. For example, the $\frac{1}{8}$ resonance horn is shown in Fig. 2.9, where these arcs are labelled *ED* and *E′D′*. Moreover, our numerical evidence indicates that *ED* is the limit of the right-hand boundaries of those horns corresponding to resonances greater that $\frac{1}{8}$. Similarly, *E′D′* is the limit of the left-hand boundaries of horns with resonances less than $\frac{1}{8}$.

In the region between the arcs *ED* and *E′D′* the hypotheses of Theorem 2.1 are satisfied and the rotation number is unique. However, in this subset of the $\frac{1}{8}$ horn the structure of $A_{(a,b)}$ is by no means simple. Indeed, $A_{(a,b)}$ can be quite strange. Figure 2.10 shows part of the attractor for $F_{(a,0)}$ with $a = 2.252\,118\,5$. Note that this value of a belongs to the segment *DD′* in Figs 2.3 and 2.9. The set in Fig. 2.10 consists of

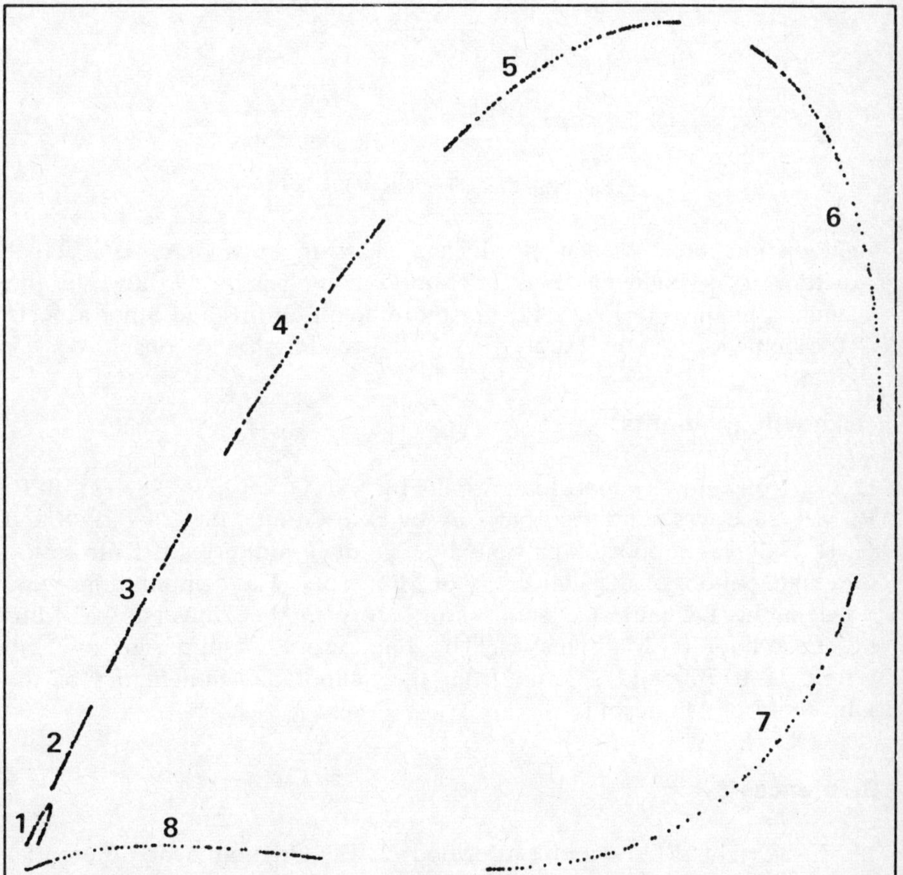

Fig. 2.10 Iterates of the map F_a for $a = 2.252\,1185$

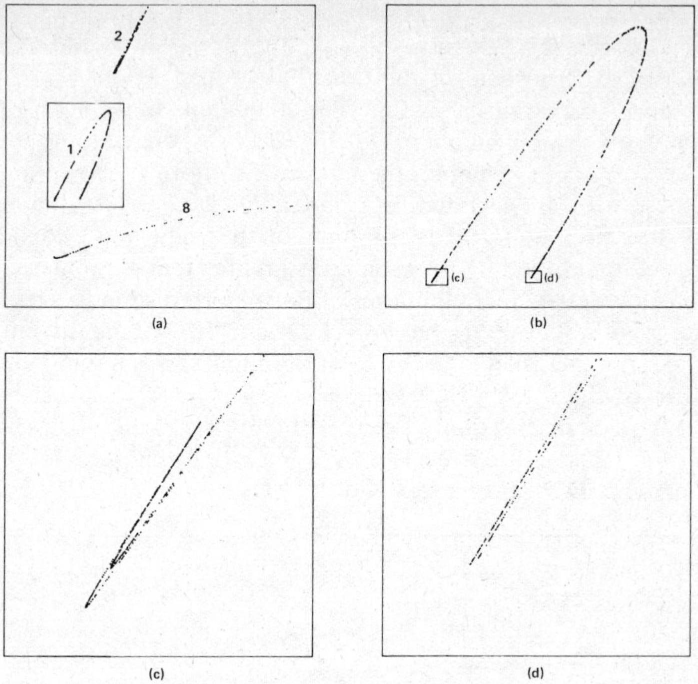

Fig. 2.11 Magnifications of Fig. 2.10

eight distinct pieces, each of which appears to be an arc. The actual structure of the eight pieces is much more complicated, as shown by the magnified pictures in Fig. 2.11. For more detail on this and other aspects of the behaviour of the family $\{F_{(a,b)}\}$, the reader should consult [5].

Acknowledgements

D. G. Aronson was supported partially by NSF Grant MSC 78–02158, R. P. McGehee was supported partially by NSF Grant MSC 79–01998 and G. R. Hall was supported partially by a grant-in-aid of research from the Graduate School of the University of Minnesota. The computations were done in the Dynamical Systems Laboratory in the University of Minnesota School of Mathematics. The Laboratory is supported by NSF Grant 81–01700 and a grant from the School of Mathematics of the University of Minnesota.

References

1. Cartwright, M. L. and Littlewood, J. E., On nonlinear differential equations of the second order, *Journal of the London Mathematical Society*, **20,** 180–189, 1945.

2. Levinson, N., A second order differential equation with singular solutions, *Annals of Mathematics*, **50,** 127–153, 1949.
3. Flaherty, J. and Hoppensteadt, F., Frequency entrainment of a forced van der Pol oscillator, *Studies in Applied Mathematics*, **58,** 5–15, 1978.
4. Levi, M., Qualitative analysis of the periodically forced relaxation oscillators, *Memoirs of the American Mathematical Society*, (244), 1981.
5. Aronson, D. G., Chory, M. A., Hall, G. R. and McGehee, R. P., Bifurcations from an invariant circle for two parameter families of maps of the plane: a computer assisted study, *Communications in Mathematical Physics*, **83,** 303–351, 1982.
6. Pounder, J. R. and Rogers, T. D., The geometry of chaos: dynamics of a nonlinear second-order difference equation, *Bulletin of Mathematical Biology*, **42,** 551–597, 1980.
7. Arnol'd, V. I., Loss of stability of self-oscillations close to resonance and versal deformations of equivariant vector fields, *Functional Analysis and its Applications*, **11**(2), 1–10, 1977.
8. Smale, S., Diffeomorphisms with many periodic points, in Cairns, S. S. (Ed.), *Differential and Combinatorial Topology*: Princeton University Press, 1965, 63–80.
9. Bowen, R., *On Axiom A Diffeomorphisms*, CBMS Regional Conference Series in Mathematics No. 35: Providence, American Mathematical Society, 1978.
10. Conley, C., Hyperbolic sets and shift automorphisms, in Moser, J. (Ed.), *Dynamical Systems: Theory and Applications*, Lecture Notes in Physics No. 38: Berlin, Springer-Verlag, 1975, 539–549.

3 Selfsimilar Turbulence Propagation from an Instantaneous Plane Source

G I Barenblatt

Propagation of turbulence from an instantaneous plane source originally concentrated in a horizontal plane layer of finite depth is considered in this chapter. The mean fluid velocity in the course of motion is supposed identically equal to zero, and the fluid is incompressible; the cases both of homogeneous and of stratified fluids are considered. Analysis of the problem is based on the turbulent energy balance equation whose closure is reached by certain Kolmogorov-type similarity hypotheses together with an assumption that the turbulence length scale is equal to a certain fixed part of actual turbulent layer depth.

The asymptotic solution to the problem under consideration appears to be a selfsimilar one, the selfsimilarity being, however, of the second kind so that the time exponent in selfsimilar variables cannot be obtained from purely dimensional considerations. A nonlinear eigenvalue problem is stated and solved to evaluate this exponent.

3.1 Statement of the problem

Consider an infinite space filled with an incompressible and at first homogeneous fluid being initially at rest. At the initial moment $t=0$ a plane turbulent layer is formed instantaneously (Fig. 3.1). At subsequent moments $t>0$ the turbulent layer expands, entraining the ambient fluid. Since the flow is shearless, turbulent energy is not generated. The turbulence is therefore dissipated into heat and decays in the course of motion.

The equation of the turbulent energy balance for the flow under consideration expresses the simple fact that the rate of turbulent energy variation at any point consists of the local contribution of the turbulent energy turbulent diffusion together with the rate of turbulent energy dissipation into heat. This equation is of the form [1]

$$\partial_t q = -\partial_z \langle q'w' + p'w'/\rho \rangle - \varepsilon_t. \qquad (3.1)$$

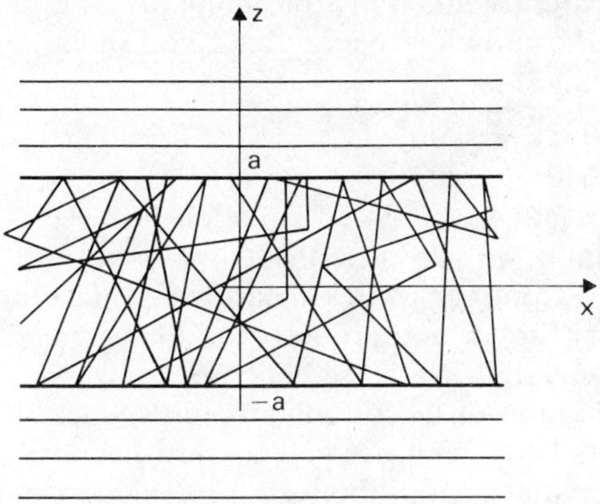

Fig. 3.1 The turbulent plane layer is formed instantaneously in unbounded fluid

In this equation $q = \langle (u'^2 + v'^2 + w'^2)/2 \rangle$ is the mean turbulent energy per unit mass; u', v', w' are the x, y, z-components of the velocity fluctuation; xyz is an orthogonal Cartesian coordinate system where the z-axis is taken along the normal to the middle plane of the layer (the turbulent energy distribution is assumed symmetric with respect to this plane); t is the time; ρ is the fluid density; q' is the turbulent energy fluctuation per unit mass; ε_t is the mean rate of turbulent energy dissipation per unit mass. The sign $\langle \ldots \rangle$ means an ensemble averaging.

Let us introduce the turbulent energy eddy diffusion coefficient K_q according to the relation

$$K_q = \frac{-\langle q'w' + p'w'/\rho \rangle}{\partial_z q}. \tag{3.2}$$

We stress that relation (3.2) for a horizontally uniform flow does not in itself represent a hypothesis, but simply introduces a new notation. According to the Kolmogorov similarity hypothesis [1, 2] we shall assume the local values of the turbulent energy eddy diffusion coefficient K_q and the mean rate of turbulent energy dissipation ε_t to be dependent only on the mean eddy size and local mean eddy energy. The mean eddy size is proportional to the external turbulence length scale l. As a measure of the eddy energy the mean turbulent energy per unit mass q may be considered. Dimensional analysis gives

$$K_q = l\sqrt{(q)}, \qquad \varepsilon_t = \frac{cq^{3/2}}{l}. \tag{3.3}$$

Here c is a constant, universal within the framework of the model accepted; its value cannot be considered as determined from experiments with shear flows. Indeed, taking K_q equal, not proportional, to $l\sqrt{(q)}$ we redefine l in comparison with the shear flow, including in its expression a factor equal to the ratio of K_q to the momentum transfer eddy coefficient.

This factor is apparently different from unity. Taking into account (3.2), (3.3), we reduce the turbulent energy balance Eq. (3.1) to the form

$$\partial_t q = \partial_z l\sqrt{(q)}\,\partial_z q - \frac{cq^{3/2}}{l}. \tag{3.4}$$

Relation (3.4) is not yet a closed equation because l—the turbulence length scale—is unknown. To close this equation let us apply the following considerations. Contrary to the linear heat conduction equation, Eq. (3.4) has the very important property of finite speed of disturbance propagation in the following sense. For an arbitrary bounded l and a finite initial distribution of the turbulent energy, such that the turbulent energy at the initial moment is equal to zero everywhere except for a certain bounded spatial region, the turbulent region where the turbulent energy is different from zero is bounded also at arbitrary moments. This property is well known for heat-conduction-type equations with the power-type nonlinearity [3, 4] and its proof is easily extended to the case of Eq. (3.4). Because of the symmetry of the problem under consideration, the turbulent region at an arbitrary moment t corresponds to $-h(t) \leqslant z \leqslant h(t)$ where $h(t)$—the turbulent layer half-depth—is a nondecreasing function of time. Note the remarkable fact that a sharp boundary between turbulent and nonturbulent regions has repeatedly been observed experimentally [5, 6].

We shall assume that the turbulence length scale l is equal to a definite fixed part α of the actual turbulent layer depth:

$$l = \alpha h(t), \tag{3.5}$$

where $\alpha < 1$ is a certain universal constant which cannot be found within the framework of the model accepted and should be determined experimentally. The constancy of the turbulence length scale within the layer at a given moment is prompted by Townsend's theory [7] concerning the governing role of large eddies in the turbulent shear flow transfer process. In principle, one of the equations for higher moments might be used instead for closure (*cf.* [8]); however, the final closure would demand some similarity hypotheses of the type (3.3), which are less justified at this stage. Relation (3.5) gives an ultimate closure of Eq. (3.4). We stress that the function $h(t)$ is not known beforehand and is to be determined in the course of solving the problem.

An initial condition should be added to Eq. (3.4). We suppose that the function giving the initial distribution of the turbulent energy is symmetrical with respect to the middle plane of the turbulent layer and is equal to zero outside a certain interval $-a \leqslant z \leqslant a$, where $2a$ is the initial depth of the turbulent layer. It is convenient to take the initial condition in the form

$$q(z,0) = \frac{Q}{a}\,u(\tau), \quad \tau = \frac{z}{a} \tag{3.6}$$

where $u(\tau)$ is an even dimensionless function of its dimensionless argument, equal identically to zero for $|\tau| \geq 1$. The quantity Q—the bulk intensity of the instantaneous turbulence source—is given by

$$Q = \int_{-a}^{a} q(z, 0)\,dz, \tag{3.7}$$

so that

$$\int_{-1}^{1} u(\tau)\,d\tau = 1. \tag{3.8}$$

Thus the problem under investigation is reduced to the construction of a solution to the Cauchy problem for the equation

$$\partial_t q = \partial_z \alpha h(t)\sqrt{(q)}\,\partial_z q - \frac{cq^{3/2}}{\alpha h(t)}, \tag{3.9}$$

which is different from zero in the region $-h(t) \leq z \leq h(t)$ and satisfies the initial condition (3.6). The exact function $h(t)$, half-depth of the turbulent layer at the moment t, is not known beforehand and should also be determined in the course of solving the problem.

3.2 Exact solution for the limiting case of lack of dissipation

Let us first consider the limiting case $c = 0$ where no dissipation of turbulent energy occurs and Eq. (3.9) takes the form

$$\partial_t q = \partial_z \alpha h(t)\sqrt{(q)}\,\partial_z q. \tag{3.10}$$

The solution of an idealized problem may be constructed for this case corresponding to the initial condition (3.6) at a finite Q and $a = 0$, i.e., to an infinitely thin instantaneous turbulence source of finite bulk intensity. Indeed, the solution of this idealized problem can depend on the governing parameters Q, t, z, α only. Dimensional analysis gives

$$q = Q^{2/3}t^{-2/3}f(\xi), \qquad \xi = \frac{z}{Q^{1/3}t^{2/3}}, \qquad h(t) = \xi_0(\alpha)Q^{1/3}t^{2/3}, \tag{3.11}$$

where ξ_0 is a certain positive constant.

Substituting (3.11) into (3.10) and assuming that $\zeta = \xi/\xi_0 = z/h(t)$, $\Phi(\zeta) = \xi_0^{-2}f(\xi)$, we obtain the equation for the function Φ:

$$\frac{d}{d\zeta}\alpha\sqrt{(\Phi)}\frac{d\Phi}{d\zeta} + \frac{2}{3}\zeta\frac{d\Phi}{d\zeta} + \frac{2}{3}\Phi = 0. \tag{3.12}$$

Because of symmetry, $\Phi(\zeta) = \Phi(-\zeta)$, $d\Phi(0)/d\zeta = 0$, and according to the definition of $h(t)$ the quantity $\Phi(1)$ is equal to zero. Integrating Eq. (3.12) under these conditions we obtain

$$\Phi(\zeta) = \begin{cases} (1-\zeta^2)^2/36\alpha^2, & (\zeta \leq 1), \\ 0, & (\zeta \geq 1). \end{cases} \tag{3.13}$$

The unknown parameter ξ_0 can be obtained from the energy conservation relation which is valid in the case of the lack of dissipation considered here:

$$E = \int_{-h(t)}^{h(t)} q(z, t)\,\mathrm{d}z = \text{const} = Q. \tag{3.14}$$

Hence we obtain the relation

$$\xi_0 = \left[\int_{-1}^{1} \Phi(\zeta)\,\mathrm{d}\zeta\right]^{-1/3} = \left(\frac{135\alpha^2}{4}\right)^{1/3}, \tag{3.15}$$

which completes the solution of the limiting problem under consideration.

3.3 Solution for the case of a finite dissipation—selfsimilar intermediate asymptotics

It seems at first glance that an analogous consideration must also be valid for the case of finite dissipation when $c \neq 0$, so that an attempt should also be made to construct a solution corresponding to $a = 0$ and $Q \neq 0$ for the case $c \neq 0$. However, it is not so. Indeed, the constant c in Eq. (3.9) is dimensionless. Therefore it seems that, because of the same dimensional arguments as in the preceding section, the solution of such a problem has to be presented in a form analogous to (3.2):

$$q = \frac{Q^{2/3}}{t^{2/3}} f(\xi), \qquad \xi = \frac{z}{Q^{1/3} t^{2/3}}, \qquad h(t) = \xi_0(\alpha, c) Q^{1/3} t^{2/3}. \tag{3.16}$$

However, this solution satisfies relation (3.14) for the bulk turbulent energy conservation which, in fact, cannot be preserved because of dissipation, which is nonzero in this case. Indeed, integrating Eq. (3.9) from $z = -h(t)$ to $z = h(t)$ and taking into account the vanishing of the turbulent energy flow at the turbulent layer boundaries, we obtain

$$\frac{\mathrm{d}E}{\mathrm{d}t} = -\frac{c}{\alpha h(t)} \int_{-h(t)}^{h(t)} q^{3/2}(z, t)\,\mathrm{d}z < 0. \tag{3.17}$$

In order to explain the paradox which has arisen, we note that at $a = 0$ the initial condition (3.6) has a singular character and is expressed, in fact, by a generalized function. Therefore the corresponding limiting solution, if it does exist, has to be a limit (if it exists and is different from zero) of the solutions to the problem corresponding to finite Q and a when a tends to zero. Let us consider therefore the general case corresponding to $a \neq 0$ and $c \neq 0$: the initial depth of the turbulent layer is different from zero and dissipation is also nonzero. The solution of this latter problem is now not a selfsimilar one: it depends on the parameters Q, t, z, α, c, a, and dimensional analysis gives

$$q = Q^{2/3} t^{-2/3} f_q(\xi, \eta, \alpha, c), \qquad h = Q^{1/3} t^{2/3} f_h(\eta, \alpha, c) \tag{3.18}$$

where $\xi = z/Q^{1/3}t^{2/3}$, as before, and $\eta = a/Q^{1/3}t^{2/3}$.

The asymptotics of the solution for turbulent layer depths much larger than the initial one, $h \gg a$, is of primary interest to us. Indeed for $h \gg a$ or (which is the same) $\eta \ll 1$, the influence of the details of the initial turbulent energy distribution has to disappear. The nonexistence of a solution corresponding to $a=0$ or $\eta=0$ demonstrates that the finite and nonzero limits of functions f_q and f_h at $\eta \to 0$ do not exist. We assume, therefore, incomplete selfsimilarity with respect to parameter η as $\eta \to 0$. This means (for a more detailed explanation of this concept, see [9]) that although the finite nonzero limits of functions f_q and f_h do not exist, at $\eta \to 0$ the power-type asymptotics of functions f_q and f_h are valid:

$$f_q = \eta^{\lambda} f\left(\frac{\xi}{\eta^{\nu}}\right), \qquad f_h = \xi_0(\alpha, c)\eta^{\sigma}. \tag{3.19}$$

so that the limiting solutions of the problem also become selfsimilar.

Here $\xi_0(\alpha, c)$ is a certain positive dimensionless function of its dimensionless arguments, and λ, ν, σ are certain constant numbers. The differential equation for the function f obtained when substituting (3.18), (3.19) into (3.9) has to be an ordinary one, therefore the conditions $\lambda = 2\nu, \sigma = \nu$ must be fulfilled. Denoting for convenience

$$\nu = \frac{(3\mu-1)}{2}, \qquad \zeta = \frac{\xi}{\xi_0} = \frac{z}{h(t)}, \qquad \Phi(\zeta) = \xi_0^{-2} f, \tag{3.20}$$

we obtain the equation

$$\frac{d}{d\zeta}\alpha\sqrt{(\Phi)}\frac{d\Phi}{d\zeta} + (1-\mu)\zeta\frac{d\Phi}{d\zeta} + 2\mu\Phi - \frac{c}{\alpha}\Phi^{3/2} = 0. \tag{3.21}$$

Because of the symmetry of the problem under consideration, $\Phi(\zeta) = \Phi(-\zeta)$, we need to construct a solution for $\zeta \geqslant 0$ only. Hence we search for a nontrivial solution of Eq. (3.20) satisfying the conditions

$$\frac{d\Phi(0)}{d\zeta} = 0, \qquad \Phi(1) = 0, \tag{3.22}$$

This solution should be continuous and should have a continuous derivative $d\Phi^{3/2}/d\zeta$; these latter conditions follow from the requirement of continuity of the turbulent energy itself and of its flow, which are proportional to Φ and $d\Phi^{3/2}/d\zeta$, respectively. In particular, from the continuity of the derivative $d\Phi^{3/2}/d\zeta$ the condition

$$\frac{d\Phi^{3/2}(1)}{d\zeta} = 0 \tag{3.23}$$

follows, which the desired solution has also to satisfy. Ultimately, the asymptotic solution that we construct has to be presented in the form

$$q = A^2 t^{-2\mu}\Phi\left(\frac{z}{At^{1-\mu}}\right), \qquad h = At^{1-\mu}, \tag{3.24}$$

where

$$A = \xi_0(\alpha, c)Q^{(1-\mu)/2}a^{(3\mu-1)/2} \tag{3.25}$$

is a constant. Substituting (3.24) into relation (3.17) we obtain

$$\mu - \frac{1}{3} = \frac{c\int_{-1}^{1} \Phi^{3/2}(\zeta)\,d\zeta}{3\alpha\int_{-1}^{1} \Phi(\zeta)\,d\zeta} > 0, \tag{3.26}$$

whence it follows that $\mu > \frac{1}{3}$ because c is positive. The condition of turbulent layer expansion (rather than contraction) clearly gives $\mu < 1$. Thus, if we let the initial depth of layer $2a$ tend to zero, we cannot (contrary to the case $c = 0$) preserve the intensity of the source Q invariant if we wish to preserve the asymptotics of the solution at large t; the intensity has to increase and moreover in such a way that the moment,

$$Qa^{(3\mu-1)/(1-\mu)}, \tag{3.27}$$

is invariant.

At small Φ the last term of the left-hand side of Eq. (3.21) is negligibly small in comparison with the preceding one and can be neglected. It is easy to show that the equation simplified in this way is invariant with respect to the transformation group

$$\Phi' = \lambda^4\Phi, \qquad \zeta' = \lambda\zeta, \tag{3.28}$$

where $\lambda > 0$ is the group parameter. Therefore the equation can be reduced to a first-order one by a standard procedure and the equation thus obtained can easily be investigated. This investigation shows that the nontrivial solution to Eq. (3.21) satisfying the conditions

$$\Phi(1) = 0, \qquad \frac{d\Phi^{3/2}(1)}{d\zeta} = 0 \tag{3.29}$$

does exist and is uniquely determined; a series expansion at $\zeta \leqslant 1$ is valid for this solution:

$$\Phi = \left(\frac{1-\mu}{2\alpha}\right)^2 (1-\zeta)^2 - \frac{1-\mu}{6\alpha^2}(1-\zeta)^3 + \ldots. \tag{3.30}$$

However, for arbitrary μ the remaining condition $d\Phi(0)/d\zeta = 0$ is in general violated. Thus, we arrive at a classical situation of a nonlinear eigenvalue problem: we have to construct a non-trivial solution to Eq. (3.21) that satisfies conditions (3.22), (3.23) and, moreover, to determine the value of parameter μ for which such a solution exists. It can be shown that the solution of this eigenvalue problem is unique; it can easily be solved numerically. The result of numerical calculation of parameter μ for different values of parameters α and c/α are presented

Table 3.1 μ-value for different values of α and c/α (numerical calculation due to E. S. Benilov)

α \ c/α	0.0	0.05	0.10	0.15	0.25	0.5	0.75	1.00	1.50	2.50	3.50
0.1	0.333	0.357	0.378	0.397	0.431	0.499	0.549	0.587	0.643	0.710	0.740
0.2	0.333	0.346	0.357	0.367	0.387	0.431	0.468	0.499	0.549	0.619	0.664

in Table 3.1 and Fig. 3.2. The solution $\Phi(\zeta)$ for the case $\alpha = 0.2$ and $c/\alpha = 0.5$ is presented in Fig. 3.3.

We note that in the selfsimilar solution constructed the constant A is left undetermined. The point is that, contrary to the case $c = 0$, the law of turbulent energy conservation at $c \neq 0$ assumes a nonintegrable form (3.17) and does not allow this constant to be determined. The only way to

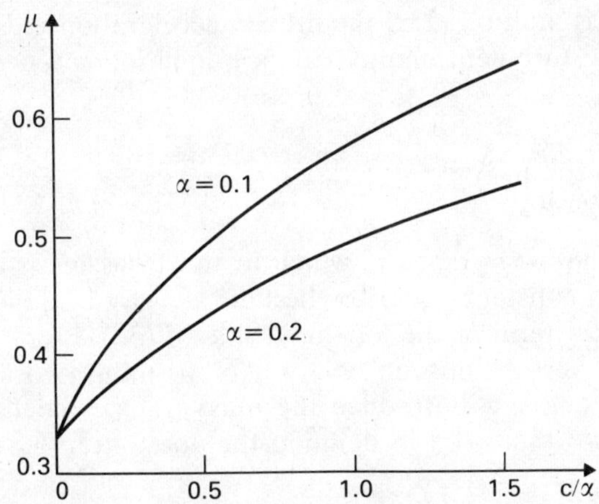

Fig. 3.2 Dependence of μ on the problem parameters α and c/α

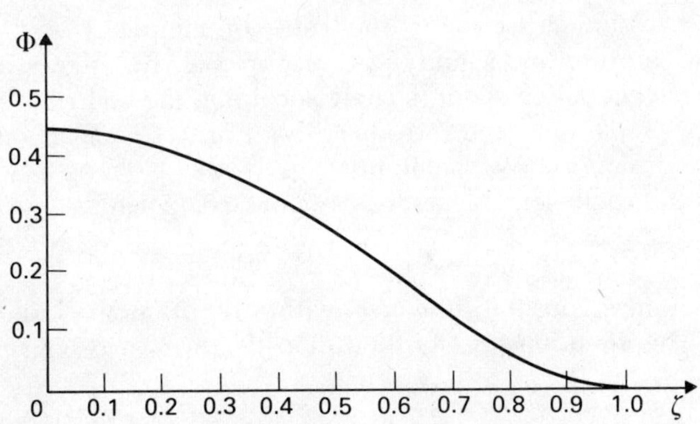

Fig. 3.3 Universal turbulent energy distribution over the turbulent layer depth, $\alpha = 0.2$, $c/\alpha = 0.5$

determine constant A that can be proposed at present is to match it asymptotically with the numerical solution of the original nonselfsimilar problem to which solution (3.24) relates (numerical calculations of the solution to the nonselfsimilar Cauchy problem, performed by E. S. Benilov, confirm that this solution rapidly arrives at asymptotics of the type (3.24), thus confirming the basic assumption of incomplete selfsimilarity).

3.4 Effect of stratification

Let us now suppose that the fluid is stratified so that there is a variation of fluid density in the direction of the strong gravity field which is directed down the z-axis (Fig. 3.1). We shall assume furthermore (the Boussinesq approximation) that the fluid density variation is negligibly small except in terms where it is multiplied by the gravity acceleration g. Under all these assumptions the turbulent energy balance equation can be written in the form:

$$\partial_t q + \partial_z\left\langle\left(q' + \frac{p'}{\rho}\right)w'\right\rangle + \varepsilon_t + \frac{\langle\rho' w'\rangle g}{\rho} = 0. \tag{3.31}$$

Hence ρ is the mean density, which in the Boussinesq approximation can be taken as constant equal to the fluid density at $z=0$ at the initial moment. The last term of the left-hand side of (3.31) represents the rate of turbulent energy consumption in performing work against the buoyancy forces. Let us introduce the mass transfer coefficient K_ρ in a way analogous to that used in defining the above transfer coefficients:

$$K_\rho = -\frac{\langle\rho' w'\rangle}{\partial_z\langle\rho\rangle} \tag{3.32}$$

We have assumed here for the sake of simplicity that only one factor—temperature or salinity—is responsible for the stratification. Otherwise we should consider, strictly speaking, the contribution of both these factors separately and introduce two transfer coefficients, for temperature and salinity. We shall now apply the Kolmogorov similarity hypothesis for coefficient K_ρ as for previous coefficients:

$$K_\rho = c_1 l\sqrt{(q)} \tag{3.33}$$

where c_1 is a new constant, universal within the framework of the model proposed. The equation for the fluid density can be written in the form

$$\partial_t s = \partial_z K_\rho \partial_z s \tag{3.34}$$

Here $s = (\langle\rho\rangle - \rho)/\rho$ is the relative density excess with respect to the reference density ρ; by assumption, the quantity s is small. The closure

(3.3)–(3.4) allows one to write Eqs (3.31), (3.34) in the form

$$\partial_t q - \partial_z l\sqrt{(q)}\,\partial_z q + \frac{cq^{3/2}}{l} - c_1 l\sqrt{(q)}(\partial_z s)g = 0, \tag{3.35}$$

$$\partial_t s - \partial_z c_1 l\sqrt{(q)}\,\partial_z s = 0. \tag{3.36}$$

The ultimate closure can be obtained if we suppose as before $l = \alpha h(t)$ where $2h(t)$ is the turbulent layer depth at the moment t.

Let the fluid be initially linearly stratified:

$$s(z, 0) = -\kappa z \tag{3.37}$$

where κ is a positive constant having the dimensions of inverse length. Then the inflow of the excess mass into the turbulent layer at the turbulent layer boundaries is equal to

$$\begin{aligned}\langle\rho' w'\rangle|_{z=\pm h(t)} &= -\rho K_\rho(\partial_z s)|_{z=\pm h(t)} \\ &= -\rho c_1 l\sqrt{(q)}(\partial_z s)|_{z=\pm h(t)} = \mp\kappa\rho h(t)\frac{\mathrm{d}h}{\mathrm{d}t}.\end{aligned} \tag{3.38}$$

Indeed, it is equal to the speed of the turbulent layer boundary propagation multiplied by the excess mass in the volume entrained by this boundary per unit time. By symmetry the excess density at $z = 0$ is equal to zero at arbitrary time:

$$s(0, t) = 0. \tag{3.39}$$

The selfsimilar excess density distribution,

$$s = \kappa A t^{1-\mu} g(\zeta), \tag{3.40}$$

satisfies all these conditions if the function $g(\zeta)$, in turn, satisfies the ordinary linear differential equation

$$\frac{\alpha c_1}{1-\mu}\frac{\mathrm{d}}{\mathrm{d}\zeta}\sqrt{(\Phi)}\frac{\mathrm{d}g}{\mathrm{d}\zeta} + \zeta\frac{\mathrm{d}g}{\mathrm{d}\zeta} - g = 0 \tag{3.41}$$

and the boundary conditions

$$g(0) = 0; \qquad \sqrt{(\Phi)}\frac{\mathrm{d}g}{\mathrm{d}\zeta} = -\frac{1-\mu}{c_1\alpha}, \quad \zeta = 1. \tag{3.42}$$

Note now that if the excess density is represented by expression (3.40) and the turbulent energy, as before, by (3.24), then the term in the turbulent energy balance equation corresponding to the rate of turbulent energy consumption in performing work against the buoyancy forces will have a time–power exponent different from the remaining terms of this equation. In fact, the buoyancy term in the turbulent energy balance equation can be represented in the form $A^2\kappa g t^{1-2\mu}G(\zeta)$, and the remaining terms in the form $A^2 t^{-(1+2\mu)}F(\zeta)$, where $G(\zeta), F(\zeta)$ are certain dimensionless functions of their dimensionless argument. Thus the ratio of the first term to the remaining ones is represented in the form $\kappa g t^2 S(\zeta)$; it is relatively small at the first stage of the motion but increases later on, reaching a value of the order of unity. When the buoyancy term becomes equal, by an order of magnitude, to the remaining terms of the turbulent energy balance equation, the selfsimilar regime considered disappears and a nonselfsimilar stage of the motion sets in. Thus the selfsimilar regime considered above can take place at the initial stage of the motion only when the rate of work against the buoyancy forces is negligibly small.

Fluid stratification can, however, have another most essential influence on turbulence expansion and decay, even at the initial stage. The point is that at the boundaries of the turbulent region part of the energy can be transferred to the internal waves excited in the nonturbulent fluid above and below the turbulent layer. The selfsimilarity of turbulence propagation and decay may be preserved only if the energy transferred to the internal waves per unit time is a constant fraction $m, 0<m<1$, of the rate of the bulk turbulent energy decay:

$$-(l\sqrt{q}\,\partial_z q)|_{z=\pm h(t)} = \mp(m/2)\frac{dE}{dt}\,. \tag{3.43}$$

Hence, instead of the condition $d\Phi^{3/2}(\pm 1)/d\zeta = 0$ of the turbulent energy flow vanishing at the turbulent layer boundaries, the condition is obtained:

$$-\alpha\left(\sqrt{(\Phi)}\frac{d\Phi}{d\zeta}\right)\bigg|_{\zeta=\pm 1} = \pm\frac{m}{2}(1-3\mu)\int_{-1}^{1}\Phi(\zeta)\,d\zeta \tag{3.44}$$

and, consequently, instead of condition (3.26) we have

$$\mu-\frac{1}{3}=\frac{(c/(1-m))\displaystyle\int_{-1}^{1}\Phi^{3/2}(\zeta)\,d\zeta}{3\alpha\displaystyle\int_{-1}^{1}\Phi(\zeta)\,d\zeta}\,. \tag{3.45}$$

The value of μ will evidently increase with increasing fraction of energy m transferred to the internal waves. Therefore turbulent energy decay will also increase with increasing m, whereas turbulent layer extension will slow down.

We have assumed in this section the symmetry of the turbulent layer extension. Such an assumption is quite natural under the condition of linear initial stratification (3.37). If, for instance, the initial stratification has an asymmetric form:

$$s(z,0)=-\kappa z,\quad z\leqslant 0;\qquad s(z,0)=-\kappa_1 z,\quad z>0, \tag{3.46}$$

then at $\kappa_1 \neq \kappa$ the turbulent layer extension will be asymmetric likewise, the upper layer boundary will move as $z=h_1(t)$, and the lower as $z=-h_2(t)$, $h_2(t)\neq h_1(t)$. The turbulent energy flow will vanish at a certain plane $z = h_3(t)$ which is to be determined, rather than at the plane $z = 0$ as before. The selfsimilarity, however, can be retained without the symmetry.

3.5 Conclusion

Simple analytical relations obtained for the laws of extension of a turbulent layer and decay of turbulence within it allow one, in principle, to check the application of a semi-empirical model of turbulence to an essentially nonsteady flow. Other properties of the solution obtained are also of interest for experimental verification. For instance, the turbulent energy near the edge of the turbulent region in a homogeneous fluid

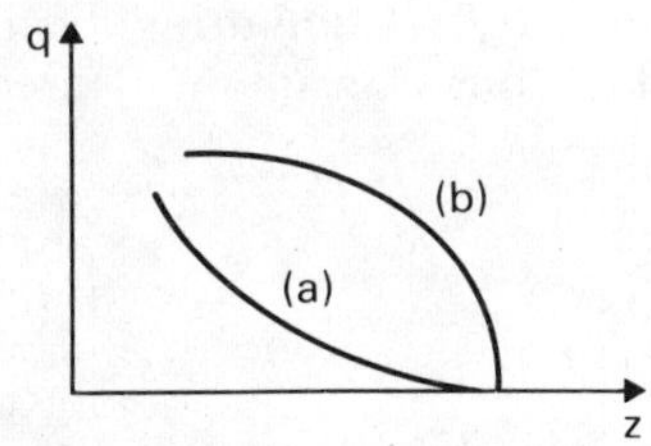

Fig. 3.4 Turbulent energy distribution near the edge of the turbulent layer, (*a*) in a homogeneous fluid and (*b*) in a stratified fluid

decreases smoothly (Fig. 3.4, Curve (a))—formulae (3.24), (3.30) give

$$q = \left(\frac{1-\mu}{2\alpha t}\right)^2 (h-z)^2, \tag{3.47}$$

whereas, near the edge of the turbulent region in a stratified fluid, formulae (3.24), (3.44) give

$$q \sim t^{-2(1+2\mu)/3}(h-z)^{2/3}, \tag{3.48}$$

so the turbulent energy vanishes abruptly (Fig. 3.4, Curve (b)).

References

1. Monin, A. S. and Yaglom, A. M., *Statistical Fluid Mechanics: Mechanics of Turbulence*, vol. 1: Cambridge, Massachusetts and London, MIT Press, 1971.
2. Kolmogorov, A. N., Equations of turbulent motion of an incompressible fluid, *Izvestija Akademii Nauk SSSR, Serija Fizicheskaya*, **6,** 56–58, 1942.
3. Zel'dovich, Ya. B. and Kompaneets, A. S., On the theory of heat conduction for temperature dependent heat conductivity, in Lukirsky,

P. I. (Ed.), *Collection of Papers Dedicated to A. F. Ioffe 70th Anniversary*, Moscow, Izd. Akademy Nauk USSR, 1950, 61–71.

4. Barenblatt, G. I., On some unsteady motions of fluid and gas in porous media, *Prikladnaya Matematika i Mekhanika*, **16,** 67–78, 1952.
5. Kovasznay, L. S. G., Kilens, C. and Blackwelder, R. F., Large-scale motion in the intermittent region of a turbulent boundary layer, *Journal of Fluid Mechanics*, **41,** 283–325, 1970.
6. Turner, J. S., *Buoyancy Effects in Fluids*: Cambridge University Press, 1973.
7. Townsend, A. A., *The Structure of Turbulent Shear Flow*: Cambridge University Press, 1976.
8. Hanjalic, K. and Launder, B. E., A Reynolds stress model of turbulence and its application to thin shear flows, *Journal of Fluid Mechanics*, **52,** 609–638, 1972.
9. Barenblatt, G. I., *Similarity, Selfsimilarity and Intermediate Asymptotics*: New York and London, Consultants Bureau, 1979.

4 Transition to Stochasticity of Viscous Flow between Rotating Spheres

Yu N Belayev and I M Yavorskaya

4.1 Theoretical and experimental background

New ideas concerning the transition to turbulence have recently emerged from the field of mathematics that embraces the qualitative theory of differential equations, topology and functional analysis. These ideas are broadly formulated in the distinguished paper by Ruelle and Takens [1] (see also [2]).

According to these authors the transition to stochasticity in a dynamical system with 'generic' (typical) behaviour becomes possible after a small number of bifurcations (no more than three) which results in the appearance in phase space of an attracting set—a 'strange attractor'—on which all phase trajectories are unstable. The fluid flows described by such trajectories appear to have continuous spectra and decaying autocorrelation functions.

As is well known, long before this it had been shown by Lorenz [3], using the system of three ordinary differential equations (obtained by rough Galerkin approximation) as a model of the thermal convection in a horizontal layer, that completely deterministic initial and boundary conditions may give a stochastic solution after only two bifurcations of the basic solutions.

These ideas presented an alternative to the Landau–Hopf picture according to which the transition to turbulence occurs as a result of a large number of bifurcations which leads to a quasiperiodic process with a countable number of degrees of freedom.

Like any new model or hypothesis, the Ruelle–Takens model needs accurate experimental tests. Such tests in the flow between cylinders [4–6] and for thermal convection in a horizontal layer [7–9] confirmed the proposed model in principle, but some important discrepancies were discovered as well. These may be explained in particular by certain nongeneric properties of the flows under consideration, connected with their high degree of symmetry—and so arises the test problem for more general flows.

In this chapter an experimental investigation of the transition to chaos in the flow between two concentric coaxially rotating spheres (spherical Couette flow, SCF) is presented. The flow is governed by three dimensionless parameters: the Reynolds number, $Re = \Omega_1 r_1^2/\nu$, the angular velocity ratio, $\omega = \Omega_2/\Omega_1$, and the relative gap width, $\delta = (r_2 - r_1)/r_1$ (indexes 1 and 2 correspond to the inner and outer spheres, respectively). In contrast to its more simple and classical analogs, plane and circular Couette flow (CCF), SCF has all three velocity components and consists of a differential rotation and a circulation in the meridional plane.

At small *Re* and where the gap is narrow the intensity of the circulation, being proportional to $Re\,\delta^3$ [10], is negligible, so that the flow is close to purely azimuthal. With wider gaps the increasing meridional circulation largely redistributes the angular momentum in a layer. This difference between narrow and wide gaps becomes more prominent at the onset of instability. In the narrow-gap case the initial instability has a centrifugal nature and the secondary flow takes the form of two Taylor-type vortices near the equator [11–14]. In the wide-gap case the first instability is connected, apparently, with the inviscid Rayleigh instability, so the resulting secondary flow appears as an azimuthal periodic wave with phase velocity dependent on gap size [15, 16]. This flow consists of n eddies in each hemisphere and a sinusoidal disturbance of the equator (Fig. 4.1); n depends on δ—the wider the gap, the smaller is n.

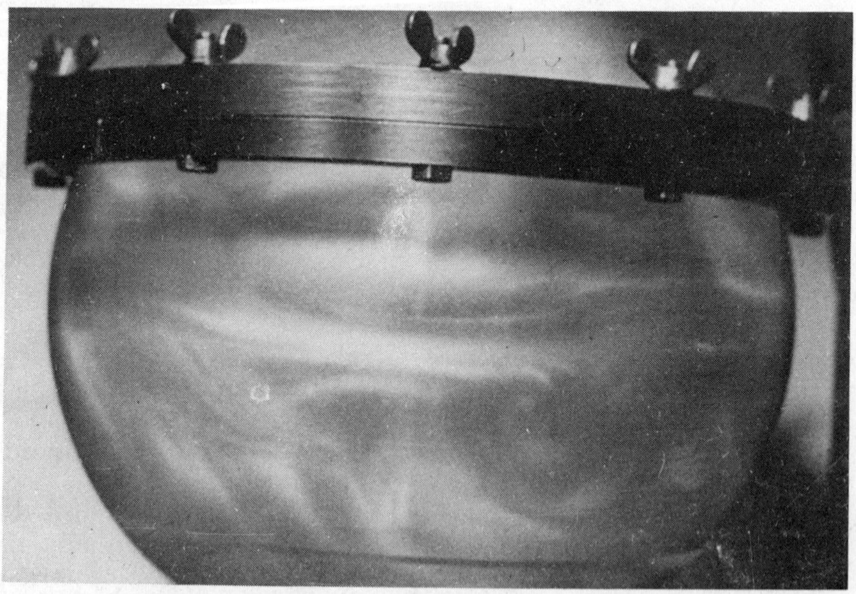

Fig. 4.1

4.2 Spectral and correlation analysis

The experiments were carried out in a rather wide-gap spherical layer ($\delta = 1.006$) in a wide range of Reynolds numbers, $50 \leqslant Re \leqslant 16\,030$, and with only the inner sphere rotating. Detailed description of the apparatus used is given elsewhere [17]: r_2 was equal to 150.08 ± 0.03 mm; radial beats were ± 0.05 mm; the angular velocity stabilization was better than 0.03 per cent. The fluid temperature was maintained constant to ± 0.05°C. Vacuum oil and its solutions in petrol were used as operating fluids.

The onset of instability and transition to turbulence in SCF were investigated by a common visualization technique (a small amount of aluminium flakes was added in the fluid) and by spectral analysis of velocity fluctuations. Visual observation was possible up to $Re \approx 2\,Re_c^{(1)}$; for larger Re it was rather difficult because of the highly complex spatial flow structure and the appearance in the flow of more than one frequency ($Re_c^{(1)}$ is the Reynolds number corresponding to the onset of instability in the basic flow). In the range $450 \leqslant Re \leqslant 900$, four spatial transitions were observed in the fluid layer. Periodic flow with four ($n = 4$) waves and phase velocity $\Omega_f/\Omega_1 = 0.155$ appeared at $Re = Re_c^{(1)}$. Then, with increasing Re, three-wave ($n = 3$), 'irregular' and two-wave ($n = 2$) sets of states were detected, with phase velocities almost independent of n. All states but the first one showed considerable hysteresis.

The film anemometer was fixed in the equatorial plane and sometimes at one pole of a spherical gap for registration of the velocity fluctuations. The anemometer maximum sensitivity diagram coincided with the radial direction. The velocity records in our experiment contained 8192 successive time intervals, the length of each data record being about 20–30 min. The velocity power spectrum was obtained using the fast Fourier transform of the digital velocity records and the velocity autocorrelation function was calculated by the usual method:

$$R(\tau) = \frac{1}{T - \tau} \int_0^{T-\tau} u(t)u(t+\tau)\,\mathrm{d}t$$

where $\tau \leqslant 0.1T$, and T is the length of the data record containing typically 500–1000 periods of the fundamental frequency. Before the onset of the first instability the basic flow was steady. Above the transition at $Re_c^{(1)}$ the spectrum consisted of instrumentally sharp peaks at ν_0 (corresponding to the frequency of the visually observed periodic wavy flow, Fig. 4.1) and its higher harmonics. Autocorrelation function decay of the order of 0.3–0.5 per cent was caused by instrumental noise. Besides the frequency ν_0 and its harmonics, the spectrum contained the frequency of the rotating sphere excited by the slightly nonideal form of the spherical surfaces. The dependence of the detected dimensionless frequencies ($\nu = \nu_{\mathrm{dim}}/\Omega_1$) on Re is shown in Fig. 4.2; the critical Reynolds

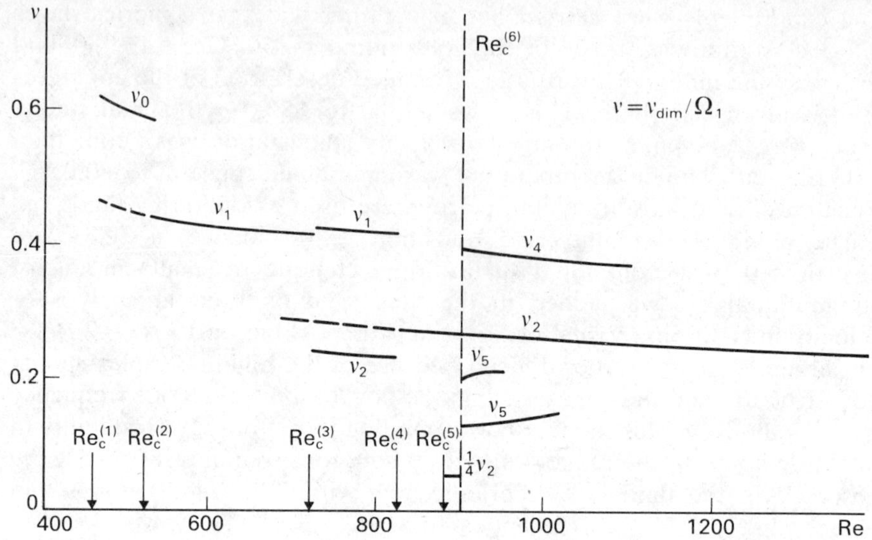

Fig. 4.2

numbers corresponding to a set of successive bifurcations are indicated by arrows.

The reorganization of the flow at $Re_c^{(2)} = 525$ was accompanied by disappearance of the frequency ν_0 and by the emergence of a new sharp frequency component at ν_1 corresponding to a visually observed periodic flow with three eddies. Next, transition to a quasiperiodic flow with two fundamental frequencies ν_1 and ν_2 occurred at $Re_c^{(3)} = 720$. Then at $Re_c^{(4)} = 820$ the frequency ν_1 disappeared and a new periodic regime with a single frequency ν_2 arose, corresponding to a periodic flow with two eddies. One more laminar flow regime occurred at $Re_c^{(5)} = 880$ when a subharmonic $\frac{1}{4}\nu_2$ with many of its higher harmonics appeared in the spectrum. It is interesting to note that at the pole the spectrum contained a single fundamental frequency $\nu_p = \frac{1}{4}\nu_2$ and its higher harmonics. Thus the first five flow regimes up to $Re = 2\, Re_c^{(1)}$ are laminar and have no more than two fundamental frequencies. Some spectra are shown in Fig. 4.3(*a*)–(*d*). The increase of the higher harmonic intensities was sometimes accompanied by a slight broadening of the spectral lines.

The first noticeable decay (about 2 per cent) of the autocorrelation function was detected at $Re = 895$ when two satellites appeared in the spectrum near the subharmonic and its odd harmonics, which was equivalent to an amplitude modulation. The dependencies of ν_p and ν_2 on *Re* were found to be different. At $Re = 895$, phase-locking of the pole and equatorial oscillations at frequencies ν_p and ν_2 no longer occurred and this possibly caused the onset of stochasticity.

With increasing *Re* at the pole and near the equator, the frequency $\frac{1}{4}\nu_2$

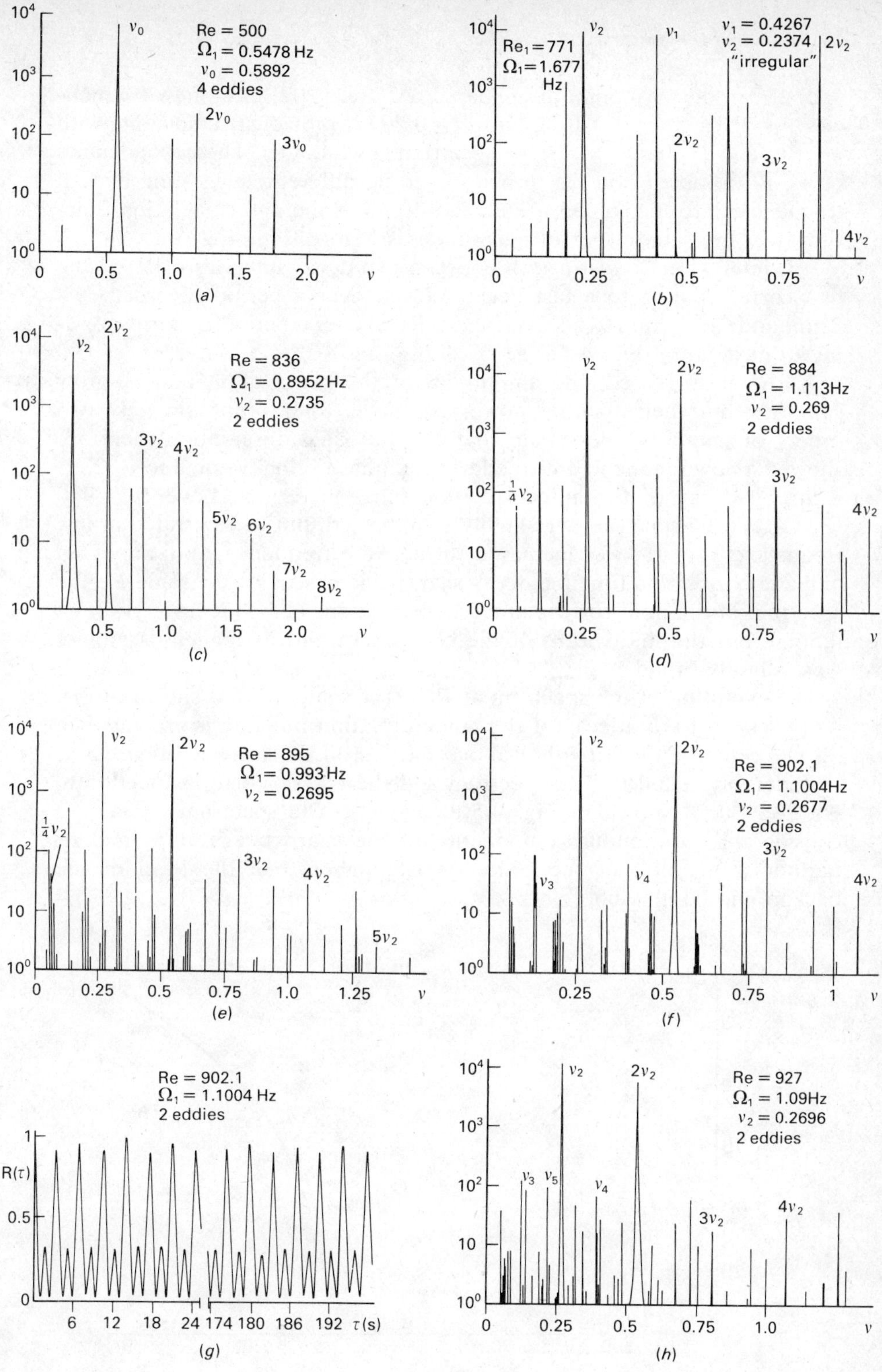

Re = 500
$\Omega_1 = 0.5478$ Hz
$\nu_0 = 0.5892$
4 eddies
ν_0
$2\nu_0$
$3\nu_0$
(a)
$Re_1 = 771$
$\Omega_1 = 1.677$ Hz
$\nu_1 = 0.4267$
$\nu_2 = 0.2374$
"irregular"
ν_2
ν_1
$2\nu_2$
$3\nu_2$
$4\nu_2$
(b)
Re = 836
$\Omega_1 = 0.8952$ Hz
$\nu_2 = 0.2735$
2 eddies
ν_2
$2\nu_2$
$3\nu_2$
$4\nu_2$
$5\nu_2$
$6\nu_2$
$7\nu_2$
$8\nu_2$
(c)
Re = 884
$\Omega_1 = 1.113$ Hz
$\nu_2 = 0.269$
2 eddies
$\frac{1}{4}\nu_2$
ν_2
$2\nu_2$
$3\nu_2$
$4\nu_2$
(d)
Re = 895
$\Omega_1 = 0.993$ Hz
$\nu_2 = 0.2695$
$\frac{1}{4}\nu_2$
ν_2
$2\nu_2$
$3\nu_2$
$4\nu_2$
$5\nu_2$
(e)
Re = 902.1
$\Omega_1 = 1.1004$ Hz
$\nu_2 = 0.2677$
2 eddies
ν_2
$2\nu_2$
ν_3
ν_4
$3\nu_2$
$4\nu_2$
(f)
Re = 902.1
$\Omega_1 = 1.1004$ Hz
2 eddies
$R(\tau)$
τ(s)
(g)
Re = 927
$\Omega_1 = 1.09$ Hz
$\nu_2 = 0.2696$
2 eddies
ν_2
$2\nu_2$
ν_3
ν_5
ν_4
$3\nu_2$
$4\nu_2$
(h)

Fig. 4.3

and all its odd harmonics disappeared. At $Re = 902$, three new frequencies $\nu_3 = 0.1363$, $\nu_4 = 0.4022$ and $\nu_5 = 0.2078$, previously coinciding with $\frac{1}{4}\nu_2$, $\frac{3}{2}\nu_2$ and $\frac{3}{4}\nu_2$, emerged in the spectrum (Fig. 4.3(*f*)). The dependencies of ν_2, ν_3, ν_4 and ν_5 on *Re* turned out to be different, suggesting that all frequencies are incommensurate. The decay of the autocorrelation function is of the same order of magnitude as before (Fig. 4.3(g)).

It is interesting to distinguish between two decay intervals of $R(\tau)$, the first corresponding to a fast decay $\Delta R(\tau_0)$ over a period of order $\tau_0 = 2\pi/\nu_2$ and the second to a slow decay $\Delta R(\tau_\infty)$ over a period τ_∞ corresponding to next 50 τ_0 (Fig. 4.4).

As *Re* is increased, the amplitudes of the peaks at ν_4 and ν_5 grow smoothly but their spectral power continues to be approximately two orders of magnitude less than that of ν_2. Visual observation therefore showed a flow regime with two eddies and phase velocity equal to $\nu_2/2$.

Strong decay of the autocorrelation function begins above $Re = 930$ (Fig. 4.4). The spectrum was highly complicated, but concentrated at low frequencies. As *Re* was increased further, the frequency ν_5 disappeared and autocorrelation function decay sharply increased. At the same *Re* the spectrum at the pole corresponded to oscillations at frequency $\approx\frac{1}{2}\nu_2$ (Fig. 4.5(*a*)), but the line widths of the component and its higher harmonics were slightly broadened.

The evolution of the spectrum at $Re = 1008$ and 1657 is shown in Fig. 4.5(*b*) and (*c*). The decay of the autocorrelation function is greater than 50 per cent (Fig. 4.5(*d*)), but above $Re = 1100$ the increase in decay is considerably reduced. The spectrum and the autocorrelation function at $Re = 3893$ are shown in Fig. 4.5(*e*) and (*f*). The spectrum contains a considerable continuous component but the sharp peaks at ν_2 and its harmonics are still of the order of 1.5 higher than the level of the background continuum.

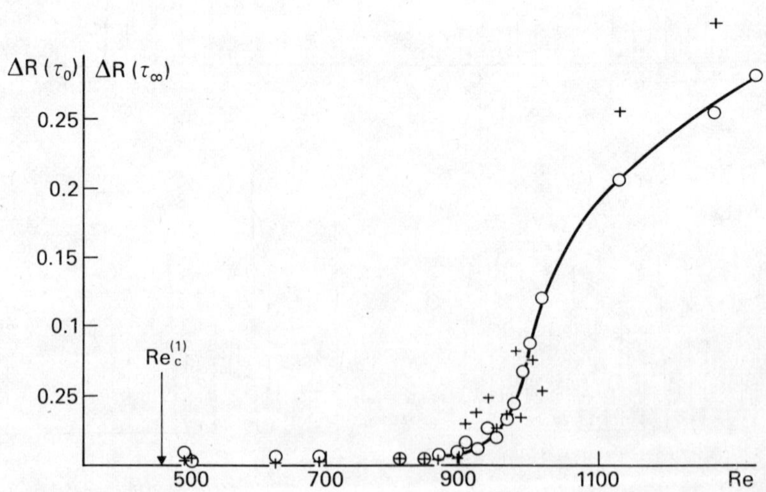

Fig. 4.4

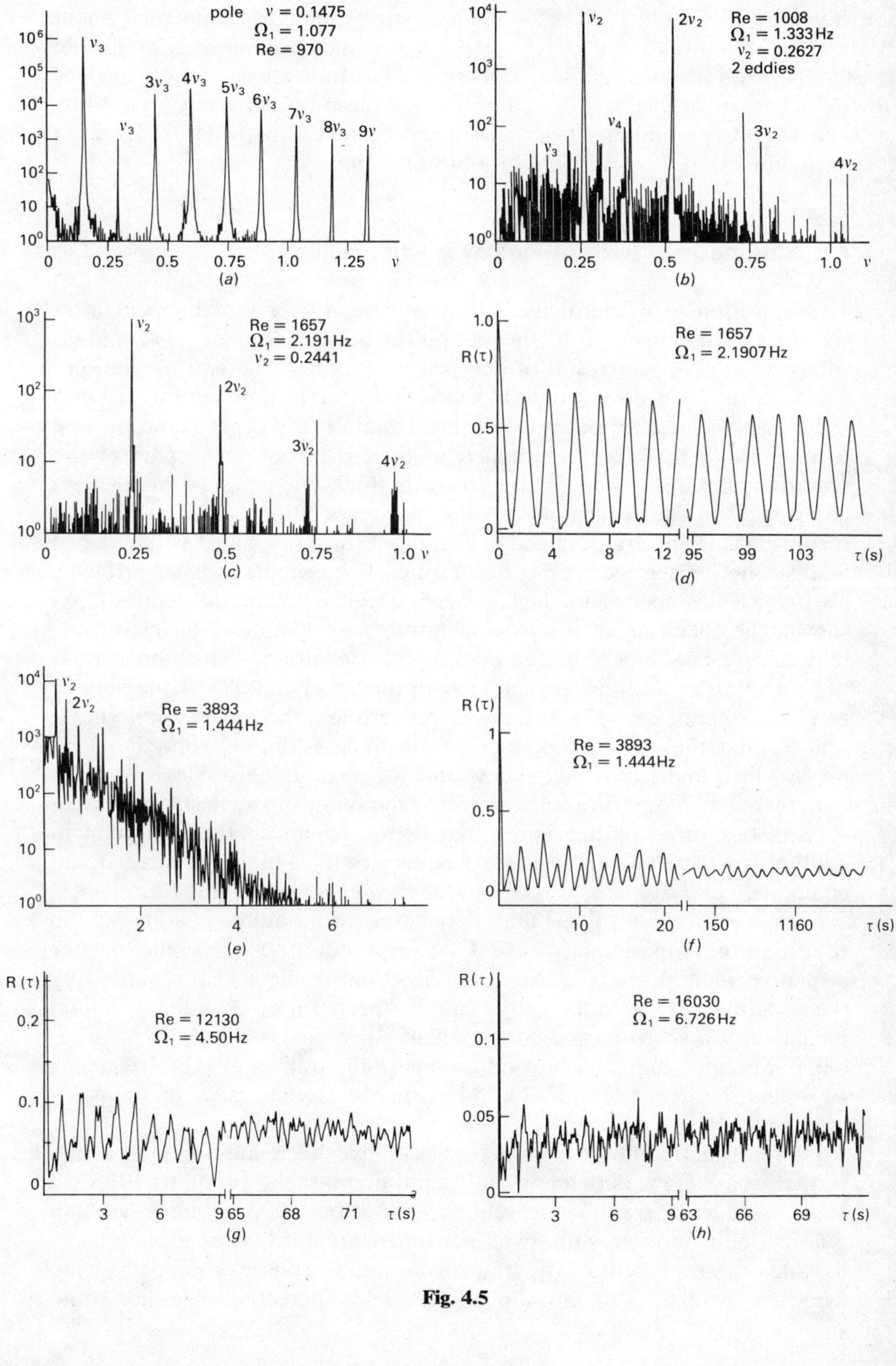

pole $\nu = 0.1475$
$\Omega_1 = 1.077$
Re = 970
(a)
Re = 1008
$\Omega_1 = 1.333$ Hz
$\nu_2 = 0.2627$
2 eddies
(b)
Re = 1657
$\Omega_1 = 2.191$ Hz
$\nu_2 = 0.2441$
(c)
Re = 1657
$\Omega_1 = 2.1907$ Hz
(d)
Re = 3893
$\Omega_1 = 1.444$ Hz
(e)
Re = 3893
$\Omega_1 = 1.444$ Hz
(f)
Re = 12130
$\Omega_1 = 4.50$ Hz
(g)
Re = 16030
$\Omega_1 = 6.726$ Hz
(h)

Fig. 4.5

The study of the spectral characteristics of SCF was carried out up to $Re = 16\,030 \approx 35\, Re_c^{(1)}$. The oscillations corresponding to the component ν_2 were monitored up to $Re = 16\,030$, but they contained a negligible part of the bulk energy. The autocorrelation function at these values of Re is presented in Fig. 4.5(g) and (h). All these results are supported by our recent preliminary measurements obtained with laser Doppler anemometers (DISA and Thermo-Systems Inc.).

4.3 Comparison with other flows—discussion

The transition to stochasticity in wide spherical gaps thus happens after six Hopf bifurcations from the periodical flow at frequency ν_2 and its subharmonic $\frac{1}{4}\nu_2$ as a result of desynchronization of the two oscillations. The bifurcations lead not only to the appearance of a new frequency but sometimes to the disappearance of those that already exist. The transition to nonperiodicity occurs gradually; at first, the continuous part of the spectrum contains only a small part of the bulk energy. New frequencies also appear in the spectrum after the first registration of stochasticity. It seems that the spectra obtained at the initial stage of chaotic behaviour of SCF strongly resemble a power spectrum of the simple Rossler attractor [18] which has special mixing properties (called 'phase coherence'). As shown, the spectrum of the Rossler attractor is composed of instrumentally sharp peaks superimposed on a broad background. The autocorrelation function of such an attractor, as in our case, consists of the sum of two components, one of which decays to zero and the other is periodic. It follows that the onset of nonperiodicity in SCF differs from the same process in cylindrical Couette flow and Rayleigh–Bénard convection.

In fact, according to recent results [5] the onset of stochasticity in CCF occurs after three bifurcations when there appears in the spectrum, in addition to two incommensurate frequencies ν_1 and ν_3, a weak broad component at a frequency $\sim\frac{1}{3}\nu_3$ and the flow becomes weakly turbulent.

It has been shown [6, 19] that, depending on the initial conditions, the transition to nonperiodicity in CCF occurs either after a periodic or after a quasiperiodic flow state. Another surprising result is also recorded by these authors: for sufficiently large aspect ratio, $\Gamma \geqslant 25$ (Γ = fluid height/gap), a sharp periodic component ν_2 re-emerges in fully continuous power spectrum well beyond the aperiodic transition. The frequency ν_2 is almost independent of Re and exists over a wide range of Reynolds numbers.

Four distinct routes to nonperiodicity have been identified in a box heated from below, depending on Prandtl number and geometry [9]. One route includes the following sequence of states: a periodic motion, a quasiperiodic motion with two incommensurate frequencies, a phase-locking or entrainment of the frequencies and the onset of nonperiodicity associated with loss of entrainment as Re is increased. Another route involves successive subharmonic bifurcations of a periodic flow. Yet another contains successive transitions to a periodic flow, to a quasiperiodic flow with two frequencies and to a quasiperiodic flow with three incommensurate frequencies and, soon after the latter transition, broadband noise begins to develop. A qualitatively different fourth route (first discovered by Libchaber and Maurer [20]) consists of the appearance of intermittent noise in a quasiperiodic state with two incommensurate frequencies.

The transition to stochasticity discussed above is something like a combination of the first two routes obtained in a convective box and involves a subharmonic bifurcation and a loss of phase-locking of the fundamental frequency ν_2 and its subharmonic.

Note added in proof

Our recent results show that six different flow regimes depending on initial conditions may exist at the same Reynolds number. In each case the transition to stochasticity occurs through its own sequence of states. Up to now, two sequences of transition to nonperiodic states at $Re = 823$ and 1075 have been considered in detail. Thus, the transition described in this chapter is one of many possible routes, and regular and nonperiodic flow regimes may exist at the same *Re*. One case of inverse transition from stochastic to regular flow regime with nondecaying correlation function was also registered when *Re* increased.

References

1. Ruelle, D. and Takens, F., On the nature of turbulence, *Communications in Mathematical Physics*, **20,** 167–192, 1971.
2. Newhouse, S., Ruelle, D. and Takens, F., Occurrence of strange axiom A attractors near quasiperiodic flows on T^m, $m \geq 3$. *Communications in Mathematical Physics*, **64,** 35–49, 1978.
3. Lorenz, E. N., Deterministic nonperiodic flow, *Journal of the Atmospheric Sciences*, **20,** 130–141, 1963.
4. Gollub, J. P. and Swinney, H. L., Onset of turbulence in a rotating fluid, *Physical Review Letters*, **35,** 927–930, 1975.
5. Fenstermacher, P. R., Swinney, H. L. and Gollub, J. P., Dynamical instabilities and the transition to chaotic Taylor vortex flow, *Journal of Fluid Mechanics*, **94,** 103–128, 1979.
6. Walden, R. W. and Donnelly, R. J., Re-emergent order of chaotic circular Couette flow, *Physical Review Letters*, **42,** 301–304, 1979.
7. Ahlers, G., Low-temperature studies of the Rayleigh–Bénard instability and turbulence, *Physical Review Letters*, **33,** 1185–1188, 1974.
8. Ahlers, G. and Walden, R. W., Turbulence near onset of convection, *Physical Review Letters*, **44,** 445–448, 1980.
9. Gollub, J. P. and Benson, S. V., Time-dependent instabilities and the

transition to turbulent convection, *Journal of Fluid Mechanics*, **100,** 449–470, 1980.

10. Yavorskaya, I. M., The influence of latitudinal temperature gradient on a spherical Couette flow, *Fluid Dynamics*, **10,** N2, 1975.
11. Khlebutin, G. N., Stability of fluid motion between rotating and stationary concentric spheres, *Fluid Dynamics*, **3,** N6, 1968.
12. Sawatzki, O. and Zierep, J., Das Stromfeld im Spalt zwischen zwei konzentrischen Kugelflächen, von denen die Innere rotiert, *Acta Mechanica*, **9,** 13–35, 1970.
13. Yavorskaya, I. M., Belayev, Yu. N. and Monakhov, A. A., Experimental study of loss of stability in a spherical Couette flow, in *Turbulentnye Techeniya* (*Turbulent flows*): Moscow, Nauka, 1977, 162–170.
14. Wimmer, M., Experiments on a viscous fluid flow between concentric rotating spheres, *Journal of Fluid Mechanics*, **78,** 317–335, 1976.
15. Yavorskaya, I. M., Belayev, Yu. N. and Monakhov, A. A., Experimental study of a spherical Couette flow, *Soviet Physics Doklady*, **221,** 1059–1062, 1975.
16. Belayev, Yu. N., Monakhov, A. A. and Yavorskaya, I. M., Stability of spherical Couette flow in thick layers when the inner sphere rotates, *Fluid Dynamics*, **13,** N2, 1978.
17. Petrov, G. I., Yavorskaya, I. M., Belayev, Yu. N., Monakhov, A. A., Slezinger, I. I., Goryainov, A. A. and Fuks, V. I., Simulation of dynamical processes in the planet atmospheres (Preprint N 255), Space Research Institute, USSR Academy of Sciences, 1975.
18. Farmer, D., Crutchfield, J., Froehling, H., Packard, N. and Shaw, R., Power spectra and mixing properties of strange attractor (preprint), University of California, Santa Cruz, 1980.
19. Donnelly, R. J., Park, K., Shaw, R. and Walden, R. W., Early non-periodic transition in Couette flow, *Physical Review Letters*, **44,** 987–989, 1980.
20. Libchaber, A. and Maurer, J., Rayleigh–Bénard experiment in liquid helium; frequency locking and the onset of turbulence, *Journal de Physique—Lettres*, **40,** L419–423, 1979.

5 Statistical Properties of Lorenz Attractors

L A Bunimovich

Dedicated to the memory of Vladimir Mikhailovich Alexeev

5.1 Introduction—ergodic theory of dissipative systems

A new branch of ergodic theory has appeared in recent years, which may be called *ergodic theory of dissipative systems.* The central concept here is the concept of strange attractors [1]. Many papers have now been published reporting various examples of strange attractors found in physical, chemical, biological and ecological problems (a recent survey of part of this activity can be found in [2]).

From the very beginning, attempts have been made to connect the concept of strange attractors with the onset of turbulence. Strange attractors have a complicated topological structure and the trajectories of the corresponding dynamical system resemble qualitatively the trajectories of fluid particles in a turbulent flow. However, this analogy is rather formal and superficial and does not shed much light on the essence of the onset of turbulence.

The question naturally arises, therefore, as to whether the behaviour of the dynamical system on a strange attractor is an adequate image of turbulence. To answer this question it is necessary to investigate the stochastical properties of the dynamical system under consideration. In the present chapter such investigation is presented for Lorenz attractors.

We first introduce several basic definitions. Let M be a compact n-dimensional manifold. It is always possible to suppose that M is an n-dimensional submanifold of an N-dimensional Euclidean space for some large N. Let us consider a time evolution $x \to S^t x$ on M. For continuous t the time evolution S^t is the *flow* obtained by integrating the differential equation $(\mathrm{d}/\mathrm{d}t)S^t x = X(S^t x)$, where X is a vector field on M. This equation means that the vector $X(x)$ belongs to the tangent space to M at the point x and the mapping $x \to X(x)$ is a smooth one.

A point $x \in M$ is called a *wandering point* if it has a neighbourhood U such that $U \cap S^t U = \varnothing$ for all sufficiently large t. Otherwise it is called *nonwandering.* The set NW of all nonwandering points is called a *nonwandering set.* The Axiom A introduced by Smale [3] for a diffeomorphism f consists of two conditions: (1) the nonwandering set $NW \subset M$ is a hyperbolic one; (2) the periodic points are everywhere dense in the set NW. (Hyperbolicity is defined by Plykin in Chapter 13, p. 282.)

Bowen and Ruelle [4] have studied the asymptotic behaviour of dynamical systems satisfying Axiom A. In this case the nonwandering set NW is the union of a finite number of compact invariant sets $(NW)_i$. For each i there exists a point $x_i \in (NW)_i$ such that its trajectory $\{S^t x_i\}$ is dense in $(NW)_i$. This property is called *topological transitivity.* The sets $(NW)_i$ are called *basic sets.* A basic set Λ is an *attractor* if it has a neighbourhood U such that $\bigcap_{t>0} S^t U = \Lambda$.

One of the principal questions in the study of strange attractors is the behaviour of the restriction of the dynamical system to the attractor. An attractor can contain a stable stationary point or a stable periodic trajectory. Such a dynamical system would not normally be considered to be stochastic. Sinai [5] introduced the following definition of the *stochastic attractor.* Let $\{S^t\}$, $0 \leqslant t < \infty$, be a dynamical system with phase space M. It is assumed that M is a compact closed Riemannian manifold. The stochastic attractor is an invariant (i.e., consisting of trajectories) closed set $\Lambda \subset M$ having the following properties:

(1) There exists a neighbourhood $U \supset \Lambda$ such that dist $(S^t x, \Lambda)$ tends to zero as $t \to \infty$ for any point $x \in U$ (dist (. . .) is the distance generated by a Riemannian metric).
(2) Let P_0 be a probability distribution on U with a density ρ_0. Then the sequence of probability distributions $S^t_* P_0$ converges weakly to an invariant distribution P which does not depend on P_0. The support of P is an attractor Λ.
(3) The probability distribution P is mixing, i.e., its autocorrelation functions $b(\tau) = \int f(S^\tau x) g(x)\, dP(x)$ tend to zero at $\tau \to \infty$.

The stochasticity of attractors in systems satisfying Axiom A (Axiom A systems) was shown in [4]. Later [6] it was proved that the attractor considered by Lorenz in his famous paper [7] is a stochastic one. As well, the following dissipative ergodic theorem was stated and proved for the Lorenz system in [6]:

Theorem 5.1. *Let $f(x)$ be a continuous function defined in a neighbourhood U of an attractor Λ and let ν be an absolutely continuous measure with support U. Then almost everywhere (with respect to ν) the following limit exists*:

$$\lim_{T\to\infty} \frac{1}{T} \int_0^T f(S^t x)\, dt = \int_\Lambda f(x)\, d\mu. \tag{5.1}$$

Here μ is an invariant absolutely continuous measure with support Λ. An analogous theorem was stated earlier for Axiom A systems [4].

5.2 The Lorenz system and its simplest properties

Consider a hydrodynamic flow in a layer of fluid of uniform depth between two parallel smooth planes, with a constant temperature difference between the upper and lower surfaces. The partial differential equations for this system (see, e.g., [8]) are well known. Lorenz [7] gave the following system of three ordinary differential equations with the help of the Galerkin procedure:

$$\begin{aligned}\frac{dx}{dt}&=-\sigma x+\sigma y,\\ \frac{dy}{dt}&=rx-y-xz,\\ \frac{dz}{dt}&=-bz+xy.\end{aligned} \tag{5.2}$$

Here b is a constant whose value depends on the dimensionless distance between the planes, σ is a Prandtl number and r is the ratio of the Rayleigh number to a certain critical value of it. Lorenz performed a qualitative investigation of this later system and its numerical simulation. System (5.2) is now known as the Lorenz system.

It is easy to verify that for the Lorenz system (5.2) the following properties hold.

(1) The divergence of the right-hand side of (5.2) is equal to $-(\sigma+1+b)$, i.e., is strictly negative. This means that any volume in the phase space of the Lorenz system shrinks exponentially with time.
(2) Let us consider the quantity $u^2(t)=x^2(t)+y^2(t)+[z(t)-r-\sigma]^2$. There exist positive constants C_1 and C_2 such that the inequality $(du/dt)\leqslant -C_1u+C_2$ holds. It follows from this inequality that any trajectory will reach the domain $\{u\leqslant 2C_2/C_1\}$ and will remain there at all subsequent times. It follows from this in particular that any solution of system (5.2) is bounded.
(3) The point O (the origin of coordinates) is a stationary point of system (5.2) for all values of parameters. If $r<1$ the point O is stable and there is no other stationary point. The point O loses its stability when the parameter r crosses the value 1. Simultaneously two new stationary points appear:

$$O_1=(\surd[b(r-1)], \surd[b(r-1)], r-1)$$
$$O_2=(-\surd[b(r-1)], -\surd[b(r-1)], r-1).$$

There are only these three stationary points in the phase space of system (5.1) when $r>1$. Points O_1 and O_2 are attracting if

$$1<r<\sigma(\sigma+b+3)(\sigma-b-1)^{-1}=r_0.$$

These points lose their stability at $r = r_0$. If $r > r_0$ then system (5.2) has no attracting stationary points.
(4) System (5.2) is invariant with respect to the transformation $x \to -x$, $y \to -y$, $z \to z$.

The limiting behaviour of system (5.2) when $r = 28$, $b = \frac{8}{3}$ and $1 \leq \sigma \leq 10$ was studied in [9] with the help of computer simulation and qualitative methods of bifurcation theory. Our consideration is based essentially on the results of this paper, which is therefore summarized briefly. A more general exposition is given by Afraimovich and Shil'nikov in Chapter 1.

(i) At point O two characteristic values belong to the left-hand semiplane of the complex plane and one to the right-hand one when $1 \leq \sigma < \sigma_1$, $\sigma_1 = 3.42$. In this case the two other stationary points O_1 and O_2 are stable focuses.

The branches of the unstable separatrix at point O will be denoted by Γ_1 and Γ_2. Then the branch $\Gamma_1(\Gamma_2)$ is attracted to $O_1(O_2)$ when $\sigma < \sigma_1$.
(ii) When $\sigma = \sigma_1$ the branches Γ_1 and Γ_2 are doubly asymptotic. This means that Γ_1 and Γ_2 belong locally to the stable separatrix of the point O. From a general point of view this situation was studied by Shil'nikov [10], who proved that two periodic hyperbolic trajectories (L_1 and L_2) appear when σ crosses the value σ_1. Moreover the branch $\Gamma_1(\Gamma_2)$ is attracted to the point $O_2(O_1)$ when $\sigma > \sigma_1$ (Fig. 5.1).
(iii) When $\sigma = \sigma_2$, $\sigma_2 \approx 5.87$, another bifurcation takes place. As a result the branch Γ_1 is attracted to a periodic motion L_2 and, symmetrically, Γ_2 to L_1.

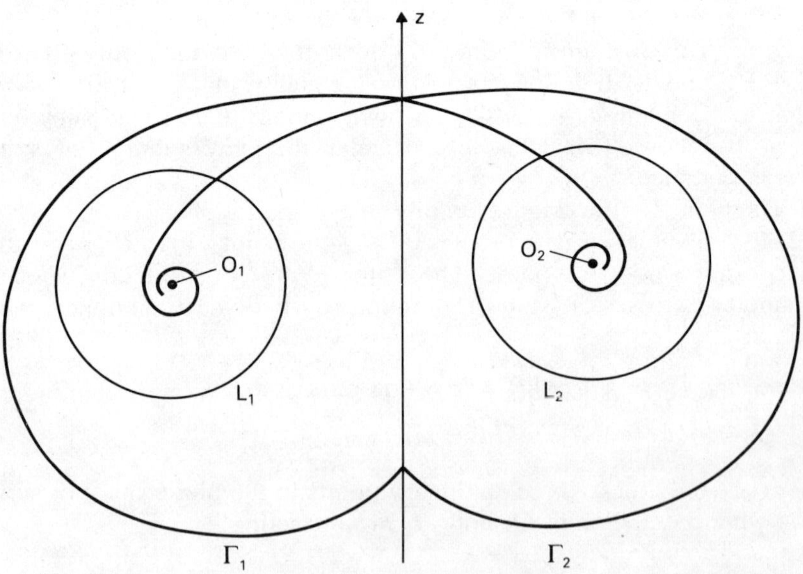

Fig. 5.1

A limiting set Λ, which is called the Lorenz attractor, appears in the phase space of system (5.2) when $\sigma \geqslant \sigma_2$. If $\sigma_2 \leqslant \sigma < \sigma_3$, $\sigma_3 \approx 6.47$, then in the phase space there are three attracting sets: the Lorenz attractor Λ and the stationary points O_1 and O_2. If parameter σ equals σ_3, then the periodical motions L_1 and L_2 shrink to the stationary points O_1 and O_2. At this moment O_1 and O_2 lose their stability. As a result, the system has a single stable limiting set when $\sigma_3 < \sigma \leqslant 10$. This set is the Lorenz attractor.

Consider the Poincaré mapping T of the plane $z = 27$ into itself. This plane contains the stationary points O_1 and O_2. Let us take any point a in this plane where $(\mathrm{d}z/\mathrm{d}t) < 0$. This means that an integral curve containing a intersects the plane $z = 27$ from above to below at this point. We continue this integral curve and take as Ta the next point at which this curve again intersects the plane $z = 27$ from above to below.

The transformation T when $\sigma_2 \leqslant \sigma \leqslant 10$ is illustrated in Fig. 5.2. Here z_1^* and z_2^* are points of intersection of L_1 and L_2 with the plane $z = 27$, respectively. It is easy to see that z_1^* and z_2^* are stationary points of T. The curve $\bar{W}$ is the intersection of the stable separatrix of the point O with the plane under consideration. The curves $W^s(z_1^*)$ and $W^s(z_2^*)$ are the intersections of the stable separatrices of the points z_1^* and z_2^* with this plane. Finally, z_1^0 and z_2^0 are the points of intersection of Γ_1 and Γ_2 with the plane $z = 27$.

We consider now the transformation T on a rectangle $\Pi_0 \subset \{z = 27\}$. The right-hand curvilinear triangle in Fig. 5.2 is the image of the top rectangle $\Pi_0^{(1)}$. By analogy, the left-hand triangle is the image of the lower rectangle $\Pi_0^{(2)}$.

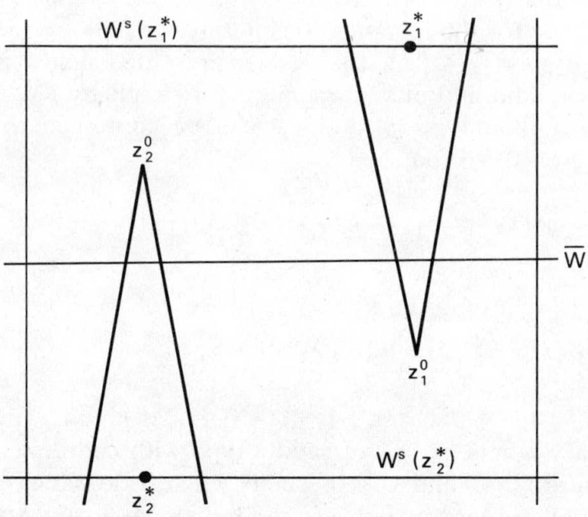

Fig. 5.2

It is easy to verify that the fixed point O has two negative (θ_1 and θ_2) and one positive (θ_3) characteristic values such that $0<-\theta_1<\theta_3<-\theta_2$ when $\sigma_2\leqslant\sigma\leqslant 10$.

With the help of computer simulations, Sinai and Vul [11] have shown that the transformation T is hyperbolic on the Lorenz attractor. This explains the appearance of statistical properties of the restriction of the dynamical system under consideration to the Lorenz attractor.

The difference between this dynamical system and the dynamical system considered in [4, 12] is connected with discontinuity of T. In this sense the transformation T is similar to dynamical systems with elastic reflections (or billiards) [13] which have an everywhere dense singular set but have, nevertheless, hyperbolic properties.

We now define axiomatically a transformation T which will be considered in what follows. This definition is based on results of many investigations in which the behaviour of a Lorenz system was simulated with the help of computers (see, for instance, [9]). We must mention, however, that there exists a rigorous result due to A. Simonov [14] showing that transformations with such properties are structurally stable and that corresponding systems of three differential equations constitute an open set in the space of all such systems. Therefore it is natural to call all attractors appearing in these systems Lorenz-type attractors. (For a more detailed exposition, see [9].)

Let Π_0 be the rectangle $\{|x|\leqslant 1, |y|\leqslant 1\}$. We shall consider a mapping T of Π_0 into itself defined by the following conditions. In the semi-open rectangle $\Pi_0^{(1)}=\{|x|\leqslant 1, 0<y\leqslant 1\}$ this mapping is defined with the help of functions $\bar{x}=f_1(x, y)$, $\bar{y}=g_1(x, y)$ of class C^2. Correspondingly, in the semi-open rectangle $\Pi_0^{(2)}=\{|x|\leqslant 1, -1\leqslant y<0\}$ the mapping T is defined by the functions $\bar{x}=f_2(x, y)$ and $\bar{y}=g_2(x, y)$ also belonging to C^2. We shall assume that $f_2(x, y)=-f_1(-x, -y)$ and $g_2(x, y)=-g_1(-x, -y)$. (This property is valid for the mapping T generated by the Lorenz system (5.2) in view of Property (4); all the results presented below are also valid without this condition, but we assume it for simplicity.) We shall assume that functions f_1 and g_1 (f_2 and g_2) can be continued to the segment $\bar{W}=\{|x|\leqslant 1,\ y=0\}$ so that

$$\lim_{y\to 0} f_1(x, y)=x_1^0 \qquad \left(\lim_{y\to 0} f_2(x, y)=x_2^0\right)$$

and

$$\lim_{y\to 0} g_1(x, y)=y_1^0 \qquad \left(\lim_{y\to 0} g_2(x, y)=y_2^0\right)$$

where $0<x_1^0<1$, $-1<y_1^0<0$, $-1<x_2^0<0$, $0<y_2^0<1$. The mapping T has two stationary points z_1^*, z_2^* of saddle type with coordinates $(x_1^*, 1)$ and $(x_2^*, -1)$ where $x_1^*>0$ and $x_2^*<0$. The characteristic values in these points are positive. The segment $y=1$ $(y=-1)$ belongs to the stable invariant manifold of the point $z_1^*(z_2^*)$.

We shall assume that the following conditions are valid [9]:

$$\left\|\frac{\partial f_1}{\partial x}\right\|<1, \quad \left\|\left(\frac{\partial g_1}{\partial y}\right)^{-1}\right\|<1,$$

$$\left\|\left(\frac{\partial g_1}{\partial y}\right)^{-1}\frac{\partial f_1}{\partial y}\right\|\left\|\frac{\partial g_1}{\partial x}\right\|<\left(1-\left\|\frac{\partial f_1}{\partial x}\right\|\right)\left(1-\left\|\left(\frac{\partial g_1}{\partial y}\right)^{-1}\right\|\right), \tag{5.3}$$

$$1-\left\|\left(\frac{\partial g_1}{\partial y}\right)^{-1}\right\|\left\|\frac{\partial f_1}{\partial x}\right\|>2\sqrt{\left[\left\|\left(\frac{\partial g_1}{\partial y}\right)^{-1}\left(\frac{\partial f_1}{\partial y}\right)\right\|\left\|\left(\frac{\partial g_1}{\partial y}\right)^{-1}\right\|\left\|\frac{\partial g_1}{\partial x}\right\|\right]}$$

where

$$\|h(x, y)\| = \max_{x,y\in\Pi_0} |h(x, y)|.$$

Analogous conditions must be valid for the functions f_2 and g_2. It is easy to see that the conditions (5.3) express hyperbolicity in the form of the existence of invariant expanding cones in the tangent space to Π_0.

In this chapter we consider stronger statistical properties of the dynamical system acting on the Lorenz attractor. In Section 5.3 we estimate the rate of decay of correlation functions. Let $h(x)$ be a sufficiently smooth, bounded function defined on the phase space of the dynamical system under consideration such that $\int_\Lambda h(x)\,d\mu=0$. We shall prove that

$$Eh(x)h(T^n x)=\int_\Lambda h(x)h(T^n x)\,d\mu\leqslant\exp(-n^\gamma)$$

for all sufficiently large n and some positive constant $\gamma<1$.

In connection with the validity of a dissipative ergodic theorem a natural question arises: what is the rate of convergence of the time averages in (5.1) to their limit? In Section 5.4, for any smooth bounded function $h(x)$ defined on the Lorenz attractor, we shall prove a theorem which is analogous to the central limit theorem of probability theory. More precisely we shall prove that the random variables

$$\sqrt{(n)}\left[\frac{1}{n}\sum_{k=0}^{n-1} h(T^k x)-\int_\Lambda h(x)\,d\mu\right]$$

have Gaussian distribution in the limit $n\to\infty$.

5.3 Symbolic dynamics for the Lorenz attractor and some auxiliary assertions

Let us consider a set of curves of class C^1 in the rectangle Π_0 with one-to-one projection on the horizontal axis. Each such curve can be considered as a smooth monotonic function $y=f(x)$ defined on some segment $[x_0, x_1]$. Let us denote by $\mathcal{B}$ the set of all such curves having the property $|f'(x)|>1$. It follows from (5.3) that $T\mathcal{B}\subset\mathcal{B}$.

For any integer $k>0$ the image $T^k\Pi_0$ is a collection of a finite number

m_k of closed sets. It is easy to see that $m_k \leq 2^k$. Two of these sets are curvilinear triangles, and they contain the saddle-points z_1^* and z_2^*, respectively. Two sides of every such triangle are the images of the vertical sides of the rectangle Π_0. The vertex where these sides intersect is the point $z_1^0=(x_1^0, y_1^0)$ (or $z_2^0=(x_2^0, y_2^0)$). The third side is a segment of the stable manifold of the point z_1^* (or z_2^*). Any other set from $T^k\Pi_0$ has the form of a spindle stretched along the vertical axis. The boundary of such a set consists of two smooth components. These components intersect in two points or *vertices*. One of these vertices is the point $T^iz_1^0$ and the other is $T^jz_2^0$, where $0 \leq i, j \leq k$ and $i \neq j$. We shall call such sets the *cells*.

A cell with vertices $T^iz_1^0$ and $T^jz_2^0$ will be called a *main cell* if it contains all other cells with the same vertices. The rank $\text{rk}(\Delta)$ of a cell Δ with these vertices is equal to the maximum of the numbers i and j. By definition the *order* of a cell Δ is the maximum of all numbers k such that $\Delta \subset T^k\Pi_0$. It is easy to see that the rank of any cell does not exceed its order and that the rank and the order are equal for main cells only. We shall say that two cells Δ_1 and Δ_2 have a right intersection if $\Delta_1 \supseteq \Delta_2$ or $\Delta_1 \subseteq \Delta_2$ and their vertices coincide.

We shall denote the length of the projection to the vertical axis of a cell Δ by $y(\Delta)$ and shall call it the *y-length* of Δ. It follows from the inequalities (5.3) that, for any cell Δ, the ratio of the y-length of $T\Delta$ to the y-length of Δ is not less than $1+v$, where the number $v>0$ does not depend on Δ. One should mention that the set $T\Delta$ can consist of two connected components. This occurs if $\Delta \cap \bar{W} \neq \emptyset$. In this case we mean by $y(T\Delta)$ the sum of the y-lengths of these two components.

Let the points z_1^0 and z_2^0 and all their images be nonperiodic points. Then it follows from (5.3) that there exist neighbourhoods $U(z_1^0)=U_1$ and $U(z_2^0)=U_2$ such that $T^iU_j \cap \bar{W} = \emptyset$ where $i=0, 1, \ldots, p$, $j=1, 2$ and $(1+v)^p = 2+\delta$, $\delta>0$. Let Δ be an arbitrary cell and $T\Delta$ consist of two connected components Δ' and Δ''. Further, let Δ' (or both Δ' and Δ'') have a small y-length l and be contained in U_1 or in U_2. Then all images $T\Delta', T^2\Delta', \ldots, T^p\Delta'$ are connected sets. Moreover if $k=\max\{i>0, T^j\Delta' \cap \bar{W} \neq \emptyset$ for all $0<j\leq i\}$ then the y-length of the longer connected component of the set $T^{k+1}\Delta'$ is not less than $l[1+(\delta/2)]$.

It follows from (5.3) that there exists such a partition η of the rectangle Π_0 into almost horizontal smooth curves that if the curves $\gamma_a \in \eta$ and $\gamma_b \in \eta$ intersect the y-axis in points $y=a$ and $y=b$ respectively, and $T\gamma_a \cap \gamma_b \neq \emptyset$, then $T\gamma_a \subset \gamma_b$. Indeed, let us consider the set $\mathcal{D}_{\bar{W}}$ which consists of such curves $\gamma_{\bar{W}}$ that $T^k\gamma_{\bar{W}} \subset \bar{W}$ for a certain $k>0$. One can deduce from (5.3) that the set $\mathcal{D}_{\bar{W}}$ is dense everywhere in Π_0. Hence this set of curves can be naturally continued to a partition of Π_0 with the desired properties. Further, from (5.3) we obtain that the length of any smooth curve from η decreases under the action of T by not less than a factor $(1+\gamma)^{-1}$, where $\gamma>0$.

We denote by γ_1^0 and γ_2^0 the curves belonging to the partition η which contain the points z_1^0 and z_2^0 respectively. A *horizontal strip* is a set which lies between two arbitrary curves from the partition η and is contained in $\Pi_0^{(1)}$ or $\Pi_0^{(2)}$. If a horizontal strip contains γ_1^0 or γ_2^0, it will be called a *boundary strip*.

We shall say that the transformation T is nontransitive if there exist boundary strips $\Pi_1' \subset \Pi_0^{(1)}$ and $\Pi_2' \subset \Pi_0^{(2)}$ and an integer $m > 0$ such that $T^m(\Pi_1' \cup \Pi_2') \subset \Pi_1' \cup \Pi_2'$. In this case the projection of the attractor Λ to the vertical axis consists of a finite number d $(d \geq 5)$ of nonoverlapping segments, i.e., it has lacunae. At first we assume that the transformation T is transitive. A sufficient (but not necessary) condition for this is the validity of the following inequalities:

$$\left\| \left(\frac{\partial g_i}{\partial y} \right)^{-1} \right\| < \frac{1}{\sqrt{2}} \qquad \text{where} \quad i = 1, 2.$$

Let us denote by μ a measure on the attractor Λ invariant with respect to T which was constructed in [6]. It was also shown in [6] that μ is absolutely continuous with respect to Lebesgue measure, which is on Λ generated by the length.

We denote by $\mathcal{R}_n$ the collection of all cells with ranks not more than n and by A_n the collection of all cells with order equal to n. It is easy to see that A_n contains not more than $2n$ cells. Let Q_n be equal to $\bigcup_{k=1}^n A_k$.

Let us take a cell $\Delta \in A_n$. We call Δ a *long cell* if $y(\Delta) > C_1(1+\delta/2)^{-n}$ where the positive constant C_1 does not depend on n and Δ. All other cells from A_n will be called *short cells*. We denote by LA_n the union of all long cells belonging to A_n, $LQ_n = \bigcup_{k=1}^n LA_k$.

We now formulate three assertions which were proved in [6]. They will be useful for us in what follows.

Lemma 5.1. *Let ν be the Lebesgue measure on the rectangle Π_0. Then*

$$\nu[T^{-n}(A_n \backslash LA_n)] < C_2 n^\beta (1+v)^{-n} \qquad \beta > 0, n > 0$$

is a sufficiently large integer and the constant C_2 does not depend on n.

The following two Lemmas are true for a transitive transformation T only:

Lemma 5.2. *For any integer $r > 0$, one can find an integer $N(r) > 0$ such that, for any cell $\Delta_1 \in \mathcal{R}_r$ and for any cell $\Delta_2 \in A_r$, the set $T^N\Delta_1 \cap \Delta_2$ is not empty for all $N \geq N(r)$. Moreover, the set $T^N\Delta_1 \cap \Delta_2$ contains a cell $\Delta_3 \subset \Delta_2$ such that $\Delta_3 \cap \Delta_2$ is a right intersection.*

It was shown in [6] that $N(r) \leq C_3 r^2$ where the constant $C_3 > 0$ does not depend on r.

Lemma 5.3. *Let $r > \mathcal{D}_1 \log_{1+(\delta/2)}(2/\mathcal{D}_2)$, where the values of the constants $\mathcal{D}_1$ and $\mathcal{D}_2$ depend only on the functions f_1, f_2, g_1, g_2 (see [6]). Then there exists such a constant $C_4 > 0$ independent of n that for any cell $\Delta \in LQ_n$ one* *can find a cell $\Delta' \in \mathcal{R}_r$ such that $T^{[C_4 n]}\Delta \cap \Delta'$ is a right intersection. (Here we denote by $[a]$ the maximal integer which does not exceed a.)*

Let us now take a cell $\Delta \in LQ_n$. It follows from Lemmas 5.2, 5.3 that there exists an integer k such that the set $T^k\Delta$ has a right intersection with all the main cells $\Delta' \in R_\eta$. Further, one can choose a number k so that

$$k \leqslant C_3 r^2 + \mathscr{D}_1 \log_{1+(\delta/2)}\left(\frac{2}{\mathscr{D}_2}\right) + C_5 n \qquad \text{where} \quad C_5 > 0$$

does not depend on n and r. Moreover, for any integer r one can find a number $\hat{r} = \hat{r}(r)$ such that $T^{\hat{r}}\Delta_1 \cap \Delta_2 \neq \varnothing$ for arbitrary cells $\Delta_1, \Delta_2 \in A_r$.

The Lorenz attractor Λ consists of a continuum of smooth curves. The end points of any such curve coincide with vertices of some main cell. Every such curve will be called a *regular component* of the attractor Λ. We shall denote by ξ the partition of Λ into regular components.

We shall call two regular components $C'_\xi, C''_\xi \in \Lambda$ *canonically isomorphic* if for every point $x \in C'_\xi$ there exists a single point $y \in C''_\xi$ and a single curve $\gamma_x \in \eta$ such that $y \in \gamma_x$ and, conversely, for every point $y \in C''_\xi$ one can define in a single way one $x \in C'_\xi$ and a $y \in \gamma_y \in \eta$ such that $x \in \gamma_y$.

Let now Δ be any cell and $C_\xi \in \xi$. We shall say that $\Delta \cap C_\xi \neq \varnothing$ is a *right intersection* if both end points of the regular component C_ξ coincide with vertices of Δ and $C_\xi \in \Delta$. Let $C'_\xi, C''_\xi \in \xi$ and let $C'_\xi \cap \Delta$ and $C''_\xi \cap \Delta$ be right intersections. Then, according to the definition of the partition η, regular components C'_ξ and C''_ξ are canonically isomorphic. We shall denote the corresponding canonical isomorphism by $\Pi : C'_\xi \to C''_\xi$.

Now we consider a new partition ζ of the attractor Λ. Every element of ζ is the set of all canonically isomorphic components of Λ, i.e., any element of ζ consists of all regular components of Λ which have a right intersection with one and the same main cell. It is clear that ζ is a countable partition. We shall denote its elements by $C_1(\zeta), C_2(\zeta), \ldots,$ in such a way that main cells with smaller order correspond to elements with smaller indices. Consequently the elements $C_1(\zeta)$ and $C_2(\zeta)$ correspond to two main cells of first order, $C_3(\zeta)$ and $C_4(\zeta)$ to two main cells of second order and so on, i.e., elements $C_{2k-1}(\zeta)$ and $C_{2k}(\zeta)$ correspond to two main cells of order k.

Let us consider an infinite matrix $\mathbb{Q} = \|q_{ij}\|$, where $q_{ij} = 1$ if $\mu[C_i(\zeta) \cap TC_j(\zeta)] > 0$ and $q_{ij} = 0$ otherwise. We shall construct a space Ω of all two-sided infinite sequences $\omega = (\ldots, \omega_{-n}, \ldots, \omega_0, \ldots, \omega_n, \ldots)$, where $\omega_i = 1, 2, \ldots$ and $q_{\omega_n \omega_{n+1}} = 1, -\infty < n < \infty$. We consider in the space of all integer numbers a discrete topology. Then Ω is a topological space with the product topology of topological spaces. In Ω a shift automorphism $\tilde{T}$ acts via the formula $\tilde{T}(\ldots, \omega_{-n}, \ldots, \omega_0, \ldots, \omega_n, \ldots) = (\ldots, \omega_{-n+1}, \ldots, \omega_1, \ldots, \omega_{n+1}, \ldots)$, i.e., every sequence is shifted by one coordinate to the left.

Consider now a mapping $\varphi : \Omega \to \Lambda$ by putting $\varphi(\omega) = \bigcap_{n=-\infty}^{\infty} T^n C_{\omega_n}(\zeta)$. It follows from the definition of the limit set (attractor) Λ that φ is defined everywhere in Ω.

With the help of the mapping φ we can consider a symbolic representation of our dynamical system. This representation is very similar to the symbolic dynamics for Axiom A systems [4, 15].

Lemma 5.4. *The mapping φ is a one-to-one mapping of the symbolic space Ω onto its image. The image $\varphi(\Omega)$ is a full measure subset of Λ.*

It is easy to see that $\varphi(\Omega)=\Lambda\setminus\bigcup_{n=-\infty}^{\infty} T^n\bar{W}$ where $T^n\bar{W}$ $(n<0)$ is the full inverse image of the set $\bar{W}$. The proof of Lemma 5.4 follows from the definition of the mapping φ and from the properties of the transformation T. It continues in the same way as the proof of Theorem 3.1 in [15].

Now we can introduce a measure $\tilde{\mu}$ into the space Ω via the formula $\tilde{\mu}(C)=\mu[\varphi(C)]$. One can obtain from Lemma 5.4 that φ is an isomorphism of the measurable spaces $(\Lambda, \mathfrak{B}, \mu)$ and $(\Omega, \tilde{\mathfrak{B}}, \tilde{\mu})$, where $\tilde{\mathfrak{B}}$ is the completion with respect to $\tilde{\mu}$ of the Borel σ-algebra in the space Ω. It follows from the equality $\varphi T=\tilde{T}\varphi$ that the measure $\tilde{\mu}$ is invariant with respect to $\tilde{T}$. The measurable space $(\Omega, \tilde{\mathfrak{B}}, \tilde{\mu})$ is called a symbolic representation of the dynamical system $(\Lambda, \mathfrak{B}, \mu)$. We shall denote by $\tilde{\xi}$ a partition of Ω induced by the partition ξ of Λ.

The following assertion is in fact a reformulation of Lemma 5.1:

Lemma 5.5. $\tilde{\mu}\{\omega:\omega_0>2n\}\leq C_2n^{\beta}(1+\delta)^{-n}$.

Let us consider conditional probabilities $\tilde{\mu}(\omega_n \mid \omega_{n-1},\dots,\omega_0,\omega_{-1},\dots)$ which exist with $\tilde{\mu}$-probability 1. Our main purpose now is to show that $\tilde{\mu}(\omega_n \mid \omega_{n-1},\dots,\omega_0,\omega_{-1},\dots)$ can be approximated by conditional probabilities with finite memory, $\tilde{\mu}(\omega_n \mid \omega_{n-1},\dots,\omega_0,\dots,\omega_{-k})$.

Now we introduce the sets

$$\begin{aligned}
\mathscr{E}_n &=\{z=(x,y)\in\Lambda:|y|<\alpha_1^n\},\\
\mathscr{F}_n &=\{z=(x,y)\in\Lambda:T^m z\notin\mathscr{E}_k,\ k=[n^{\alpha_2}]\ \text{ for }\ |m|\leq n\\
&\qquad \text{and}\quad T^r z\notin\mathscr{E}_r\quad\text{for}\quad r>n\},\\
K_n &=\{z=(x,y)\in\Lambda:\mu[\mathscr{F}_n\mid C_\xi(z)]>1-\alpha_3^{n^{1/2}}\}
\end{aligned}$$

where $0<\alpha_1,\ \alpha_2,\ \alpha_3<1,\ \alpha_1<\alpha_3$.

It is easy to see that

$$\mu(\mathscr{E}_n)<\alpha_1^n,\quad \mu(\mathscr{F}_n)\leq 2n\alpha_1^{[n^{\alpha_2}]}+\frac{\alpha_1^n}{1-\alpha_1},\quad \mu(K_n)<\frac{2n}{1-\alpha_1}\left(\frac{\alpha_1}{\alpha_3}\right)^{[n^{\alpha_2}]}.$$

The corresponding sets $\tilde{\mathscr{E}}_n$, $\tilde{\mathscr{F}}_n$ and $\tilde{K}_n$ in the symbolic space Ω have analogous properties. It is easy to see that the partition $\tilde{\xi}$ of the space Ω corresponds to the partition ξ of Λ. The partition $\tilde{\xi}$ appears when we fix all ω_i, $-\infty<i\leq 0$.

Lemma 5.6. *Let the elements C'_ξ, C''_ξ correspond to the sequences ω_0, $\omega_{-1},\dots,\omega_{-n+1},\omega'_{-n},\omega'_{-n-1},\dots$ and $\omega_0,\omega_{-1},\dots,\omega_{-n+1},\omega''_{-n},\omega''_{-n-1},\dots$,* *respectively, and* C'_{ξ}, $C''_{\xi} \in \varphi(K_n)$. *Then there exist such* α_1, α_2, α_3 *and* α_4, $0<\alpha_4<1$ *that* $\sum_{\omega_1} |\tilde{\mu}(\omega_1 \mid C'_{\xi}) - \tilde{\mu}(\omega_1 \mid C''_{\xi})| < \alpha_4^{n^{1/2}}$.

We now formulate the property which is an analogy of the Doeblin condition in the theory of ordinary Markov chains (see, for instance, [16]).

Let us consider the conditional probabilities

$$q_1(\omega_{n+1}, \ldots, \omega_{2n}) = \tilde{\mu}(\omega_{n+1}, \ldots, \omega_{2n} \mid \omega'_0, \ldots, \omega'_{-n+1})$$

and

$$q_2(\omega_{n+1}, \ldots, \omega_{2n}) = \tilde{\mu}(\omega_{n+1}, \ldots, \omega_{2n} \mid \omega''_0, \ldots, \omega''_{-n+1}).$$

After that we have the probability distributions q_1, q_2 on the space of words $\omega_{n+1}, \ldots, \omega_{2n}$ under different conditions $\omega'_0, \ldots, \omega'_{-n+1}$ and $\omega''_0, \ldots, \omega''_{-n+1}$.

Lemma 5.7. *Let* $\varphi(\omega'_i) \subset LQ_{n^{1/2}}$, $\varphi(\omega''_i) \subset LQ_{n^{1/2}}$. *There exists a constant* λ, $0<\lambda<1$, *such that for all large enough* n

$$\operatorname{var}(q_1, q_2) = \frac{1}{2} \sum_{\omega_{n+1}, \ldots, \omega_{2n}} |q_1(\omega_{n+1}, \ldots, \omega_{2n}) - q_2(\omega_{n+1}, \ldots, \omega_{2n})| \leq \lambda.$$

The proof of this Lemma follows the proof of an analogous assertion in [17]. The following considerations are quite similar to [18].

Let $n_1 < n$ be chosen. Now we shall transform the probability distribution $\tilde{\mu}(\omega_{-n_1}, \ldots, \omega_{n+n_1})$. Denote by G_n the set of all sequences $(\omega_{-n_1}, \ldots, \omega_{n+n_1})$ satisfying the conditions $C_{\omega_i}(\tilde{\xi}) \in LA_n$ for all $-n_1 \leq i \leq n+n_1$ and

$$\tilde{\mu}(\mathcal{F}_{n_1} \cap (\omega_i, \ldots, \omega_j)) \geq (1 - \tilde{\mu}^{1/2}(\bar{\mathcal{F}}_{n_1}))\tilde{\mu}(\omega_i, \ldots, \omega_j)$$

where $\bar{\mathcal{F}}_{n_1} = \Omega \backslash \mathcal{F}_{n_1}$ and $-n_1 \leq i$, $j \leq n+n_1$. It follows from Chebyshev's inequality that

$$\tilde{\mu}(G_n) \geq 1 - n^2 \tilde{\mu}^{1/2}(\bar{\mathcal{F}}_{n_1}).$$

Lemma 5.8. *Let* $(\omega_{-n_1}, \ldots, \omega_{n+n_1}) \in G_n$. *Then there exists a number* λ_1, $0<\lambda_1<1$, *such that*

$$[1 - \tilde{\mu}^{1/2}(\bar{\mathcal{F}}_{n_1})]^{2n}(1+\lambda_1^{n_1})^{-n}$$
$$\leq \frac{\tilde{\mu}(\omega_{-n_1} \ldots \omega_{n+n_1})}{\tilde{\mu}(\omega_{-n_1} \ldots \omega_{n_1}) \prod_{i=n_1+1}^{n+n_1} \tilde{\mu}(\omega_i \mid \omega_{i-1} \ldots \omega_{i-2n_1-1})}$$
$$\leq (1+\lambda_1^{n_1})^n [1 - \tilde{\mu}^{1/2}(\bar{\mathcal{F}}_{n_1})]^{-2n}$$

Proof. According to the formula of complete probability we have

$$\tilde{\mu}(\omega_{-n_1} \ldots \omega_{n+n_1}) = \tilde{\mu}(\omega_{-n_1} \ldots \omega_{n_1}) \prod_{i=n_1+1}^{n+n_1} \tilde{\mu}(\omega_i \mid \omega_{i-1} \ldots \omega_{-n_1}).$$

Now we must estimate the fraction

$$I_i = \frac{\tilde{\mu}(\omega_i \mid \omega_{i-1} \dots \omega_{i-2n_1-1})}{\tilde{\mu}(\omega_i \mid \omega_{i-1} \dots \omega_{-n_1})}.$$

Let us consider the other fraction

$$I_i^{(0)} = \frac{\int \tilde{\mu}(\mathscr{F}_{n_1} \cap \omega_i \mid C_{T^{i-1}\zeta})\, d\nu^{(1)}}{\int \tilde{\mu}(\mathscr{F}_{n_1} \cap \omega_i \mid C_{T^{i-1}\zeta})\, d\nu^{(2)}}$$

where $\nu^{(1)}(\nu^{(2)})$ is the induced probability distribution on the space of $C_{T^{i-1}\zeta} \in (\omega_{i-1} \dots \omega_{-n_1})(C_{T^{i-1}\zeta} \in (\omega_{i-1} \dots \omega_{i-2n_1-1}))$. We fix

$$C^{(0)}_{T^{i-1}} \in (\omega_{i-1} \dots \omega_{i-2n_1-1}) \subset (\omega_{i-1} \dots \omega_{-n_1})$$

and rewrite $I_i^{(0)}$ as follows:

$$I_i^{(0)} = \frac{\int [\tilde{\mu}(\omega_i \cap \mathscr{F}_{n_1} \mid C_{T^{i-1}\zeta}) / \tilde{\mu}(\omega_i \cap \mathscr{F}_{n_1} \mid C^{(0)}_{T^{i-1}\zeta})]\, d\nu^{(1)}}{\int [\tilde{\mu}(\omega_i \cap \mathscr{F}_{n_1} \mid C_{T^{i-1}\zeta}) / \tilde{\mu}(\omega_i \cap \mathscr{F}_{n_1} \mid C^{(0)}_{T^{i-1}\zeta})]\, d\nu^{(2)}}$$

It is easy to see that the sets $(\omega_i \cap \mathscr{F}_{n_1}) \cap C_{T^{i-1}\zeta}$ and $(\omega_i \cap \mathscr{F}_{n_1}) \cap C^{(0)}_{T^{i-1}\zeta}$ are canonically isomorphic for arbitrary $C_{T^{i-1}\zeta}$, $C^{(0)}_{T^{i-1}\zeta}$. The absolute value of the difference between the corresponding density and 1 is not more than $\lambda_1^{n_1}$ for some λ_1, $0 < \lambda_1 < 1$. Hence

$$\left| \frac{\tilde{\mu}(\omega_i \cap \mathscr{F}_{n_1} \mid C_{T^{i-1}\zeta})}{\tilde{\mu}(\omega_i \cap \mathscr{F}_{n_1} \mid C^{(0)}_{T^{i-1}\zeta})} - 1 \right| < \lambda_1^{n_1}$$

and $(1+\lambda_1^{n_1})^{-2} \leq I_i^{(0)} \leq (1+\lambda_1^{n_1})^2$.

Therefore

$$\frac{\tilde{\mu}(\omega_{-n_1} \dots \omega_{n+n_1})}{\tilde{\mu}(\omega_{-n_1} \dots \omega_{n_1}) \prod_{i=n_1+1}^{n+n_1} \tilde{\mu}(\omega_i \mid \omega_{i-1} \dots \omega_{i-2n_1-1})} = \prod_{i=n_1+1}^{n+n_1} I_i$$

$$\leq [1 - \tilde{\mu}^{1/2}(\bar{\mathscr{F}}_{n_1})]^{-2n} \prod_{i=n_1+1}^{n+n_1} I_i^{(0)} \leq (1+\lambda_1^{n_1})^n [1 - \tilde{\mu}^{1/2}(\bar{\mathscr{F}}_{n_1})]^{-2n}. \quad \blacksquare$$

In the same way one can estimate the fraction under consideration from below.

5.4 Estimation of the correlation function decay

Let a function $h(\omega)$ defined in Ω be bounded and let there exist a constant λ_2, $0 < \lambda_2 < 1$ such that for all sufficiently large n one can find functions $h_n(\omega)=h_n(\omega_{-n}\dots\omega_n)$ such that $\sup_{\omega\in\Omega}|h(\omega)-h_n(\omega)|\leqslant\lambda_2^n$ and $\int_\Omega h_n\,\mathrm{d}\tilde\mu=0$. Then the following theorem holds:

Theorem 5.2. *If $Eh=\int_\Omega h\,\mathrm{d}\tilde\mu=0$ then $|Eh(\tilde T^n\omega)h(\omega)|\leqslant\exp(-n^{\gamma_1})$ for all sufficiently large n where a constant γ_1, $0<\gamma_1<1$ does not depend on n.*

Proof. Let the numbers n and n_1, $n_1<n$ be fixed. It is easy to see that

$$|Eh(\tilde T^n\omega)h(\omega)-Eh_{n_1}(\tilde T^n\omega)h_{n_1}(\omega)|\leqslant\text{const }\lambda^{n_1}.$$

We have

$$Eh_{n_1}(\tilde T^n\omega)h_{n_1}(\omega)=\int_\Omega h_{n_1}(\tilde T^n\omega)h_{n_1}(\omega)\,\mathrm{d}\tilde\mu$$

$$=\sum_{\omega_{-n_1}\dots\omega_{n+n_1}} h_{n_1}(\omega_{-n_1+n}\dots\omega_{n_1+n})h_{n_1}(\omega_{-n_1}\dots\omega_{n_1})\tilde\mu(\omega_{-n_1}\dots\omega_{n_1+n}).$$

According to Lemma 5.8 the problem reduces to investigation of the expression

$$\sum h_{n_1}(\omega_{-n_1}\dots\omega_{n_1})h_{n_1}(\omega_{n-n_1}\dots\omega_{n+n_1})\tilde\mu(\omega_{-n_1}\dots\omega_{n_1})$$
$$\times\prod_{i=n_1+1}^{n+n_1}\tilde\mu(\omega_i\mid\omega_{i-1}\dots\omega_{i-2n_1-1}).$$

The numbers

$$\mu^{(1)}(\omega_{-n_1}\dots\omega_{n+n_1})=\tilde\mu(\omega_{-n_1}\dots\omega_{n_1})\prod_{i=n_1+1}^{n+n_1}\tilde\mu(\omega_i\mid\omega_{i-1}\dots\omega_{i-2n_1-1})$$

define the probability distribution on the space of all sequences $(\omega_{-n_1}\dots\omega_{n+n_1})$ which is a homogeneous Markov chain with the memory $2n_1+1$. Its stationary probability distribution is equal to $\{\tilde\mu(\omega_{-n_1}\dots\omega_{n_1})\}$ while transition probabilities are $\tilde\mu(\omega_i\mid\omega_{i-1}\dots\omega_{i-2n_1-1})$. The space of states $\mathcal{M}$ of this Markov chain consists of all words $(\omega_{-n_1}\dots\omega_{n_1})$.

We introduce the subset $V_{n_1}\subset\mathcal{M}$ which is defined by the following conditions:

(i) $C_{\omega_i}(\zeta)\subset LA_{2n_1}$, $-n_1\leqslant i\leqslant n_1$;
(ii) $\tilde\mu[(\omega_{-n_1}\dots\omega_{n_1})\cap\mathcal{F}_{n_1}]\geqslant[1-\tilde\mu^{1/2}(\bar{\mathcal{F}}_{n_1})]\tilde\mu(\omega_{-n_1}\dots\omega_{n_1})$

A transition $(\omega_{-n_1}\dots\omega_{n_1})\to(\omega_{-n_1+1}\dots\omega_{n_1+1})$ will be called admissible if $\tilde\mu[(\omega_{-n_1}\dots\omega_{n_1+1})\cap\mathcal{F}_{n_1}]\geqslant[1-\tilde\mu^{1/2}(\bar{\mathcal{F}}_{n_1})]\tilde\mu[(\omega_{-n_1}\dots\omega_{n_1+1})]$.

According to this definition, for all states $(\omega_{-n_1}\dots\omega_{n_1})\in V_{n_1}$ the conditional probability of admissible transitions is not less than $\tilde\mu^{1/2}(\mathcal{F}_{n_1})$.

We now consider the subset H_n of all sequences $(\omega_{-n_1}\dots\omega_{n+n_1})$ for which $(\omega_{i-n_1}\dots\omega_{i+n_1})\in V_{n_1}$, $0\leqslant i\leqslant n$ and all transitions $(\omega_{i-n_1}\dots\omega_{i+n_1})\to(\omega_{i+1-n_1}\dots\omega_{i+1+n_1})$ are admissible. We define on H_n a new probability distribution $\bar{\mu}$:

$$\bar{\mu}(\omega_{-n_1}\ldots\omega_{n_1+n})=\mathscr{D}^{-1}\tilde{\mu}(\omega_{-n_1}\ldots\omega_{n_1})\prod_{i=n_1+1}^{n+n_1}\tilde{\mu}(\omega_i\mid\omega_{i-1}\ldots\omega_{i-2n_1-1})$$

for all $(\omega_{-n_1}\ldots\omega_{n+n_1})\in H_n$.

Here $\mathscr{D}$ is a normalized factor:

$$\mathscr{D}=\sum_{\omega_{-n_1}\ldots\omega_{n+n_1}}\tilde{\mu}(\omega_{-n_1}\ldots\omega_{n_1})\prod_{i=n_1+1}^{n+n_1}\tilde{\mu}(\omega_i\mid\omega_{i-1}\ldots\omega_{i-2n_1-1}).$$

It follows from Chebyshev's inequality that

$$\tilde{\mu}(H_n)\geq 1-2(n+2n_1)\tilde{\mu}^{1/2}(\mathscr{F}_{n_1})-(2n_1+n)\lambda_1^{2n_1}.$$

Now in view of Lemma 5.8 we find that the probability distributions $\bar{\mu}$ and $\tilde{\mu}$ are very close to each other on the set $V_n\cap H_n$.

The probability distribution $\bar{\mu}$ is a nonhomogeneous Markov chain with the memory $2n_1+1$. Its transition probabilities have the form

$$\bar{\mu}(\omega_i\mid\omega_{i-1}\ldots\omega_{i-2n_1-1})=\frac{\tilde{\mu}(\omega_i\mid\omega_{i-1}\ldots\omega_{i-2n_1-1})}{\sum_{\omega_i\ldots\omega_{n+n_1}}\prod_{j=i}^{n+n_1}\tilde{\mu}(\omega_j\mid\omega_{j-1}\ldots\omega_{j-2n_j-1})}$$

$$\times\sum_{\omega_{i+1}\ldots\omega_n}\prod_{j=i+1}^{n+n_1}\tilde{\mu}(\omega_j\mid\omega_{j-1}\ldots\omega_{j-2n_1-1}).$$

Both sums are taken over the numbers $(\omega_{j-2n}\ldots\omega_j)\in V_n$ and all transitions are admissible. From the definition of V_n we have

$$\left|\frac{\bar{\mu}(\omega_i\mid\omega_{i-1}\ldots\omega_{i-2n_1-1})}{\tilde{\mu}(\omega_i\mid\omega_{i-1}\ldots\omega_{i-2n_1-1})}-1\right|\leq 1-[1-\tilde{\mu}(\bar{\mathscr{F}}_{n_1})]^{2(n+2n_1+1)}.$$

In the same way one can obtain that for any m the variation

$$\tfrac{1}{2}\sum|\bar{\mu}(\omega_{3(2n_1+1)+m+1}\ldots\omega_{4(2n_1+1)+m}\mid\omega'_{-n_1}\ldots\omega'_{n_1})$$

$$-\bar{\mu}(\omega_{3(2n_1+1)+m+1}\ldots\omega_{4(2n_1+1)+m}\mid\omega''_{-n_1}\ldots\omega''_{n_1})|$$

differs from the analogous variation for $\tilde{\mu}$ by a number whose absolute value does not exceed $2\{1-[1-\tilde{\mu}^{1/2}(\bar{\mathscr{F}}_{n_1})]^{4(2n_1+1)}\}$. Now from Lemma 5.7 we find that the last sum is not more than λ_1.

We have

$$Eh_{n_1}(\tilde{T}^n\omega)h_{n_1}(\omega)=\int_\Omega h_{n_1}(\tilde{T}^n\omega)h_{n_1}(\omega)\,d\tilde{\mu}$$

$$=\sum_{\omega_{-n_1}\ldots\omega_{n+n_1}}h_{n_1}(\omega_{-n_1+n}\ldots\omega_{n_1+n})h_{n_1}(\omega_{-n_1}\ldots\omega_{n_1})\tilde{\mu}(\omega_{-n_1}\ldots\omega_{n+n_1}).$$

If the summation in the last expression is performed only over values ω_i, $-n_1\leq i\leq n+n_1$ for which all $\omega_i\leq n_1$ then, according to Lemma 5.5, the total is altered by not more than a factor equal to const $n\exp(-n_1)$. We denote by $\sum'$ a sum over all ω_i, $-n_1 \leq i \leq n+n_1$ for which $\omega_i \leq n_1$, so we can write

$$\sum{}' h_{n_1}(\omega_{-n_1+n}\ldots\omega_{n_1+n})h_{n_1}(\omega_{-n_1}\ldots\omega_{n_1})\tilde{\mu}(\omega_{-n_1}\ldots\omega_{n+n_1})$$

$$=\sum{}' h_{n_1}(\omega_{-n_1+n}\ldots\omega_{n_1+n})h_{n_1}(\omega_{-n_1}\ldots\omega_{n_1})\tilde{\mu}(\omega_{-n_1}\ldots\omega_{2n_1-1})$$

$$\times\prod_{n_1<i\leq n_1+n}\tilde{\mu}(\omega_i\mid\omega_{i-1}\ldots\omega_{-n_1}).$$

By analogy with the ordinary ergodic theorem for Markov chains we have

$$\sum{}' |\tilde{\mu}(\omega_{-n_1+n}\ldots\omega_{n_1+n})-\tilde{\mu}(\omega_{-n_1+n}\ldots\omega_{n_1+n}\mid\omega_{-n_1}\ldots\omega_{n_1})|$$

$$\leq\lambda^{n/n_1}+\text{const}\; n\lambda_5^n$$

where $0<\lambda_5<1$.

Consequently

$$\sum{}' h_{n_1}(\omega_{-n_1}\ldots\omega_{n_1})\tilde{\mu}(\omega_{-n_1}\ldots\omega_{n+n_1})\tilde{\mu}(\omega_{-n_1+n}\ldots\omega_{n_1+n}\mid\omega_{-n_1}\ldots\omega_{n_1})$$

$$\times h_{n_1}(\omega_{-n_1+n}\ldots\omega_{n_1+n})=\sum{}' h_{n_1}(\omega_{-n_1}\ldots\omega_{n_1})\tilde{\mu}(\omega_{-n_1}\ldots\omega_{n_1})$$

$$\times\tilde{\mu}(\omega_{-n_1+n}\ldots\omega_{n_1+n})h_{n_1}(\omega_{-n_1+n}\ldots\omega_{n_1+n})+\text{r.t.},$$

where $|\text{r.t.}|\leq\text{const}(\lambda^{n_1}+\lambda^{n/n_1})$, $0<\lambda<1$. Now, combining all the foregoing inequalities and putting $n_1=[\sqrt{n}]$, we get the assertion of Theorem 5.2. ■

5.5 The central limit theorem

Consider a function $h(\omega)$ for which all conditions formulated in the beginning of Section 5.4 hold. It follows from Theorem 5.2 that $\sum_{-\infty}^{\infty}|E(\tilde{T}^n h\cdot h)|<\infty$ and $\sigma(h)=\sum_{-\infty}^{\infty}E(\tilde{T}^n h\cdot h)<\infty$ where an expectation E is taken with respect to the measure $\tilde{\mu}$. We assume that $\sigma(h)>0$.

Theorem 5.3. *Let $Eh=0$. Then for any numbers a, b, $a<b$, the limiting relation holds:*

$$\lim_{n\to\infty}\tilde{\mu}\left\{\omega: a<\frac{1}{\sqrt{(\sigma(h)n)}}\sum_{k=0}^{n-1}h(\tilde{T}^k\omega)<b\right\}=\frac{1}{\sqrt{(2\pi)}}\int_a^b\exp(-u^2/2)\,du.$$

Proof. The assertion of Theorem 5.3 means that the probability distribution of the normalized sum $1/[\sqrt{(\sigma(h)n)}]\sum_{k=0}^{n-1}h(\tilde{T}^k\omega)$ converges weakly to the standard Gaussian distribution.

Consider now a characteristic function

$$\varphi_n(\lambda)=E\exp\left[\mathrm{i}\frac{\lambda}{\sqrt{(\sigma(h)n)}}\sum_{k=0}^{n-1}h(\tilde{T}^k\omega)\right].$$

We must show that $\varphi_n(\lambda)\xrightarrow[n\to\infty]{}\exp(-\lambda^2/2)$, where λ varies in any compact set. We divide the segment $I=[0,n-1]$ into nonoverlapping subintervals $I=I_1^{(1)}\cup I_1^{(2)}\cup I_2^{(1)}\cup I_2^{(2)}\cup\ldots\cup I_{m-1}^{(2)}\cup I_m^{(1)}$ in such a way that the length $|I_i^{(1)}|$ of any subinterval $I_i^{(1)}$ (except possibly the last one) equals $[n^{\gamma_1}]$ and the length $|I_j^{(2)}|$ of any subinterval $I_j^{(2)}$ equals $[n^{\gamma_2}]$, where $0<\gamma_2<\gamma_1<1$, $\gamma_1>\gamma_2+\frac{1}{2}$ and $|I_m^{(1)}|\leq[n^{\gamma_1}]$. Hence we have

$$\varphi_n(\lambda)=E\exp\left\{\mathrm{i}\frac{\lambda}{\sqrt{(\sigma(h)n)}}\left[\sum_{s=1}^{m}\sum_{k\in I_s^{(1)}}h(\tilde{T}^k\omega)+\sum_{s=1}^{m-1}\sum_{k\in I_s^{(2)}}h(\tilde{T}^k\omega)\right]\right\}.$$

It is easy to see that

$$\left|\sum_{s=1}^{m-1}\sum_{k\in I_s^{(2)}}h(\tilde{T}^k\omega)\right|\leq\text{const}\ mn^{\gamma_2}=\text{const}\ n^{1-\gamma_1+\gamma_2}.$$

Therefore

$$\sup\left|\frac{1}{\sqrt{n}}\sum_{s=1}^{m-1}\sum_{k\in I_s^{(2)}}h(T^k\omega)\right|\xrightarrow[n\to\infty]{}0$$

and the limiting behaviour of the function $\varphi_n(\lambda)$ is the same as for the expression

$$\psi_n^{(1)}(\lambda)=E\exp\left\{\mathrm{i}\frac{\lambda}{\sqrt{(\sigma(h)n)}}\sum_{s=1}^{m}\sum_{k\in I_s^{(1)}}h(\tilde{T}^k\omega)\right\}.$$

We now show that the function $\psi_n^{(1)}(\lambda)$ is close to

$$\left[E\exp\left\{\mathrm{i}\frac{\lambda}{\sqrt{(\sigma(h)n)}}\sum_{k\in I_1^{(1)}}h(\tilde{T}^k\omega)\right\}\right]^m,$$

i.e., to the characteristic function of the sum of m independent random variables distributed as the random variable $\sum_{k\in I_1^{(1)}}h(\tilde{T}^k\omega)$. Let us take $\gamma_3<\gamma_2$, $n_1=n^{\gamma_3}$ and consider the function $h_{n_1}(\omega)$. We have

$$\left|\sum_{s=1}^{m}\sum_{k\in I_s^{(1)}}[h(\tilde{T}^k\omega)-h_{n_1}(\tilde{T}^k\omega)]\right|\leq\text{const}\ \lambda_1^{n_1}.$$

Therefore we can replace the function $\psi_n^{(1)}(\lambda)$ by a function

$$\psi_n^{(2)}(\lambda)=E\exp\left\{\mathrm{i}\frac{\lambda}{\sqrt{(\sigma(h)n)}}\sum_{s=1}^{m}\sum_{k\in I_s^{(1)}}h_{n_1}(\tilde{T}^k\omega)\right\}.$$

The function

$$\exp\left\{\mathrm{i}\frac{\lambda}{\sqrt{(\sigma(h)n)}}\sum_{s=1}^{m}\sum_{k\in I_s^{(1)}}h_{n_1}(\tilde{T}^k\omega)\right\}$$

depends on the variables $\omega_{-n_1}, \ldots, \omega_{n+n_1}$ only. In the expression of the function $\psi_n^{(2)}(\lambda)$ we restrict the summation to sequences for which $\omega_i \leqslant 2[n_1^{1/2}] = 2n_2$. According to Lemma 5.1 the set of all such sequences has a probability not less than $1-(n+2n_1)\lambda_1^{n_2}$.

Thus we can consider the function

$$\psi^{(3)}(\lambda) = \sum_{\{\omega_{-n_1}\ldots\omega_{n+n_1}\}} \exp\left\{i\frac{\lambda}{\sqrt{(\sigma(h)n)}}\sum_{s=1}^{m}\sum_{k\in I_s^{(1)}} h_{n_1}(\tilde{T}^k\omega)\right\}\tilde{\mu}(\omega_{-n_1}\ldots\omega_{n+n_1}).$$

Now we replace the probability distribution $\tilde{\mu}$ by the Markov probability distribution $\bar{\mu}$ constructed in Section 5.4 and consider the function

$$\psi_n^{(4)}(\lambda) = \sum \exp\left\{i\frac{\lambda}{\sqrt{(\sigma(h)n)}}\sum_{s=1}^{m}\sum_{k\in I_s^{(1)}} h_{n_1}(\tilde{T}^k\omega)\right\}\bar{\mu}(\omega_{-n_1}\ldots\omega_{n+n_1}).$$

It follows from the results of the previous section that $|\psi_n^{(4)}(\lambda) - \psi_n^{(3)}(\lambda)| \xrightarrow[n\to\infty]{} 0$ uniformly in λ. So we can write for $m = n - [n^{\gamma_1}] - [n^{\gamma_2}]$

$$\psi_n^{(4)}(\lambda) = \sum \exp\left\{i\frac{\lambda}{\sqrt{(\sigma(h)n)}}\sum_{s=1}^{m-1}\sum_{k\in I_s^{(1)}} h_{n_1}(\tilde{T}^k\omega)\right\}\bar{\mu}(\omega_{-n_1}\ldots\omega_{m+n_1})$$

$$\times \sum_{\omega_j, n-[n^{\gamma_1}]-n_1\leqslant j\leqslant n+n_1} \exp\left\{i\frac{\lambda}{\sqrt{(\sigma(h)n)}}\sum_{k\in I_m^{(1)}} h_{n_1}(\tilde{T}^k\omega)\right\}$$

$$\times \bar{\mu}(\omega_{n-[n^{\gamma_1}]-n_1}\ldots\omega_{n+n_1} \mid \omega_{m-n_1}\ldots\omega_{m+n_1}).$$

The same arguments as in Section 5.4 show that

$$\left|\sum \exp\left\{i\frac{\lambda}{\sqrt{(\sigma(h)n)}}\sum h_{n_1}(\tilde{T}^k\omega)\right\}\bar{\mu}(\omega_{n-[n^{\gamma_1}]-n_1}\ldots\varphi_{m-n_1}\ldots\omega_{m+n_1})\right.$$

$$\left. -\sum \exp\left\{i\frac{\lambda}{\sqrt{(\sigma(h)n)}}\sum_{k\in I_m^{(1)}} h_{n_1}(\tilde{T}^k\omega)\right\}\bar{\mu}(\omega_{n-[n^{\gamma_1}]-n_1}\ldots\omega_{n+n_1})\right|$$

$$\leqslant \lambda_3^{n[\gamma_1]/n_1} + n^2\tilde{\mu}(\bar{\mathscr{F}}_{n_1}).$$

Hence the absolute value of the difference

$$\sum \exp\left\{i\frac{\lambda}{\sqrt{(\sigma(h)n)}}\sum_{k\in I_m^{(1)}} h_{n_1}(\tilde{T}^k\omega)\right\}\bar{\mu}(\omega_{n-[n^{\gamma_1}]-n_1}\ldots\omega_{n+n_1})$$

$$-\sum \exp\left\{i\frac{\lambda}{\sqrt{(\sigma(h)n)}}\sum_{k\in I_m^{(1)}} h_{n_1}(\tilde{T}^k\omega)\right\}\tilde{\mu}(\omega_{n-[n^{\gamma_1}]-n_1}\ldots\omega_{n+n_1})$$

is sufficiently small. As a result, we find that the difference

$$\left|\psi_n^{(4)}(\lambda) - \prod_{r=1}^{m} E\exp\left\{i\frac{\lambda}{\sqrt{(\sigma(h)n)}}\sum_{k\in I_r^{(1)}} h_{n_1}(\tilde{T}^k\omega)\right\}\right|$$

tends to zero uniformly in λ.

It is easy to see that the variance

$$E\Big[\sum_{k\in I_s^{(1)}} h_{n_1}(\tilde{T}^k\omega)\Big]^2 \sim \sigma(h)n^{\gamma_1}$$

and that

$$E\Big[\sum_{s=1}^{m}\sum_{k\in I_s^{(1)}} h_{n_1}(\tilde{T}^k\omega)\Big]^2 \sim \sigma(h)n_1 m \sim \sigma(h)n \quad \text{as } n\to\infty.$$

Thus to prove Theorem 5.3 it is sufficient to check Lindeberg's condition (see [19]). In view of Chebyshev's inequality it is sufficient to estimate the fourth moment

$$E\Big[\sum_{k\in I_s^{(1)}} h_{n_1}(\tilde{T}^k\omega)\Big]^4 = \sum_{k_1,k_2,k_3,k_4} E[h_{n_1}(\tilde{T}^{k_1}\omega)h_{n_1}(\tilde{T}^{k_2}\omega)h_{n_1}(\tilde{T}^{k_3}\omega)h_{n_1}(\tilde{T}^{k_4}\omega)].$$

Without loss of generality we can assume that $k_1<k_2<k_3<k_4$.

It is convenient to consider three special cases:

(1) There exists some ε, $0<\varepsilon<\gamma_2-\gamma_3$, so that $\max_{1\le i\le 3}(k_{i+1}-k_i)\le n_1^{1+\varepsilon}$. In this case

$$|Eh_{n_1}(\tilde{T}^{k_1}\omega)h_{n_1}(\tilde{T}^{k_2}\omega)h_{n_1}(\tilde{T}^{k_3}\omega)h_{n_1}(\tilde{T}^{k_4}\omega)|\le \text{const}.$$

(2) For $i_0=1$ or $i_0=3$ the difference $k_{i_0+1}-k_{i_0}>n_1^{1+\varepsilon}$.

Let us consider, for certainty, the case when $i_0=3$. Making use of the formula for conditional probabilities, we can write

$$\begin{aligned}&Eh_{n_1}(\tilde{T}^{k_1}\omega)h_{n_1}(\tilde{T}^{k_2}\omega)h_{n_1}(\tilde{T}^{k_3}\omega)h_{n_1}(\tilde{T}^{k_4}\omega)\\ &\quad=E\{E[h_{n_1}(\tilde{T}^{k_4}\omega)\mid \omega_i,\, i\le k_3+n_1]\}h_{n_1}(\tilde{T}^{k_1}\omega)h_{n_1}(\tilde{T}^{k_2}\omega)h_{n_1}(\tilde{T}^{k_3}\omega).\end{aligned}$$

From the ergodic theorem for Markov chains and the condition $Eh_{n_1}=0$ we obtain, for the case when all $\omega_i\le\sqrt{(n_1)}$,

$$|E[h_{n_1}(\tilde{T}^{k_4}\omega)\mid \omega_i,\, i\le k_1+n_1]|\le \lambda_5^{\text{const}(n/n_1)}.$$

In the opposite case the expectation under consideration is bounded from above by some constant, but the probability of a corresponding condition is not more than const $\lambda_1^{n_1}$. The general inequality for this case takes the following form:

$$|Eh_{n_1}(\tilde{T}^{k_1}\omega)h_{n_1}(\tilde{T}^{k_2}\omega)h_{n_1}(\tilde{T}^{k_3}\omega)h_{n_1}(\tilde{T}^{k_4}\omega)|\le \text{const}(\lambda_5^{n/n_1}+n\lambda_1^{n_1}).$$

(3) $k_3-k_2>n_1^{1+\varepsilon}$, but the other differences are not more than $n_1^{1+\varepsilon}$. A total number of such cases is not more than const $n^2n_1^2$. Therefore we can use the following simple estimation:

$$|E(h_{n_1}(\tilde{T}^{k_1}\omega)h_{n_1}(\tilde{T}^{k_2}\omega)h_{n_1}(\tilde{T}^{k_3}\omega)h_{n_1}(\tilde{T}^{k_4}\omega))|\le \text{const}.$$

Finally, we have

$$E\left(\sum_{k\in I_s^{(1)}} h_{n_1}(\tilde{T}^k\omega)\right)^4 \leqslant \text{const}\ n^2 n_1^2 \leqslant \text{const}\ |I_s^{(1)}|^{\tilde{\gamma}}$$

for some $\tilde{\gamma}$, $0<\tilde{\gamma}\leqslant 2$. Lindeberg's condition follows from Chebyshev's inequality. ■

We have considered only the case when the points z_1^0 and z_2^0 and all their images are aperiodic. It is easy to see, however, that if they or their images are periodic then the proofs of Lemmas 5.6, 5.7 and, correspondingly, of Theorems 5.2, 5.3 become simpler since in this case every cell is long.

Consider now the case when the transformation T is not transitive. In this case the attractor Λ can be represented as the union of a finite number of sets:

$$\Lambda = \Lambda_1 \cup \ldots \cup \Lambda_k, \quad \text{where} \quad k \geqslant 3$$

and

$$T^s\Lambda_i = \begin{cases} \Lambda_{i+s} & \text{if} \quad i+s \leqslant k \\ \Lambda_{i+s-k} & \text{if} \quad i+s > k. \end{cases}$$

This means that the spectrum of the dynamical system under consideration contains a cyclic component [6]. We shall denote by θ the partition of the attractor Λ into the sets Λ_i, $1\leqslant i\leqslant k$. Let us consider the factor transformation T/θ of the factor space Λ/θ. It is easy to see that for the factor system $(\Lambda/\theta, \mathfrak{B}/\theta, \mu_\theta, T/\theta)$ Theorems 5.2 and 5.3 hold. This case is the same as the case when the points z_1^0 and z_2^0 (or their images) are periodic. In fact all the cells of all ranks for the factor system have one and the same y-length.

Finally, we should mention that at present stochasticity is proved rigorously for attractors in systems with Axiom A and in the Lorenz system only. This is because, for the dynamical systems induced on these attractors, the properties of hyperbolicity are valid almost everywhere. For attractors of more general type in Euclidean spaces this property does not hold (see Chapter 13 by Plykin, where all possible types of hyperbolic attractors are discussed). The situation here is very similar to transformations of the unit segment into itself. For expanding (i.e., uniformly hyperbolic) transformations, stochasticity was proved some time ago. Recent results of Yakobson [20] for one-dimensional transformations of more general type (for instance, $x \to ax(1-x)$ (mod. 1), $0<a\leqslant 4$) give the hope of proving rigorous theorems concerning the statistical properties of attractors of more general type than mentioned above.

Acknowledgements

I express my sincere gratitude to Dr Ja. B. Pesin who has read the whole text and made many useful comments.

I am greatly indebted to Professor Ja. G. Sinai for useful discussions and to Dr D. Szasz for his cooperation in translating the text into English.

I am grateful to Ms E. Pasternak for her help in preparation of the manuscript.

Finally, I would like to express my appreciation to Professor G. I. Barenblatt for his encouragement and some helpful comments and suggestions.

References

1. Ruelle, D. and Takens, F., On the nature of turbulence, *Communications in Mathematical Physics*, **20,** 167–192, 1971.
2. Rabinovich, M. I., Stochastic auto-oscillations and turbulence, *Russian Physical Surveys*, **125,** 123–165, 1978.
3. Smale, S., Differential dynamical systems, *Bulletin of the American Mathematical Society*, **73,** 747–817, 1967.
4. Bowen, R. and Ruelle, D., The ergodic theory of Axiom A flows, *Inventiones Mathematicae*, **29,** 181–202, 1975.
5. Sinai, Ja. G., Stochasticity of dynamical systems, in Gaponov-Grekhov, A. V. (Ed.), *Nonlinear Waves*: Moscow, Nauka, 1979, 192–212.
6. Bunimovich, L. A. and Sinai, Ja. G., Stochasticity of the attractor in the Lorenz model, in Gaponov-Grekhov, A. V. (Ed.), *Nonlinear Waves*: Moscow, Nauka, 1979, 212–226.
7. Lorenz, E., Deterministic nonperiodic flow, *Journal of the Atmospheric Sciences*, **20,** 130–144, 1963.
8. Saltzman, B., Finite amplitude free convection as an initial value problem, *Journal of the Atmospheric Sciences*, **19,** 329–341, 1962.
9. Afraimovich, V. S., Bykov, V. V. and Shil'nikov, L. P., On the origin and structure of the Lorenz attractor, *Doklady Akademii Nauk USSR*, **243,** 334–336, 1977.
10. Shil'nikov, L. P., The generation of a periodic motion from a trajectory which is doubly asymptotic to a saddle-type equilibrium state, *Mathematical Sbornik*, **73,** 747–817, 1967.
11. Sinai, Ja. G. and Vul, E. B., Hyperbolicity conditions for the Lorenz model, *Physica*, **2D,** 3–7, 1980.
12. Anosov, D. V., Geodesic flows on closed Riemannian manifolds of negative curvature, *Trudy of V. A. Steklov Mathematical Institute Akademii Nauk USSR*, No. 90: Moscow, 1967.
13. Sinai, Ja. G., Dynamical systems with elastic reflections, *Russian Mathematical Surveys*, **XXV,** 141–192, 1970.
14. Simonov, A. A., Investigation of bifurcations in some dynamical systems by the methods of symbolic dynamics, *Doklady Akademy Nauk USSR*, **240,** 1297–1300, 1978.
15. Sinai, Ja. G., Markov partitions and U-diffeomorphisms, *Functional Analysis and Applications*, **2,** 45–68, 1968.

16. Gnedenko, B. V. and Kolmogorov, A. N., *Limit Theorems for Sums of Independent Random Variables*: Moscow, Gostekhizdat, 1947.
17. Bunimovich, L. A. and Sinai, Ja. G., Markov partitions for dispersed billiards, *Communications in Mathematical Physics*, **78,** 247–280, 1980.
18. Bunimovich, L. A. and Sinai, Ja. G., Statistical properties of Lorenz gas with periodic configuration of scatters, *Communications in Mathematical Physics*, **78,** 479–497, 1981.
19. Ibragimov, J. A. and Linnik, Yu. V. *Independent and Stationary Connected Variables*: Moscow, Nauka, 1965.
20. Yakobson, M. V., Invariant measures of one-parameter families of one-dimensional mappings absolutely continuous with respect to dx, *Russian Mathematical Surveys*, **35,** 215–216, 1980.

6 An Example of Direct Bifurcation into a Turbulent State

F H Busse

6.1 Introduction

The transition from laminar to turbulent flow without intermediate states of periodic or quasiperiodic flow is a common phenomenon in plane parallel shear flows. It is caused by subcritical finite amplitude instability, i.e., the lower branch of an inverted bifurcation from the laminar state is unstable and the instability leads to a discontinuous jump from one stationary state to another stationary (in the statistical sense) state. Sometimes this is called 'snap-through instability' [1].

This chapter is concerned with the different phenomenon of a continuous transition from a motionless state to a statistically stationary state of turbulent flow. This kind of transition occurs in a fluid layer heated from below and rotating about a vertical axis, as shown in Fig. 6.1. When the critical Rayleigh number for onset of instability is exceeded, small amplitude convection occurs that exhibits random features both in its spatial and its time-dependence. Since the equations describing the motion are weakly nonlinear, the mathematical analysis can be carried out without expensive numerical computations and insights can be gained into the nature of the motion that are not readily accessible in other kinds of turbulent flow.

The result that all small amplitude solutions for convection in a rotating layer are unstable, in contrast to the nonrotating case, was first derived by Küppers and Lortz [2]. The evolution in time of the convection motion was considered by Busse and Clever [3]. The importance of the interaction of patches of convection rolls was recognized only when experimental observations became available [4, 5]. These steps are reviewed below and an interpretation of the observed phenomena in terms of the statistical limit cycle concept is given. Although the phenomenon of weakly nonlinear turbulence in a rotating layer heated from below appears to be unique, it can serve as a model for more general types of turbulence, as will be discussed in the conclusion.

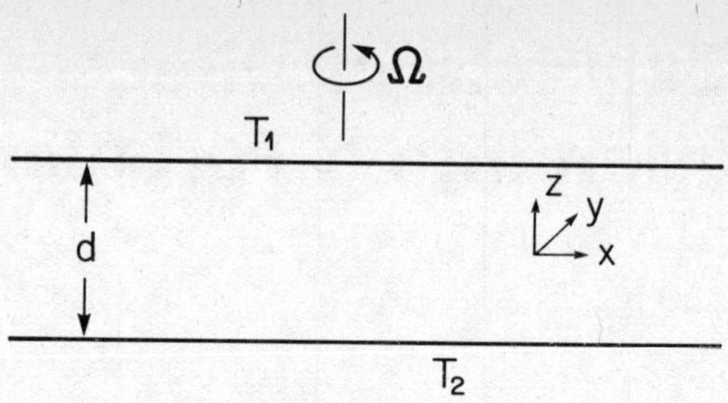

Fig. 6.1 Geometrical configuration of the rotating convection layer

6.2 Mathematical formulation

For the dimensionless description of the problem the depth d of the layer, the thermal diffusion time d^2/κ, and the temperature difference across the layer divided by the Rayleigh number $(T_2 - T_1)/R$ are used as scales for length, time and temperature, respectively. The Navier–Stokes equations for the velocity vector $\mathbf{v}$ and the heat equation for the deviation θ of the temperature from the static state can thus be written in the form

$$P^{-1}\left(\frac{\partial}{\partial t} + \mathbf{v}\cdot\nabla\right)\mathbf{v} + \tau\boldsymbol{\lambda}\times\mathbf{v} = -\nabla\pi + \boldsymbol{\lambda}\theta + \nabla^2\mathbf{v} \tag{6.1a}$$

$$\nabla\cdot\mathbf{v} = 0 \tag{6.1b}$$

$$\left(\frac{\partial}{\partial t} + \mathbf{v}\cdot\nabla\right)\theta = R\boldsymbol{\lambda}\cdot\mathbf{v} + \nabla^2\theta \tag{6.1c}$$

where $\boldsymbol{\lambda}$ is the unit vector in the vertical direction opposite to gravity and the dimensionless external parameters are defined by

$$R = \frac{\gamma g(T_2 - T_1)d^3}{\kappa\nu}, \qquad P = \frac{\nu}{\kappa}, \qquad \tau = \frac{\Omega d^2}{\nu}. \tag{6.2}$$

Here γ, g, ν and κ are the coefficient of thermal expansion, the acceleration of gravity, kinematic viscosity and thermal diffusivity, respectively. Ω is the angular velocity of rotation about the vertical axis through the centre of the layer. For simplicity, a circular layer of radius r_0 is assumed such that

$$d \ll r_0 \ll g/\Omega^2 \tag{6.3}$$

is satisfied. The second inequality permits neglect of the centrifugal force in Eqs (6.1). For simplicity, the limit $P \to \infty$ will be assumed since no qualitatively new aspects seem to appear for finite values of P [6, 7].

The linearized version of Eqs (6.1) admits solutions of the form

$$v_z = f(z, \alpha, \tau) \sum_{n=-N}^{N} C_n \exp\{i\mathbf{k}_n \cdot \mathbf{r}\} \tag{6.4}$$

where **r** is the position vector and z is its vertical component. The horizontal vectors $\mathbf{k}_n$ are arbitrary except for their absolute value, $|\mathbf{k}_n| = \alpha$. For notational convenience the definitions

$$C_{-n} = C_n^+, \qquad \mathbf{k}_{-n} = -\mathbf{k}_n$$

are used which ensure a real expression (6.4). The equation for $f(z, \alpha, \tau)$ together with the boundary conditions at $z = \pm\frac{1}{2}$ represent an eigenvalue problem for $R(\alpha, \tau)$. At a finite value α_c this function reaches its minimum $R_c(\tau)$ at which the onset of convection occurs.

Since a sufficiently large horizontal extent of the layer is assumed such that the conditions at the side boundaries become irrelevant, the choice of the coefficients C_n and the directions of the vectors $\mathbf{k}_n$ are not constrained as far as the linear analysis is concerned. The parameter N may tend to infinity. The nonlinear terms of the equations, however, impose conditions that must be satisfied by the coefficients C_n for a given set of vectors $\mathbf{k}_n$. The solvability conditions arise when the full equations are expanded in terms of powers of the amplitude of convection

$$\varepsilon \equiv \overline{\left(\sum_{n=-N}^{N} |C_n|^2 \right)^{1/2}} \tag{6.5}$$

where the bar indicates a time average. We allow for a slow time-dependence of the coefficients. At Rayleigh numbers slightly above the critical value R_c the time derivative will be at most of the order of $R - R_c$ for solutions of the form (6.4). Since the solvability conditions in the order ε^2 of Eqs (6.1) do not provide any constraints, the analysis must be carried to the order ε^3 of the problem where the conditions

$$M \frac{\mathrm{d}}{\mathrm{d}t} C_i = \left\{ (R - R_c)K - \frac{1}{2} \sum_{m=-N}^{N} T_{im} |C_m|^2 \right\} C_i \qquad i = 1, \ldots, N \tag{6.6}$$

result. The notation follows that of [2] except for the addition of the time derivative on the left-hand side. The matrix elements T_{im} obey the symmetry relationships

$$T_{im} = T_{i-m} = T_{-im} \tag{6.7}$$

but they do not satisfy $T_{im} = T_{mi}$ as in the nonrotating case [8]; this is because they depend on $\boldsymbol{\lambda} \cdot \mathbf{k}_i \times \mathbf{k}_m$ as well as on $\mathbf{k}_i \cdot \mathbf{k}_m$. For regular distributions of **k**-vectors, for which the angle between neighbouring **k**-vectors is a constant, Eqs (6.6) are solved by

$$|C_1|^2 = |C_2|^2 = \ldots = |C_N|^2 = \frac{\varepsilon^2}{2N}. \tag{6.8}$$

By superimposing infinitesimal disturbances onto the steady solutions of Eqs (6.6), Küppers and Lortz [2] found that all steady solutions except for the two-dimensional solution, $N = 1$, are unstable as in the nonrotating case. But the novel result of [2] is that the two-dimensional solution is also unstable if τ^2 exceeds a critical value τ_c^2. The value τ_c^2 decreases generally with Prandtl number [6, 7]. For $P=\infty$ and stress-free boundaries, $\tau_c^2=2285$.

6.3 Initial value problem

The paradoxical situation created by the result of Küppers and Lortz is that no other solution is available to take the place of the unstable roll solution. Oscillatory solutions can occur only for $P<1$, and since the growing disturbances themselves assume the form of rolls they do not provide an indication of a new stable state. To investigate the realized form of convection, the initial value problem of Eqs (6.6) has been solved numerically [3]. While up to $N=60$ modes have been included in the original equations, the essential features can actually be seen in the case $N=3$. Since the Küppers–Lortz instability assumes the form of rolls with an axis in a direction of nearly 60° in the sense of rotation with respect to the given steady rolls, it is sufficient to choose the three **k**-vectors satisfying $\mathbf{k}_1+\mathbf{k}_2+\mathbf{k}_3=0$. Using the definitions

$$t^*=\frac{2tK(R-R_c)}{M}, \quad g_i=\frac{|C_i|^2\, T_{11}}{2K(R-R_c)}, \quad \beta=\frac{T_{12}}{T_{11}}, \quad \gamma=\frac{T_{13}}{T_{11}} \tag{6.9}$$

Eqs (6.6) can be written, in the case $N=3$,

$$\begin{aligned}
\dot{g}_1&=g_1(1-g_1-\beta g_2-\gamma g_3)\\
\dot{g}_2&=g_2(1-g_2-\beta g_3-\gamma g_1)\\
\dot{g}_3&=g_3(1-g_3-\beta g_1-\gamma g_2)
\end{aligned} \tag{6.10}$$

where the dot indicates the differentiation with respect to t^*. A typical solution is shown in Fig. 6.2. Starting with the steady solution $g_1=1$ and small values for the disturbances g_2, g_3, the former grows to replace g_1

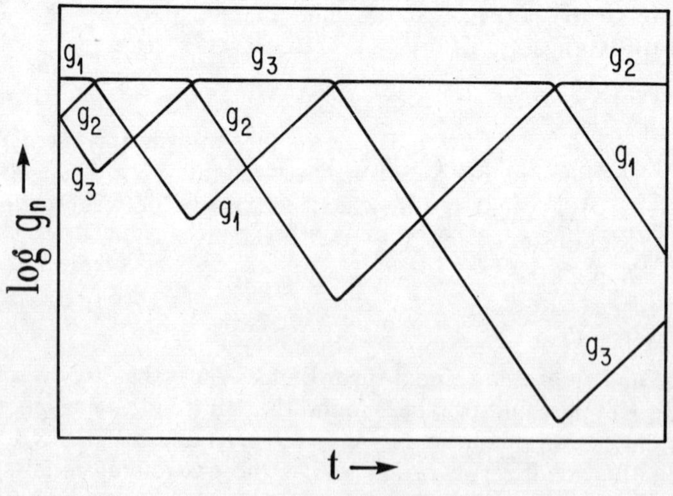

Fig. 6.2 Solution of Eqs (6.10)

while g_3 decays until g_2 has replaced g_1. Since the disturbance g_3 now starts growing from a lower level, it takes longer to replace g_2 as the nearly steady solution. Thus a cyclical solution with ever-increasing cycle period results.

The mathematical properties of Eqs (6.10) have been studied in the context of population biology by May and Leonard [9]. The particular case of interest in the present context is the regime $\gamma+\beta>2$ and $\gamma<1$ or $\beta<1$ for which all fixpoints are unstable. As the authors demonstrate, a surface $f(g_1, g_2, g_3)=\text{const.}$ spanned by the three fixpoints $g_n=1$, $g_m=0$ for $m\neq n$, is approached rapidly from arbitrary initial conditions. Within this surface the solution path wanders from the neighbourhood of one unstable fixpoint to that of the next. As the path proceeds it approaches ever more closely the lines $f(g_1, g_2, 0)=f(0, g_1, g_2)=f(g_1, 0, g_3)=\text{const.}$, connecting the three fixpoints as shown in Fig. 6.3(a). This behaviour of

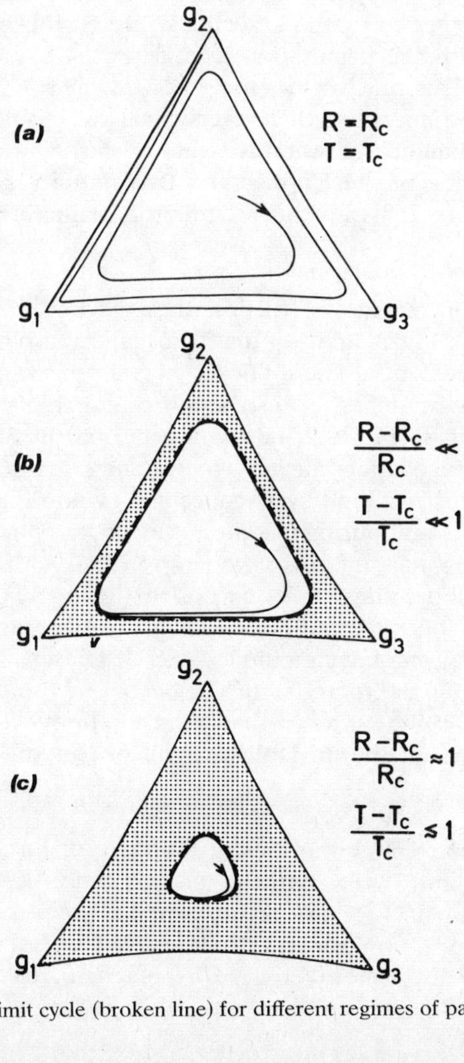

Fig. 6.3 Statistical limit cycle (broken line) for different regimes of parameters $T=\tau^2$ and R

the solutions of Eqs (6.10) is not generic. Small modifications of Eqs (6.10) lead to quite different behaviour [10]. Experimental observations [4, 5] show that spatial inhomogeneities of the disturbances are primarily responsible for the discrepancy between the physically realized convection flow and the asymptotic behaviour predicted by Eqs (6.10). This discrepancy can be understood in terms of the concept of the statistical limit cycle.

6.4 Statistical limit cycle

The evolution of the Küppers–Lortz instability is only incompletely described by the solution shown in Fig. 6.2. Since the initial amplitude of the disturbances is generally not constant throughout the layer, as assumed in that solution, phase lags develop in different parts of the layer. While rolls characterized by the **k**-vector $\mathbf{k}_3$ are established in one part of the layer, rolls with the orientation $\mathbf{k}_2$ still persist in other parts of the layer and so on. This patchy structure soon reaches a statistical equilibrium in which the typical length L over which rolls are well correlated is determined by a balance between the time of horizontal diffusion, L^2/κ, and the growth time of the Küppers–Lortz instability. Since the latter is proportional to $(R-R_c)(\tau^2-\tau_c^2)d^2/\kappa$, the approximate relationship

$$L/d \propto (R-R_1)^{1/2}(\tau^2-\tau_c^2)^{1/2} \tag{6.11}$$

holds. The solution of Eqs (6.10) describes the observations locally if a statistically varying lower limit on the disturbance amplitudes g_n is added as a constraint representing the influence of rolls on the other orientations in the neighbouring patches. The solutions then exhibit the same cyclical character as shown in Fig. 6.2, but with a period fluctuating statistically about a mean value depending on $R-R_c$ and on $\tau^2-\tau_c^2$ only. We are thus led to the statistical limit cycle concept shown in Fig. 6.3. Because of the influence of neighbouring patches, the lines connecting the points $g_n=1$, $g_m=0$, $m \neq n$, can never be approached more closely than a certain distance depending on L, except in the singular case $R=R_c$ or $\tau^2=\tau_c^2$ when L tends to infinity. However, experimental noise will prevent the realization of such a limit. As L decreases, the statistical limit cycle shrinks until the patch size L becomes comparable with the wavelength of the rolls, $2\pi/\alpha$. At that point the limit cycle has contracted to a small circle about the unstable fixpoint of Eqs (6.10),

$$g_1=g_2=g_3=(1+\beta+\gamma)^{-1}, \tag{6.12}$$

which describes convective motions in the form of hexagonal cells.

The statistical limit cycle character of convection is borne out by the observations shown in Fig. 6.4. At a particular point of the layer, three orientations of rolls are exchanged in a cyclical fashion with a well-defined period in the time average. The patch to the lower left of the

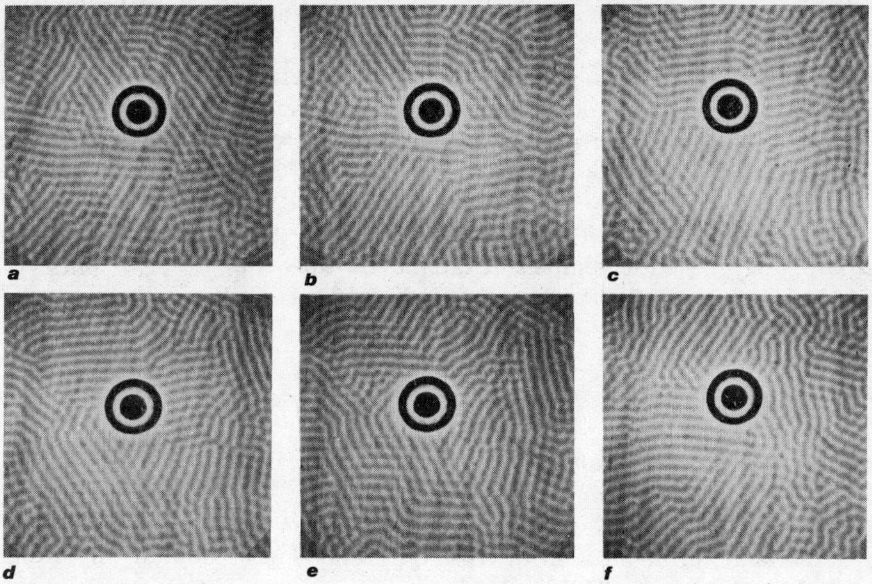

Fig. 6.4 Development in time of the horizontal structure of convection in a rotating layer heated from below (after [4])

circle in Fig. 6.4(*a*) is thus reproduced approximately in Fig. 6.4(*f*). However, since the interaction of patches does not permit a persistently recurrent regular pattern, a strictly periodic solution of the problem is not possible.

As the Rayleigh number is increased the patch size decreases and finally the replacement of the three rolls by one another happens so rapidly that all three rolls are present at the same time with nearly the same amplitude. The resulting hexagonal cells are of two kinds, those with rising motion in the centre and those with descending motion in the centre. The fact that both kinds occur with equal probability already indicates that their origin is entirely different from that of hexagonal convection in nonrotating layers with temperature-dependent properties [11]. The description by the statistical limit cycle picture fits the observations well, including the remaining cyclical fluctuation of the hexagonal cells.

6.5 Discussion

The fluid system considered above is remarkable in that it exhibits characteristic features of turbulence in the case of small amplitudes of motion. Random properties enter into both its spatial and time dependence such that the correlation is characterized by a single wave number with respect to its horizontal dependence. There is no energy cascade associated with this kind of turbulence.

Convection in a rotating layer owes its unique properties to the horizontal isotropy of the layer. However, anisotropies of fluid systems tend to become less important as a fully developed turbulent state is approached. The number of different solutions of the Navier–Stokes equations that are approximately equivalent increases at the same time. The turbulent fluid state may be characterized by many such solutions, all of which are unstable so that the actually realized state wanders from the neighbourhood of one of the solutions to that of the next. Convection in a rotating layer is the simplest example of this kind of description of turbulence.

References

1. Joseph, D. D., *Stability of Fluid Motions*: Berlin, Heidelberg and New York, Springer-Verlag, 1976.
2. Küppers, G. and Lortz, D., Transition from laminar convection to thermal turbulence in a rotating fluid layer, *Journal of Fluid Mechanics*, **35,** 609–620, 1969.
3. Busse, F. H. and Clever, R. M., Nonstationary convection in a rotating system, in Müller, U., Roesner, K. G. and Schmidt, B. (Eds), *Recent Developments in Theoretical and Experimental Fluid Mechanics*: Berlin, Heidelberg and New York, Springer-Verlag, 1979, 376–385.
4. Busse, F. H. and Heikes, K. E., Convection in a rotating layer: a simple case of turbulence, *Science*, **208,** 173–175, 1980.
5. Heikes, K. E. and Busse, F. H., Weakly nonlinear turbulence in a rotating convection layer, *Nonlinear Dynamics, Annals of the New York Academy of Sciences*, **357,** 28–36, 1980.
6. Küppers, G., The stability of steady finite amplitude convection in a rotating fluid layer, *Physics Letters*, **32A,** 7–8, 1970.
7. Clever, R. M. and Busse, F. H., Nonlinear properties of convection rolls in a horizontal layer rotating about a vertical axis, *Journal of Fluid Mechanics*, **94,** 609–627, 1979.
8. Schlüter, A., Lortz, D. and Busse, F., On the stability of steady finite amplitude convection, *Journal of Fluid Mechanics* **23,** 129–144, 1965.
9. May, R. M. and Leonard, W. J., Nonlinear aspects of competition between three species, *SIAM Journal of Applied Mathematics*, **29,** 243–253, 1975.
10. Coste, J., Peyraud, J. and Coullet, P., Asymptotic behaviors in the dynamics of competing species, *SIAM Journal of Applied Mathematics*, **36,** 516–543, 1979.
11. Busse, F. H., The stability of finite amplitude cellular convection and its relation to an extremum principle, *Journal of Fluid Mechanics*, **30,** 625–649, 1967.

7 Universal Behaviour in Nonlinear Systems

M J Feigenbaum

7.1 Introduction

There exist in nature processes that can be described as complex or chaotic and processes that are simple or orderly. Technology attempts to create devices of the simple variety: an idea is to be implemented, and various parts executing orderly motions are assembled. For example, cars, aircraft, radios and clocks are all constructed from a variety of elementary parts each of which, ideally, implements one ordered aspect of the device. Technology also tries to control or minimize the impact of seemingly disordered processes, such as the complex weather patterns of the atmosphere, the myriad whorls of turmoil in a turbulent fluid, the erratic noise in an electronic signal and other such phenomena. It is the complex phenomena that interest us here.

When a signal is noisy, its behaviour from moment to moment is irregular and has no simple pattern of prediction. However, if we analyse a sufficiently long record of the signal we may find that signal amplitudes occur within narrow ranges a definite fraction of the time. Analysis of another record of the signal may reveal the same fraction. In this case, the noise can be given a statistical description. This means that while it is impossible to say what amplitude will appear next in succession, it is possible to estimate the probability or likelihood that the signal will attain some specified range of values. Indeed, for the last hundred years disorderly processes have been taken to be statistical (one has given up asking for a precise causal prediction), so that the goal of a description is to determine what the probabilities are, and from this information to determine various behaviours of interest—for example, how air turbulence modifies the drag on an aircraft.

We know that perfectly definite causal and simple rules can have statistical (or random) behaviours. Thus, modern computers possess 'random number generators' that provide the statistical ingredient in a simulation of an erratic process. However, this generator does nothing more than shift the decimal point in a rational number whose repeating block is suitably long. Accordingly, it is possible to predict what the nth generated number will be. Yet, in a list of successive generated numbers there is such a seeming lack of order that all statistical tests will confer upon the numbers a pedigree of randomness. Technically, the term 'pseudorandom' is used to indicate this nature.

One now may ask whether the various complex processes of nature themselves might not be merely pseudorandom, with the full import of randomness, which is untestable, a historic but misleading concept. Indeed our purpose here is to explore this possibility. What will prove altogether remarkable is that some very simple schemes to produce erratic numbers behave identically to some of the erratic aspects of natural phenomena. More specifically, there is now cogent evidence that the problem of how a fluid changes over from smooth to turbulent flow can be solved through its relation to the simple scheme described in this chapter. Other natural problems that can be treated in the same way are the behaviour of a population from generation to generation and the noisiness of a large variety of mechanical, electrical and chemical oscillators. Also, there is now evidence that various Hamiltonian systems—those subscribing to classical mechanics, such as the solar system—can come under this discipline.

The feature common to these phenomena is that, as some external parameter (temperature, for example) is varied, the behaviour of the system changes from simple to erratic. More precisely, for some range of parameter values, the system exhibits an orderly *periodic* behaviour; that is, the system's behaviour reproduces itself every period of time T. Beyond this range, the behaviour fails to reproduce itself after T seconds; it almost does so, but in fact it requires two intervals of T to repeat itself. That is, the period has doubled to $2T$. This new periodicity remains over some range of parameter values until another critical parameter value is reached, after which the behaviour almost reproduces itself after $2T$ but, in fact, it now requires $4T$ for reproduction. This process of successive period doubling recurs continually (with the range of parameter values for which the period is 2^nT becoming successively smaller as n increases) until, at a certain value of the parameter, it has doubled *ad infinitum*, so that the behaviour is no longer periodic.

Period doubling is, then, a characteristic route for a system to follow as it changes over from simple periodic to complex aperiodic motion. All the phenomena mentioned above exhibit period doubling. In the limit of aperiodic behaviour, there is a unique and hence universal solution common to all systems undergoing period doubling. This fact implies remarkable consequences. For a given system, if we denote by Λ_n the value of the parameter at which its period doubles for the nth time, we find that the values Λ_n converge to Λ_∞ (at which the motion is aperiodic) *geometrically* for large n. This means that

$$\Lambda_\infty - \Lambda_n \propto \delta^{-n} \qquad (7.1)$$

for a fixed value of δ (the rate of onset of complex behaviour) as n becomes large. Put differently, if we define

$$\delta_n \equiv \frac{\Lambda_{n+1}-\Lambda_n}{\Lambda_{n+2}-\Lambda_{n+1}}, \tag{7.2}$$

δ_n (quickly) approaches the constant value δ. (Typically, δ_n will agree with δ to several significant figures after just a few period doublings.) What is quite remarkable (beyond the fact that there is always a geometric convergence) is that, for all systems undergoing this period doubling, the value of δ is *predetermined* at the universal value [1, 2]

$$\delta = 4.669\,2016\ldots. \tag{7.3}$$

Thus, this definite number must appear as a natural rate in oscillators, populations, fluids, and all systems exhibiting a period-doubling route to turbulence. In fact, most measurable properties of any such system in this aperiodic limit now can be determined in a way that essentially bypasses the details of the equations governing each specific system, because the theory of this behaviour is universal over such details. That is, so long as a system possesses certain *qualitative* properties that enable it to undergo this route to complexity, its *quantitative* properties are determined. (This result is analogous to the results of the modern theory of critical phenomena, where a few qualitative properties of the system undergoing a phase transition, notably the dimensionality, determine universal critical exponents. Indeed at a formal level the two theories are identical in that they are fixed-point theories, and the number δ, for example, can be viewed as a critical exponent.) Accordingly, it is sufficient to study the simplest system exhibiting this phenomenon to comprehend the general case.

7.2 Functional iteration

A random number generator is an example of a simple iteration scheme that has complex behaviour. Such a scheme generates the next pseudorandom number by a definite transformation upon the present pseudorandom number. In other words, a certain function is re-evaluated successively to produce a sequence of such numbers. Thus, if f is the function and x_0 is a starting number (or 'seed'), then $x_0, x_1, \ldots, x_n, \ldots$, where

$$x_1 = f(x_0), \quad x_2 = f(x_1), \quad \ldots \quad x_{n+1} = f(x_n), \quad \ldots \tag{7.4}$$

is the sequence of generated pseudorandom numbers. That is, they are generated by *functional iteration*. The nth element in the sequence is

$$x_n = f\{f[\ldots f(f(x_0))\ldots]\} \equiv f^n(x_0), \tag{7.5}$$

where n is the total number of applications of f. ($f^n(x)$ is not the nth power of $f(x)$; it is the nth *iterate* of f.) A property of iterates worthy of mention is

$$f^n[f^m(x)] = f^m[f^n(x)] = f^{m+n}(x), \tag{7.6}$$

since each expression is simply $m+n$ applications of f. It is understood that

$$f^0(x) = x. \tag{7.7}$$

It is also useful to have a symbol, $\circ$, for functional iteration (or composition), so that

$$f^n \circ f^m = f^m \circ f^n = f^{m+n}. \tag{7.8}$$

Now f^n in Eq. (7.5) is itself a definite and computable function, so that x_n as a function of x_0 is known in principle.

If the function f is linear as, for example,

$$f(x) = ax \tag{7.9}$$

for some constant a, it is easy to see that

$$f^n(x) = a^n x, \tag{7.10}$$

so that, for this f,

$$x_n = a^n x_0 \tag{7.11}$$

is the solution of the recurrence relation defined in Eq. (7.4),

$$x_{n+1} = ax_n. \tag{7.12}$$

Should $|a| < 1$, then x_n geometrically converges to zero at the rate $1/a$. This example is special in that the linearity of f allows for the explicit computation of f^n.

We must choose a nonlinear f to generate a pseudorandom sequence of numbers. If we choose for our nonlinear f

$$f(x) = a - x^2, \tag{7.13}$$

then it turns out that f^n is a polynomial in x of order 2^n. This polynomial rapidly becomes unmanageably large; moreover, its coefficients are polynomials in a of order up to 2^{n-1} and become equally difficult to compute. Thus even if $x_0 = 0$, x_n is a polynomial in a of order 2^{n-1}. These polynomials are nontrivial, as can be surmised from the fact that for certain values of a the sequence of numbers generated for almost all starting-points in the range $(a - a^2, a)$ possess all the mathematical properties of a random sequence. To illustrate this, let us consider the iterates of a similar system in two dimensions:

$$x_{n+1} = y_n - x_n^2 \qquad y_{n+1} = a - x_n. \tag{7.14}$$

Analogous to Eq. (7.4), a starting coordinate pair (x_0, y_0) is used in Eq. (7.14) to determine the next coordinate (x_1, y_1). Equation (7.14) is reapplied to determine (x_2, y_2) and so on. For some initial points, all iterates lie along a definite elliptic curve, whereas for others the iterates are distributed 'randomly' over a certain region. It should be obvious that no explicit formula will account for the vastly rich behaviour that is possible. That is, while the iteration scheme of Eq. (7.14) is trivial to specify, its nth iterate as a function of (x_0, y_0) is unavailable. Put differently, applying the simplest of nonlinear iteration schemes to itself sufficiently many times can create vastly complex behaviour. Yet, precisely because the same operation is reapplied, it is conceivable that only a select few self-consistent patterns might emerge where the consistency is determined by the key notion of iteration and not by the particular function performing the iterates. These self-consistent patterns do occur in the limit of infinite period doubling and in a well-defined intricate organization that can be determined *a priori* amidst immense complexity.

7.3 Fixed-point behaviour of functional iterations

Let us now make a direct onslaught on Eq. (7.13) to see what it possesses. We want to know the behaviour of the system after many iterations. As we already know, high iterates of f rapidly become very complicated. One way in which this growth can be prevented is to have the first iterate of x_0 be precisely x_0 itself. Generally, this is impossible. Rather this condition determines possible x_0's. Such a self-reproducing point is called a *fixed point* of f. The sequence of iterates is then $x_0, x_0, x_0, \ldots$ so that the behaviour is static or, if viewed as periodic, it has period 1.

It is elementary to determine the fixed points of Eq (7.13). For future convenience we shall use a modified form of Eq. (7.13) obtained by a translation in x and some redefinitions:

$$f(x) = 4\lambda x(1-x), \tag{7.15}$$

so that as λ is varied, $x=0$ is always a fixed point. Indeed, the fixed-point condition for Eq. (7.15),

$$x^* = f(x^*) = 4\lambda x^*(1-x^*), \tag{7.16}$$

gives as the two fixed points

$$x^* = 0, \qquad x_0^* = 1 - \tfrac{1}{4}\lambda. \tag{7.17}$$

The maximum value of $f(x)$ in Eq. (7.15) is attained at $x=\frac{1}{2}$ and is equal to λ. Also, for $\lambda > 0$ and x in the interval $(0, 1)$, $f(x)$ is always positive. Thus, if λ is anywhere in the range $[0, 1]$, then any iterate of any x in $(0, 1)$ is also always in $(0, 1)$. Accordingly, in all that follows we shall consider only values of x and λ lying between 0 and 1. By Eq. (7.16) for $0 \leqslant \lambda < \frac{1}{4}$, only $x^*=0$ is within range, whereas for $\frac{1}{4} \leqslant \lambda \leqslant 1$, both fixed points are within the range. For example, if we set $\lambda = \frac{1}{2}$ and we start at the fixed point $x_0^* = \frac{1}{2}$ (that is, we set $x_0 = \frac{1}{2}$), then

$$x_1 = x_2 = \ldots = \tfrac{1}{2};$$

similarly, if $x_0 = 0$,

$$x_1 = x_2 = \ldots = 0$$

and the problem of computing the nth iterate is obviously trivial.

What if we choose an x_0 that is not at a fixed point? The easiest way to see what happens is to perform a graphical analysis. We graph $y = f(x)$ together with $y = x$. Where the lines intersect we have $x = y = f(x)$, so that the intersections are precisely the fixed points. Now, if we choose an x_0 and plot it on the x-axis, the ordinate of $f(x)$ at x_0 is x_1. To obtain x_2, we must transfer x_1 to the x-axis before reapplying f. Reflection through the straight line $y = x$ accomplishes precisely this operation. Altogether, to iterate an initial x_0 successively,

(1) move *vertically* to the graph of $f(x)$;
(2) move *horizontally* to the graph of $y = -x$; and
(3) repeat steps 1, 2, etc.

Figure 7.1 depicts this process for $\lambda = \frac{1}{2}$. The two fixed points are circled, and the first several iterates of an arbitrarily chosen point x_0 are shown. What should be obvious is that if we start from any x_0 in (0, 1) ($x = 0$ and $x = 1$ excluded), upon continued iteration x_n will converge to the fixed point at $x = \frac{1}{2}$. No matter how close x_0 is to the fixed point at $x = 0$, the iterates diverge away from it. Such a fixed point is termed *unstable.*

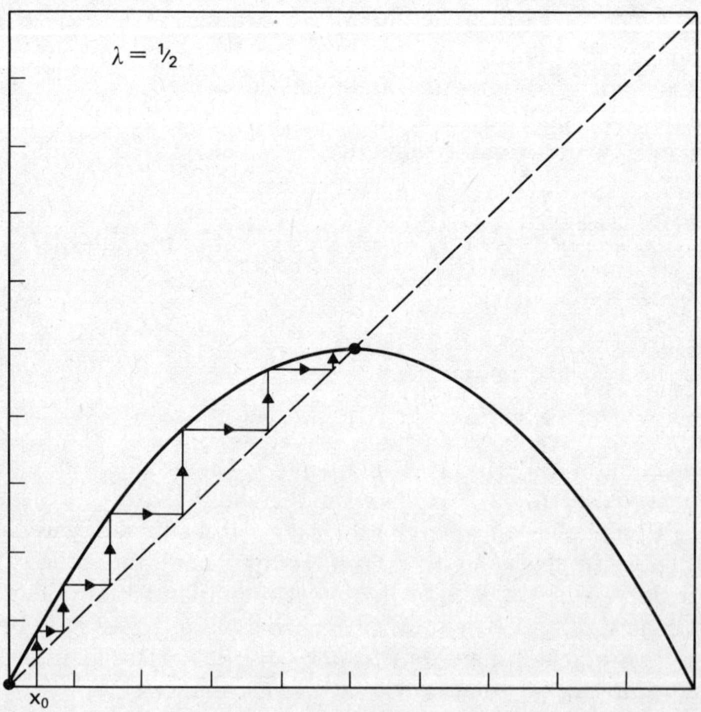

Fig. 7.1 Iterates of x_0 at $\lambda = 0.5$

Alternatively, for almost all x_0 near enough to $x=\frac{1}{2}$ (in this case, all x_0 in (0, 1)), the iterates converge towards $x=\frac{1}{2}$. Such a fixed point is termed *stable* or is referred to as an *attractor* of period 1.

Now, if we are not concerned with the transient behaviour of the iterates of x_0, but only with some regular behaviour that will emerge eventually, then knowledge of the stable fixed point at $x=\frac{1}{2}$ satisfies our concern for the eventual behaviour of the iterates. In this restricted sense of eventual behaviour, the existence of an attractor determines the solution independently of the initial condition x_0 provided that x_0 is within the *basin of attraction* of the attractor; that is, that it is attracted. The attractor satisfies Eq. (7.16), which is explicitly independent of x_0. This condition is the basic theme of universal behaviour: if an attractor exists, the eventual behaviour is independent of the starting-point.

What makes $x=0$ unstable, but $x=\frac{1}{2}$ stable? The reader should be able to convince himself that $x=0$ is unstable because the slope of $f(x)$ at $x=0$ is greater than 1. Indeed, if x^* is a fixed point of f and the derivative of f at x^*, $f'(x^*)$, is smaller than 1 in absolute value, then x^* is stable. If $|f'(x^*)|$ is greater than 1, then x^* is unstable. Also, only stable fixed points can account for the eventual behaviour of the iterates of an arbitrary point.

We now must ask, 'For what values of λ are the fixed points attracting?' By Eq. (7.15), $f'(x)=4\lambda(1-2x)$ so that

$$f'(0)=4\lambda \tag{7.18}$$

and

$$f'(x_0^*)=2-4\lambda. \tag{7.19}$$

For $0<\lambda<\frac{1}{4}$, only $x^*=0$ is stable. At $\lambda=\frac{1}{4}$, $x_0^*=0$ and $f'(x_0^*)=1$. For $\frac{1}{4}<\lambda<\frac{3}{4}$, x^* is unstable and x_0^* is stable, while at $\lambda=\frac{3}{4}$, $f'(x_0^*)=-1$ and x_0^* also has become unstable. Thus, for $0<\lambda<\frac{3}{4}$, the eventual behaviour is known.

7.4 Period 2 from the fixed point

What happens to the system when λ is in the range $\frac{3}{4}<\lambda<1$, where there are no attracting fixed points? We shall see that as λ increases slightly beyond $\lambda=\frac{3}{4}$, f undergoes period doubling. That is, instead of having a stable cycle of period 1 corresponding to one fixed point, the system has a stable cycle of period 2; that is, the cycle contains two points. Since these two points are fixed points of the function f^2 (f applied twice) and since stability is determined by the slope of a function at its fixed points, we must now focus on f^2. First, we examine a graph of f^2 at λ just below $\frac{3}{4}$. Figures 7.2(a) and (b) show f and f^2, respectively, at $\lambda=0.7$.

To understand Fig. 7.2(b), observe first that, since f is symmetric about its maximum at $x=\frac{1}{2}$, f^2 is also symmetric about $x=\frac{1}{2}$. Also, f^2 must have

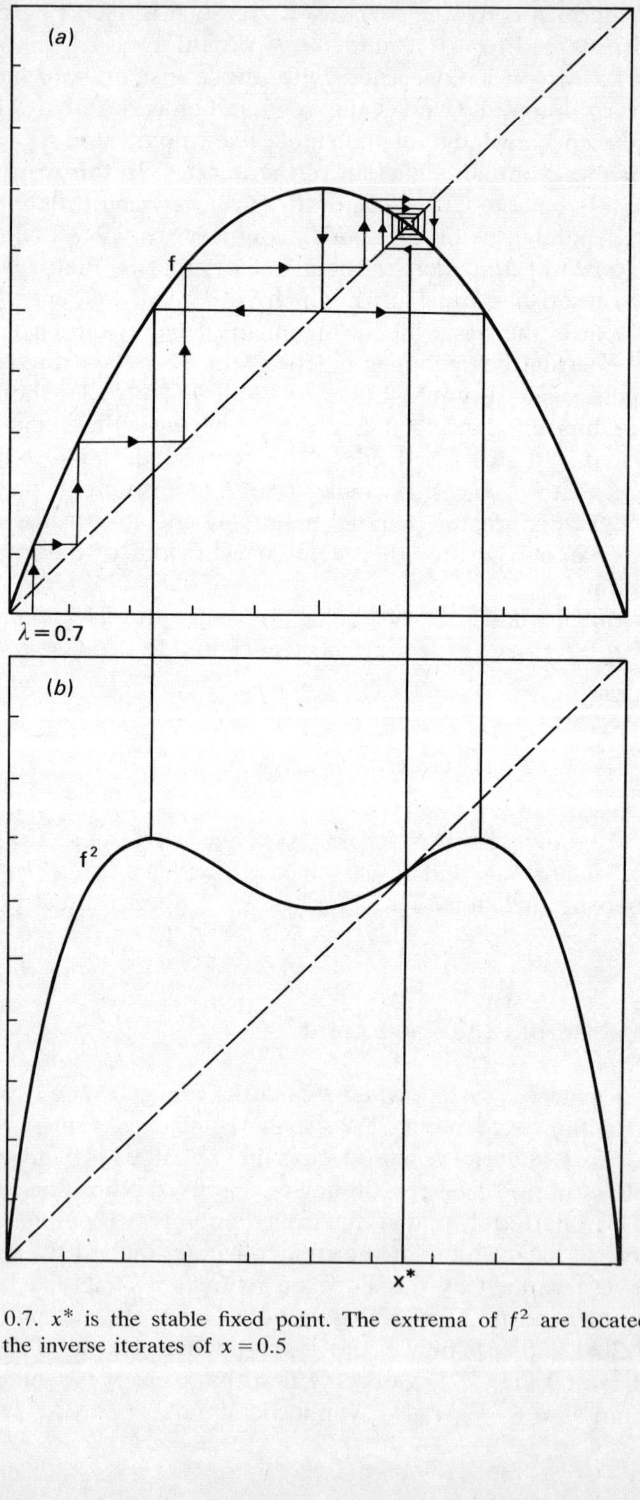

Fig. 7.2 $\lambda = 0.7$. x^* is the stable fixed point. The extrema of f^2 are located in (a) by constructing the inverse iterates of $x = 0.5$

a fixed point whenever f has, because the second iterate of a fixed point is still that same point. The main ingredient that determines the period-doubling behaviour of f as λ increases is the relationship of the slope of f^2 to the slope of f. This relationship is a consequence of the chain rule. By definition

$$x_2 = f^2(x_0) \qquad \text{where} \qquad x_1 = f(x_0), \quad x_2 = f(x_1).$$

We leave it to the reader to verify by the chain rule that

$$f^{2\prime}(x_0) = f'(x_0)f'(x_1) \tag{7.20}$$

and

$$f^{n\prime}(x_0) = f'(x_0)f'(x_1)\ldots f'(x_{n-1}), \tag{7.21}$$

an elementary result that determines period doubling. If we start at a fixed point of f and apply Eq. (7.20) to $x_0 = x^*$, so that $x_2 = x_1 = x^*$, then

$$f^{2\prime}(x^*) = f'(x^*)f'(x^*) = [f'(x^*)]^2. \tag{7.22}$$

Since at $\lambda = 0.7$, $|f'(x^*)| < 1$, it follows from Eq. (7.22) that

$$0 < f^{2\prime}(x^*) < 1.$$

Also, if we start at the extremum of f, so that $x_0 = \frac{1}{2}$ and $f'(x_0) = 0$, it follows from Eq. (7.21) that

$$f^{n\prime}(\tfrac{1}{2}) = 0 \tag{7.23}$$

for all n. In particular, f^2 is extreme (and a minimum) at $\frac{1}{2}$. Also, by Eq. (7.20), f^2 will be extreme (and a maximum) at the x_0 that will iterate under f to $x = \frac{1}{2}$, since then $x_1 = \frac{1}{2}$ and $f'(x_1) = 0$. These points, the inverses of $x = \frac{1}{2}$, are found by going vertically down along $x = \frac{1}{2}$ to $y = x$ and then horizontally to $y = f(x)$. (Reverse the arrows in Fig. 7.1, and see Fig. 7.2(a).) Since f has a maximum, there are two horizontal intersections and, hence, the two maxima of Fig. 7.2(b). *The ability of f to have complex behaviours is precisely the consequence of its double-valued inverse,* which is in turn a reflection of its possession of an extremum. A monotone f, one that always increases, always has simple behaviours, whether or not the behaviours are easy to compute. A linear f is always monotone. The f's we are concerned with always fold over and so are strongly nonlinear. This folding nonlinearity gives rise to universality. Just as linearity in any system implies a definite method of solution, folding nonlinearity in any system also implies a definite method of solution. In fact folding nonlinearity in the aperiodic limit of period doubling in any system is solvable, and many systems, such as various coupled nonlinear differential equations, possess this nonlinearity.

To return to Fig. 7.2(b), as $\lambda \to \frac{3}{4}$ and the maximum value of f increases to $\frac{3}{4}$, $f'(x^*) \to -1$ and $f^{2\prime}(x^*) \to +1$. As λ increases beyond $\frac{3}{4}$, $|f'(x^*)| > 1$ and $f^{2\prime}(x^*) > 1$, so that f^2 must develop two new fixed points beyond those of f; that is, f^2 will cross $y = x$ at two more points. This transition is

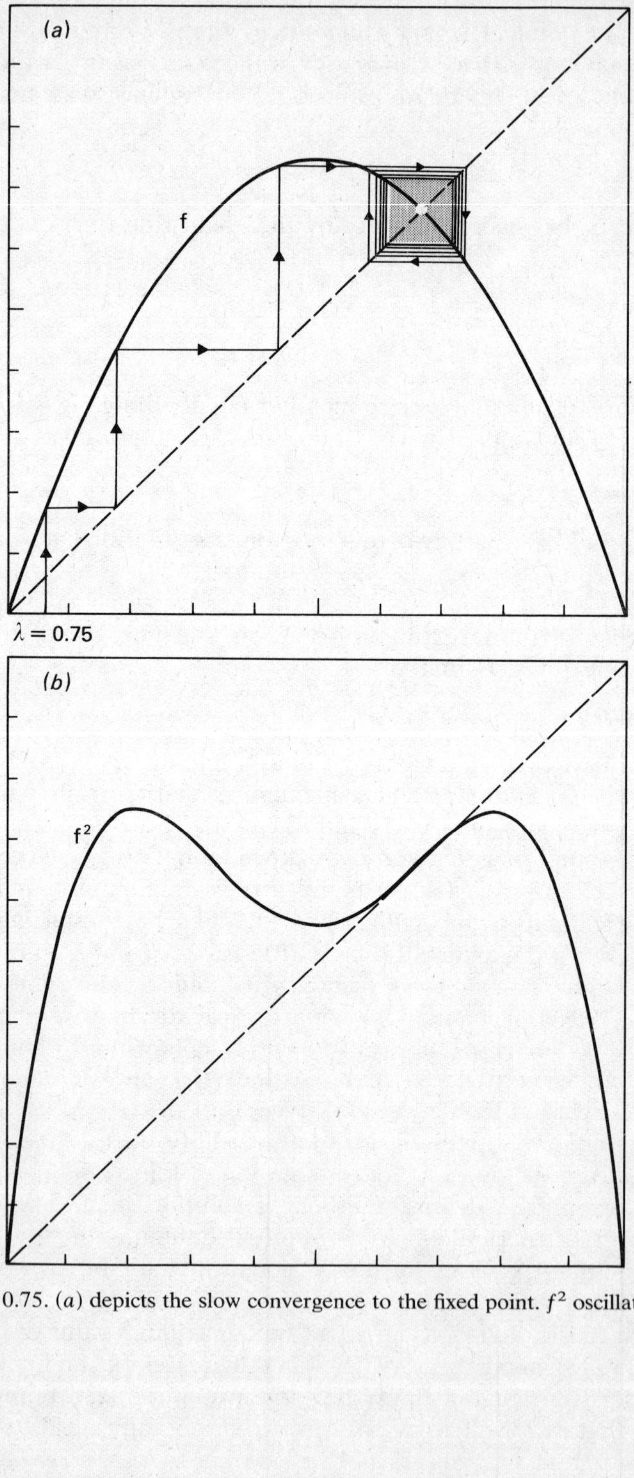

Fig. 7.3 $\lambda = 0.75$. (*a*) depicts the slow convergence to the fixed point. f^2 oscillates about the fixed point

depicted in Figs 7.3(a) and (b) for f and f^2, respectively, at $\lambda = 0.75$, and similarly in Figs 7.4(a) and (b) at $\lambda = 0.785$. (Observe the exceptionally slow convergence to x^* at $\lambda = 0.75$, where iterates approach the fixed point not geometrically, but rather with deviations from x^* inversely proportional to the square root of the number of iterations.) Since x_1^* and x_2^*, the new fixed points of f^2, are not fixed points of f, it must be that f sends one into the other:

$$x_1^* = f(x_2^*) \quad \text{and} \quad x_2^* = f(x_1^*).$$

Such a pair of points, termed a 2-cycle, is depicted by the limiting unwinding circulating square in Fig. 7.4(a). Observe in Fig. 7.4(b) that the slope of f^2 is in excess of 1 at the fixed point of f and so is an unstable fixed point of f^2, while the two new fixed points have slopes smaller than 1, and so are stable; that is, every two iterates of f will have a point attracted toward x_1^* if it is sufficiently close to x_1^* or toward x_2^* if it is sufficiently close to x_2^*. This means that the sequence under f,

$$x_0, x_1, x_2, x_3, \ldots,$$

eventually becomes arbitrarily close to the sequence

$$x_1^*, x_2^*, x_1^*, x_2^*, \ldots,$$

so that this is stable 2-cycle, or an attractor of period 2. Thus, we have observed for Eq. (7.15) the first period doubling as the parameter λ has increased.

There is a point of paramount importance to be observed; namely, f^2 has the same slope at x_1^* and at x_2^*. This point is a direct consequence of Eq. (7.20), since if $x_0 = x_1^*$, then $x_1 = x_2^*$, and vice versa, so that the product of the slopes is the same. More generally, if $x_1^*, x_2^*, \ldots, x_n^*$ is an n-cycle so that

$$x_{r+1}^* = f(x_r^*), \quad r = 1, 2, \ldots, n-1$$

and

$$x_1^* = f(x_n^*), \tag{7.24}$$

then each is a fixed point of f^n with identical slopes:

$$x_r^* = f^n(x_r^*), \quad r = 1, 2, \ldots, n \tag{7.25}$$

and

$$f^{n\prime}(x_r^*) = f'(x_1^*) \ldots f'(x_n^*). \tag{7.26}$$

From this observation will follow period doubling *ad infinitum*.

As λ is increased further, the minimum at $x = \frac{1}{2}$ will drop as the slope of f^2 through the fixed point of f increases. At some value of λ, denoted by λ_1, $x = \frac{1}{2}$ will become a fixed point of f^2. Simultaneously, the right-hand maximum will also become a fixed point of f^2. (By Eq. (7.26), both

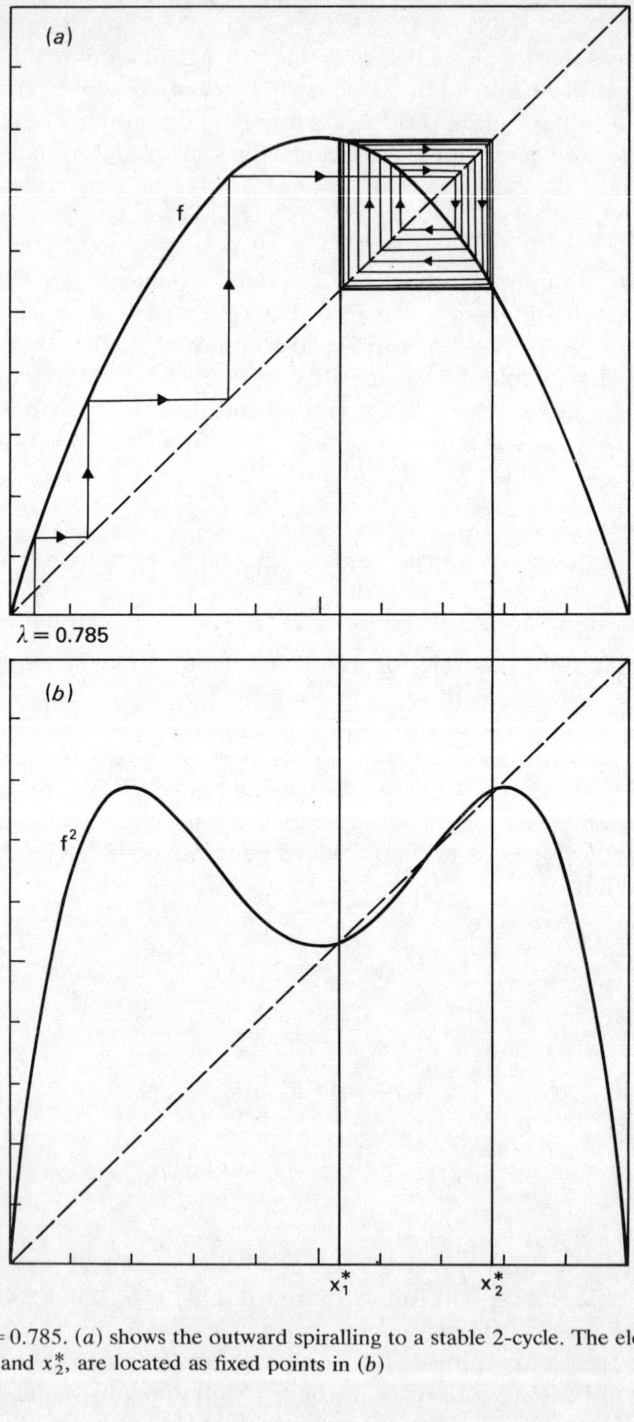

Fig. 7.4 $\lambda = 0.785$. (*a*) shows the outward spiralling to a stable 2-cycle. The elements of the 2-cycle, x_1^* and x_2^*, are located as fixed points in (*b*)

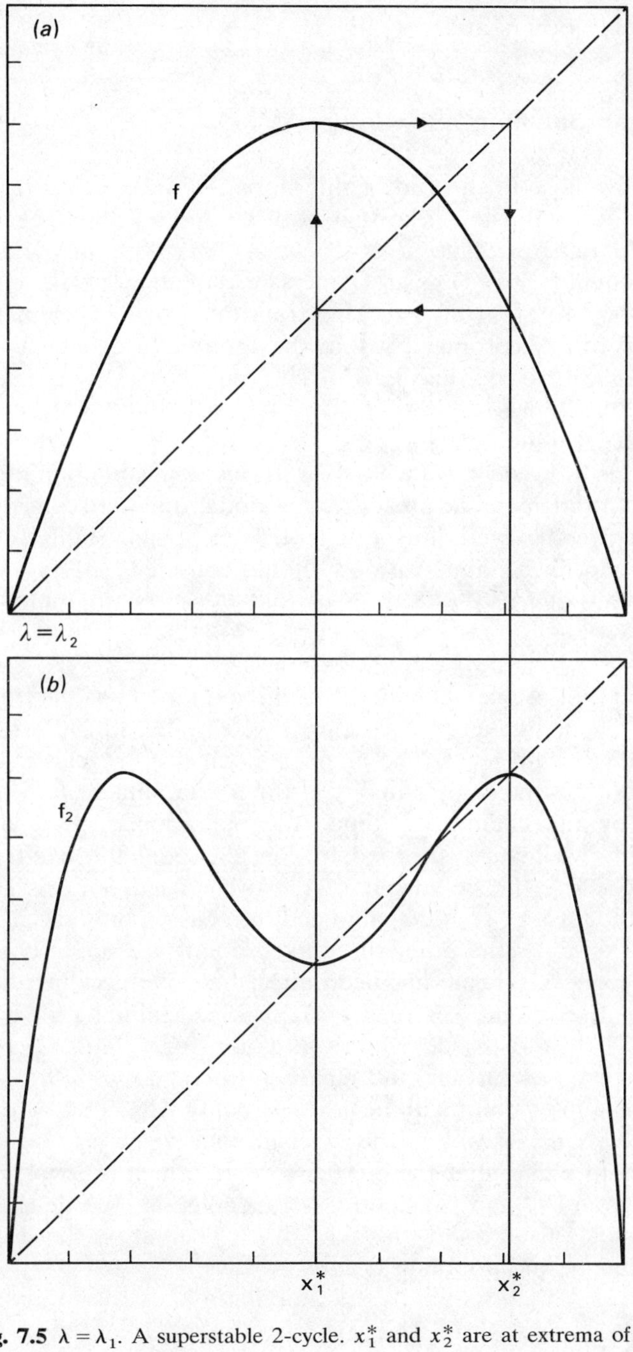

Fig. 7.5 $\lambda = \lambda_1$. A superstable 2-cycle. x_1^* and x_2^* are at extrema of f^2

elements of the 2-cycle have slope 0.) Figures 7.5(a) and (b) depict the situation that occurs at $\lambda = \lambda_1$ [3].

7.5 Period doubling *ad infinitum*

We are now close to the end of this story. As we increase λ further, the minimum drops still lower so that both x_1^* and x_2^* have negative slopes. At some parameter value, denoted by Λ_2, the slope at both x_1^* and x_2^* becomes equal to -1. Thus at Λ_2 the same situation has developed for f^2 as developed for f at $\Lambda_1 = \frac{3}{4}$. This transitional case is depicted in Figs 7.6(a) and (b). Accordingly, just as the fixed point of f at Λ_1 issued into being a 2-cycle, so too does each fixed point of f^2 at Λ_2 create a 2-cycle, which in turn is a 4-cycle of f. That is, we have now encountered the second period doubling.

The manner in which we were able to follow the creation of the 2-cycle at Λ_1 was to anticipate the presence of period 2, and so to consider f^2, which would resolve the cycle into a pair of fixed points. Similarly, to resolve period 4 into fixed points we now should consider f^4. Beyond being the fourth iterate of f, Eq. (7.8) tells us that f^4 can be computed from f^2:

$$f^4 = f^2 \circ f^2.$$

From this point, we can abandon f itself, and take f^2 as the 'fundamental' function. Then, just as f^2 was constructed by iterating f with itself, we now iterate f^2 with itself. The manner in which f^2 reveals itself as being an iterate of f is the slope equality at the fixed points of f^2, which we saw imposed by the chain rule. Since the operation of the chain rule is 'automatic', we actually needed to consider only the fixed point of f^2 nearest to $x = \frac{1}{2}$; the behaviour of the other fixed point is slaved to it. Thus, at the level of f^4, we again need to focus on only the fixed point of f^4 nearest to $x = \frac{1}{2}$: the other three fixed points are similarly slaved to it. Thus, a recursive scheme has been unearthed. We now increase λ to λ_2, so that the fixed point of f^4 nearest to $x = \frac{1}{2}$ is again at $x = \frac{1}{2}$ with slope 0. Figures 7.7(a) and (b) depict this situation for f^2 and f^4, respectively. When λ increases further, the maximum of f^4 at $x = \frac{1}{2}$ now moves up, developing a fixed point with negative slope. Finally, at Λ_3 when the slope of this fixed point (as well as the other three) is again -1, each fixed point will split into a pair giving rise to an 8-cycle, which is now stable. Again, $f^8 = f^4 \circ f^4$, and f^4 can be viewed as fundamental. We define λ_3 so that $x = \frac{1}{2}$ again is a fixed point, this time of f^8. Then at Λ_4 the slopes are -1, and another period doubling occurs. Always,

$$f^{2^{n+1}} = f^{2^n} \circ f^{2^n}, \tag{7.27}$$

Provided that a constraint on the range of λ does not prevent it from decreasing the slope at the appropriate fixed point past -1, this doubling must recur *ad infinitum*.

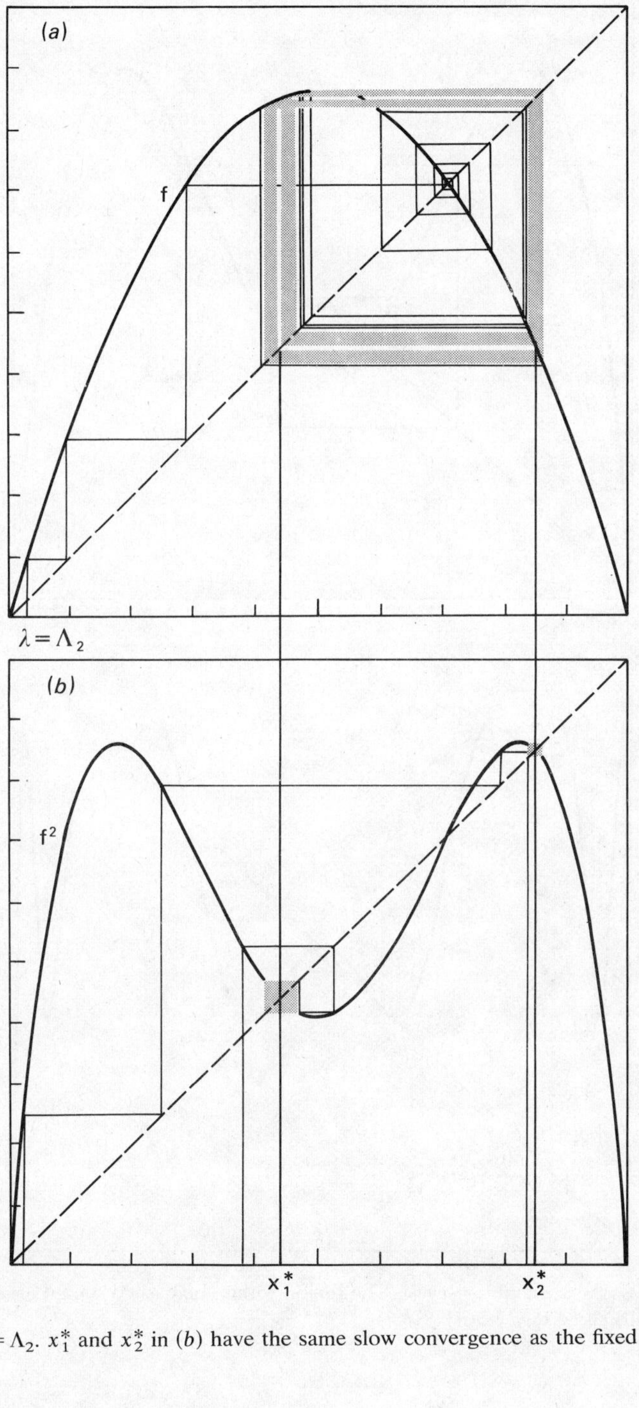

Fig. 7.6 $\lambda = \Lambda_2$. x_1^* and x_2^* in (b) have the same slow convergence as the fixed point in Fig. 7.3(a)

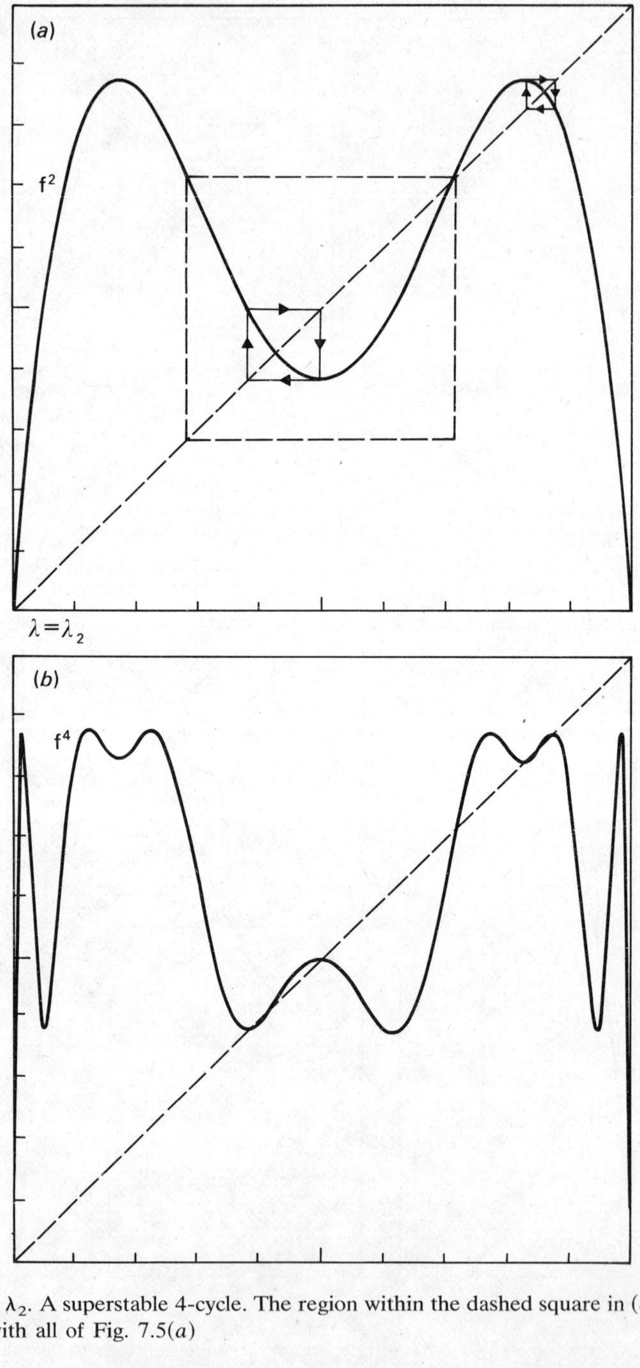

Fig. 7.7 $\lambda = \lambda_2$. A superstable 4-cycle. The region within the dashed square in (*a*) should be compared with all of Fig. 7.5(*a*)

Basically, the mechanism that f^{2^n} uses to period double at Λ_{n+1} is the same mechanism that $f^{2^{n+1}}$ will use to double at Λ_{n+2}. The function $f^{2^{n+1}}$ is constructed from f^{2^n} by Eq. (7.27), and similarly $f^{2^{n+2}}$ will be constructed from $f^{2^{n+1}}$. Thus, there is a definite operation that, by acting on functions, creates functions; in particular, the operation acting on f^{2^n} at Λ_{n+1} (or better, f^{2^n} at λ_n) will determine $f^{2^{n+1}}$ at λ_{n+1}. Also, since we need to keep track of f^{2^n} only in the interval including the fixed point of f^{2^n} closest to $x=\frac{1}{2}$ and since this interval becomes increasingly small as λ increases, the part of f that generates this region is also the restriction of f to an increasingly small interval about $x=\frac{1}{2}$. (Actually, slopes of f at points farther away also matter, but these merely set a 'scale', which will be eliminated by a rescaling.) The behaviour of f away from $x=\frac{1}{2}$ is immaterial to the period-doubling behaviour, and in the limit of large n only the nature of f's maximum can matter. This means that in the infinite period-doubling limit, all functions with a quadratic extremum will have identical behaviour ($f''(\frac{1}{2})\neq 0$ is the generic circumstance). Therefore, the operation on functions will have a stable fixed point in the space of functions, which will be the common universal limit of high iterates of any specific function [2].

To determine this universal limit we must enlarge our scope vastly, so that the role of the starting-point, x_0, will be played by an arbitrary function; the attracting fixed point will become a universal function obeying an equation implicating only itself. The role of the function in the equation $x_0=f(x_0)$ now must be played by an operation that yields a new function when it is performed upon a function. In fact, the heart of this operation is the functional composition of Eq. (7.27). If we can determine the exact operator and actually can solve its fixed-point problem, we shall understand why a special number, such as δ of Eq. (7.3), has emerged independently of the specific system (the starting function) we have considered.

7.6 The universal limit of high iterates

In this section we sketch the solution to the fixed-point problem. In Fig. 7.7(a) a dashed square encloses the part of f^2 that we must focus on for all further period doublings. This square should be compared with the unit square that comprises all of Fig. 7.5(a). If the Fig. 7.7(a) square is reflected through $x=\frac{1}{2}$, $y=\frac{1}{2}$ and then magnified so that the circulation squares of Figs 7.4(a) and 7.5(a) are of equal size, we will have in each square a piece of a function that has the same kind of maximum at $x=\frac{1}{2}$ and falls to zero at the right-hand lower corner of the circulation square. Just as f produced this second curve of f^2 in the square as λ increased from λ_1 to λ_2, so too will f^2 produce another curve, which will be similar to the other two when it has been magnified suitably and reflected twice.

Figure 7.8 shows this superposition for the first five such functions; at

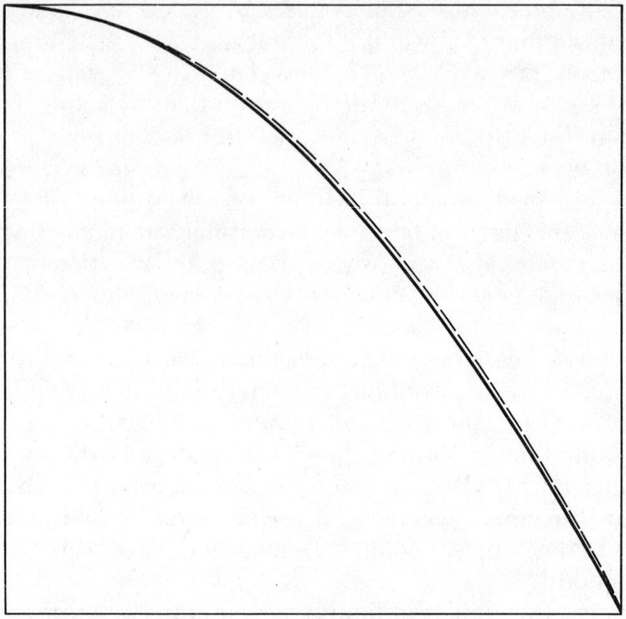

Fig. 7.8 Superposition of the suitably magnified dotted squares of $f^{2^{n-1}}$ (as in Figs 7.5(a), 7.7(a), . . .)

the resolution of the figure, observe that the last three curves are coincident. Moreover, the scale reduction that f^2 will determine for f^4 is based solely on the functional composition, so that if these curves for f^{2^n}, $f^{2^{n+1}}$, converge (as they obviously do in Fig. 7.8), the scale reduction from level to level will converge to a definite constant. But the width of each circulation square is just the distance between $x=\frac{1}{2}$ when it is a fixed point of f^{2^n} and the fixed point of f^{2^n} next nearest to $x=\frac{1}{2}$ (Figs 7.7(a) and (b)). That is, asymptotically, *the separation of adjacent elements of period-doubled attractors is reduced by a constant value from one doubling to the next.* Also, from one doubling to the next, this next nearest element alternates from one side of $x=\frac{1}{2}$ to the other. Let d_n denote the algebraic distance from $x=\frac{1}{2}$ to the nearest element of the attractor cycle of period 2^n, in the 2^n-cycle at λ_n. A positive number a scales this distance down in the 2^{n+1}-cycle at λ_{n+1} [1, 2]:

$$\frac{d_n}{d_{n+1}} \sim -a. \tag{7.28}$$

But since rescaling is determined only by functional composition, there is some function that composed with itself will reproduce itself reduced in scale by $-\alpha$. The function has a quadratic maximum at $x=\frac{1}{2}$, is symmetric about $x=\frac{1}{2}$ and can be scaled by hand to equal 1 at $x=\frac{1}{2}$. Shifting coordinates so that $x=\frac{1}{2}\to x=0$, we have

$$-\alpha g[g(x/\alpha)]=g(x). \tag{7.29}$$

Substituting $g(0)=1$, we have

$$g(1)=-\frac{1}{\alpha}. \tag{7.30}$$

Accordingly, Eq. (7.29) is a definite equation for a function g depending on x through x^2 and having a maximum of 1 at $x=0$. There is a unique smooth solution to Eq. (7.29), which determines

$$\alpha=2.502\,907\,875\ldots. \tag{7.31}$$

Knowing α, we can predict through Eq. (7.28) a definite scaling law binding on the iterates of any scheme possessing period doubling. The law has, indeed, been amply verified experimentally. By Eq. (7.29) we see that the relevant operation upon functions that underlies period doubling is functional composition followed by magnification, where the magnification is determined by the fixed-point condition of Eq. (7.29) with the function g the fixed point in this space of functions. However, Eq. (7.29) does not describe a stable fixed point because we have not incorporated in it the parameter increase from λ_n to λ_{n+1}. Thus, g is not the limiting function of the curves in the circulation squares, although it is intimately related to that function.

The full theory is described in the next section. Here we merely state that we can determine the limiting function and thereby can *determine the location of the actual elements of limiting* 2^n*-cycles.* We also have established that g is an unstable fixed point of functional composition, where the rate of divergence away from g is precisely δ of Eq. (7.3) and so is computable. Accordingly, there is a full theory that determines, in a precise quantitative way, the aperiodic limit of functional iterations with an unspecified function f [4].

7.7 Some details of the full theory

Returning to Eq (7.28), we are in a position to describe theoretically the universal scaling of high-order cycles and the convergence to a universal limit. Since d_n is the distance between $x=\frac{1}{2}$ and the element of the 2^n-cycle at λ_n nearest to $x=\frac{1}{2}$ and since this nearest element is the 2^{n-1} iterate of $x=\frac{1}{2}$ (which is true because these two points were coincident before the nth period doubling began to split them apart), we have

$$d_n=f^{2^{n-1}}(\lambda_n,\tfrac{1}{2})-\tfrac{1}{2}. \tag{7.32}$$

For future work it is expedient to perform a coordinate translation that moves $x=\frac{1}{2}$ to $x=0$. Thus, Eq. (7.32) becomes

$$d_n=f^{2^{n-1}}(\lambda_n,0). \tag{7.33}$$

Equation (7.28) now determines that the rescaled distances,

$$r_n \equiv (-\alpha)^n d_{n+1},$$

will converge to a definite finite value as $n \to \infty$. That is,

$$\lim_{n\to\infty} (-\alpha)^n f^{2^n}(\lambda_{n+1}, 0) \tag{7.34}$$

must exist if Eq. (7.28) holds.

However, from Fig. 7.8 we know something stronger than Eq. (7.34). When the nth iterated function is magnified by $(-\alpha)^n$, it converges to a definite function. Equation (7.34) is the value of this function at $x=0$. After the magnification, the convergent functions are given by

$$(-\alpha)^n f^{2n}[\lambda_{n+1}, x/(-\alpha)^n].$$

Thus,

$$g_1(x) \equiv \lim_{n\to\infty} (-\alpha)^n f^{2^n}[\lambda_{n+1}, x/(-\alpha)^n] \tag{7.35}$$

is the limiting function inscribed in the square of Fig. 7.8. The function $g_1(x)$ is, by the argument of the restriction of f to increasingly small intervals about its maximum, the universal limit of all iterates of all f's with a quadratic extremum. Indeed, it is numerically easy to ascertain that g_1 of Eq. (7.35) is always the same function independent of the f in Eq. (7.32).

What is this universal function good for? Figure 7.5(a) shows a crude approximation of g_1 ($n=0$ in the limit of Eq. (7.35)), while Fig. 7.7(a) shows a better approximation ($n=1$). In fact, the extrema of g_1 near the fixed points of g_1 support circulation squares each of which contains two points of the cycle. (The two squares shown in Fig. 7.7(a) locate the four elements of the cycle.) That is, g_1 determines the location of elements of high-order 2^n-cycles near $x=0$. Since g_1 is universal, we now have the amazing result that the location of the actual elements of highly doubled cycles is universal. The reader might guess this is a very powerful result. Figure 7.9 shows g_1 out to x sufficiently large to have eight circulation squares, and hence locates the fifteen elements of a 2^n-cycle nearest to $x=0$. Also, the universal value of the scaling parameter α, obtained numerically, is

$$\alpha = 2.502\,907\,875\ldots. \tag{7.36}$$

Like δ, α is a number that can be measured (through an experiment that observes the d_n of Eq. (7.28)) in any phenomenon exhibiting period doubling.

If g_1 is universal, then of course its iterate g_1^2 also is universal. Figure 7.7(b) depicts an early approximation to this iterate. In fact, let us define a new universal function g_0, obtained by scaling g_1^2:

$$g_0(x) \equiv -\alpha g_1^2(-x/\alpha). \tag{7.37}$$

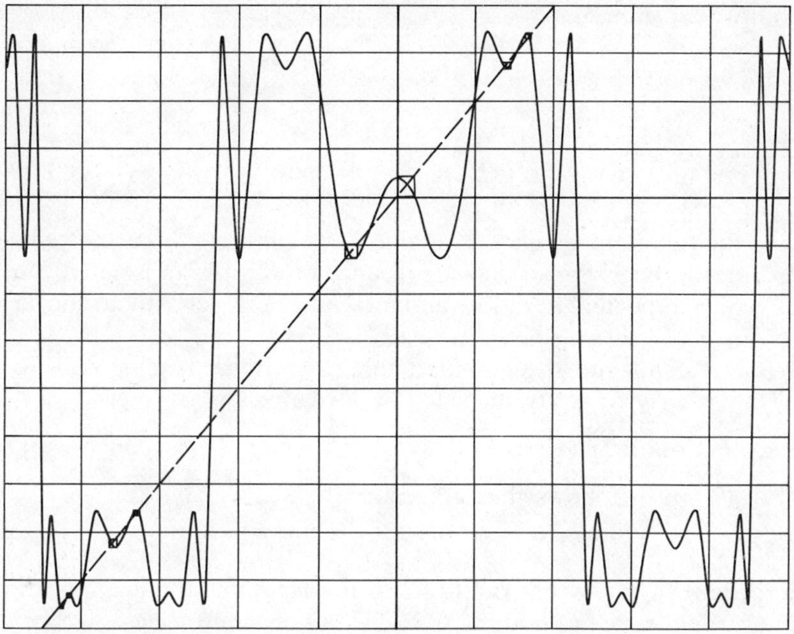

Fig. 7.9 The function g_1. The squares locate cycle elements

(Because g_1 is universal and the iterates of our quadratic function are all symmetric in x, both g_1 and g_0 are symmetric functions. Accordingly, the minus sign within g_1^2 can be dropped with impunity.) From Eq. (7.35), we now can write

$$g_0(x) = \lim_{n\to\infty} (-\alpha)^n f^{2^n}[\lambda_n, x/(-\alpha)^n]. \tag{7.38}$$

(We introduced the scaling of Eq. (7.37) to provide one power of α per period doubling, since each successive iterate of f^{2^n} reduces the scale by α.)

In fact, we can generalize Eqs (7.35) and (7.38) to a family of universal functions g_r:

$$g_r(x) = \lim_{n\to\infty} (-\alpha)^n f^{2^n}[\lambda_{n+1}, x/(-\alpha)^n]. \tag{7.39}$$

To understand this, observe that g_0 locates the cycle elements as the fixed points of g_0 at extrema; g_1 locates the same elements by determining two elements per extremum. Similarly, g_r determines 2^r elements about each extremum near a fixed point of g_r. Since each f^{2^n} is always magnified by $(-\alpha)^n$ for each r, the scales of all g_r are the same. Indeed, g_r for $r > 1$ looks like g_1 of Fig. 7.9, except that each extremum is slightly higher, to accommodate a 2^r-cycle. Since each extremum must grow by convergently small amounts to accommodate higher and higher 2^r-cycles, we are led to conclude that

$$g(x) = \lim_{r\to\infty} g_r(x) \tag{7.40}$$

must exist. By Eq. (7.39),

$$g(x) = \lim_{n\to\infty} (-\alpha)^n f^{2^n}[\lambda_\infty, x/(-\alpha)^n]. \tag{7.41}$$

Unlike the functions g_r, $g(x)$ is obtained as a limit of f^{2^n}'s at a fixed value of λ. Indeed, this is the special significance of λ_∞; it is an isolated value of λ at which repeated iteration and magnification lead to a convergent function.

We now can write the equation that g satisfies. Analogously to Eq. (7.37), it is easy to verify that all g_r are related by

$$g_{r-1}(x) = -\alpha g_r[g_r(-x/\alpha)]. \tag{7.42}$$

By Eq. (7.40), it follows that g satisfies

$$g(x) = -\alpha g[g(x/\alpha)]. \tag{7.43}$$

The reader can verify that Eq. (7.43) is invariant under a magnification of g. Thus, the theory has nothing to say about absolute scales. Accordingly, we must fix this by hand by setting

$$g(0) = 1. \tag{7.44}$$

Also, we must specify the nature of the maximum of g at $x = 0$ (for example, quadratic). Finally, since g is to be built by iterating $a - x^2$, it must be both smooth and a function of x through x^2. With these specifications, Eq. (7.43) has a unique solution. By Eqs (7.44) and (7.43),

$$g(0) = 1 = -\alpha g[g(0)] = -\alpha g(1),$$

so that

$$\alpha = -1/g(1). \tag{7.45}$$

Accordingly, Eq. (7.43) determines α together with g.

Let us comment on the nature of Eq. (7.43), a so-called functional equation. Because g is smooth, if we know its value at a finite number of points we know its value to some approximation on the interval containing these points by any sufficiently smooth interpolation. Thus, to some degree of accuracy, Eq. (7.43) can be replaced by a finite coupled system of nonlinear equations. Exactly, then, Eq. (7.43) is an infinite-dimensional nonlinear vector equation. Accordingly, we have obtained the solution to one-dimensional period doubling through our infinite-dimensional, explicitly universal problem. Equation (7.43) must be infinite-dimensional because it must keep track of the infinite number of cycle elements demanded of any attempt to solve the period-doubling problem. Rigorous mathematics for equations like Eq. (7.43) is just beyond the boundary of present mathematical knowledge.

At this point, we must determine two items. First, where is δ? Second, how do we obtain g_1, the real function of interest for locating cycle elements? The two problems are part of one question. Equation (7.42) is itself an iteration scheme. However, unlike the elements in Eq. (7.4), the elements acted on in Eq. (7.42) are functions. The analog of the function of f in Eq. (7.4) is the operation in function space of functional composition followed by a magnification. If we call this operation T, and an element of the function space ψ, Eq. (7.42) gives

$$T[\psi](x) = -\alpha\psi^2(-x/\alpha). \tag{7.46}$$

In terms of T, Eq. (7.42) now reads

$$g_{r-1} = T[g_r], \tag{7.47}$$

and Eq. (7.43) reads

$$g = T[g]. \tag{7.48}$$

Thus, g is precisely the fixed point of T. Since g is the limit of the sequence g_r, we can obtain g_r for large r by linearizing T about its fixed point g. Once we have g_r in the linear regime, the exact repeated application of T by Eq. (7.47) will provide g_1. Thus, we must investigate the stability of T at the fixed point g. However, it is obvious that T is unstable at g: for a large enough r, g_r is a point arbitrarily close to the fixed point g; by Eq. (7.47), successive iterates of g_r under T move away from g.

How unstable is T? Consider a one-parameter family of functions f_λ, which means a 'line' in the function space. For each f there is an isolated parameter value λ_∞ for which repeated applications of T lead to convergence towards g (Eq. (7.41)). Now, the function space can be 'packed' with all the lines corresponding to the various f's. The set of all the points on these lines specified by the respective λ_∞'s determines a 'surface' having the property that repeated applications of T to any point on it will converge to g. This is the surface of stability of T (the 'stable manifold' of T through g). But through each point of this surface issues out the corresponding line, which is one-dimensional since it is governed by a single parameter, λ. Accordingly, T is unstable in only one direction in function space. Linearized about g, this line of instability can be written as the one-parameter family

$$f_\lambda(x) = g(x) - \lambda h(x), \tag{7.49}$$

which passes through g (at $\lambda = 0$) and deviates from g along the unique direction h. But f_λ is just one of our transformations (Eq. (7.4))! Thus, as we vary λ, f_λ will undergo period doubling, doubling to a 2^n-cycle at Λ_n. By Eq. (7.41), λ_∞ for the family of functions f_λ in Eq. (7.49) is

$$\lambda_\infty = 0. \tag{7.50}$$

Thus, by Eq. (7.1)

$$\lambda_n \sim \delta^{-n}. \tag{7.51}$$

Since applications of T by Eq. (7.47) iterate in the opposite direction (diverge away from g), it now follows that the rate of instability of T along h must be precisely δ.

Accordingly, we find δ and g_1 in the following way. First, we must linearize the operation T about its fixed point g. Next, we must determine the stability directions of the linearized operator. Moreover, we expect there to be precisely one direction of instability. Indeed, it turns out that infinitesimal deformations (conjugacies) of g determine stable directions, while a unique unstable direction, h, emerges with a stability rate (eigenvalue) precisely the δ of Eq. (7.3). Equation (7.49) at λ_r is precisely g_r for asymptotically large r. Thus g_r is known asymptotically, so that we have entered the sequence g_r and can now, by repeated use of Eq. (7.47), step down to g_1. All the ingredients of a full description of high-order 2^n-cycles now are at hand and evidently are universal.

Although we have said that the function g_1 universally locates cycle elements near $x=0$, we must understand that it does not locate all cycle elements. This is possible because a finite distance of the scale of g_1 (for example, the location of the element nearest to $x=0$) has been magnified by α^n for n diverging. Indeed, the distances from $x=0$ for all elements of a 2^n-cycle, 'accurately' located by g_1, are reduced by $-\alpha$ in the 2^{n+1}-cycle. However, it is obvious that some elements have no such scaling: because $f(0)=a_n$ in Eq. (7.13), and $a_n \to a_\infty$, which is a definite nonzero number, the distance from the origin of the element of the 2^n-cycle farthest to the right certainly has not been reduced by $-\alpha$ at each period doubling. This suggests that we must measure locations of elements on the far right with respect to the farthest right point. If we do this, we can see that these distances scale by α^2, since they are the images through the quadratic maximum of f at $x=0$ of elements close to $x=0$ scaling with $-\alpha$. In fact, if we image g_1 through the maximum of f (through a quadratic conjugacy), then we shall indeed obtain a new universal function that locates cycle elements near the right-most element. The correct description of a highly doubled cycle now emerges as one of universal local clusters [2].

We can state the scope of universality for the location of cycle elements precisely. Since $f(\lambda_1, x)$ exactly locates the two elements of the 2^1-cycle, and since $f(\lambda_1, x)$ is an approximation to g_1 ($n=0$ in Eq. (7.35)), we evidently can locate both points exactly by appropriately scaling g_1. Next, near $x=0$, $f^2(\lambda_2, x)$ is a better approximation to g_1 (suitably scaled). However, in general, the more accurately we scale g_1 to determine the smallest 2-cycle elements, the greater is the error in its determination of the right-most elements. Again, near $x=0$, $f^4(\lambda_3, x)$ is a still better approximation to g_1. Indeed, the suitably scaled g_1 now can determine several points about $x=0$ accurately, but determination of the right-most elements is still worse. In this fashion, it follows that g_1, suitably scaled, can determine 2^r-points of the 2^n cycle near $x=0$ for $r \ll n$. If we focus on the neighbourhood of one of these 2^r points at some definite distance from $x=0$, then by Eq. (7.35) the larger the n, the larger the scaled distance of this region from $x=0$, and so the poorer the approximation of the location of fixed points in it by g_1. However, just as we can construct the version of g_1 that applies at the right-most cycle element, we also can construct the version of g_1 that applies at this chosen neighbourhood. Accordingly, the universal description is set through an acceptable tolerance: if we 'measure' f^{2^n} at some definite n, then we can use the actual location of the elements as foci for 2^n versions of g_1, each applicable at one such point. For all further period doubling, we determine the new cycle elements through the g_1's.

In summary, the more accurately we care to know the locations of arbitrarily high-order cycle elements, the more parameters we must measure (namely, the cycle elements at some chosen order of period doubling). This is the sense in which the universality theory is asymptotic. Its ability to have serious predictive power is the fortunate consequence of the high convergence rate δ (~4.67). Thus, typically after the first two or three period doublings, this asymptotic theory is already accurate within several percent. If a period-doubling system is measured in its 4- or 8-cycle, its behaviour throughout and symmetrically beyond the period-doubling regime also is determined to within a few percent.

To make precise dynamical predictions, we do not have to construct all the local versions of g_1; all we really need to know is the local scaling everywhere along the attractor. The scaling is $-\alpha$ at $x=0$ and α^2 at the right-most element. But what is it at an arbitrary point? We can determine the scaling law if we order elements not by their location on the x-axis, but rather by their order as iterates of $x=0$. Because the time sequence in which a process evolves is precisely this ordering, the result will be of immediate and powerful predictive value. It is precisely this scaling law that allows us to compute the spectrum of the onset of turbulence in period-doubling systems [5].

What must we compute? First, just as the element in the 2^n-cycle nearest to $x=0$ is the element halfway around the cycle from $x=0$, the element nearest to an arbitrarily chosen element is precisely the one halfway around the cycle from it. Let us denote by $d_n(m)$ the distance between the mth cycle element (x_m) and the element nearest to it in a 2^n-cycle. (The d_n of Eq. (7.28) is $d_n(0)$.) As just explained,

$$d_n(m) = x_m - f^{2^{n-1}}(\lambda_n, x_m). \tag{7.52}$$

However, x_m is the mth iterate of $x_0 = 0$. Recalling from Eq. (7.6) that powers commute, we find

$$d_n(m) = f^m(\lambda_n, 0) - f^m[\lambda_n, f^{2^{n-1}}(\lambda_n, 0)]. \tag{7.53}$$

Let us, for the moment, specialize to m of the form 2^{n-r}, in which case

$$\begin{aligned} d_n(2^{n-r}) &= f^{2^{n-r}}(\lambda_n, 0) - f^{2^{n-r}}[\lambda_n, f^{2^{n-1}}(\lambda_n, 0)] \\ &= f^{2^{n-r}}(\lambda_{(n-r)+r}, 0) - f^{2^{n-r}}[\lambda_{(n-r)+r}, f^{2^{n-1}}(\lambda_n, 0)]. \end{aligned} \tag{7.54}$$

For $r \ll n$ (which can still allow $r \gg 1$ for n large), we have, by Eq. (7.39),

$$d_n(2^{n-r}) \sim (-\alpha)^{-(n-r)}\{g_r(0) - g_r[(-\alpha)^{n-r} f^{2^{n-1}}(\lambda_n, 0)]\}$$

or

$$d_n(2^{n-r}) \sim (-\alpha)^{-(n-r)}\{g_r(0) - g_r[(-\alpha)^{-r+1} g_1(0)]\}. \tag{7.55}$$

The object we want to determine is the local scaling at the mth element, that is, the ratio of nearest separations at the mth iterate of $x = 0$, at successive values of n. That is, if the scaling is called σ,

$$\sigma_n(m) \equiv \frac{d_{n+1}(m)}{d_n(m)}. \tag{7.56}$$

(Observe by Eq. (7.28), the definition of α, that $\sigma_n(0) \sim (-\alpha)^{-1}$.) Specializing again to $m = 2^{n-r}$, where $r \ll n$, we have by Eq. (7.55)

$$\sigma(2^{n-r}) \sim \frac{g_{r+1}(0) - g_{r+1}[(-\alpha)^{-r} g_1(0)]}{g_r(0) - g_r[(-\alpha)^{-r+1} g_1(0)]}. \tag{7.57}$$

Finally, let us rescale the axis of iterates so that all 2^{n+1} iterates are within a unit interval. Labelling this axis by t, the value of t of the mth element in a 2^n-cycle is

$$t_n(m) = m/2^n. \tag{7.58}$$

In particular, we have

$$t_n(2^{n-r}) = 2^{-r}. \tag{7.59}$$

Defining σ along the t-axis naturally as

$$\sigma(t_n(m)) \sim \sigma_n(m) \qquad \text{as } n \to \infty,$$

we have by Eqs (7.57) and (7.59),

$$\sigma(2^{-r-1}) = \frac{g_{r+1}(0) - g_{r+1}[(-\alpha)^{-r} g_1(0)]}{g_r(0) - g_r[(-\alpha)^{-r+1} g_1(0)]}. \tag{7.60}$$

It is not much more difficult to obtain σ for all t. This is done first for rational t by writing t in its binary expansion:

$$t_{r_1 r_2 r_3 \ldots} = 2^{-r_1} + 2^{-r_2} + \ldots.$$

In the 2^n-cycle approximation we require σ_n at the $2^{n-r_1} + 2^{n-r_2} + \ldots$ iterate of the origin. But, by Eq. (7.8),

$$f^{2^{n-r_1} + 2^{n-r_2} + \ldots} = f^{2^{n-r_1}} \circ f^{2^{n-r_2}} \ldots.$$

It follows by manipulations identical to those that led from Eq. (7.54) to

Eq. (7.60) that σ at such values of t is obtained by replacing the individual g_r terms in Eq. (7.60) by appropriate iterates of various g_r's.

There is one last ingredient to the computation of σ. We know that $\sigma(0) = -\alpha^{-1}$. We also know that $\sigma_n(1) \sim \alpha^{-2}$. But, by Eq. (7.59),

$$t_n(1) = 2^{-n} \rightarrow 0.$$

Thus σ is discontinuous at $t = 0$, with $\sigma(0-\varepsilon) = -\alpha^{-1}$ and $\sigma(0+\varepsilon) = \alpha^{-2}(\varepsilon \rightarrow 0^+)$. Indeed, since $x_{2^{n-r}}$ is always very close to the origin, each of these points is imaged quadratically. Thus Eq. (7.60) actually determines $\sigma(2^{-r-1} - \varepsilon)$, while $\sigma(2^{-r-1} + \varepsilon)$ is obtained by replacing each numerator and denominator g_r by its square. The same replacement also is correct for each multi-g_r term that figures into σ at the binary expanded rationals [6].

Altogether, we have the following results. $\sigma(t)$ can be computed for all t, and it is universal since its explicit computation depends only upon the universal functions g_r. σ is discontinuous at all the rationals. However, it can be established that the larger the number of terms in the binary expansion of a rational t, the smaller the discontinuity of σ. Lastly, as a finite number of iterates leaves t unchanged as $n \rightarrow \infty$, σ must be continuous except at the rationals. Figure 7.10 depicts $1/\sigma(t)$. Despite the pathological nature of σ, the reader will observe that basically it is constant half the time at α^{-1} and half the time at α^{-2} for $0 < t < \frac{1}{2}$. In a succeeding approximation, it can be decomposed in each half into two slightly different quarters, and so forth. (It is easy to verify from Eq. (7.52) that σ is periodic in t of period 1, and has the symmetry

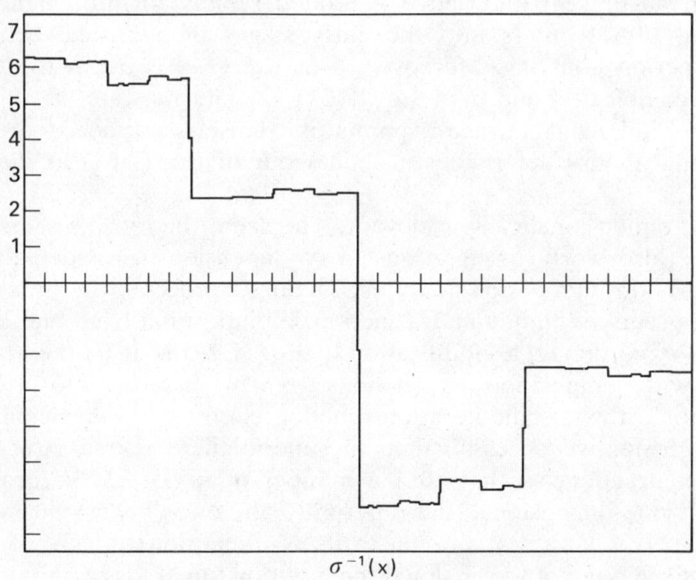

Fig. 7.10 The trajectory scaling function: $\sigma(x + \frac{1}{2}) = -\sigma(x)$

$\sigma(t+\frac{1}{2})=-\sigma(t)$. Accordingly, we have paid attention to its first half $0<t<\frac{1}{2}$.) With σ we are at last finished with one-dimensional iterates *per se*.

7.8 Universal behaviour in higher dimensional systems

So far we have discussed iteration in one variable; Eq. (7.15) is the prototype. Equation (7.14), an example of iteration in two dimensions, has the special property of preserving areas. A generalization of Eq. (7.14),

$$x_{n+1}=y_n-x_n^2$$

and

$$y_{n+1}=a+bx_n \tag{7.61}$$

with $|b|<1$, contracts areas. Equation (7.61) is interesting because it possesses a so-called *strange attractor.* This means an attractor (as before) constructed by folding a curve repeatedly upon itself (Fig. 7.11) with the consequent property that two initial points very near to one another are, in fact, very far from each other when the distance is measured along the folded attractor, which is the path they follow upon iteration. This means that after some iteration they will soon be far apart in actual distance as well as when measured along the attractor. This general mechanism gives a system highly sensitive dependence upon its initial conditions and a truly statistical character: since very small differences in initial conditions are magnified quickly, unless the initial conditions are known to infinite precision, all present knowledge is eroded rapidly to future ignorance. Now, Eq. (7.61) enters into the early stages of statistical behaviour through period doubling. Moreover, δ of Eq. (7.3) is *again* the rate of onset of complexity, and α of Eq. (7.31) is again the rate at which the spacing of adjacent attractor points is vanishing. Indeed, the one-dimensional theory determines all behaviour of Eq. (7.61) in the onset regime.

In fact, dimensionality is irrelevant. The same theory, the same numbers, etc., also work for iterations in N dimensions, provided that the system goes through period doubling. The basic process, wherever period doubling occurs *ad infinitum*, is functional composition from one level to the next. Accordingly, a modification of Eq. (7.29) is at the heart of the process, with composition on functions from N dimensions to N dimensions. Should the specific iteration function contract N-dimensional volumes (a dissipative process), then in general there is one direction of slowest contraction, so that after a number of iterations the process is effectively one-dimensional. Put differently, the one-dimensional solution to Eq. (7.29) is always a solution to its N-dimensional analog. It is the relevant fixed point of the analog if the iteration function is contractive [7].

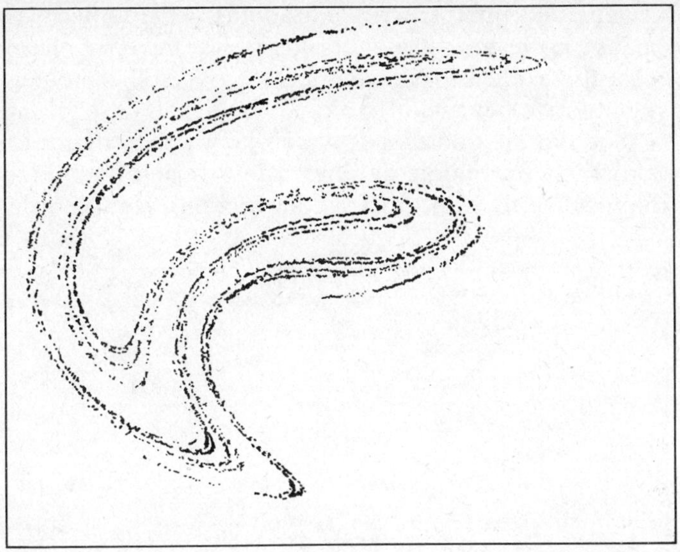

Fig. 7.11 The plotted points lie on the 'strange attractor' of Duffing's equation

7.9 Universal behaviour in differential systems

The next step of generalization is to include systems of differential equations. A prototypic equation is Duffing's oscillator, a driven damped anharmonic oscillator,

$$\ddot{x} + k\dot{x} + x^3 = b \sin 2\pi t. \tag{7.62}$$

The periodic drive of period 1 determines a natural time step. Figure 7.12(a) depicts a period 1 attractor, usually referred to as a *limit cycle*. It is an attractor because, for a range of initial conditions, the solution to Eq. (7.62) settles down to the cycle. It is period 1 because it repeats the same curve in every period of the drive. Figures 7.12(b) and (c) depict attractors of periods 2 and 4 as the friction or damping constant k in Eq. (7.62) is reduced systematically. The parameter values $k = \lambda_0, \lambda_1, \lambda_2, \ldots,$ are the damping constants corresponding to the most stable 2^n-cycle in analogy to the λ_n of the one-dimensional functional iteration. Indeed, this oscillator's period doubles (at least numerically) *ad infinitum*. In fact, by $k = \lambda_5$, the δ_3 of Eq. (7.2) has converged to 4.69. Why is this?

Instead of considering the entire trajectories, as shown in Fig. 7.12, let us consider only where the trajectory point is located every 1 period of the drive. The 1-cycle then produces only one point, while the 2-cycle produces a pair of points, and so forth. This *time-one map* (if the trajectory point is $(x, \dot{x})$ now, where is it one period later?) is by virtue of the differential equation a smooth and invertible function in two dimensions. Qualitatively, it looks like the map of Eq. (7.61). In the present state of mathematics, little can be said about the analytic behaviour of time-one maps; however, since our theory is universal, it makes no difference that the explicit form is not known. We still can determine the complete quantitative behaviour of Eq. (7.62) in the onset regime where the motion tends to aperiodicity. If we already know, by measurement, the precise form of the trajectory after a few period doublings, we can compute the form of the trajectory as the friction is reduced throughout

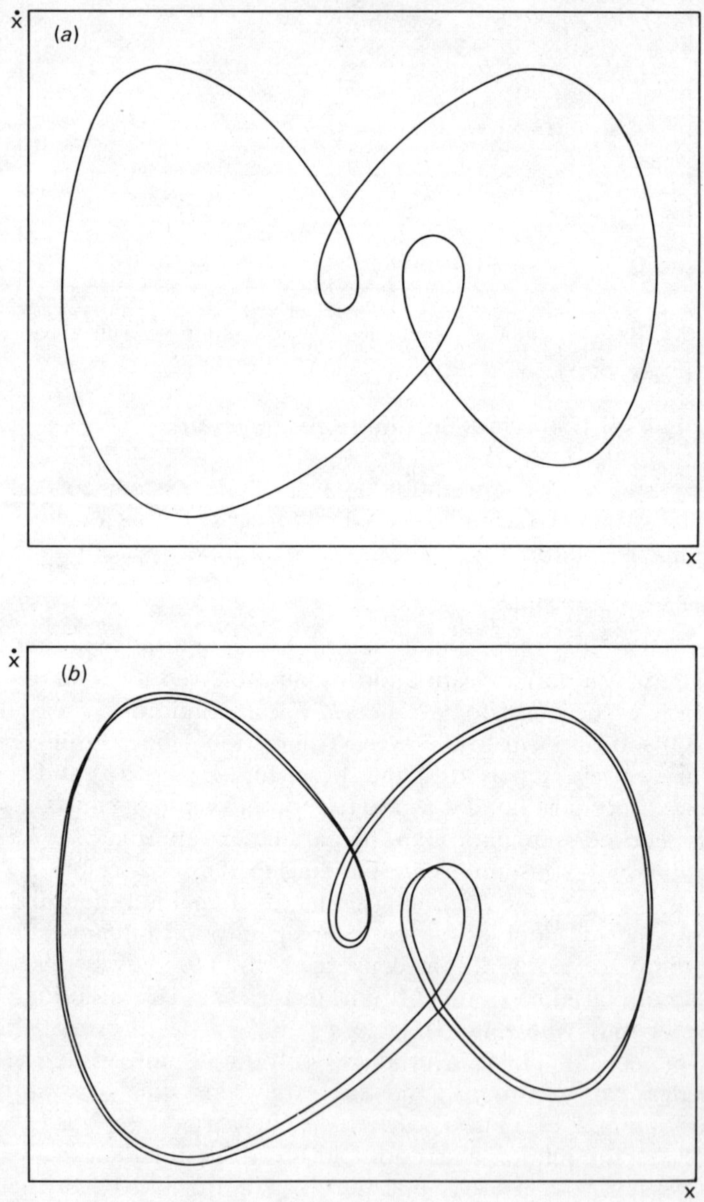

Fig. 7.12 (*a*) The most stable 1-cycle of Duffing's equation in phase space $(x, \dot{x})$
(*b*) The most stable 2-cycle of Duffing's equation—observe that it is two displaced copies of (*a*)
(*c*) The most stable 4-cycle of Duffing's equation—observe that the displaced copies of (*b*) have either a broad or a narrow separation

the region of onset of complexity by carefully using the full power of the universality theory to determine the spacings of elements of a cycle.

Let us see how this works in some detail. Consider the time-one map of the Duffing's oscillator in the superstable 2^n-cycle. In particular, let us focus on an element at which the scaling function σ (Fig. 7.10) has the value σ_0, and for which the next iterate of this element also has the scaling σ_0. (The element is not at a big discontinuity of σ.) It is then intuitive that if we had taken our time-one examination of the trajectory at values of time displaced from our first choice, we would have seen the same scaling σ_0 for this part of the trajectory. That is, the differential equations will extend the map-scaling function continuously to a function along the entire trajectory so that, if two successive time-one elements have scaling σ_0, then the entire stretch of trajectory over this unit time interval has scaling σ_0. In the last section, we were motivated to construct σ as a function of t along an interval precisely towards this end.

To implement this idea, the first step is to define the analog of d_n. We require the spacing between the trajectory at time t and at time $T_n/2$ where the period of the system in the 2^n-cycle is

$$T_n \cong 2^n T_0. \tag{7.63}$$

That is, we define

$$d_n(t) \equiv x_n(t) - x_n\left(t + \frac{T_n}{2}\right) \tag{7.64}$$

(There is a d for each of the N variables for a system of N differential equations.) Since σ was defined as periodic of period 1, we now have

$$d_{n+1}(t) \sim \sigma\left(\frac{t}{T_{n+1}}\right) d_n(t). \tag{7.65}$$

The content of Eq. (7.65), based on the n-dependence arising solely through the T_n in σ, and not on the detailed form of σ, already implies a strong scaling prediction, in that the ratio

$$\frac{d_{n+1}(t)}{d_n(t)},$$

when plotted with t scaled so that $T_n = 1$, is a function independent of n. Thus if Eq. (7.65) is true for some σ, whatever it may be, then knowing $x_n(t)$, we can compute $d_n(t)$ and, from Eq. (7.65), $d_{n+1}(t)$. As a consequence of periodicity, Eq. (7.64) for $n \to n+1$ can be solved for $x_{n+1}(t)$ (through a Fourier transform). That is, if we have measured any chosen coordinate of the system in its 2^n-cycle, we can compute its time-dependence in the 2^{n+1}-cycle. Because this procedure is recursive, we can compute the coordinate's evolution for all higher cycles through the infinite period-doubling limit. If Eq. (7.65) is true and σ not known, then by measurement at a 2^n-cycle and at a 2^{n+1}-cycle, σ could be constructed from Eq. (7.65), and hence all higher order doublings would again be determined. Accordingly, Eq. (7.65) is a very powerful result. However, we know much more. The universality theory tells us that period doubling is universal and that there is a unique function σ which, indeed, we have computed in the previous section. Accordingly, by measuring $x(t)$ in some chosen 2^n-cycle (the higher the n, the greater the number of effective parameters to be determined empirically, and the more precise are the predictions), we now can compute the entire evolution of the system on its route to turbulence.

How well does this work? The empirically determined σ (for Eq. (7.62)) of Eq. (7.65) is shown for $n = 3$ in Fig. 7.13(a) and $n = 4$ in Fig. 7.13(b). The figures were constructed by plotting the ratios of d_{n+1} and d_n scaled respective to $T = 16$ in Fig. 7.13(a) and $T = 32$ in Fig. 7.13(b). Evidently the scaling law Eq. (7.65) is being obeyed. Moreover, on the same graph Fig. 7.14 shows the empirical σ for $n = 4$ and the recursion theoretical σ of Fig. 7.10. The reader should observe the detail-by-detail agreement of the two. In fact, if we use Eq. (7.65) and the theoretical σ with $n = 2$ as empirical input, the $n = 5$ frequency spectrum agrees with the empirical $n = 5$ spectrum to within 10%. (The $n = 4$ determines $n = 5$ to within 1%.) Thus the asymptotic universality theory is correct and is already well obeyed, even by $n = 2$ [5].

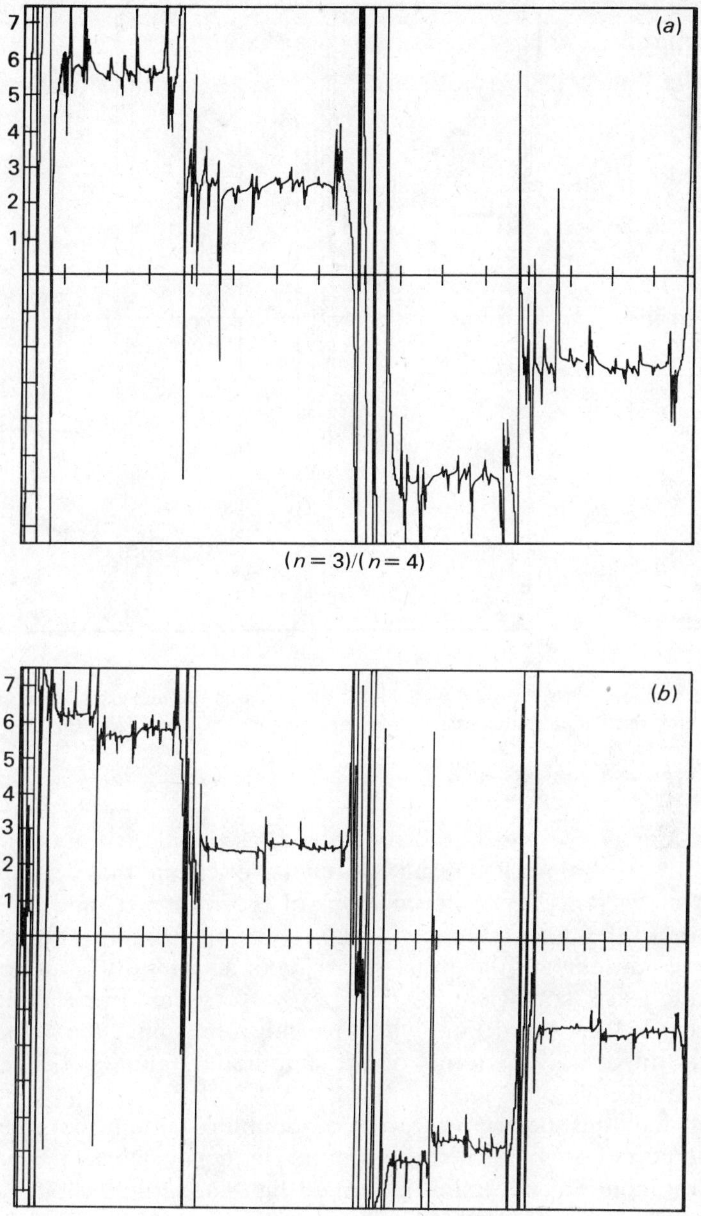

Fig. 7.13 (*a*) Ratio of nearest copy separations in the 8-cycle and the 16-cycle for Duffing's equation
(*b*) As (*a*), but for the 16-cycle and the 32-cycle. Here the axis is twice as compressed

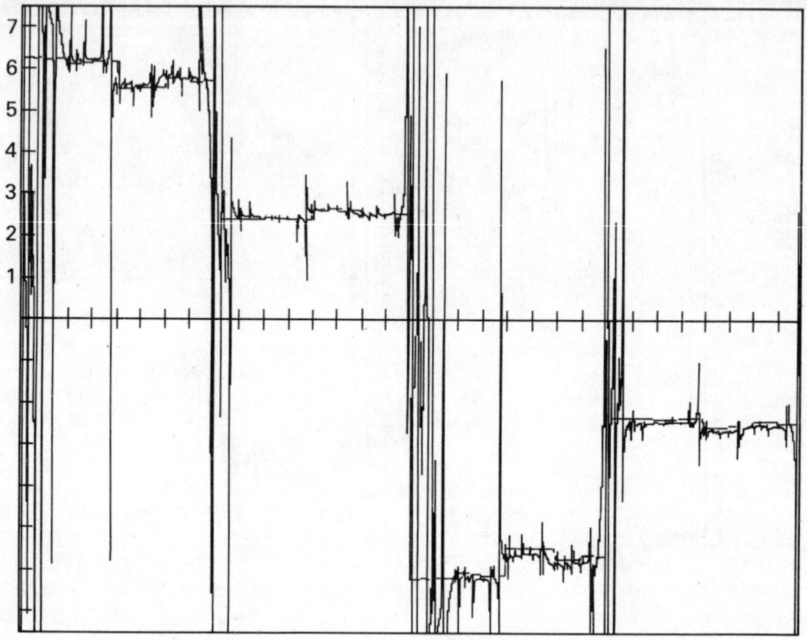

Fig. 7.14 Figure 7.13(*b*) overlaid with Fig. 7.10 compares the universal scaling function σ with the empirically determined scaling of nearest copy separations from the 16-cycle to the 32-cycle for Duffing's equation

Equations (7.64) and (7.65) are solved, as mentioned above, through Fourier transforming. The result is a recursive scheme that determines the Fourier coefficients of $x_{n+1}(t)$ in terms of those of $x_n(t)$ and the Fourier transform of the (known) function $\sigma(t)$. To employ the formula accurately requires knowledge of the entire spectrum of x_n (amplitude and phase) to determine each coefficient of x_{n+1}. However, the formula enjoys an approximate local prediction, which roughly determines the amplitude of a coefficient of x_{n+1} in terms of the amplitudes (alone) of x_n near the desired frequency of x_{n+1}.

What does the spectrum of a period-doubling system look like? Each time the period doubles, the fundamental frequency halves; period doubling in the continuum version is termed half-subharmonic bifurcation, a typical behaviour of coupled nonlinear differential equations. Since the motion almost reproduces itself every period of the drive, the amplitude at this original frequency is high. At the first subharmonic halving, spectral components of the odd halves of the drive frequency come in. On the route to aperiodicity they saturate at a certain amplitude. Since the motion more nearly reproduces itself every two periods of drive, the next saturated subharmonics, at the odd fourths of the original frequency, are smaller still than the first ones, and so on, as each set of odd 2^nths

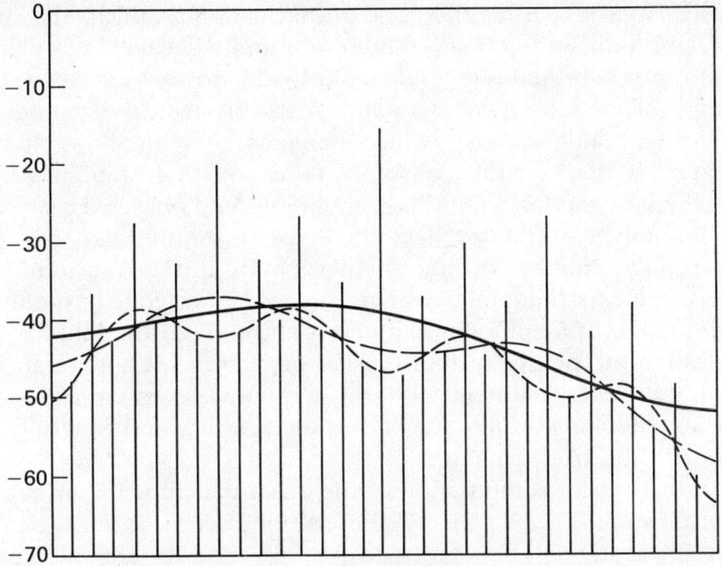

Fig. 7.15 Subharmonic spectrum of Duffing's equation in the 32-cycle
····· Interpolation of the odd 32nd subharmonics
- - - - Interpolation of the odd 16th subharmonics, lowered by 8.2 dB
— — — Interpolation of the odd 8th subharmonics, lowered by 16.4 dB
——— Interpolation of the odd 4th subharmonics, lowered by 24.6 dB

comes into being. A crude approximate prediction of the theory is that whatever the system, the saturated amplitudes of each set of successively lower half-frequencies define a smooth interpolation located 8.2 dB below the smooth interpolation of the previous half-frequencies. (This is shown in Fig. 7.15 for Eq. (7.62).) After subharmonic bifurcations *ad infinitum*, the system is now no longer periodic; it has developed a continuous broad spectrum down to zero frequency with a definite internal distribution of the energy. That is, the system emerges from this process having developed the beginnings of broadband noise of a determined nature. This process also occurs in the onset of turbulence in a fluid.

2.10 Onset of turbulence

The existing idea of the route to turbulence is Landau's 1941 theory. The idea is that a system becomes turbulent through a succession of instabilities, where each instability creates a new degree of freedom (through an indeterminate phase) of a time-periodic nature with the frequencies successively higher and incommensurate (not harmonics); because the resulting motion is the superposition of these modes, it is quasiperiodic.

In fact, it is experimentally clear that quasiperiodicity is incorrect. Rather, to produce the observed noise of rapidly decaying correlation the spectrum must become continuous (broadband noise) down to zero frequency. The defect can be eliminated through the production of successive half-subharmonics, which then emerge as an allowable route to turbulence. If the general idea of a succession of instabilities is maintained, the new modes do not have indeterminate phases. However, only a small number of modes need be excited to produce the required spectrum. (The number of modes participating in the transition is, as of now, an open experimental question.) Indeed, knowledge of the phases of a small number of amplitudes at an early stage of period doubling suffices to determine the phases of the transition spectrum. What is important is that a purely causal system can and does possess essentially statistical properties. Invoking *ad hoc* statistics is unnecessary and generally incompatible with the true dynamics.

A full theoretical computation of the onset demands the calculation of successive instabilities. The method used traditionally is perturbative. We start at the static solution and add a small time-dependent piece. The fluid equations are linearized about the static solution, and the stability of the perturbation is studied. To date, only the first instability has been computed analytically. Once we know the parameter value (for example, the Rayleigh number) for the onset of this first time-varying instability, we must determine the correct form of the solution after the perturbation has grown large beyond the linear regime. To this solution we add a new time-dependent perturbative mode, again linearized (now about a time-varying, nonanalytically available solution) to discover the new instability. To date, the second step of the analysis has been performed only numerically. This process, in principle, can be repeated again and again until a suitably turbulent flow has been obtained. At each successive stage, the computation grows successively more intractable.

However, it is just at this point that the universality theory solves the problem; it works only after enough instabilities have entered to reach the asymptotic regime. Since just two such instabilities already serve as a good approximate starting-point, we need only a few parameters for each flow to empower the theory to complete the hard part of the infinite cascade of more complex instabilities.

Why should the theory apply? The fluid equations make up a set of coupled field equations. They can be spatially Fourier-decomposed to an infinite set of coupled ordinary differential equations. Since a flow is viscous, there is some smallest spatial scale below which no significant excitation exists. Thus, the equations are effectively a finite coupled set of nonlinear differential equations. The number of equations in the set is completely irrelevant. The universality theory is generic for such a dissipative system of equations. Thus it is possible that the flow exhibits period doubling. If it does, then our theory applies. However, to prove that a given flow (or any flow) actually should exhibit doubling is well beyond present understanding. All we can do is experiment.

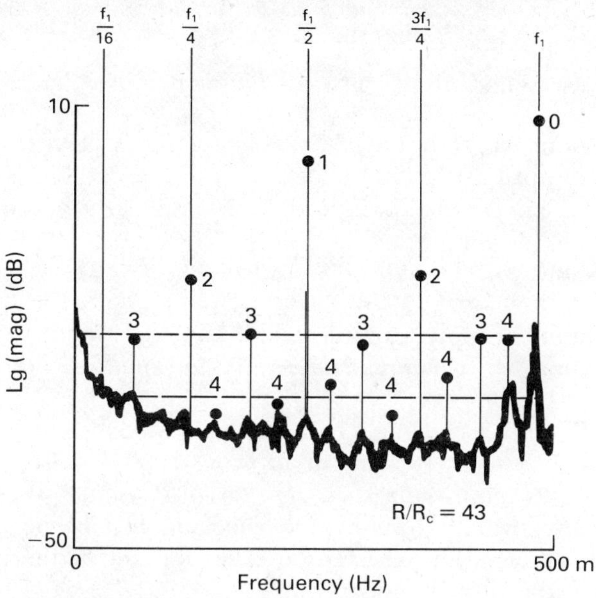

Fig. 7.16 Experimental spectrum of a convecting fluid at its transition to turbulence (after Libchaber and Maurer). The broken lines result from dropping a horizontal line down through the odd 4th subharmonics (labelled 2) by 8.2 and 16.4 dB

Figure 7.16 depicts the experimentally measured spectrum of a convecting liquid helium cell at the onset of turbulence [8]. The system displays measurable period doubling through four or five levels; the spectral components at each set of odd half-subharmonics are labelled with the level. With $n=2$ taken as asymptotic, the dotted lines show the crudest interpolations implied for the $n=3$, $n=4$ component. Given the small amount of amplitude data, the interpolations are perforce poor, while ignorance of higher odd multiples prevents construction of any significant interpolation at the right-hand side. Accordingly, to do the crudest test, the farthest right-hand amplitude was dropped, and the oscillations were smoothed away by averaging [9]. The experimental results, −8.3 dB and −8.4 dB, are in surprisingly good agreement with the theoretical 8.2.

From this good experimental agreement and the many period doublings as the clincher, we can be confident that the measured flow has made its transition according to our theory. A measurement of δ from its fundamental definition would, of course, be altogether convincing. (Experimental resolution is insufficient at present. [10]) However, if we work backwards, we find that the several percent agreement in 8.2 dB is an experimental observation of α in the system to the same accuracy. Thus, the present method has provided a theoretical calculation of the actual dynamics in a field where such a feat has been impossible since the construction of the Navier–Stokes equations. In fact, the scaling law Eq. (7.65) transcends these equations, and applies to the true equations, whatever they may be.

References

1. *Annual Report of the Theoretical Division, Los Alamos*, LA-6816-PR, 1976.
2. Feigenbaum, M., *Journal of Statistical Physics*, **19,** 25, 1978 (first circulated April, 1976).
3. See May, R., *Nature*, **261,** 459, 1976 (an early account of these properties).
4. Feigenbaum, M., *Journal of Statistical Physics*, **21,** 669, 1979 (first circulated November, 1976).
5. Feigenbaum, M., *Physics Letters*, A **74,** 375, 1979.
6. Feigenbaum, M., *Communications in Mathematical Physics*, **77,** 65, 1980.
7. Collet, P. and Eckmann, J-P., *Iterated Maps on the Interval as Dynamical Systems*, Birkhaüser, Boston, 1980.
8. Libchaber, A. and Maurer, J., *Journal de Physique*, **41,** C3, 1980.
9. Feigenbaum, M., in *Nonlinear Phenomena in Chemical Dynamics*, Springer-Verlag, Berlin, 1981 (a fuller account of the power spectrum than that in [5] and [6]).
10. Since the time of writing, new and more accurate fluid experiments have also measured δ. See:
    Giglio, M., Muzzati, S. and Perini, U., *Physical Review Letters*, **47,** 243, 1981.
    Libchaber, A., Laroche, C. and Fauve, S., *Journal de Physique*, **43,** L211, 1982.

8 Asymptotic Numerical Analysis for the Navier–Stokes Equations

C Foias and R Temam

8.1 Introduction

Up to now the numerical analysis of the Navier–Stokes equations has been limited to the study of the approximation of time-dependent solutions on a finite interval of time or to the approximation of stationary solutions (*cf.*, among many others, [1–4]). In the presence of a turbulent flow driven by a steady excitation, a different type of problem naturally arises: the study of the long-term behaviour of the solutions.

The purpose of this chapter is to show that for the two-dimensional Navier–Stokes equations and, under some circumstances, for the three-dimensional Navier–Stokes equations, there is a theoretical basis for determining the qualitative long-term behaviour of a fluid by the study of a finite number of adequate modes. A typical result is the following. Let W be a finite-dimensional subspace of the natural function space V. If W satisfies a certain condition, then the behaviour for $t \to \infty$ of a solution u of the Navier–Stokes equation is completely determined by the behaviour for $t \to \infty$ of *its projection* on W.

Several results of this type are derived in this chapter. While this kind of problem has already been discussed in [5], our present interest was aroused and inspired by the questions, conjectures and ideas of O. P. Manley and Y. M. Treve [6–8], with whom we have had fruitful discussions and correspondence.

8.2 Notation and recapitulation of results

Let Ω be a bounded domain of $\mathbb{R}^l$, $l = 2$ or 3, and let Γ be its boundary. We assume that Γ is a manifold of dimension $l-1$, of class $\mathscr{C}^4$ with a finite number of connected components, and that Ω is locally located on one side of Γ. We shall first consider the initial value problem for the

Navier–Stokes equations:

$$\frac{\partial u}{\partial t} - \nu\,\Delta u + (u \cdot \nabla)u + \nabla p = f \quad \text{in} \quad \Omega \tag{8.1}$$

$$\nabla u = 0 \quad \text{in} \quad \Omega \tag{8.2}$$

$$u = 0 \quad \text{on} \quad \Gamma \tag{8.3}$$

$$u\,|_{t=0} = u_0, \tag{8.4}$$

where $\nu > 0$ is the kinematic viscosity; $u = (u_1, u_2)$ or (u_1, u_2, u_3) is the velocity, a vector-valued function of $x \in \Omega$ and $t \geqslant 0$; p is the pressure, $p = p(x, t)$; and f represents the external body force per unit of mass.

All that follows applies to the case where (8.3) is replaced by a nonhomogeneous boundary condition that corresponds more closely to real physical situations (Couette–Taylor flow, etc.). We will refrain from treating this case to avoid purely technical difficulties.

We denote by $L^2(\Omega)$ the space of square integrable functions on Ω and by $H_0^1(\Omega)$ the Sobolev space made up of the functions which are in $L^2(\Omega)$ together with their first derivatives and which vanish on Γ. We put, for u in $L^2(\Omega)$ or $L^2(\Omega)^l$ (u in $H_0^1(\Omega)$ or $H_0^2(\Omega)^l$, respectively),

$$|u|^2 = \int_\Omega |u(x)|^2\,\mathrm{d}x, \qquad \left(\text{resp. } \|u\|^2 = \int_\Omega |\nabla u(x)|^2\,\mathrm{d}x\right).$$

The space $L^2(\Omega)^l$ admits classically an orthogonal decomposition of the form $H \oplus G$, where

$$G = \{v = \nabla q,\ q \in H^1(\Omega)\}$$

and its orthogonal H satisfies

$$H = \{v \in L^2(\Omega)^l,\ \operatorname{div} v = 0,\ v \cdot n\,|_\Gamma = 0\},$$

n being the unit outward normal of Γ. Let also

$$V = \{v \in H_0^1(\Omega)^l,\ \operatorname{div} v = 0\}.$$

It is clear that H and V are Hilbert spaces for the norms $|\cdot|$ and $\|\cdot\|$, and their corresponding scalar products. While $\frac{1}{2}|v|^2$ is equal to the kinetic energy of the fluid with velocity v (the density $\rho = 1$), we recall that for $v \in V$, $\|v\|$ also reduces to a physical quantity:

$$\|v\|^2 = \int_\Omega |\operatorname{curl} v(x)|^2\,\mathrm{d}x.$$

Let Π denote the orthogonal projection of $L^2(\Omega)^l$ onto H and define the operators A and B by

$$Au = -\Pi\,\Delta u, \quad \text{for } u \in D(A) = V \cap H^2(\Omega)^l, \tag{8.5}$$

$$B(u, v) = \Pi[(u \cdot \nabla)v)], \quad \text{for } u,\ v \in D(A). \tag{8.6}$$

Then A is a self-adjoint operator in H with an orthonormal basis

$\{w_m\}_{m\geq 1}$ of eigenvectors, such that

$$Aw_m = \lambda_m w_m, \quad m \geq 1, 0 < \lambda_1 \leq \lambda_2 \leq \ldots, \tag{8.7}$$

$\lambda_m \to +\infty$ as $m \to +\infty$ and

$$V = D(A^{1/2}), \qquad \|u\| = |A^{1/2}u| \quad \text{for } u \in V. \tag{8.8}$$

The operator A is an isomorphism from $D(A)$ onto H and from V onto V' (the dual of V which one can identify as a superspace of H). We recall that B is a compact mapping from $D(A)\times V$ or $V\times D(A)$ into H and from $V\times V$ into V'. Furthermore, we have the estimates

$$\|B(u, v)\|_{V'} \leq \begin{cases} \sqrt{2}\,|u|^{1/2}\,\|u\|^{1/2}\,|v|^{1/2}\,\|v\|^{1/2} & \text{if } l=2 \\ 6\,|u|^{1/4}\,\|u\|^{3/4}\,|v|^{1/4}\,\|v\|^{3/4} & \text{if } l=3 \end{cases} \tag{8.9}$$

$$|B(u, v)| \leq c_0 \begin{cases} |u|^{1/2}\,|Au|^{1/2}\,\|v\| \\ |u|^{1/2}\,\|u\|^{1/2}\,\|v\|^{1/2}\,|Av|^{1/2} \end{cases} \tag{8.10}$$

if $l=2$; if $l=2$ or 3,

$$|B(u, v)| \leq c_1 \begin{cases} |Au|\,\|v\| \\ \|u\|\,|Av|. \end{cases} \tag{8.11}$$

The constants c_0 and c_1 depend only on Ω but are not easy to determine since they involve the norm of the operator A^{-1}, and Sobolev constants. All the above results can be found, for instance, in [4, 9].

We can now recall the functional formulation of (8.1)–(8.4): this is the functional differential equation

$$\frac{du}{dt} + \nu Au + B(u, u) = f, \qquad u|_{t=0} = u_0, \tag{8.12}$$

where we shall assume for simplicity $u_0 \in V$ and

$$\left.\begin{aligned} &f \text{ is continuous and bounded from } [0, \infty) \text{ into } H, \\ &f' = \frac{df}{dt} \text{ belongs to } L^2_{\text{loc}}\,(0, \infty; V'). \end{aligned}\right\} \tag{8.13}$$

It is well known that, if $l=2$, there exists a unique function u such that

$$u \in \mathscr{C}([0, \infty); H) \cap L^2_{\text{loc}}([0, \infty); V), \tag{8.14}$$

satisfying Eq. (8.12) in V' (*cf.*, for instance, [4]). Moreover, u is actually a continuous function from $(0, \infty)$ into $D(A)$, which is bounded in $D(A)$ as $t \to +\infty$. We set

$$\|u(t)\| \leq c_2 = c_2(|u_0|, [f], 1/\nu, c_0, \lambda_1), \quad \text{for } t \geq 0, \tag{8.15}$$

$$|Au(t)| \leq c_3 = c_3(|u_0|, [f], 1/\nu, c_0, \lambda_1, 1/\alpha) \quad \text{for } t \geq \alpha > 0, \tag{8.16}$$

where

$$[f] = \sup_{0<t<\infty} |f(t)|. \tag{8.17}$$

The estimate (8.15) is given in [5]; the more recent estimate (8.16) is given in [10] (*cf.* also [11]).

If $l=3$, there exists a weak solution of (8.12) bounded in H. Such a solution may or may not be bounded in V. We will only consider such a solution if it is bounded in V:

$$\|u(t)\| \leqslant R \quad \text{for } t \geqslant 0. \tag{8.18}$$

In this case it follows also from [10–12] that $u(t)$ belongs to $D(A)$ as $t \to \infty$:

$$|Au(t)| \leqslant c_4(R, [f], 1/\nu, c_1, \lambda_1, 1/\alpha) \quad \text{for } t \geqslant \alpha > 0. \tag{8.19}$$

In the following section we will let $l=2$ or 3, and consider solutions of (8.12) which are uniformly bounded in V on $[0, \infty)$; the existence of $R < \infty$ is an assumption if $l=3$, and is automatic if $l=2$ ($R=c_2$, *cf.* (8.15)). See Remark 8.2 at the end of the chapter.

8.3 Approximation in the subspaces V_m

8.3.1 An inequality

For $m \in \mathbb{N}$, we denote by V_m the space spanned by the eigenfunctions $w_1, \ldots, w_m$, of A, and by P_m the projector in V, H or V' onto V_m. Finally, $Q_m = I - P_m$. It is easy to see that

$$\|\varphi\| \leqslant \lambda_{m+1}^{1/2} |\varphi|, \forall \varphi \in V_m, \tag{8.20}$$

$$|\psi| \leqslant \lambda_{m+1}^{-1/2} \|\psi\|, \forall \psi \in Q_m V. \tag{8.21}$$

Let $u(\cdot)$ and $v(\cdot)$ be solutions of the equations

$$u' + \nu Au + B(u, u) = f \qquad t > 0, u(0) = u_0, \tag{8.22}$$

$$v' + \nu Av + B(v, v) = g \qquad t > 0, v(0) = v_0, \tag{8.23}$$

where $v_0 \in V$ and g satisfies the same assumption (8.13) as f. We set

$$w = u - v, \quad p_m = P_m w, \quad q_m = Q_m w, \quad e = f - g, \quad e_m = Q_m e.$$

Then

$$q_m' + \nu A q_m + Q_m B(v, w) + Q_m B(w, u) = e_m$$

and consequently, recalling that $(B(\varphi, \psi), \theta) = -(B(\varphi, \theta), \psi)$ for θ, φ, ψ in V,

$$\frac{1}{2}\frac{\mathrm{d}}{\mathrm{d}t}|q_m|^2 + \nu \|q_m\|^2 = -(B(v, p_m), q_m) - (B(p_m, u), q_m)$$
$$-(B(q_m, u), q_m) + (e_m, q_m). \tag{8.24}$$

Taking into account (8.11), the right-hand side of (8.24) can be bounded by

$$\lambda_1^{-1/2} |e_m| \|q_m\| + c_1(|Au| + |Av|) |p_m| \|q_m\| + c_1 |Au| |q_m| \|q_m\|$$

which (for $\varepsilon > 0$ arbitrary)

$$\leq \frac{\varepsilon\nu}{2}\|q_m\|^2 + \frac{\lambda_1^{-1}}{2\varepsilon\nu}|e_m|^2 + \frac{\varepsilon\nu}{2}\|q_m\|^2 + \frac{c_1^2}{2\varepsilon\nu}(|Au|+|Av|)^2|p_m|^2$$
$$+c_1\lambda_{m+1}^{-1/2}|Au|\,\|q_m\|^2$$

which, by (8.16), (8.19),

$$\leq \varepsilon\nu\|q_m\|^2 + \frac{\lambda_1^{-1}}{2\varepsilon\nu}|e_m|^2 + \frac{2c_1^2c_4^2}{\varepsilon\nu}|p_m|^2 + c_1c_4\lambda_{m+1}^{-1/2}\|q_m\|^2$$

for $t \geq \alpha$. (Here c_i, c_i', c_i'', . . . , denote various positive constants. The c_i's are the same all the time; c_i', c_i'', . . . , may represent different quantities at different places.)

We have then proved the following:

Lemma 8.1. *If m is large enough, so that*

$$\lambda_{m+1} > \frac{c_1^2c_4^2}{\nu^2}, \tag{8.25}$$

then, for $t \geq \alpha > 0$,

$$\frac{\mathrm{d}}{\mathrm{d}t}|q_m|^2 + \nu'\|q_m\|^2 \leq \frac{\lambda_1^{-1}}{\varepsilon\nu}|e_m|^2 + \frac{4c_1^2c_4^2}{\varepsilon\nu}|p_m|^2,$$

$$\nu' = \nu - c_1c_4\lambda_{m+1}^{-1/2} > 0, \qquad \varepsilon = \frac{1}{2} - \frac{c_1c_4\lambda_{m+1}^{-1/2}}{2\nu}. \tag{8.26}$$

(Here we need $\lambda_{m+1} \geq (36c_1^2/\nu^2)\,(\sup_{t\geq\alpha}|Au(t)|^2)$ and actually it is sufficient to have $\lambda_{m+1} > (c_1^2/\nu^2)\limsup_{t\to\infty}|Au(t)|^2$.)

8.3.2 The main result

We introduce the following weak mode of convergence: we will say that $\varphi(t)$ converges essentially to 0 as $t\to\infty$ (we write $\varphi(t) \xrightarrow{\text{c.e.}} 0$) if:

$$\left.\begin{array}{l}\text{there exists } K>0 \text{ such that, for every } \delta>0, \text{ there exists } t_\delta \\ \text{such that, for } t\geq t_\delta, \\ \text{meas}\,\{\tau\in(t,t+1), |\varphi(\tau)|\geq\delta\}\leq K\delta.\end{array}\right\} \tag{8.27}$$

It is easy to see that if $\varphi(t)\to 0$ for $t\to\infty$, then $\varphi(t)\xrightarrow{\text{c.e.}}0$ for $t\to\infty$. Also, if $\varphi\in L^1(0,\infty)$, then $\varphi(t)\xrightarrow{\text{c.e.}}0$ as $t\to\infty$.

Theorem 8.1. *We assume that $l=2$ or that $l=3$ and that u and v are solutions of (8.22) and (8.23) uniformly bounded in V. We assume also* *that* (8.25) *is satisfied*:

(1) *If* $|P_m[u(t)-v(t)]|\to 0$, $|(I-P_m)[f(t)-g(t)]|\to 0$, *for* $t\to\infty$, *then*

$$|(I-P_m)[u(t)-v(t)]|\to 0 \text{ for } t\to\infty, \tag{8.28}$$

$$|u(t)-v(t)|\to 0 \text{ for } t\to\infty. \tag{8.29}$$

(2) *If* $|P_m[u(t)-v(t)]|^2 \xrightarrow{\text{c.e.}} 0$, $|(I-P_m)[f(t)-g(t)]|^2 \xrightarrow{\text{c.c.}} 0$, *for* $t\to\infty$, *then* (8.28) *still holds but, instead of* (8.29),

$$|u(t)-v(t)|^2 \xrightarrow{\text{c.e.}} 0, \quad \text{for } t\to\infty. \tag{8.30}$$

Proof. We infer from (8.21) and (8.26) that

$$\frac{\mathrm{d}}{\mathrm{d}t}|q_m|^2+\nu'\lambda_{m+1}^{1/2}|q_m|^2 \leqslant \frac{\lambda_1^{-1}}{\varepsilon\nu}|e_m|^2+\frac{4c_1^2c_4^2}{\varepsilon\nu}|p_m|^2,$$

whence, for $t\geqslant t_0\geqslant\alpha$,

$$|q_m(t)|^2\leqslant|q_m(t_0)|^2\exp[-\nu'\lambda_{m+1}^{1/2}(t-t_0)] + \int_{t_0}^{t}\left(\frac{\lambda_1^{-1}}{\varepsilon\nu}|e_m(\tau)|^2+\frac{4c_1^2c_4^2}{\varepsilon\nu}|p_m(\tau)|^2\right)\exp[-\nu'\lambda_{m+1}^{1/2}(t-\tau)]\,\mathrm{d}\tau. \tag{8.31}$$

There exist K, $K'>0$ and, for every $\delta>0$, there exists t_δ such that, for $t\geqslant t_\delta$,

$$\operatorname{meas}\{\tau\in(t,t+1), |P_m[u(\tau)-v(\tau)]|^2\geqslant\delta\}\leqslant K\delta$$
$$\operatorname{meas}\{\tau\in(t,t+1), |P_m[f(\tau)-g(\tau)]|^2\geqslant\delta\}\leqslant K'\delta.$$

For fixed δ we take $t\geqslant t_0+M$, $t_0\geqslant\max(t_\delta,\alpha)$, M a fixed integer. Then (8.31) implies that

$$\begin{aligned}|q_m(t)|^2\leqslant{}& c(u,v)\exp[-\nu'\lambda_{m+1}^{1/2}(t-t_0)]+\delta\left(\frac{\lambda_1^{-1}}{\varepsilon\nu}+\frac{4c_1^2c_4^2}{\nu}\right)\\ &\times\int_{t-M}^{t}\exp[-\nu'\lambda_{m+1}^{1/2}(t-\tau)]\,\mathrm{d}\tau+\left[\frac{\lambda_1^{-1}}{\varepsilon\nu}c(f,g)+\frac{4c_1^2c_4^2}{\varepsilon\nu}c(u,v)\right]\\ &\times\left\{\int_{t_0}^{t-M}\exp[-\nu'\lambda_{m+1}^{1/2}(t-\tau)]\,\mathrm{d}\tau+2\delta M\right\}\\ &\qquad+M\delta\left[K'\frac{\lambda_1^{-1}}{\varepsilon\nu}c(f,g)+K\frac{4c_1^2c_4^2}{\nu}c(u,v)\right]\end{aligned}$$

where

$$c(u,v)=\sup_{t\geqslant0}|u(t)-v(t)|,\qquad c(f,g)=\sup_{t\geqslant0}|f(t)-g(t)|.$$

Therefore, as $t \to \infty$,

$$\limsup_{t\to\infty} |q_m(t)|^2 \le \delta\left(\frac{\lambda_1^{-1}}{\varepsilon\nu} + \frac{4c_1^2c_4^2}{\varepsilon\nu}\right)\left(\frac{1-\exp(-\nu'\lambda_{m+1}^{1/2}M)}{\nu'\lambda_{m+1}^{1/2}}\right)$$
$$+\left[\frac{\lambda_1^{-1}}{\varepsilon\nu}c(f,g) + \frac{4c_1^2c_4^2}{\varepsilon\nu}c(u,v)\right]\left(\frac{\exp(-\nu'\lambda_{m+1}^{1/2}M)}{\nu'\lambda_{m+1}^{1/2}}\right)$$
$$+M\delta\left[K'\frac{\lambda_1^{-1}}{\varepsilon\nu}c(f,g) + K\frac{4c_1^2c_4^2}{\nu}c(u,v)\right].$$

We let $\delta \to 0$ and then $M \to \infty$, and we obtain (8.28). ■

Remark 8.1. Theorem 8.1 (1) is contained in [5]; the proof is reproduced here for the sake of completeness.

8.4 Approximation in a general subspace

In applications, utilization of the basis $\{w_m\}_{m\ge1}$ formed by the eigenvectors of A is not practical since these functions are not easy to determine. We therefore now show how Theorem 8.1 can be extended when V_m is replaced by a general finite-dimensional subspace W of V.

8.4.1 Assumptions

Let W be a finite-dimensional subspace of V and let $P(W)$ be the projector in H onto W, $Q(W) = I - P(W)$. Since $P(W)$ is not a projector in V, it may happen that $((\varphi, \psi)) \ne 0$ if $\varphi \in W$ and $\psi \in V$, $P(W)\psi = 0$. However, one can show (*cf.* Lemma 8.3 below) that there exists $\rho(W)$, $0 \le \rho(W) < 1$ such that

$$|((\varphi, \psi))| \le \rho(W)\,\|\varphi\|\,\|\psi\|, \quad \forall\varphi \in W, \forall\psi \in V, P(W)\psi = 0. \tag{8.32}$$

We associate also with W the two numbers $\lambda(W)$, $\mu(W)$:

$$\lambda(W) = \inf\{\|\varphi\|^2, \varphi \in V, P(W)\varphi = 0, |\varphi| = 1\},$$
$$\mu(W) = \sup\{\|\psi\|^2, \psi \in W, |\psi| = 1\},$$

so that

$$|\varphi| \le \lambda(W)^{-1/2}\|\varphi\|, \quad \forall\varphi \in V, P(W)\varphi = 0, \tag{8.33}$$
$$\|\psi\| \le \mu(W)^{1/2}|\psi|, \forall\psi \in W. \tag{8.34}$$

When it is not necessary to mention the dependence on W, we write simply P, Q, ρ, λ, μ, instead of $P(W)$, etc.

8.4.2 An inequality

We consider, as in Section 8.3, the solutions $u(\cdot)$, $v(\cdot)$ of (8.22), (8.23) and we put

$$w = u - v, \quad p = Pw, \quad q = Qw, \quad e = f - g.$$

We have

$$q' + \nu QAw + QB(v, w) + QB(w, u) = Qe$$

and, taking the scalar product in H with q, we get

$$\frac{1}{2}\frac{\mathrm{d}}{\mathrm{d}t}|q|^2 + \nu\,\|q\|^2 = -(B(v, p), q) - (B(p, u), q)$$
$$-(B(q, u), q) + (Qe, q) - \nu((p, q)).$$

Using (8.11) and (8.32)–(8.34) we find that the right-hand side of this inequality is less than

$$\lambda_1^{-1/2}\,|Qe|\,\|q\| + \nu\rho\mu^{1/2}\,|p|\,\|q\| + c_1(|Au| + |Av|)\,|p|\,\|q\| + c_1\,|Au|\,|q|\,\|q\|$$

which, by (8.16) (8.19),

$$\leqslant \frac{\varepsilon\nu}{3}\|q\|^2 + \frac{3\lambda_1^{-1}}{4\varepsilon\nu}|Qe|^2 + \frac{\varepsilon\nu}{3}\|q\|^2 + \frac{3\nu\rho^2}{\varepsilon}\mu\,|p|^2 + \frac{\varepsilon\nu}{3}\|q\|^2$$
$$+\frac{3c_1^2}{4\varepsilon\nu}(|Au| + |Av|)^2\,|p|^2 + c_1c_4\lambda^{-1/2}\,\|q\|^2 \quad \text{for } t \geqslant \alpha > 0$$

$$\leqslant \varepsilon\nu\,\|q\|^2 + \frac{3\lambda_1^{-1}}{4\varepsilon\nu}|Qe|^2 + \frac{3}{\varepsilon}\left(\frac{c_1^2c_4^2}{\nu} + \nu\rho^2\mu\right)|p|^2 + c_1c_4\lambda^{-1/2}\,\|q\|^2$$

for $t \geqslant \alpha > 0$.

If

$$\lambda(W) > \frac{c_1^2c_4^2}{\nu^2}, \tag{8.35}$$

then we set

$$\nu' = \nu - c_1c_4\lambda(W)^{-1/2} > 0, \qquad \varepsilon = \frac{1}{2} - \frac{c_1c_4\lambda(W)^{-1/2}}{2\nu}, \tag{8.36}$$

and we have established:

Lemma 8.2. *If* (8.35) *holds, then for* $t \geqslant \alpha > 0$,

$$\frac{\mathrm{d}}{\mathrm{d}t}|q|^2 + \nu'\,\|q\|^2 \leqslant \frac{3\lambda_1^2}{2\varepsilon\nu}|Qe|^2 + \frac{6}{\varepsilon}\left(\frac{c_1^2c_4^2}{\nu} + \nu\rho^2\mu\right)|p|^2, \quad \nu', \varepsilon \text{ as in } (8.36). \tag{8.37}$$

8.4.3 Statement of the result

As in Section 8.3.2, we have:

Theorem 8.2. *We assume that* $l = 2$ *or that* $l = 3$ *and that* u *and* v *are solutions of* (8.22) (8.23) *uniformly bounded in* V. *Let* W *be a finite-*

dimensional subspace of V such that (3.4) *is satisfied*:

(1) *If* $|P[u(t)-v(t)]| \to 0, |(I-P)[f(t)-g(t)]| \to 0$ *for* $t \to \infty$, *then*

$$|(I-P)[u(t)-v(t)]| \to 0 \quad \text{for } t \to \infty, \tag{8.38}$$

$$|u(t)-v(t)| \to 0 \quad \text{for } t \to \infty. \tag{8.39}$$

(2) *If* $|P[u(t)-v(t)]|^2 \xrightarrow{\text{c.e.}} 0$, $|(I-P)[f(t)-g(t)]|^2 \xrightarrow{\text{c.e.}} 0$ *for* $t \to \infty$, *then* (8.38) *holds and, instead of* (8.39),

$$|u(t)-v(t)| \xrightarrow{\text{c.e.}} 0, \text{ for } t \to \infty. \tag{8.40}$$

The proof, starting from (8.37), is essentially the same as that of Theorem 8.1.

In the rest of this section we give examples and show that (8.32) is always satisfied while assumption (8.35) is satisfied, if W is 'sufficiently large'.

8.4.4 Assumption (8.32)

Lemma 8.3. *Under the assumptions of Section* 8.4.1 *there exists* $\rho = \rho(W)$, $0 \leq \rho < 1$ *such that* (8.32) *holds.*

Proof. If (8.32) were not true, we could find two sequences $\{\varphi_j\}_{j\geq 1}$, $\{\psi_j\}_{j\geq 1}$, $\varphi_j \in W$, $\psi_j \in V$, $P\psi_j = 0$, such that

$$\|\varphi_j\|\,\|\psi_j\| \geq |((\varphi_j, \psi_j))| \geq \left(1-\frac{1}{j}\right)\|\varphi_j\|\,\|\psi_j\|.$$

Setting

$$\varphi_j' = \frac{\varphi_j}{\|\varphi_j\|}, \quad \psi_j' = \frac{\psi_j}{\|\psi_j\|},$$

we find

$$1 \geq |((\varphi_j', \psi_j'))| \geq \left(1-\frac{1}{j}\right). \tag{8.41}$$

We can extract a subsequence (still denoted by j) such that φ_j' converges to some limit φ, $\|\varphi\| = 1$, $\varphi \in W$ (W has finite dimension), and ψ_j' converge weakly in V to ψ, $\psi \in V$, $\|\psi\| \leq 1$, $P\psi = 0$. At the limit, (8.41) gives

$$|((\varphi, \psi))| = 1, \quad \|\varphi\| = 1, \quad \|\psi\| \leq 1,$$

so that $\|\psi\| = 1$, $\psi = K\varphi \neq 0$, in contradiction to $P\psi = 0$. ■

8.4.5 Assumption (8.35)—Example

We consider the following situation which is classical in the numerical analysis of partial differential equations and in particular of the Navier–Stokes equations (*cf.* [4], Chap. I, §4):

We are given a family $\{W_h\}_{h\in\mathcal{H}}$, of finite-dimensional subspaces of V. The set of indices $\mathcal{H}$ is arbitrary but is equipped with a concept of a limit, which we denote for simplicity $\lim_{h\to 0}$ (a filter $\mathcal{F}$ with a denumerable basis: $\lim_{h\to 0}$ means, roughly speaking, $\lim_{\mathcal{F}}$). For example, in the Galerkin method, $\mathcal{H}=1/\mathbb{N}$, $h=1/m$, and we pass to the limit $m\to\infty$, $h\to 0$. For finite element methods (*cf.* [4]), $\mathcal{H}$ is a family of regular triangulations of the domain Ω, and we let the diameter of the largest triangle go to 0.

The main assumption on the spaces W_h is the following:

$$\forall\varphi\in V,\qquad \inf_{\psi\in W_h}\|\varphi-\psi\|\to 0 \quad\text{as}\quad h\to 0. \tag{8.42}$$

In the case of an increasing sequence of subspaces W_m of V (Galerkin method), assumption (8.42) means simply that

$$\bigcup_{m\in\mathbb{N}} W_m \text{ is dense in } V. \tag{8.43}$$

Then we see that assumption (8.35) is satisfied for 'sufficiently small' h.

Lemma 8.4. *Under assumption* (8.42),

$$\lim_{h\to 0}\lambda(W_h)=+\infty, \tag{8.44}$$

and (8.35) *is satisfied for sufficiently small h.*

Proof. The proof consists in establishing the following more precise statement:

For every integer m there exists h_m and, for $h\leqslant h_m$, $\lambda(W_h)\geqslant\lambda_m$. (8.45)

For given m and $\delta>0$, assumption (8.42), written with $\varphi=w_j$, $j=1,\ldots,m$, shows that there exists h_m such that

$$\inf_{\psi\in W_h}\|w_j-\psi\|\leqslant\delta,\quad \text{for } j=1,\ldots,m \text{ and for every } h<h_m.$$

Thus for every $h\leqslant h_m$, there exist $\tilde{w}_1,\ldots,\tilde{w}_m$ in W_h, with $\|w_j-\tilde{w}_j\|\leqslant\delta$. Therefore if $\varphi\in V$, $[I-P(W_h)]\varphi=0$, we have

$$\begin{aligned}
\|\varphi\|^2&=\|P_m\varphi\|^2+\|(I-P_m)\varphi\|^2\\
&\geqslant\lambda_{m+1}\,|(I-P_m)\varphi|^2+\lambda_1\,|P_m\varphi|^2\\
&\geqslant\lambda_{m+1}\,|\varphi|^2-(\lambda_{m+1}-\lambda_1)\,|P_m\varphi|^2\\
&=\lambda_{m+1}\,|\varphi|^2-(\lambda_{m+1}-\lambda_1)\sum_{j=1}^{m}(\varphi,\,w_j-\tilde{w}_j)^2\\
&\geqslant[\lambda_{m+1}-(\lambda_{m+1}-\lambda_1)\lambda_1^{-1}m\delta^2]\,|\varphi|^2.
\end{aligned}$$

This implies that

$$\lambda(W_h) \geq \lambda_{m+1} - (\lambda_{m+1} - \lambda_1)\lambda_1^{-1} m\delta^2$$

and the result follows by letting $\delta \to 0$. ■

It is also useful for later purposes to establish:

Lemma 8.5. *Under assumption* (8.42),

$$\lim_{h\to 0} \mu(W_h) = +\infty. \tag{8.46}$$

Proof. Because of (8.42), for every $\psi \in V$ and for every h there exists $\psi_h \in W_h$, with

$$\lim_{h\to 0} \|\psi - \psi_h\| = 0.$$

Because of (8.34), $\|\psi_h\| \leq \mu(W_h)^{1/2} |\psi_h|$, and if the family $\mu(W_h)$ did not converge to $+\infty$ for $h \to 0$, $\lim_{h\to 0} \inf \mu(W_h) \leq K < \infty$, we would have at the limit: $\|\psi\| \leq K |\psi|$, $\forall \psi \in V$; this is impossible. ■

8.5 Time-periodic solutions

The notation is the same as in Section 8.4. If assumption (8.35) is satisfied, we infer from Lemma 8.2 that there exist two positive constants, $c_5 > 0$, $\eta > 0$, independent of u and v, such that

$$|(I-P)[u(t)-v(t)]|^2 \leq c_5\{\exp[-\eta(t-t_0)] \tag{8.47}$$
$$+ \int_{t_0}^{t} \exp[-\eta(t-\tau)][|P(u(\tau)-v(\tau))|^2 + |(I-P)(f(\tau)-g(\tau))|^2]\,d\tau\},$$

for $t \geq t_0 \geq \alpha > 0$.

We can now prove the following:

Theorem 8.3. *We assume that $l = 2$ or that $l = 3$ and that u is a solution of* (8.22) *uniformly bounded in V. Let W be a finite-dimensional subspace of V such that* (8.35) *is satisfied. Assume, moreover, that there exist periodic functions $f_\infty(\cdot)$, $p_\infty(\cdot)$ with values in H and W and period $T > 0$, such that*

$$\lim_{t\to\infty} |f(t) - f_\infty(t)| = \lim_{t\to\infty} |Pu(t) - p_\infty(t)| = 0. \tag{8.48}$$

Then there exists a periodic solution u_∞ with period T of the equation

$$u'_\infty + \nu A u_\infty + B(u_\infty, u_\infty) = f_\infty, \tag{8.49}$$

such that

$$\lim_{t\to\infty} \|u(t) - u_\infty(t)\| = 0. \tag{8.50}$$

Proof. Apply (8.47) to u and v, $v(t)=u(t+jT)$. Then for $lT\geqslant t_0$ we obtain, at time $t+lT$,

$$\begin{aligned}|(I-P)(u(t+lT)-u[t+(j+l)T)]|^2\leqslant c_5\Big\{&\exp[-\eta(t+lT-t_0)]\\ &+\int_{t_0}^{t+lT}\exp[-\eta(t+lT-\tau)]\times[|P(u(\tau)-u(\tau+jT))|^2\\ &+|(I-P)(f(\tau)-f(\tau+jT))|^2]\,d\tau\Big\}.\end{aligned}\tag{8.51}$$

For $\varepsilon>0$ given, let $t_\varepsilon\geqslant t_0$ be such that, for $t\geqslant t_\varepsilon$,

$$|Pu(t)-p_\infty(t)|\leqslant\varepsilon,\qquad |f(t)-f_\infty(t)|\leqslant\varepsilon.$$

Then from (8.51) we obtain

$$\begin{aligned}|(I-P)[u(t+lT)-u(t+(j+l)T)]|^2&\leqslant c_5\exp[-\eta(t+lT-t_0)]\\ &+c_5c(u,f)\int_{t_0}^{t_\varepsilon}\exp[-\eta(t+lT-\tau)]\,d\tau\\ &+8c_5\varepsilon^2\int_{t_\varepsilon}^{t+lT}\exp[-\eta(t+lT-\tau)]\,d\tau,\end{aligned}$$

so that

$$\sup_{t\geqslant t_0}|(I-P)[u(t+lT)-u(t+(l+j)T)]|^2\leqslant c_\varepsilon\exp(-\eta lT)+(c_5\varepsilon^2/\eta)$$

for every $j\geqslant 0$ and $l\geqslant t_0/T$. Therefore $\{(I-P)u(t+lT)\}_l$ is a Cauchy sequence in the space of continuous-bounded functions from $[t_0,\infty)$ into $(I-P)W$. Thus there exists a continuous-bounded function u_∞ from $[t_0,\infty)$ into H, such that

$$u(t+lT)\rightarrow u_\infty(t)\text{ as }l\rightarrow\infty,\text{ in }H,\text{ uniformly in }t\text{ on }(t_0,\infty).\tag{8.52}$$

Since $|Au(t+lT)|\leqslant c_4$ for all $t\geqslant\alpha$, $l\geqslant 1$, we see that $|Au_\infty(t)|\leqslant c_4$ for all $t\geqslant\alpha$, and

$$u(t+lT)\rightarrow u_\infty(t)\quad\text{in}\quad V,\quad\forall t\geqslant\alpha.\tag{8.53}$$

It is then easy to see that u_∞ is a solution bounded in V of (8.49), and u_∞ is periodic of period T by virtue of (8.52). The convergence

$$\lim_{t\rightarrow\infty}|u(t)-u_\infty(t)|=0\tag{8.54}$$

follows immediately from (8.52). The convergence (8.50) follows from (8.54) and the fact that $|Au(t)-Au_\infty(t)|$ remains bounded (by $2c_4$) as $t\rightarrow\infty$. ■

We then deduce the following result for stationary solutions:

Theorem 8.4. *The assumptions are similar to those of Theorem 8.3. We* *assume that there exist $f_\infty \in H$ and $p_\infty \in W$ such that*

$$|f(t)-f_\infty| \to 0, |Pu(t) \to p_\infty| \to 0, \quad \text{for } t \to \infty. \tag{8.55}$$

It follows that there exists $u_\infty \in D(A)$ such that

$$\|u(t)-u_\infty\| \to 0 \quad \text{for } t \to \infty, \tag{8.56}$$

where u_∞ is a (stationary) solution of the Navier–Stokes equation

$$\nu A u_\infty + B(u_\infty, u_\infty) = f_\infty. \tag{8.57}$$

Proof. We apply Theorem 8.3 with $T>0$ fixed, arbitrary, and we obtain (8.49) (8.50). Now $T>0$ can be chosen arbitrarily small and, since u_∞ must be independent of T in (8.50), we conclude that u_∞ has period 0, i.e., u_∞ is independent of t.

8.6 Comment on the Galerkin approximation

We now assume that the dimension $l=2$. For simplicity, we restrict ourselves to a Galerkin approximation of Navier–Stokes equations based on the spaces V_m, i.e., the family w_j of eigenfunctions of the Stokes problem. We will show that if m is sufficiently large, the behaviour as $t \to \infty$ of the Galerkin approximation u_m is completely determined by the behaviour, as $t \to \infty$, of a certain number m_* of its modes, i.e., of $P_{m_*}u_m$, $m_* < m$. *This number m_* does not depend on m.*

8.6.1 Galerkin approximation

For fixed m, the Galerkin approximation u_m of the solution u of (8.22) is defined by

$$\begin{cases} u'_m + \nu A u_m + P_m B(u_m, u_m) = P_m f, \quad t>0, \\ u_m(0) = P_m u_0. \end{cases} \tag{8.58}$$

It is classical practice to derive *a priori* estimates independent of m on u_m; for example, for every $t \geq s \geq 0$,

$$|u_m(t)|^2 + \nu \int_s^t \|u_m(\sigma)\|^2 \, d\sigma \leq |u_m(s)|^2 + \frac{\lambda_1^{-1}(t-s)}{\nu}[f]^2 \tag{8.59}$$

and

$$|u_m(t)|^2 \leq |u_0|^2 \exp(-\nu\lambda_1 t) + \left(\frac{1-\exp(-\nu\lambda_1 t)}{\nu^2 \lambda_1}\right)[f]^2. \tag{8.60}$$

The following *a priori* estimate is satisfied by u_m:

Lemma 8.6. *$\|u_m(t)\|$ is bounded independently of m and t for $t \geq \alpha > 0$, $m \geq 0$.*

Proof. Taking the scalar product of (8.58) with Au_m, we obtain

$$\frac{1}{2}\frac{\mathrm{d}}{\mathrm{d}t}\|u_m\|^2+\nu\,|Au_m|^2=+(f,Au_m)-(B(u_m,u_m),Au_m).$$

Because of (8.10) (8.17), the right-hand side of this expression is majorized by

$$[f]\,|Au_m|+c_0\,|u_m|^{1/2}\,\|u_m\|\,|Au_m|^{3/2}$$
$$\leqslant\frac{\nu}{4}|Au_m|^2+\frac{1}{\nu}[f]^2+\frac{\nu}{4}|Au_m|^2+c'\,|u_m|^2\,\|u_m\|^4.$$

Therefore, with (8.60),

$$\frac{\mathrm{d}}{\mathrm{d}t}\|u_m\|^2+\nu\,|Au_m|^2\leqslant c''(1+\|u_m\|^2)^2 \tag{8.61}$$

and, for $0\leqslant s\leqslant t$, we can show by integration that

$$(1+\|u_m(t)\|^2)\leqslant(1+\|u_m(s)\|^2)\exp\left[c''\int_s^t(1+\|u_m(\sigma)\|^2)\,\mathrm{d}\sigma\right].$$

If $t\geqslant\alpha>0$, we integrate in s from $t-\alpha$ to t and we find that

$$c''\alpha(1+\|u_m(t)\|^2)\leqslant\exp\left\{c''\int_{t-\alpha}^t[(1+\|u_m(\sigma)\|^2)\,\mathrm{d}\sigma]-1\right\}.$$

Using (8.59), we see that the right-hand side of this inequation is bounded by a constant depending on α but independent of t and m, and the Lemma follows. ■

8.6.2 Behaviour as $t\to\infty$ of the Galerkin approximation

Let v_m be the Galerkin approximation of the solution v of (8.23):

$$\begin{aligned}&v_m'+\nu Av_m+P_mB(v_m,v_m)=P_mg,\quad t>0,\\&v_m(0)=P_mv_0,\end{aligned} \tag{8.62}$$

and, as before, u_m denotes the Galerkin approximation of the solution u of (8.22).

We put for $m_*\leqslant m$,

$$w_m=u_m-v_m,\qquad p_{m_*}=P_{m_*}w_m,\qquad q_{m_*}=Q_{m_*}w_m,$$
$$e=f-g,\quad e_{m_*}=Q_{m_*}e.$$

Then

$$\frac{\mathrm{d}q_{m_*}}{\mathrm{d}t}+\nu Aq_{m_*}+Q_{m_*}P_mB(v_m,w_m)+Q_{m_*}P_mB(w_m,u_m)=Q_{m_*}P_me$$

and consequently

$$\frac{1}{2}\frac{\mathrm{d}}{\mathrm{d}t}|q_{m*}|^2+\nu\,\|q_{m*}\|^2=-(B(v_m,w_m),P_mq_{m*})-(B(w_m,u_m),P_mq_{m*})$$
$$+(Q_{m*}e,P_mq_{m*}).$$

The right-hand side is equal to

$$(Q_{m*}e,P_mq_{m*})-(B(v_m,p_{m*}),P_mq_{m*})-(B(v_m,(I-P_m)q_{m*}),P_mq_{m*})$$
$$-(B(p_{m*},u_m),P_mq_{m*})-(B(q_{m*},u_m),P_mq_{m*}),$$

and, because of (8.10) and Lemma 8.6, this quantity is bounded by

$$\begin{aligned}&|Q_{m*}e|\,|q_{m*}|+c'\,|v_m|^{1/2}\,\|v_m\|^{1/2}\,|p_{m*}|^{1/2}\,\|p_{m*}\|^{1/2}\,\|q_{m*}\|\\&\quad+c'\,|v_m|^{1/2}\,\|v_m\|^{1/2}\,|q_{m*}|^{1/2}\,\|q_{m*}\|^{3/2}\\&\quad+c'\,|p_{m*}|^{1/2}\,\|p_{m*}\|^{1/2}\,\|u_m\|\,\|q_{m*}\|\\&\quad+c'\,|q_{m*}|^{1/2}\,\|q_{m*}\|^{3/2}\,\|u_m\|\\&\quad\leqslant\lambda_1^{1/2}\,|Q_{m*}e|\,\|q_{m*}\|+c''\lambda_{m*+1}^{1/4}\,|p_{m*}|\,\|q_{m*}\|+c''\lambda_{m*+1}^{-1/4}\,\|q_{m*}\|^2.\end{aligned}$$

If

$$\lambda_{m*+1}>\left(\frac{c''}{\nu}\right)^4,\tag{8.63}$$

we set

$$\nu'=2(\nu-c''\lambda_{m+1}^{-1/4})>0,\qquad \varepsilon=\frac{1}{2}-\frac{c''\lambda_{m+1}^{-1/4}}{2\nu}>0,\tag{8.64}$$

and we bound the last quantity by

$$\frac{\varepsilon\nu}{2}\|q_{m*}\|^2+\frac{\lambda_1^{-1}}{2\varepsilon\nu}|Q_{m*}e|^2+\frac{\varepsilon\nu}{2}\|q_{m*}\|^2+\frac{(c'')^2\lambda_{m*+1}^{1/4}}{2\varepsilon\nu}|p_{m*}|^2+c''\lambda_{m*+1}^{-1/4}\|q_{m*}\|^2.$$

We now find that

$$\frac{\mathrm{d}}{\mathrm{d}t}|q_{m*}|^2+\nu'\,\|q_{m*}\|^2\leqslant\frac{\lambda_1}{\epsilon\nu}|Q_{m*}e|^2+\frac{c'''\lambda_{m*+1}^{1/2}}{\epsilon\nu}|p_{m*}|^2.\tag{8.65}$$

As for Theorem 8.1, we obtain:

Theorem 8.5. *We assume that $l=2$ and that $m\geqslant m_*$, where m_* is sufficiently large that* (8.63) *is satisfied.*

(1) *If*

$$|P_{m*}(u_m(t)-v_m(t))|\to 0,\qquad |(I-P_m)(f(t)-g(t))|\to 0$$

for $t\to\infty$, then

$$|(I-P_{m*})(u_m(t)-v_m(t))|\to 0\quad \textit{for } t\to 0,\tag{8.66}$$
$$|u_m(t)-v_m(t)|\to 0\quad \textit{for } t\to\infty.\tag{8.67}$$

(2) *If*

$$|P_{m*}(u_m(t)-v_m(t))|^2 \xrightarrow{\text{c.e.}} 0, \qquad |(I-P_{m*})(f(t)-g(t))|^2 \xrightarrow{\text{c.e.}} 0,$$

for $t\to\infty$, *then* (8.66) *holds and, instead of* (8.67),

$$|u_m(t)-v_m(t)|^2 \xrightarrow{\text{c.e.}} 0, \quad \textit{for } t\to\infty. \tag{8.68}$$

Remark 8.2. The full implications of this theorem will be realized if we can relate the behaviour, for $t\to\infty$, of the Galerkin approximation u_m of u, to the behaviour for $t\to\infty$ of u itself. This question will be considered in a subsequent work. However, Theorem 8.5 indicates that it may be useless to take too many modes for the long-term Galerkin approximation of flow.

References

1. Chorin, A. J., *Lectures on Turbulence Theory*: Boston, Publish or Perish, 1975 (includes Bibliography).
2. Girault, V. and Raviart, P. A., *Finite Element Approximation of the Navier–Stokes Equations*, Lecture Notes in Mathematics, Vol. 749: Berlin, Springer-Verlag, 1979.
3. Orszag, S., Lectures on the statistical theory of turbulence, in Balian, R. and Peube, J. L. (Eds), *Fluid Dynamics, Les Houches* 1973: New York, Gordon and Breach, 1974, 235–374.
4. Temam, R., *Navier–Stokes Equations, Theory and Numerical Analysis*, 2nd edn: Amsterdam, North-Holland, 1979.
5. Foias, C. and Prodi, G., Sur le comportement global des solutions nonstationnaires des équations de Navier–Stokes en dimension 2, *Rendiconti del Seminario Matematico della Università di Padova*, **39,** 1–34, 1967.
6. Manley, O. P. and Treve, Y. M., Minimum number of modes in approximate solutions to equations of hydrodynamics, *Physics Letters*, **82A,** 88, 1981.
7. Treve, Y. M. and Manley, O. P., A possible strange attractor in MHD flow, in Laval, G. and Gresillon, D. (Eds), *Intrinsic Stochasticity in Plasmas*: Orsay, Les Editions de Physique, 1979, 393.
8. Treve, Y. M., A numerical test of the reliability of Galerkin approximations to the solutions of the Navier–Stokes equations, *Journal of Computational Physics*, **40,** 2, 1981.
9. Lions, J. L., *Quelques Méthodes de Résolution des Problèmes aux Limites Nonlinéaires*: Paris, Dunod, 1969.
10. Guillopé, C., Comportement à l'infini des solutions des équations de Navier–Stokes et propriétés des ensembles fonctionnels invariants (ou attracteurs), *Annales de l'Institut Fourier* (to appear).

11. Temam, R., Behaviour at time $t=0$ of the solutions of semilinear evolution equations, *Journal of Differential Equations*, **43,** 73–91, 1982.
12. Foias, C. and Temam, R., Some analytic and geometric properties of the evolution Navier–Stokes equations, *Journal de Mathématiques Pures et Appliquées*, **58,** 339–368, 1979.

9 Recent Experiments on the Transition to Turbulent Convection

J P Gollub

9.1 Introduction

The literature on hydrodynamics contains hundreds of experimental articles on convection dating back at least to the experiments of Bénard in 1900 [1]. While some of these were concerned with the problem of the transition to turbulence, a fundamental understanding of this phenomenon was not achieved, for a variety of reasons. Experiments have used relatively primitive measurement techniques that do not permit turbulent and nonturbulent states to be reliably distinguished and quantitatively studied. The thermal control achieved in many of the early experiments was inadequate. Furthermore, experiments were hindered by the lack of an adequate conceptual structure for describing the transition to turbulence.

This situation has changed dramatically since the early 1970s. Quantitative analysis of the secondary instabilities of convection rolls using Galerkin expansions (and visual observation of these instabilities) by Busse *et al.* [2] has certainly brought us closer to an understanding of the onset of turbulent motion. The hypothesis of Ruelle and Takens [3] and Lorenz [4] that turbulence near its onset might be described as a strange attractor in a low-dimensional dynamical system has also given considerable impetus to recent experiments.

In addition to these theoretical advances, new techniques have spawned a generation of experiments that are much more quantitative than earlier ones. Major advances include: the utilization of laser Doppler techniques, which permit the measurement of local fluid velocities without perturbing the fluid significantly; cryogenic techniques, which permit sensitive measurement of the time-dependent heat flux through a convecting fluid; and computer processing of experimental data to obtain power spectra of exceedingly high resolution.

In this chapter some recent experiments are reviewed that utilize these new techniques. We discuss only experiments which are related to the onset of time-dependent convection, as opposed to the Rayleigh–Bénard instability itself. The period covered by this review is approximately 1974–1981, but is certainly not exhaustive. Rather, we highlight experiments that have some bearing on the relevance of nonlinear dynamics to the problem of turbulence. A more general and extensive review of hydrodynamic instabilities and the transition to turbulence is available elsewhere [5].

9.2 New methods

9.2.1 Cryogenic techniques

Ahlers *et al.* [6–9] have shown that there are many advantages in studying convective instabilities and turbulence in liquid helium above the lambda point, where it is a normal classical fluid. Temperature resolution of one part in 10^7 can be achieved, along with a very high degree of spatial uniformity in the boundary temperatures. Furthermore, the time-dependent heat flux through a fluid layer can be easily measured because the heat capacity of the boundary can be kept small compared with that of the liquid, a condition that cannot be approached at room temperature. Finally, the Prandtl number σ (ratio of the kinematic viscosity to the thermal diffusivity) can be varied over the range $0.6<\sigma<1.5$ by varying the mean working temperature.

Heat flux measurements are performed by imposing a steady heat current q through the cell and measuring the temperature change, ΔT, of the bottom plate while holding the temperature of the top plate constant. Although the experiments are not conducted at strictly constant Rayleigh number, the variations are small, so that the physical phenomena are believed not to be modified by the constant heat flux condition. Libchaber and Maurer [10–13] have shown that precise local temperature measurements can also be obtained at low temperatures with the aid of several carbon bolometers embedded in the upper boundary of the convection cell. This technique permits both time variations and the spatial structure of the flow to be determined.

9.2.2 Laser Doppler velocimetry

Laser light scattering provides a highly precise and nonperturbative probe of the local fluid velocity. The method has been extensively applied to studies of convective instabilities and turbulence by Gollub *et al.* [14–19] and by Bergé *et al.* [20–23]. The velocity is measured in a volume defined by the intersection of two focused laser beams. Typical dimensions of this volume are small compared with the spatial scale of the velocity variations. The angular deflections of the light scattered from the two beams toward the photomultiplier are different, leading to different Doppler shifts for the two beams. The output of the photomultiplier, which is a square-law detector, contains a signal at the difference frequency, an easily measured (but noisy) audio signal whose frequency is proportional to one component of the local fluid velocity. By locking a high precision oscillator to the noisy Doppler signal, velocities as small as 10^{-4} cm/s can be measured in real time.

By mounting the fluid system on computer-controlled translation stages, Gollub and Benson [14] were able to map the velocity in a horizontal plane when it was time-independent. An extension of this method by Gollub and Steinman [19] allowed maps of the spatial structure of the velocity field to be made even in time-dependent regimes, provided that the fluctuations were sufficiently slow.

9.2.3 Computer control and on-line data acquisition

Laboratory computer techniques have permitted a qualitative advance in experimental studies of the chaotic behaviour of fluid systems. The long time-scales and large quantities of data preclude manual experimentation. Numerical fast Fourier transform methods have been essential in interpreting the data. Finally, automation permits a programmed sequence of changes in the Rayleigh number to be executed repetitively.

9.3 Small aspect ratio convection

The nature of the transition to turbulence is profoundly affected by the relative magnitude of the horizontal and vertical dimensions of the fluid layer, commonly known as the aspect ratio. This quantity has been defined for rectangular layers as the ratio of the largest horizontal dimension to the fluid depth, while it has been customary to use the radius-to-depth ratio for cylindrical layers.

Small aspect ratio layers have been found to behave like systems with a few degrees of freedom near the onset of turbulence, in the sense that they show types of time-dependence that are also seen in simple dynamical systems. These phenomena include quasiperiodicity, phase-locking and subharmonic bifurcations. Large aspect ratio layers, on the other hand, do not show such behaviour. In this section, we summarize some of the experiments on small aspect ratio convection.

One important discovery is the observation that there are a number of stable convection patterns in cells with small aspect ratio, depending on the past history or initial conditions of the system. A given convection cell may contain either two or three rolls, for example. More surprisingly, Gollub and Benson [14] also found that there are several variants of two-roll convection patterns, both of which are stable. Each one has a distinct sequence of instabilities leading to turbulence as R is varied.

9.3.1 Quasiperiodicity and phase-locking

A number of groups [8, 11, 18, 21] have detected quasiperiodic flows with several incommensurate frequencies. In some cases these incommensurate oscillations phase-lock with each other. If a quasiperiodic state is described by a trajectory lying on a torus in phase space, then the phenomenon of phase-locking produces a limit cycle on the torus. An example of this process as described by Gollub and Benson [14] follows.

The flow first becomes time-dependent by entering a strictly periodic regime at R_1 (Fig. 9.1(a)). At a higher Rayleigh number, R_2, a second frequency component begins to grow. The interaction of these two oscillations causes the spectrum to consist of sharp peaks at the fundamental frequencies f_1 and f_2, and all of their low-order sums and differences of the form $f = m_1 f_1 + m_2 f_2$. The presence of these mixing components in the spectrum indicates that the time-dependent processes

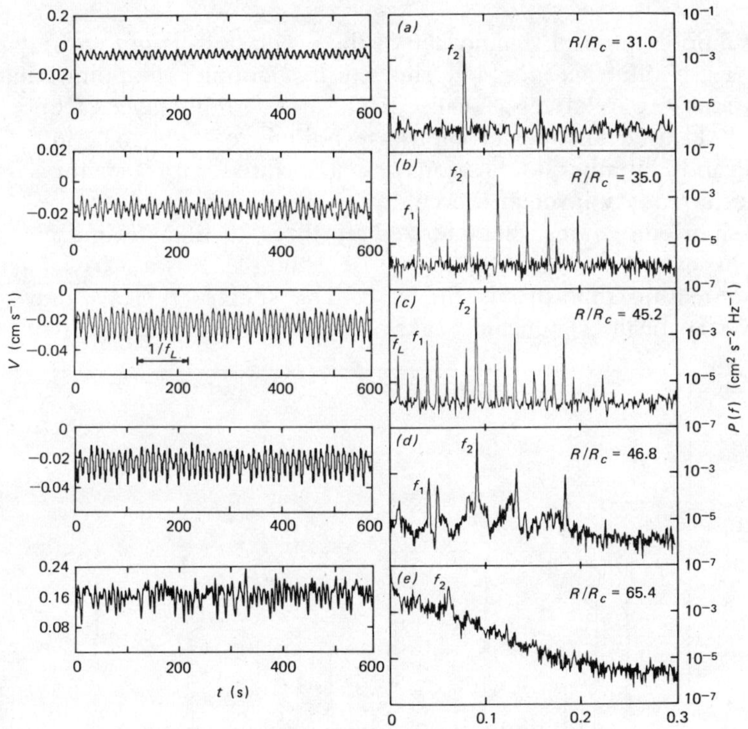

Fig. 9.1 Velocity records and power spectra showing a sequence of bifurcations leading to turbulent convection at small aspect ratio [14].
(a) Limit cycle
(b) Motion on a torus, with two incommensurate frequencies (and sums and differences)
(c) Phase-locking on the torus at the frequency ratio $f_2/f_1 = \frac{9}{4}$
(d) Weakly nonperiodic flow
(e) Strongly nonperiodic motion at high Rayleigh number

are strongly nonlinear. The ratio f_2/f_1 decreases smoothly with increasing R. Therefore, the two frequencies are generally incommensurate. At higher R, they phase-lock so that f_2/f_1 is the ratio of two small integers (typically $\frac{7}{3}$ or $\frac{9}{4}$). The spectrum (Fig. 9.1(c)) then consists of an array of spikes at all multiples of the lowest commensurate frequency of f_1 and f_2. The phase-locking persists only over a relatively narrow range in R/R_c. When f_2/f_1 begins to change again, the spectral lines become broadband (Fig. 9.1(d)) indicating the onset of nonperiodicity at R_t. At high R/R_c the line structure of the spectrum is no longer present. Near the onset of noise, the linewidth of the spectral peaks is a linear function of $(R-R_t)$.

These observations are a clear example of the production of chaos in a system that has many degrees of freedom, only a few of which are excited initially. Quite similar phenomena, especially the relationship between the onset of chaos and the loss of phase-locking, have been seen in systems of coupled electronic oscillators by Gollub, Romer and Socolar [24].

The fluid dynamical significance of the oscillations is not entirely clear. However, Gollub *et al.* [17] showed by Doppler mapping that the oscillations are relatively localized within the fluid layer. Dubois and Bergé [21] used an interferometric technique to show that, in fluids at high Prandtl number, oscillations are associated with structures in the thermal boundary layer at the cold plate.

Quasiperiodic states characterized by three distinct frequencies were found in several cases by Gollub and Benson [14]. It is not trivial actually to demonstrate that this is the case. The spectra typically have over twenty statistically significant peaks, as shown in Fig. 9.2, but they can all

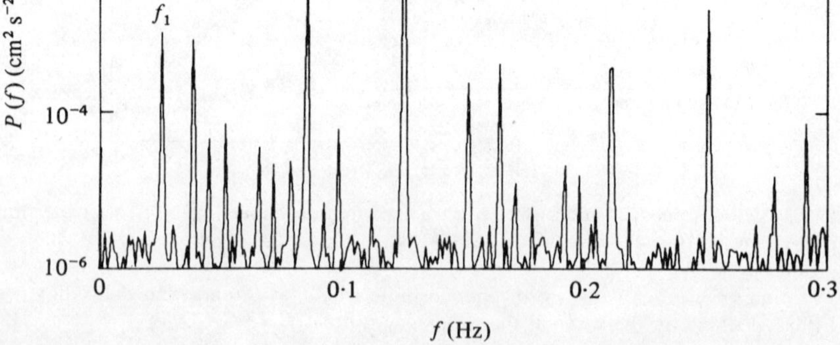

Fig. 9.2 Spectrum showing motion on a three-torus [14]. All peaks are linear combinations of the three frequencies f_1, f_2 and f_3. This is believed to be the largest number of apparently incommensurate frequencies found for a convective system

be fitted to linear combinations of three basic frequencies to within about one part in 10^4. Comparable fits cannot be obtained using two independent frequencies. These facts, combined with the observation that the ratios of the three frequencies vary smoothly with R, provide strong evidence that they are incommensurate with each other. Thus this state can be represented by a 3-torus in phase space. Nonperiodicity evolves from this state at higher R.

9.3.2 Subharmonic bifurcations

Libchaber and Maurer [12] and Gollub *et al.* [17] showed that nonperiodicity was obtained after several successive subharmonic bifurcations for certain combinations of aspect ratio and Prandtl number. This behaviour is at least qualitatively similar to that shown by nonlinear mappings of the unit interval of the form $x_{n+1}=f(x_n)$, where the function $f(x)$ has a single parabolic maximum in the unit interval.

A particularly beautiful and reproducible set of spectra showing subharmonic bifurcations was obtained recently by Giglio, Musazzi and Perini [25] using a laser beam deflection technique that responds to the average of the vertical temperature gradient along a horizontal line in a small aspect ratio cell containing water. Successive increments in R cause additional peaks to appear in the spectrum at subharmonics (and odd multiples) of the lowest frequency previously present in the spectrum. These phenomena are illustrated in Fig. 9.3. The onset Rayleigh numbers are predicted to follow a geometric series after many subharmonic bifurcations:

$$(R_{n+1}-R_n)/(R_{n+2}-R_{n+1}) \rightarrow 4.6692\ldots.$$

Data for the transition points seem to be consistent with this prediction, but the comparison is not yet precise. When R exceeds the accumulation point R_∞ of the sequence, the subharmonic bifurcations are predicted to disappear successively, forming a sequence above R_∞ which is the mirror image of that below R_∞. This feature of the subharmonic route to turbulence has been observed by Libchaber [13].

9.3.3 Intermittency

An interesting route to turbulence in which periodic or quasiperiodic oscillations are interrupted by turbulent bursts has been observed by Maurer and Libchaber [11] at low Prandtl number, and by Bergé, Dubois, Manneville and Pomeau [20] at high Prandtl number. An example of a time record for this process is shown in Fig. 9.4. The duration and frequency of the turbulent bursts increases with R until the motion is purely chaotic. Intermittent oscillations are known to occur in the Lorenz model, so this type of process may be another example of low-dimensional chaotic behaviour.

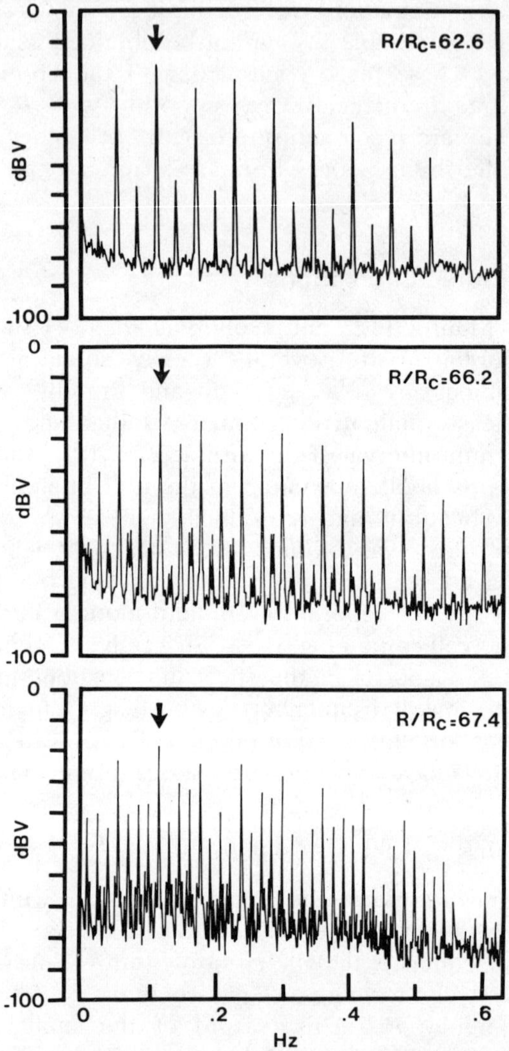

Fig. 9.3 Spectrum showing successive subharmonic bifurcations [25]. The fundamental frequency is indicated by the arrow

9.3.4 Modulation experiments

In order to examine the question of whether external noise plays any role in the onset of turbulent convection at small aspect ratio, Gollub and Steinman [16] injected spatially uniform but broadband temperature fluctuations into the lower boundary temperature. The magnitude of the applied noise was sufficient to give a root mean square fluctuation in R of about $0.3R_c$ in the range $20R_c < R < 50R_c$. The main observations were

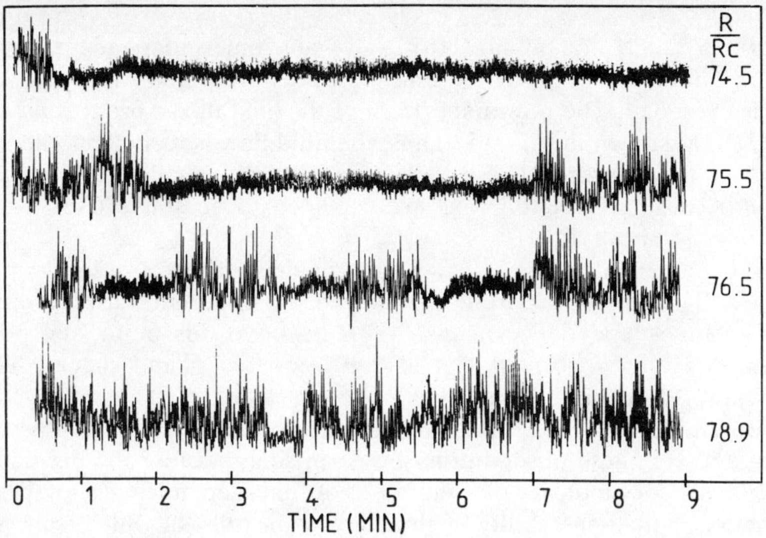

Fig. 9.4 Time records showing intermittent turbulence [11]. The motion is quasiperiodic between the noisy bursts

as follows:

(1) In the periodic regime preceding the onset of turbulence, the external noise causes slow fluctuations that can be represented by a Fourier series with time-dependent amplitudes and phases. The spectral peaks are slightly broadened and a low-level broad background is introduced.
(2) In the turbulent regime ($R < 41R_c$), the external noise produces no statistically significant change in the velocity power spectra.
(3) The transition Rayleigh number is not measurably depressed by the external noise.

These observations indicate that, under nominally time-independent boundary conditions, stochastic driving forces can probably be ignored in constructing models of the transition to turbulence at small aspect ratio.

9.4 Large aspect ratio convection

9.4.1 Heat flux experiments

The aspect ratio dependence of the transition process was discovered by Ahlers and Behringer [7] in heat flux experiments on cylindrical cells of convecting liquid helium. At small aspect ratio ($\Gamma = 2$), the time-dependent phenomena are similar to those obtained by local measurements and described in Section 9.3. However, when Γ is 4.7 or larger, no periodic or quasiperiodic state is observed. Instead, the time-dependence is nonperiodic at its onset.

At $\Gamma = 57$, it is not possible to resolve any time-independent regime. Rather, the motion seems to be time-dependent and nonperiodic as soon as R_c is exceeded. The power spectrum of the heat flux is broadband even at $1.27R_c$, as shown in Fig. 9.5. Thus, the fluid flow is turbulent whenever the velocity is nonzero. These data pose a serious challenge to stability theory, because convection rolls are expected to be stable over a finite interval in R above R_c.

The behaviour for $\Gamma = 4.7$, an intermediate aspect ratio, is more complex. The initial experiments showed an apparent onset of (nonperiodic) time-dependence near $R = 2R_c$. Above this point, the power spectra $P(f)$ are broadband, flat at low frequencies, and shaped like a power law f^{-n} at high frequencies, with $n = 4.0 \pm 0.2$. The spectra (Fig. 9.6) are similar to those of white noise passed through a double pole low-pass filter. These observations are surprising, because the predictions of stability theory indicate that the onset of time-dependence should arise from an oscillatory instability of the convection rolls, in which transverse waves propagate along the roll axis. This instability is expected to be manifested as a periodic oscillation in the heat flux beginning at (3 to 5)R_c rather than a much slower nonperiodic motion starting at about $2R_c$. However, the effects of this instability were identified in the form of a broad hump in the power-law tail of the spectra at higher R.

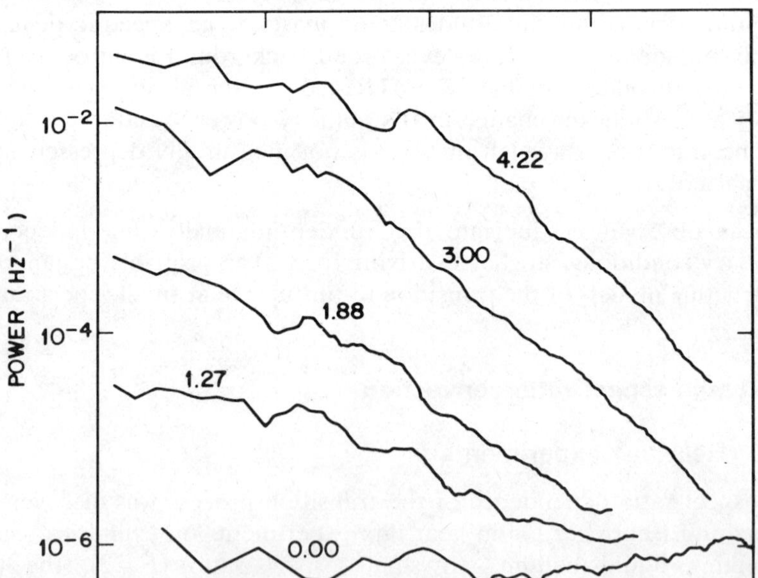

Fig. 9.5 Spectra of time-dependent heat transport in a cylindrical cell of aspect ratio 57 showing nonperiodicity at only $1.27R_c$ [8]

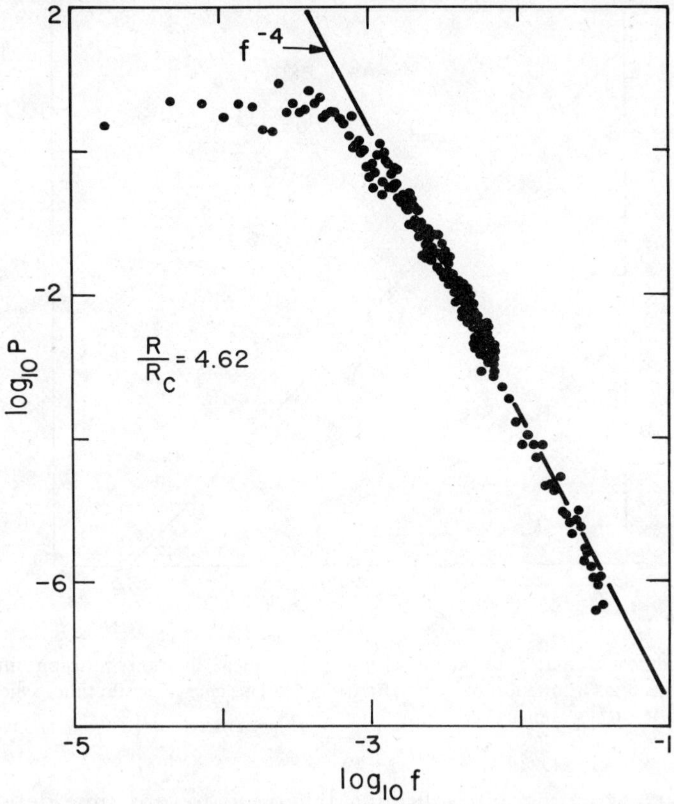

Fig. 9.6 Spectrum of the time-dependent heat transport at an intermediate aspect ratio (4.7) showing a flat region at low frequencies and a power-law tail at high frequencies [8]

More recent experiments on the same cell by Ahlers and Walden [9] showed even more puzzling phenomena. The flow is not in fact time-independent below $2R_c$ as had been previously believed. Rather, there is a very low-level random time-dependence with occasional randomly spaced major peaks in the effective thermal conductivity. As R is reduced below $2R_c$, these events become increasingly rare. Their frequency n is well represented by the equation $n = n_0 \exp(-\Delta_0/\varepsilon)$, where $\varepsilon = R/(R_c - 1)$, as shown in Fig. 9.7. Ahlers and Walden suggested that this behaviour is similar to that shown by a randomly forced mechanical system moving in a double well potential, with the parameter Δ_0 playing the role of an activation energy. Thus, the experiments clearly suggest that a stochastic explanation of the onset of turbulent convection at moderate and large aspect ratios may be required. The power-law falloff of the spectra also supports a stochastic onset hypothesis, since numerical studies of low-dimensional strange attractors [26] show exponential rather than power-law forms for the high-frequency behaviour.

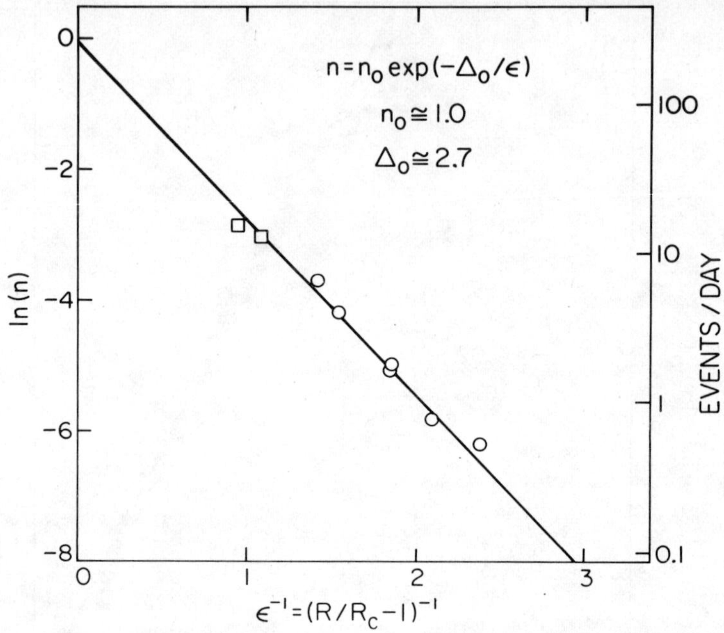

Fig. 9.7 Number of major turbulent events per vertical thermal diffusion time, on a logarithmic scale, as a function of $\varepsilon^{-1} = (R/R_c - 1)^{-1}$. The characteristic time-scale diverges exponentially as R_c is approached [9]

Heat flux experiments confirming the trend toward time-dependence near R_c at large aspect ratio have been performed recently by Behringer *et al.* [27]. At a radius-to-height ratio of 7.87 and Prandtl number 0.6, this group found broadband turbulence above $1.32R_c$.

9.4.2 Laser Doppler experiments

Gollub and Steinman [19] extended their laser Doppler experiments to a rectangular cell of aspect ratio 30 in order to make a comparison with the provocative heat flux experiments of Ahlers and Behringer [7, 8] and Ahlers and Walden [9].

Gollub and Steinman found that the large aspect ratio system is characterized by very long transients, comparable to the horizontal thermal diffusion time from one edge of the layer to the other (about 10^5 s). In the interval $R_c < R < 5R_c$, the fluid often eventually attains a time-independent steady state, but only after a day or two. Therefore, extreme patience is required in order to make the simplest dynamical distinction: the presence or absence of intrinsic time-dependence.

The observed steady-state flow is not the uniform one predicted by stability theory. Rather, the rolls preferentially align with their axes perpendicular to all of the lateral walls of the cell. The resulting pattern is splayed and contains defects where a roll ends. An example of such a

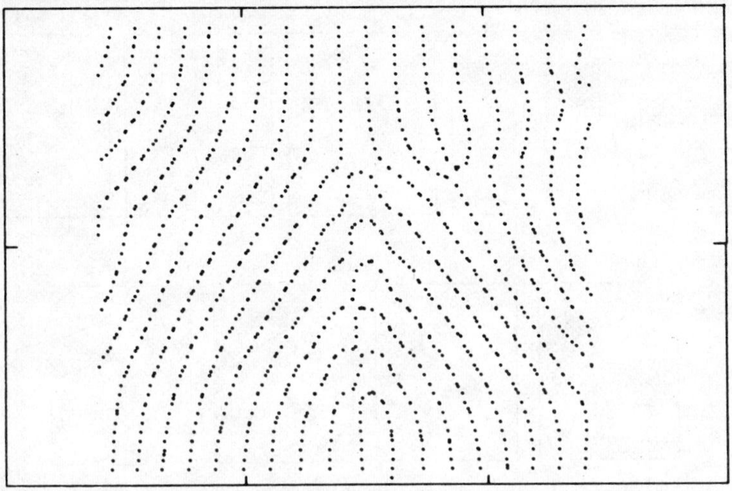

Fig. 9.8 Doppler map showing the structure of the velocity field above the midplane of a large rectangular cell at $R/R_c = 4.3$. The dots mark the roll boundaries, where the flow is entirely vertical (out of or into the page). The pattern contains defects because the rolls align perpendicular to the lateral boundaries, but does not change over several days [19]

pattern is shown in Fig. 9.8, which was obtained by mapping the Doppler shift in a horizontal plane above the centre of the cell. The dots mark the locations of the roll boundaries, where the velocity is entirely vertical.

Stable flows were only observed below $5R_c$ and generally contained two defects, the minimum number needed to allow perpendicular alignment with all lateral boundaries. On the other hand, flows containing many defects could be created by decreasing R from a value above $15R_c$. These patterns are not strictly time-independent, and apparently have extra degrees of freedom that permit gradual evolution of the pattern even after a few days.

All of the structures were found to 'melt' at about $5R_c$. The local velocity begins to fluctuate slowly, with statistics quite similar to those found by Ahlers and Behringer in heat flux measurements. At $5R_c$, the fluctuations are sufficiently slow that a portion of the flow field can be mapped in a time *shorter* than the characteristic time for the fluctuations. Therefore the space and time structure may be followed in detail by repetitive laser Doppler imaging. A sequence of Doppler scans taken at 10 minute intervals is shown in Fig. 9.9. Shaded and unshaded regions are rolls of opposite vorticity (direction of circulation). At the boundaries between these regions, the fluid moves vertically (out of or into the page). In (*b*) a defect *p* enters the field of view from the left. In (*c*) it merges with a roll *q* of similar vorticity. This process forms a defect of opposite polarity (shaded) which is then expelled from the field of view. These and other observations showed that the convective rolls are unstable with respect to deformations above about $R_1 \equiv 5R_c$. The observed onset is

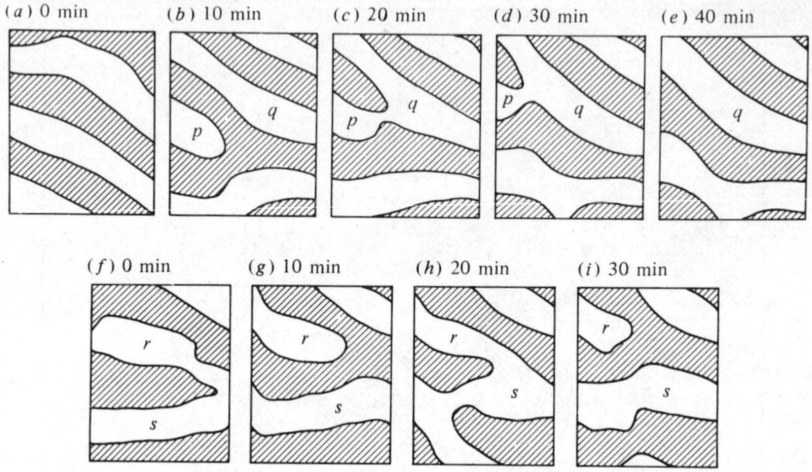

Fig. 9.9 Two sequences showing a portion of the velocity field in a large rectangular cell at intervals of 10 minutes, at $R/R_c = 5$. Shaded and unshaded regions have opposite vorticity. The rolls are clearly unstable, and the onset of this phenomenon appears to coincide with the predicted onset of the skewed varicose instability [19]

consistent with the predicted onset of the 'skewed varicose' instability of Busse and Clever [2]. While the linear stability theory does not predict any time-dependence, it is possible that nonlinear effects would cause this instability to evolve into chaotic time-dependence. Recent numerical simulations of the full Boussinesq equations by Siggia and Zippelius [28] support this view.

The transition at R_1 resembles a melting transition in two respects. First, the lattice of Rayleigh–Bénard rolls was found to be mobile only above R_1. Second, a process akin to annealing was often observed. If R is reduced (quenched) from about $15R_c$ to a value below R_1, the structure usually contains many defects. However, if the system is then held just above R_1 for a day or so, many of the defects disappear.

9.4.3 High Prandtl number experiments

Bergé *et al.* [23] have found many of the phenomena described in Section 9.3.2 at much higher Prandtl numbers (of the order of 100). Provided that the side walls have a thermal conductivity comparable to that of the fluid and are well coupled to the copper plates, these workers also found that the axes of the rolls align preferentially in a direction perpendicular to the lateral boundaries. The onset of time-dependence in flow structures containing defects is associated with a structural change known as the cross-roll instability [29]. Thus, it seems likely that the onset of obvious broadband noise in large aspect ratio convection is associated with identifiable structural instabilities at both low and high Prandtl numbers. However, the extremely slow time-dependence first observed by Ahlers and Walden closer to R_c may have a different origin.

Experiments by Wesfreid and Croquette [30] have shown that the phase variable which describes the position of the convective rolls obeys a diffusion equation, and they have measured the diffusion coefficient. The concept of phase diffusion may prove useful in understanding the time-dependence close to R_c.

9.5 Conclusion

Recent experimental advances have allowed a number of distinctions to be made that are not possible with conventional methods. Computer-assisted experimentation permits quantitative data to be obtained over long periods of time, so that the onset of time-dependence can be reliably determined. Power spectral analysis permits the experimenter to distinguish between complex oscillations and chaotic motion. Cryogenic techniques now provide thermal transport measurements of exceedingly high sensitivity. Finally, laser Doppler methods are beginning to allow simultaneous studies of the space and time structure of the velocity field. One major conclusion from the work of many experimental groups is that low-dimensional nonlinear models are most fruitful for qualitatively describing the time-dependent phenomena seen in fluid layers of small lateral extent. However, there are many experimental observations for which no quantitative explanation yet exists.

References

1. Bénard, H., *Revue Générale des Sciences Pures et Appliquées*, **12,** 1261 and 1309, 1900.
2. See especially Busse, F. H. and Clever, R. M., Instabilities of convection rolls in a fluid of moderate Prandtl number, *Journal of Fluid Mechanics*, **91,** 319, 1979.
3. Ruelle, D. and Takens, F., On the nature of turbulence, *Communications in Mathematical Physics*, **20,** 167, 1971.
4. Lorenz, E. N., Deterministic nonperiodic flow, *Journal of the Atmospheric Sciences*, **20,** 130, 1963.
5. Swinney, H. L. and Gollub, J. P. (Eds), *Hydrodynamic Instabilities and the Transition to Turbulence*, Topics in Applied Physics, Vol. 45: Berlin, Heidelberg and New York, Springer-Verlag, 1981.
6. Ahlers, G., Low temperature studies of the Rayleigh–Bénard instability and turbulence, *Physical Review Letters*, **33,** 1185, 1974.
7. Ahlers, G. and Behringer, R. P., Evolution of turbulence from the Rayleigh–Bénard instability, *Physical Review Letters*, **40,** 712, 1978.
8. Ahlers, G. and Behringer, R. P., The Rayleigh–Bénard instability and the evolution of turbulence, *Progress of Theoretical Physics, Supplement*, **64,** 186, 1978.

9. Ahlers, G. and Walden, R. W., Turbulence near onset of convection, *Physical Review Letters,* **44,** 445, 1980.
10. Libchaber, A. and Maurer, J., Local probe in a Rayleigh–Bénard experiment in liquid helium, *Journal de Physique—Lettres,* **39,** L-369, 1978.
11. Maurer, J. and Libchaber, A., Effect of the Prandtl number on the onset of turbulence in liquid ^{4}He, *Journal de Physique—Lettres,* **41,** L-515, 1980.
12. Libchaber, A. and Maurer, J., Une expérience de Rayleigh–Bénard de géométrie réduite: multiplication, accrochage et démultiplication de fréquences, *Journal de Physique,* **41,** C3-51, 1980.
13. Libchaber, A. in *Nonlinear Phenomena at Phase Transitions and Instabilities* (Proceedings of a Nato Advanced Study Institute, Geilo, Norway, March 1981): New York, Plenum, 1982.
14. Gollub, J. P. and Benson, S. V., Many routes to turbulent convection, *Journal of Fluid Mechanics,* **100,** 449, 1980.
15. Gollub, J. P. and Benson, S. V., Chaotic response to periodic perturbation of a convecting fluid, *Physical Review Letters,* **41,** 948, 1978.
16. Gollub, J. P. and Steinman, J. F., External noise and the onset of turbulent convection, *Physical Review Letters,* **45,** 551, 1980.
17. Gollub, S. V., Benson, S. V. and Steinman, J. F., A subharmonic route to turbulent convection, *Annals of the New York Academy of Sciences,* **357,** 22, 1980.
18. Gollub, J. P. and Benson, S. V., Phase locking in the oscillations leading to turbulence, in Haken, H. (Ed.), *Pattern Formation by Dynamic Systems and Pattern Recognition*: Berlin, Springer-Verlag, 1979.
19. Gollub, J. P. and Steinman, J. F., Doppler imaging of the onset of turbulent convection, *Physical Review Letters,* **47,** 505, 1981; Gollub, J. P., McCarriar, A. R. and Steinman, J. F., Convective pattern evolution and secondary instabilities, *Journal of Fluid Mechanics,* **125,** 259, 1982.
20. Bergé, P., Dubois, M., Manneville, P. and Pomeau, Y., Intermittency in Rayleigh-Bénard convection, *Journal de Physique—Lettres,* **41,** L-341, 1980.
21. Dubois, M. and Bergé, P., Experimental evidence for the oscillators in a convective biperiodic regime, *Physics Letters,* **76A,** 53, 1980.
22. Dubois, M. and Bergé, P., Experimental study of the velocity field in Rayleigh–Bénard convection, *Journal of Fluid Mechanics,* **85,** 641, 1978.
23. Bergé, P., Rayleigh–Bénard convection in high Prandtl number fluid, in Haken, H. (Ed.), *Synergetics,* Proceedings of the International Symposium on Synergetics, Schloss Elmau, 1981: Berlin, Springer-Verlag, 1981.
24. Gollub, J. P., Romer, E. J. and Socolar, J. E., Trajectory divergence

for coupled relaxation oscillators: measurements and models, *Journal of Statistical Physics*, **23,** 321, 1980.

25. Giglio, M., Musazzi, S. and Perini, U., Transition to turbulence via a reproducible sequence of period doubling bifurcations, *Physical Review Letters*, **47,** 243, 1981.
26. Greenside, H., Ahlers, G., Hohenberg, P. C. and Walden, R. W., A simple stochastic model for the onset of turbulence in Rayleigh–Bénard convection, *Physica*, **5D,** 322–334, 1982.
27. Behringer, R. P., Agoste, C., Jan, J. S. and Shaumeyer, J. N., Time-dependent Rayleigh–Bénard convection and instrumental attenuation, *Physics Letters*, **80A,** 273, 1980; Behringer, R. P., Shaumeyer, J. N., Jan, J. S., Agosta, C. and Clark, C. A., Time-dependent convection preceded by a discrete transition (preprint).
28. Siggia, E. and Zippelius, A., Pattern selection in Rayleigh–Bénard convection near threshold, *Physical Review Letters*, **47,** 855, 1981.
29. Busse, F. H., Nonlinear properties of thermal convection, *Reports on Progress in Physics*, **41,** 1929, 1978.
30. Wesfreid, J. E. and Croquette, V., Forced phase diffusion in Rayleigh–Bénard convection, *Physical Review Letters*, **45,** 634, 1980.

10 Homoclinic Orbits, Subharmonics and Global Bifurcations in Forced Oscillations

B D Greenspan and P J Holmes

10.1 Introduction—a physical example

In this chapter we develop a method for the analysis of systems of the form

$$\dot{x} = f(x) + \varepsilon g(x; t), \tag{10.1}$$

where $x = \begin{pmatrix} u \\ v \end{pmatrix} \in \mathbb{R}^2$ and

$$f = \begin{pmatrix} f_1(u, v) \\ f_2(u, v) \end{pmatrix}, \qquad g = \begin{pmatrix} g_1(u, v; t) \\ g_2(u, v; t) \end{pmatrix}$$

are sufficiently smooth (bounded) functions and g is T-periodic in t. In examples of interest g will generally also contain parameters

$$g_\mu(x, t) = g(x; t; \mu) \quad \mu \in \mathbb{R}^m. \tag{10.2}$$

The unperturbed system,

$$\dot{x} = f(x), \tag{10.3}$$

is assumed to be integrable. In particular, we shall concentrate on the Hamiltonian case, in which a real valued function $H: \mathbb{R}^2 \mapsto \mathbb{R}$ exists and

$$f_1(u, v) = \frac{\partial H}{\partial v}, \qquad f_2(u, v) = -\frac{\partial H}{\partial u}. \tag{10.4}$$

The results we describe here can be generalized to multidimensional systems and even to infinite-dimensional Hamiltonian systems arising as evolution equations in studies of partial differential equations (*cf.* [1]), and to the study of multidegree of freedom Hamiltonian systems with perturbations which do not depend explicitly on time (*cf.* [2–4]).

Our basic assumption, made precise in Section 10.3, is that the unperturbed system (10.3) possesses a nondegenerate (hyperbolic) saddle-point

p_0 with a homoclinic orbit $\Gamma^0 = \{g^0(t) \mid t \in \mathbb{R}\}$ such that

$$\lim_{t\to+\infty} q^0(t) = \lim_{t\to-\infty} q^0(t) = p_0. \tag{10.5}$$

We further assume that the interior of Γ^0 is filled with a one-parameter family of periodic orbits $\Gamma^\alpha = \{q^\alpha(t) \mid \alpha \in (-1, 0),\ t \in [0, T_\alpha]\}$ where T_α is the period of $q^\alpha(t)$,

$$\lim_{\alpha\to 0} q^\alpha(t) = q^0(t), \tag{10.6}$$

and $\lim_{\alpha\to-1} q^\alpha(t) = q^{-1}$ is an elliptic fixed point or centre (Fig. 10.1). The homoclinic orbit can be replaced by a homoclinic cycle connecting several saddle-points. Such orbits and cycles filled with periodic motions occur naturally in the phase planes of single degree of freedom nonlinear Hamiltonian systems; the pendulum,

$$\ddot{\theta} + \sin\theta = 0, \tag{10.7}$$

and Duffing's equation with negative linear stiffness,

$$\ddot{y} - y + y^3 = 0, \tag{10.8}$$

provide examples. We select Duffing's equation as our major example in subsequent sections because of its importance in mechanics, which we now briefly indicate. For more details, see [1, 5–8].

Consider a slender elastic rod, clamped at one end in a rigid framework and constrained to move in a plane (Fig. 10.2). Two magnets attached to the frame as indicated cause the beam to buckle either to the left or right, the central position being unstable. The whole framework is now moved sinusoidally, so that the beam vibrates under 'inertial' excitation. For small excitation amplitudes, periodic motions are observed about either stable equilibrium, but as the amplitude is increased an apparently sudden transition to a 'chaotic snap-through' motion is observed. In this state, the beam oscillates irregularly about first one and then the other equilibrium. An example of such a motion, measured from a strain gauge at the beam's root, is shown in Fig. 10.3.

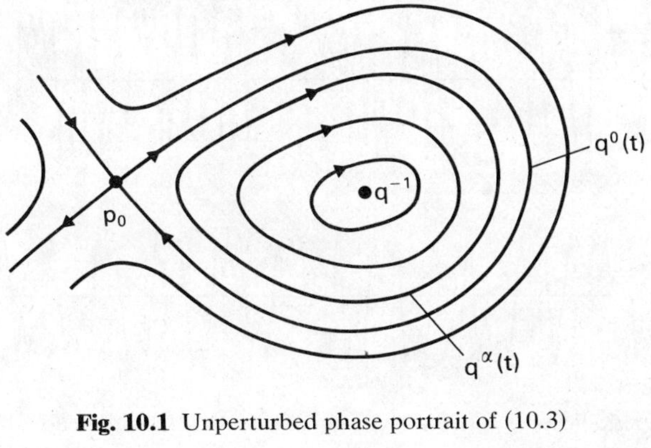

Fig. 10.1 Unperturbed phase portrait of (10.3)

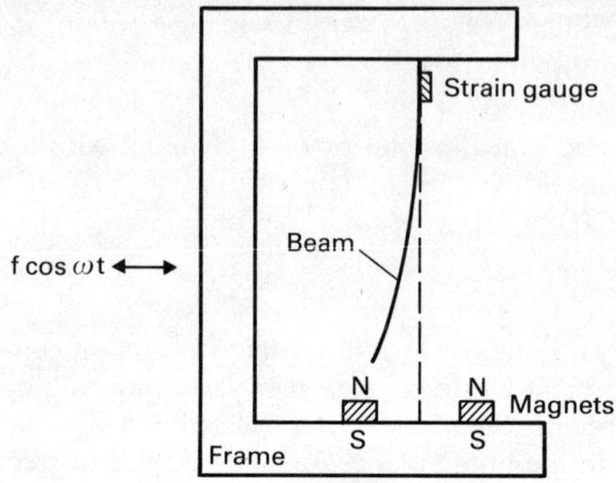

Fig. 10.2 Magnetoelastic beam [6]

As shown by Moon and Holmes [6], the nonlinear partial differential equations of an inextensible elastic rod in a nonuniform magnetic field can be truncated in a single-mode Galerkin approximation to yield Duffing's equation,

$$\ddot{y} - y + y^3 = 0, \tag{10.9}$$

where $y = y(t)$ is the (nondimensional) amplitude of the first mode of vibration and dissipation is ignored. On the introduction of weak dissipation (aerodynamic damping) and periodic forcing, one obtains

$$\ddot{y} - y + y^3 = \varepsilon(\gamma \cos \omega t - \delta\dot{y}), \tag{10.10}$$

which is the equation to be studied in Sections 10.4–10.6.

In Section 10.2 we introduce some basic concepts from the qualitative theory of dynamical systems and then, in Section 10.3, state and sketch the proofs of the results on perturbed integrable systems. In Section 10.4 we apply these results to Duffing's equation and in Section 10.5

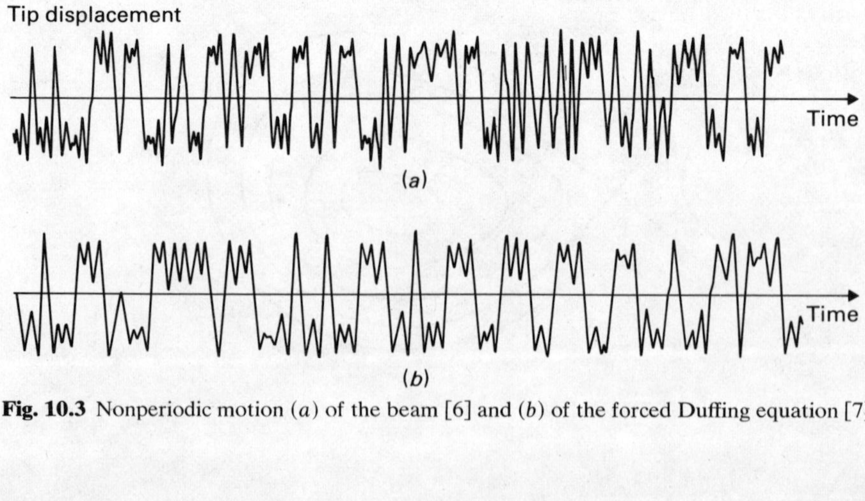

Fig. 10.3 Nonperiodic motion (*a*) of the beam [6] and (*b*) of the forced Duffing equation [7]

we review more of the relevant abstract theory. Finally, in Section 10.6 we bring the abstract theory back to bear on Duffing's equation in an attempt to understand irregular motions such as that of Fig. 10.3. In this chapter, although we assume some familiarity with the qualitative theory of dynamical systems and bifurcation theory, we do attempt to introduce or review most of the important aspects necessary for our study. Background material on differential dynamics may be found in Chillingworth [9], and on bifurcation theory and nonlinear analysis in Iooss and Joseph [10] and Chow and Hale [11]. The latter reference, and Chow, Hale and Mallet-Paret [12] contain many of the results of Section 10.3, but presented from a different viewpoint. The book of Andronov, Vitt and Khaiken [13] continues to provide the best background in nonlinear oscillations.

10.2 Poincaré maps and invariant manifolds

We consider the periodically forced system

$$\dot{x} = f(x) + \varepsilon g(x, t); \qquad x \in \mathbb{R}^2, \quad g(x, t) = g(x, t+T), \tag{10.11}$$

which can be rewritten as an autonomous system defined on the 'toroidal' phase space $\mathbb{R}^2 \times S^1$:

$$\left.\begin{aligned} \dot{x} &= f(x) + \varepsilon g(x, \theta) \\ \dot{\theta} &= 1 \end{aligned}\right\} \quad (x, \theta) \in \mathbb{R}^2 \times S^1, \tag{10.12}$$

where $S^1 = \mathbb{R}/T$, the circle of length T. This naturally copes with the T-periodicity of the vector field. We next define a *cross-section* for the flow,

$$\Sigma^{t_0} = \{(x, \theta) \in \mathbb{R}^2 \times S^1 \mid \theta = t_0 \in [0, T)\} \tag{10.13}$$

and a first return or *Poincaré map*,

$$P_\varepsilon^{t_0} : \Sigma^{t_0} \to \Sigma^{t_0}, \tag{10.14}$$

obtained by following orbits $[x(t+t_0), \theta(t+t_0)]$ based at points $[x(t_0), t_0] \in \Sigma^{t_0}$ to their next intersection with Σ^{t_0}. We have

$$P_\varepsilon^{t_0}[x(0)] = \pi \cdot [x(T+t_0), T+t_0] \tag{10.15}$$

where $[x(t+t_0), t+t_0]$ is the solution of (10.12) based at $[x(t_0), t_0]$ and π denotes projection onto the first factor. When $\varepsilon = 0$, the unperturbed Poincaré map $P_0 = P_0^{t_0}$ is identical on every section Σ^{t_0} in view of the invariance of $\dot{x} = f(x)$ under time translations. P_0 is given simply by the time T flow map of the unperturbed equation

$$P_0[x(0)] = x^0(T), \tag{10.16}$$

where $x^0(t)$ is the unperturbed solution of (10.11) based at $x(0)$. The perturbed maps $P_\varepsilon^{t_0}$ differ on different sections but any two are diffeomorphic and $P_\varepsilon^{t_0+nT} = (P_\varepsilon^{t_0})^n$.

The Poincaré map captures the dynamics of (10.11) in the sense that T-periodic orbits of (10.11) correspond to fixed points of $P_\varepsilon^{t_0}$, mT-periodic subharmonics correspond to periodic cycles of period m and stability types correspond. The study of fixed and periodic points of the two-dimensional map $P_\varepsilon^{t_0}$ is generally easier than that of the corresponding periodic motions in the three-dimensional flow of (10.12).

Suppose that the map P (the sub- and superscripts will be dropped where they are not explicitly needed) has a hyperbolic fixed point p. This implies that the linearized map $DP(p)$ has eigenvalues λ_i with modulus $|\lambda_i| \neq 1$, $i = 1, 2$: a sink, a saddle or a source. The stable manifold theorem (*cf.* [9, 14, 15], Appendix A) asserts that, in a neighbourhood $U(p)$, there are smooth local stable manifolds $W^s_{\text{loc}}(p)$, $W^u_{\text{loc}}(p)$ tangent to the eigenspaces E^s, E^u of the linearized problem

$$\dot{\xi} = DP(p)\xi, \tag{10.17}$$

and of dimension s, u ($s + u = 2$ here) where s (resp. u) is the number of eigenvalues with moduli less than 1 (resp. greater than 1). These manifolds are simply defined as the sets of points asymptotic to p under forward (resp. backward) iterations of P,

$$W^s_{\text{loc}}(p) = \{x \in U(p) \mid P^n(x) \in U(p), \forall n \geq 0 \text{ and } P^n(x) \to p \text{ as } n \to \infty\},$$
$$W^u_{\text{loc}}(p) = \{x \in U(p) \mid P^{-n}(x) \in U(p), \forall n \geq 0 \text{ and } P^{-n}(x) \to p \text{ as } n \to \infty\},$$

and global manifolds are obtained by iteration of the local manifolds backwards (resp. forwards):

$$W^s(p) = \bigcup_{n \geq 0} P^{-n}[W^s_{\text{loc}}(p)],$$

$$W^u(p) = \bigcup_{n \geq 0} P^{n}[W^u_{\text{loc}}(p)].$$

Of course, if both eigenvalues lie inside (resp. outside) the unit circle, then $W^u(p)$ (resp. $W^s(p)$) is empty.

We note that while locally they are smooth submanifolds of $\mathbb{R}^2$, $W^s(p)$ and $W^u(p)$ are not necessarily submanifolds globally, since they may wind back and forth and accumulate on themselves, as they do in the specific examples to follow (Fig. 10.4).

It is important to realize that the orbit of a map $\{P^n(x)\}$ is a sequence of points, rather than a curve as in the case of orbits of flows arising from vector fields. Each of the curves $W^s(p)$, $W^u(p)$ in Fig. 10.4 contains a one-parameter family of such orbits. Uniqueness of solutions implies that the stable (or unstable) manifolds $W^s(p_1)$, $W^s(p_2)$ of two distinct points cannot intersect, nor can such a manifold intersect itself.

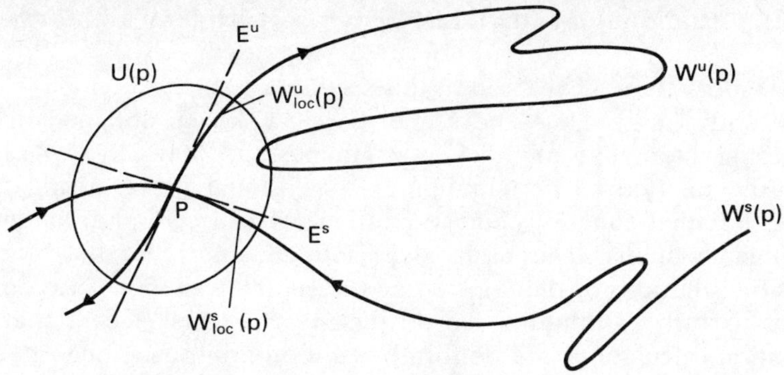

Fig. 10.4 Stable and unstable manifolds

A *homoclinic point* q to a hyperbolic saddle-point p is a point whose orbit approaches p under both forward and backward iterates of P:

$$\lim_{n\to\infty} P^n(q) = \lim_{n\to\infty} P^{-n}(q) = p,$$

thus $q \in W^u(p) \cap W^s(p)$. When two plane curves intersect at a point they generally do so transversely, so that their tangent vectors span $\mathbb{R}^2$. Such a point $q \in W^u(p) \pitchfork W^s(p)$ is called a *transversal homoclinic point* and will play an important role in our analysis. The existence of one such point immediately implies the existence of infinitely many, lying on the orbit $\{P^n(q)\}$. Moreover, the λ-Lemma ([16, 17] and Appendix) implies that $W^u(p)$ accumulates upon itself, and $W^s(p)$ accumulates upon itself, leading to the complicated structure of Fig. 10.5.

In Section 10.5 we discuss some implications of such homoclinic orbits, but first we develop a method for their detection in specific examples.

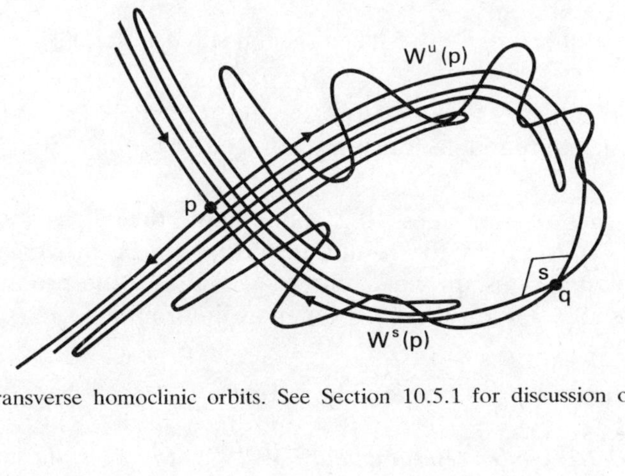

Fig. 10.5 Transverse homoclinic orbits. See Section 10.5.1 for discussion of the role of rectangle S

10.3 Perturbations of integrable systems—Melnikov's method

The basic ideas to be developed in this section are due to Melnikov [18]. More recently Chow, Hale and Mallet-Paret [12] have obtained similar results using alternative methods, and Holmes and Marsden [1–4] have applied the method to certain infinite-dimensional flows arising from partial differential equations and to multidegree of freedom autonomous Hamiltonian systems. The basic idea is to make use of the globally computable solutions of the unperturbed *integrable* system in the computation of perturbed solutions. To do this we must first ensure that the perturbation calculations are uniformly valid on arbitrarily long or semi-infinite time intervals.

First we make our assumptions precise. We consider systems of the form

$$\dot{x} = f(x) + \varepsilon g(x, t); \qquad x = \begin{pmatrix} u \\ v \end{pmatrix} \in \mathbb{R}^2, \tag{10.18}$$

where

$$f = \begin{pmatrix} f_1(x) \\ f_2(x) \end{pmatrix}, \qquad g = \begin{pmatrix} g_1(x, t) \\ g_2(x, t) \end{pmatrix}$$

are sufficiently smooth (C^r, $r \geq 2$) and bounded on bounded sets and g is T-periodic in t. For simplicity, we assume that the unperturbed system is Hamiltonian with $f_1 = \partial H/\partial v$, $f_2 = -\partial H/\partial u$. The non-Hamiltonian case is considered by Melnikov [18] and Holmes [19]. Specific assumptions on the unperturbed flow are:

Assumption 10.1. For $\varepsilon = 0$, (10.18) possesses a homoclinic orbit $q^0(t)$ to a hyperbolic saddle-point p_0.

Assumption 10.2. Let $\Gamma^0 = \{q^0(t) \mid t \in \mathbb{R}\} \cup \{p_0\}$. The interior of Γ^0 is filled with a continuous family of periodic orbits $q^\alpha(t)$, $\alpha \in (-1, 0)$. Letting

$$d(x, \Gamma) = \inf_{q \in \Gamma} |x - q| \qquad \text{we have} \qquad \lim_{\alpha \to 0} \sup_{t \in \mathbb{R}} d[q^\alpha(t), \Gamma] = 0.$$

Assumption 10.3. Let $h_\alpha = H[q^\alpha(t)]$ and T_α be the period of $q^\alpha(t)$. Then T_α is a differentiable function of h_α and $dT_\alpha/dh_\alpha > 0$ inside Γ^0.

We note that Assumptions 10.2 and 3 imply that $T_\alpha \to \infty$ monotonically as $\alpha \to 0$. Many of the results to follow can be proved under less restrictive assumptions. In what follows we indicate the proofs but omit some details. See Greenspan [20] for more information.

10.3.1 Bifurcations to homoclinic orbits

Lemma 10.1. *Under Assumptions* 10.1–3, *for* ε *sufficiently small*

(10.18) *has a unique hyperbolic periodic orbit* $\gamma_\varepsilon(t) = p_0 + 0(\varepsilon)$. *Correspondingly, the Poincaré map* $P_\varepsilon^{t_0}$ *has a unique hyperbolic saddle-point* $p_\varepsilon^{t_0} = p_0 + 0(\varepsilon)$.

Proof. This is a straightforward application of the implicit function theorem, our assumptions implying that $DP_0(p_0)$ does not contain 1 in its spectrum and hence that $Id - DP_0(p_0)$ is invertible and there is a smooth curve of fixed points in (x, ε) space passing through $(p_0, 0)$. ■

Lemma 10.2. *The local stable and unstable manifolds* $W^s_{\text{loc}}(\gamma_\varepsilon)$, $W^u_{\text{loc}}(\gamma_\varepsilon)$ *of the perturbed periodic orbit are* C^r *close to those of the unperturbed periodic orbit* $p_0 \times S^1$. *Moreover, orbits* $q^s_\varepsilon(t, t_0)$, $q^u_\varepsilon(t, t_0)$ *lying in* $W^s(\gamma_\varepsilon)$, $W^u(\gamma_\varepsilon)$ *and based on* Σ^{t_0} *can be expressed as follows, with uniform validity in the indicated time intervals:*

$$\begin{aligned} q^s_\varepsilon(t, t_0) &= q^0(t - t_0) + \varepsilon q^s_1(t, t_0) + 0(\varepsilon^2), \quad t \in [t_0, \infty); \\ q^u_\varepsilon(t, t_0) &= q^0(t - t_0) + \varepsilon q^u_1(t, t_0) + 0(\varepsilon^2), \quad t \in (-\infty, t_0]. \end{aligned} \tag{10.19}$$

Proof. The existence of the perturbed manifolds follows from the invariant manifold theory of Hirsch, Pugh and Shub [15], *cf.* Hartman [14]. We fix a ν-neighbourhood $(0 \leq \varepsilon \ll \nu \ll 1)$ U_ν of p_0, inside which the local perturbed manifolds and their tangent spaces are ε-close to those of the unperturbed flow (or Poincaré map). A standard Gronwall estimate shows that perturbed orbits starting within $0(\varepsilon)$ of $q^0(0)$ remain within $0(\varepsilon)$ of $q^0(t - t_0)$ for finite times and hence that one can follow any such orbit from an arbitrary point $q^0(0)$ on Γ^0 *outside* U_ν to the boundary of U_ν, at, say, $t = t_1$. Once in U_ν, if the perturbed orbit q^s_ε is selected to lie in $W^s(\gamma_\varepsilon)$, then its behaviour is governed by the exponential contraction associated with the linearized system. Straightforward estimates then show that

$$|q^s_\varepsilon(t, t_0) - q^0(t - t_0)| = 0(\varepsilon) \qquad \text{for} \qquad t \in (t_1, \infty).$$

Reversing time, one obtains a similar result for $q^u_\varepsilon(t, t_0)$. ■

Sanders [21] was the first to work out the asymptotics in detail. This Lemma implies that solutions lying in the stable manifold are *uniformly* approximated by the solution q^s_1 of the first variational equation:

$$\dot{q}^s_1(t, t_0) = Df[q^0(t - t_0)]q^s_1(t, t_0) + g[q^0(t - t_0), t]. \tag{10.20}$$

A similar expression holds for $q^u_1(t, t_0)$. Note that the initial time, t_0, appears explicitly since solutions of the perturbed systems are not invariant under arbitrary translations in time.

We next define the *separation* of the manifolds $W^u(p^{t_0}_\varepsilon)$, $W^s(p^{t_0}_\varepsilon)$ on the section Σ^{t_0} at the point $q^0(0)$ as

$$d(t_0) = q^u_\varepsilon(t_0) - q^s_\varepsilon(t_0), \tag{10.21}$$

where $q^u_\varepsilon(t_0) \stackrel{\text{def}}{=} q^u_\varepsilon(t_0, t_0)$, $q^s_\varepsilon(t_0) \stackrel{\text{def}}{=} q^s_\varepsilon(t_0, t_0)$ are the unique points on

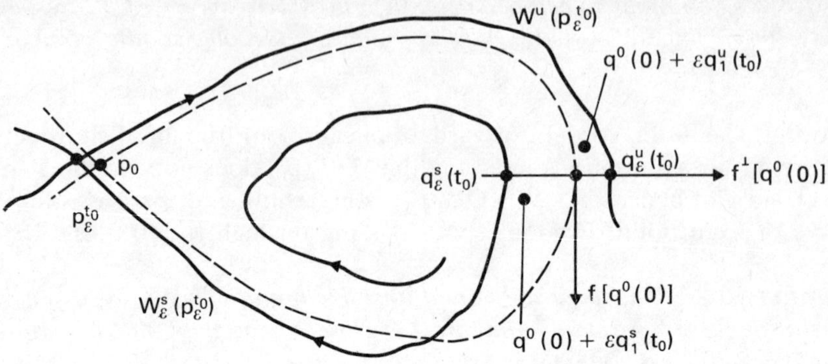

Fig. 10.6 Perturbed manifolds and the distance function

$W^u(p_\varepsilon^{t_0})$, $W^s(p_\varepsilon^{t_0})$ 'closest' to $p_\varepsilon^{t_0}$ and lying on the normal $f^\perp[q^0(0)]=\{-f_2[q^0(0)], f_1[q^0(0)]\}^T$ to Γ at $q^0(0)$. The C^r closeness of the manifolds to Γ, and Lemma 10.2 then imply that

$$d(t_0)=\varepsilon\frac{f(q_0)\wedge[q_1^u(t_0)-q_1^s(t_0)]}{|f[q_0(0)]|}+0(\varepsilon^2). \tag{10.22}$$

Here the wedge product is defined $a\wedge b=a_1b_2-a_2b_1$ and $f\wedge(q_1^u-q_1^s)$ is the projection of $q_1^u-q_1^s$ onto $f^\perp$ (Fig. 10.6).

Finally, we define the *Melnikov function*:

$$M(t_0)=\int_{-\infty}^{\infty} f[q^0(t-t_0)]\wedge g[q^0(t-t_0), t]\,dt. \tag{10.23}$$

Theorem 10.1. *If $M(t_0)$ has simple zeros and maxima and minima of* $0(1)$, *then, for $\varepsilon>0$ sufficiently small, $W^u(p_\varepsilon^{t_0})$ and $W^s(p_\varepsilon^{t_0})$ intersect transversely. If $M(t_0)$ remains bounded away from zero then $W^u(p_\varepsilon^{t_0})\cap W^s(p_\varepsilon^{t_0})=\varnothing$.*

Proof. Consider the time-dependent distance function

$$\Delta(t, t_0)=f[q^0(t-t_0)]\wedge[q_1^u(t, t_0)-q_1^s(t, t_0)]\stackrel{\text{def}}{=}\Delta^u(t, t_0)-\Delta^s(t, t_0) \tag{10.24}$$

and note that $d(t_0)=\varepsilon\Delta(t_0, t_0)/|f[q^0(0)]|+0(\varepsilon^2)$, from (10.22). We compute

$$\frac{\mathrm{d}}{\mathrm{d}t}\Delta^s(t, t_0)=Df[q^0(t-t_0)]\dot{q}_0(t-t_0)\wedge q_1^s(t, t_0)+f[q^0(t-t_0)]\wedge\dot{q}_1^s(t, t_0).$$

Using (10.20) and the fact that $\dot{q}_0=f(q_0)$ this yields

$$\begin{aligned}\Delta^s&=Df(q^0)f(q^0)\wedge q_1^s+f(q^0)\wedge[Df(q^0)q_1^s+g(q^0, t)]\\&=\text{trace }Df(q^0)\,\Delta^s+f(q^0)\wedge g(q^0, t).\end{aligned} \tag{10.25}$$

But, since f is Hamiltonian, trace $Df\equiv0$. Integrating (10.25) from t_0 to ∞

we have

$$\Delta^s(\infty, t_0) - \Delta^s(t_0, t_0) = \int_{t_0}^{\infty} f[q^0(t-t_0)] \wedge g[q^0(t-t_0), t]\,dt. \tag{10.26}$$

However $\Delta^s(\infty, t_0) = \lim_{t\to\infty} f[q^0(t-t_0)] \wedge q_1^s(\infty, t_0)$ and $\lim_{t\to\infty} q^0(t-t_0) = p_0$, so that $\lim_{t\to\infty} f[q^0(t-t_0)] = 0$ while $q_1^s(\infty, t_0)$ is bounded, from Lemma 10.2; thus $\Delta^s(\infty, t_0) = 0$ and (10.26) gives us a formula for $\Delta^s(t_0, t_0)$. A similar calculation gives

$$\Delta^u(t_0, t_0) = \int_{-\infty}^{t_0} f[q^0(t-t_0)] \wedge g[q^0(t-t_0), t]\,dt, \tag{10.27}$$

and addition of (10.26) and (10.27) and use of (10.22) yields

$$d(t_0) = \frac{\varepsilon M(t_0)}{\|f[q^0(0)]\|} + 0(\varepsilon^2). \tag{10.28}$$

Since $\|f[q^0(0)]\| = 0(1)$, $M(t_0)$ provides a good measure of the separation of the manifolds at $q^0(0)$ on Σ^{t_0}. In particular, if $M(t_0)$ oscillates about zero with maxima and minima independent of ε, then $q_\varepsilon^u(t_0) - q_\varepsilon^s(t_0)$ must change its orientation with respect to $f^\perp[q^0(0)]$ as t^0 varies and hence $W^u(p_\varepsilon^{t_0})$ and $W^s(p_\varepsilon^{t_0})$ intersect. If the zeros are simple $((dM/dt_0) \neq 0)$, then it follows that the intersections are transversal. Conversely, if no zeros exist, then q_ε^u and q_ε^s retain the same orientation and hence the manifolds do not intersect. ■

Remark 10.1. We note that $M(t_0)$ is T-periodic in t_0, as it should be, since the maps $P_\varepsilon^{t_0}$ and $P_\varepsilon^{t_0+T}$ are identical and thus $d(t_0) = d(t_0+T)$. In computing $M(t_0)$ we are effectively standing at a fixed point $q^0(0)$ on a moving cross-section Σ^{t_0} and watching the perturbed manifolds oscillate as t_0 varies. In his analysis, Greenspan [20] fixes the section and moves the base point $q^0(0)$.

Remark 10.2. If the perturbation g is derived from a (time-dependent) Hamiltonian function G, $g_1 = \partial G/\partial v$, $g_2 = -\partial G/\partial u$, then we have

$$M(t_0) = -\int_{-\infty}^{\infty} \{H[q^0(t-t_0)], G[q^0(t-t_0), t]\}\,dt, \tag{10.29}$$

where $\{\cdot, \cdot\}$ denotes the Poisson bracket. This is a natural formula if one recalls that the first variation of the unperturbed Hamiltonian energy, H, will be obtained by integrating its evolution equation

$$\dot{H} = \{H, G\} \tag{10.30}$$

along the unperturbed orbit $q^0(t-t_0)$, *cf.* Arnold [22].

Remark 10.3. If $g = g(x)$ is not explicitly time-dependent then we have, using Green's theorem,

$$\int_{-\infty}^{\infty} f[q^0(t-t_0)] \wedge g[q^0(t-t_0)]\, dt = \int_{-\infty}^{\infty} (f_1 g_2 - f_2 g_1)\, dt$$

$$= \int [g_2(u^0, v^0)\dot{u}^0 - g_1(u^0, v^0)\dot{v}^0]\, dt$$

$$= \int_{\Gamma} g_2(u, v)\, du - g_1(u, v)\, dv$$

$$= \int_{\mathrm{int}\,\Gamma} \operatorname{trace} Dg(x)\, dx \qquad (10.31)$$

Thus the formula obtained in Andronov *et al.* [23] is a special (planar) case of the more general Melnikov function which describes the 'splitting' of the perturbed saddle separatrices.

We now turn to the case in which the perturbation $g = g(x, t; \mu)$ depends upon parameters $\mu \in \mathbb{R}^p$. For simplicity, we take $p = 1$.

Corollary 10.1. *Consider the parameterized family $\dot{x} = f(x) + \varepsilon g(x, t; \mu)$, $\mu \in \mathbb{R}$ and let Assumptions* 10.1–3 *hold. Suppose that the Melnikov function $M(t_0, \mu)$ has a quadratic zero*

$$M(\tau, \mu_b) = \frac{\partial M}{\partial t_0}(\tau, \mu_b) = 0 \quad \textit{but} \quad \frac{\partial^2 M}{\partial t_0^2}(\tau, \mu_b) \neq 0 \quad \textit{and} \quad \frac{\partial M}{\partial \mu}(\tau, \mu_b) \neq 0.$$

Then $\mu_B = \mu_b + 0(\varepsilon)$ is a bifurcation value for which quadratic homoclinic tangencies occur in the family of systems.

Proof. By hypothesis, using (10.22), we have

$$d(t_0, \mu) = \varepsilon\{\alpha(\mu - \mu_b) + \beta(t_0 - \tau)^2\} + 0(\varepsilon\, |\mu - \mu_b|^2) + 0(\varepsilon^2) \qquad (10.32)$$

where we have expanded in a Taylor series about $(t_0, \mu) = (\tau, \mu_b)$ and α, β are finite constants. Taking ε sufficiently small we find that $d(t_0, \mu)$ has a quadratic zero with respect to t_0 for some μ_B near μ_b, and hence that $W^u(p_\varepsilon^\tau)$, $W^s(p_\varepsilon^\tau)$ have a quadratic tangency near $q^0(0)$ on Σ^τ. ■

This result is important, since it permits us to verify in specific examples one of the hypotheses of Newhouse's theorem [24] on wild hyperbolic sets (see Section 10.5).

We next turn to the perturbations of the periodic motions q^α inside Γ.

10.3.2 Bifurcations of subharmonics

Once more we start with a perturbation Lemma:

Lemma 10.3. *Let $q^\alpha(t - t_0)$ be a periodic orbit of the unperturbed system based on Σ^{t_0} with period T_α. Then there exists a perturbed orbit $q_\varepsilon^\alpha(t, t_0)$, not* *necessarily periodic, which can be expressed as*

$$q_\varepsilon^\alpha(t, t_0) = q^\alpha(t - t_0) + \varepsilon q_1^\alpha(t, t_0) + 0(\varepsilon^2) \tag{10.33}$$

uniformly in $t \in (t_0, t_0 + T_\alpha)$, *for* ε *sufficiently small and all* $\alpha \in (-1, 0)$.

Proof. The proof relies heavily upon the geometrical structure of the perturbed stable and unstable manifolds established in Lemma 10.2. We again fix a neighbourhood U_ν and take a curve of initial conditions $q^\alpha(0) \in \Sigma^{t_0}$ not lying in U_ν and with $\lim_{\alpha \to 0} q^\alpha(0) = q^0(0)$. Any orbit $q^\alpha(t - t_0)$ starting on such a curve takes a finite time to reach the boundary of U_ν as t increases or decreases and hence we have, say,

$$|q_\varepsilon^\alpha(t, t_0) - q^\alpha(t - t_0)| = 0(\varepsilon) \quad \text{for} \quad t \in (t_0 - t_1, t_0 + t_2).$$

Once in U_ν, the perturbed and unperturbed orbits may take arbitrarily long to pass through and exit, since, as $\alpha \to 0$, they pass arbitrarily close to the saddle-point p_0 (or to γ_ε). However, for fixed $\varepsilon = \varepsilon_0$ we can find a set of orbits *lying sufficiently close to the stable and unstable manifolds* which remain within $0(\varepsilon)$ of those manifolds until they enter a $c\varepsilon$ neighbourhood $U_{c\varepsilon}$ of p, which is chosen to contain γ_ε (Fig. 10.7). This implies that we can extend our estimate uniformly to $t \in (t_0 - t_3, t_0 + t_4)$, where $t_3 + t_4$ is the time required for the unperturbed orbit q^α to pass from the boundary of $U_{c\varepsilon}$ and return to it. It remains to check that $q_\varepsilon^\alpha(t, t_0)$ and $q^\alpha(t - t_0)$ remain within $0(\varepsilon)$ during the arbitrarily long passage through $U_{c\varepsilon}$. This follows from the 'shear' on the flow near p; one has to check that q_ε^α exits from $U_{c\varepsilon}$ 'at the correct time' and since orbits passing near γ_ε take arbitrarily long to exit, and those near the boundary of $U_{c\varepsilon}$ arbitrarily short times, at least one orbit can be found for any given (unperturbed) time of passage. It follows that an initial condition

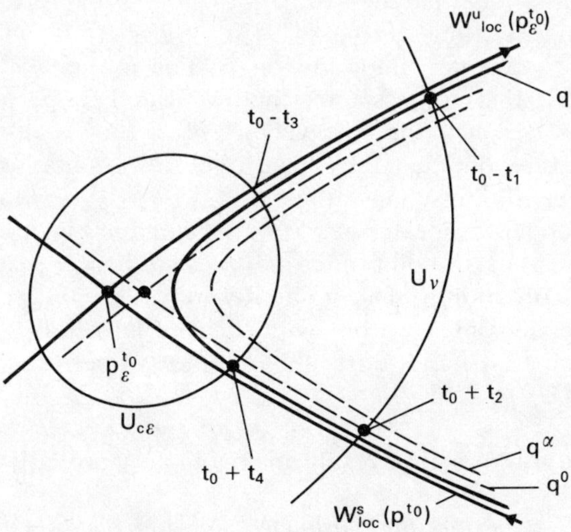

Fig. 10.7 Orbits in U_ν and $U_{c\varepsilon}$

$q_\varepsilon^\alpha(t_0) \in \Sigma^{t_0}$ can be picked within $0(\varepsilon)$ of $q^\alpha(0)$ and $q^0(0)$ such that the orbit $q_\varepsilon^\alpha(t_0)$ will remain within $0(\varepsilon)$ of $q^\alpha(t-t_0)$ (and $q_\varepsilon^s(t-t_0)$) until it reaches $U_{c\varepsilon}$. It will then 'transfer its allegiance' to $q_\varepsilon^u(t-t_0)$ until it once more reaches an ε-neighbourhood of $q^\alpha(0)$. Throughout it remains within $0(\varepsilon)$ of $q^\alpha(t-t_0)$. This takes care of orbits with unperturbed periods T_α *larger* than some $T'_\alpha = T'_\alpha(\varepsilon_0)$ depending on ε_0. For orbits with periods shorter than $T'_\alpha(\varepsilon_0)$ a standard Gronwall estimate ensures ε-closeness. Then we have our result for all T_α and $\varepsilon = \varepsilon_0$. But since f and g are C^r the solution will vary smoothly in ε and thus the result holds for all $0 < \varepsilon \leq \varepsilon_0$. For details see [20]. ■

We next define the subharmonic Melnikov function. Letting $q^\alpha(t-t_0)$ be a periodic orbit of period mT/n, with m and n relatively prime, we set

$$M^{m/n}(t_0) = \int_0^{mT} f[q^\alpha(t-t_0)] \wedge g[q^\alpha(t-t_0), t]\, dt. \tag{10.34}$$

Theorem 10.2. *If $M^{m/n}(t_0)$ has simple zeros and maxima and minima of $0(1)$, and $dT_\alpha/dh_\alpha \neq 0$, then for $0 < \varepsilon \leq \varepsilon(n)$, (10.18) has a subharmonic orbit of period mT. If $n = 1$ then the result is uniformly valid in $0 < \varepsilon \leq \varepsilon_0$.*

Proof. A calculation similar to that of Theorem 10.1 shows that

$$\begin{aligned} f[q^\alpha(0)] \wedge [q_\varepsilon^\alpha(t_0+mT, t_0) - q_\varepsilon^\alpha(t_0, t_0)] \\ = \varepsilon \int_0^{mT} f[q^\alpha(t-t_0)] \wedge g[q^\alpha(t-t_0), t]\, dt + 0(\varepsilon^2). \end{aligned} \tag{10.35}$$

Thus if $M^{m/n}(t_0)$ has a simple zero then there is a perturbed orbit $q_\varepsilon^\alpha(t, t_0)$ which leaves $q_\varepsilon^\alpha(t_0)$ and returns to Σ^{t_0} at $q_\varepsilon^\alpha(t_0+mT)$ with the vector $q_\varepsilon^\alpha(t_0+mT) - q_\varepsilon^\alpha(t_0) \subset \Sigma^{t_0}$ parallel to $f[q^\alpha(0)]$; i.e., the projection onto $f^\perp[q^\alpha(0)]$ vanishes. Letting $M^\beta(t_0) = \int_0^{mT} f(q^\beta) \wedge g(q^\beta, t)\, dt$ for β near α, it is clear that M^β depends smoothly on β. For $\beta < \alpha$, $T_\beta < T_\alpha$ and for $\beta > \alpha$, $T_\beta > T_\alpha$; it therefore follows that we can find perturbed orbits $q_\varepsilon^{\beta_1}, q_\varepsilon^{\beta_2}$, $\beta_1 < \alpha < \beta_2$, such that the vectors $q_\varepsilon^{\beta_i}(t_0+mT) - q_\varepsilon^{\beta_i}(t_0) \subset \Sigma^{t_0}$ are parallel to $f[q^{\beta_i}(0)]$, but that they have opposite orientations. Thus the curve of initial conditions connecting $q_\varepsilon^{\beta_1}(t_0)$ to $q_\varepsilon^{\beta_2}(t_0)$ is mapped back to the section Σ^{t_0} under m iterates of $P_\varepsilon^{t_0}$, as indicated in Fig. 10.8. It follows that $(P_\varepsilon^{t_0})^m$ has a fixed point near $q^\alpha(0)$, and hence that there is a subharmonic of order m/n. The nonuniformity for $n > 1$ arises because Lemma 10.3 applies only to orbits with times of length $T_\alpha = mT$ making one pass through $U_\nu(p)$ and 'ultrasubharmonics' of period mT/n make n passes through $U_\nu(p)$. ■

We also have a bifurcation result analogous to Corollary 10.1:

Corollary 10.2. *Consider the parameterized family $\dot{x} = f(x) + \varepsilon g(x, t; \mu)$, $\mu \in \mathbb{R}$, and let Assumptions 10.1–3 hold. Suppose that $M^{m/n}(t_0, \mu)$ has a*

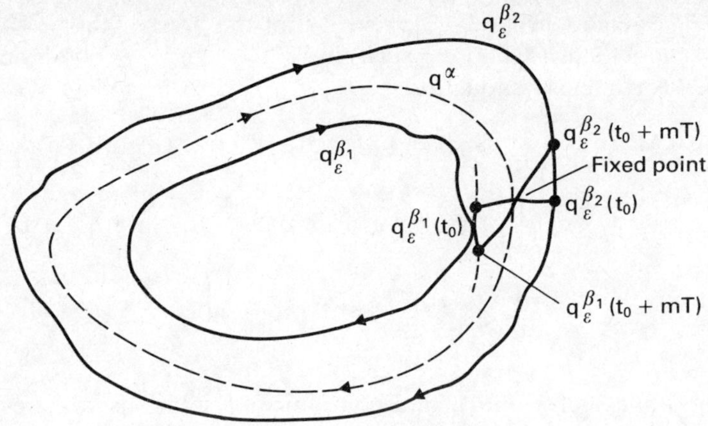

Fig. 10.8 Proof of Theorem 10.2

quadratic zero

$$M^{m/n} = \frac{\partial M^{m/n}}{\partial t_0} = 0; \qquad \frac{\partial^2 M^{m/n}}{\partial t_0^2}, \frac{\partial M^{m/n}}{\partial \mu} \neq 0 \quad at \quad \mu = \mu_b.$$

Then $\mu_{m/n} = \mu_b + 0(\varepsilon)$ is a bifurcation value at which saddle-nodes occur.

Proof. The proof is identical to that of Corollary 10.1. We will consider the stability types of the orbits created in this bifurcation in Section 10.3.3. ■

The next result is a generalization of one obtained by Chow *et al.* [12]. It implies that the homoclinic bifurcation is the limit of a countable sequence of subharmonic saddle-node bifurcations.

Theorem 10.3. *Let $M^{m/1}(t_0) = M^m(t_0)$, then*

$$\lim_{m \to \infty} M^m(t_0) = M(t_0). \tag{10.36}$$

Proof. We must show that the integral

$$M^m(t_0) = \int_{-mT/2}^{mT/2} f[q^\alpha(t-t_0)] \wedge g[g^\alpha(t-t_0), t]\, dt \tag{10.37}$$

converges to

$$M(t_0) = \int_{-\infty}^{\infty} f[q^0(t-t_0)] \wedge g[q^0(t-t_0), t]\, dt, \tag{10.38}$$

as $\alpha \to 0$ and $m \to \infty$. (Note that the periodicity of $M^m(t_0)$ implies that we can change the limits from $0 \to mT$ to $-(mT/2) \to (mT/2)$.) Letting

$$\Gamma^\alpha = \{q^\alpha(t) \mid t \in [0, T_\alpha)\} \qquad \text{and} \qquad \Gamma^0 = \{q^0(t) \mid t \in \mathbb{R}\} \cup \{p_0\},$$

we select a neighbourhood $U_\nu(p)$ such that the arc lengths of Γ^0, $\Gamma^\alpha \cap U_\nu(p)$ are less than ν. Choose τ such that $q^0(-\tau)$ and $q^0(\tau)$ both lie within U_ν. Then for α close enough to zero, $q^\alpha(\pm\tau)$ also lie in U_ν. We have

$$\begin{aligned} M(t_0) - M^m(t_0) = & \left[\int_{-\tau}^{\tau} f(q^0)\wedge g(q^0, t)\,dt - \int_{-\tau}^{\tau} f(q^\alpha)\wedge g(q^\alpha, t)\,dt\right] \\ & + \left[\int_{-\infty}^{-\tau} f(q^0)\wedge g(q^0, t)\,dt + \int_{\tau}^{\infty} f(q^0)\wedge g(q^0, t)\,dt \right. \\ & \left. - \int_{-mT/2}^{-\tau} f(q^\alpha)\wedge g(q^\alpha, t)\,dt - \int_{\tau}^{mT/2} f(q^\alpha)\wedge g(q^\alpha, t)\,dt\right]. \end{aligned} \tag{10.39}$$

The smoothness of f and g and continuity of solutions with respect to initial conditions implies that, given $\nu > 0$, there is an $\alpha' < 0$ such that, for $\alpha \in (\alpha', 0)$ the first [bracketed] term of (10.39) is less than ν. Clearly, as $\alpha' \to 0^-$, $m \to \infty$. The second term may be expressed as the difference between two integrals over arcs of Γ^0 and Γ^α. Using the arc-length increment $ds = \sqrt{(\dot{u}^{02} + \dot{v}^{02})}\,dt = |f(q^0)|\,dt$ on Γ^0 and $ds = |f(q^\alpha)|\,dt$ on Γ^α, the second term becomes

$$\oint_{q^0(-\tau)}^{q^0(\tau)} f(q^0)\wedge g(q^0, t)\frac{ds}{|f(q^0)|} - \oint_{q^\alpha(-\tau)}^{q^\alpha(\tau)} f(q^\alpha)\wedge g(q^\alpha, t)\frac{ds}{|f(q^\alpha)|}. \tag{10.40}$$

Our assumptions on g imply that

$$\sup_{\substack{q \in B_\nu(p) \\ t \in \mathbb{R}}} |g(q^\alpha, t)| = K < \infty,$$

and thus that the second term is bounded by $2K\nu$. Hence, for $\alpha \in (\alpha', 0)$,

$$|M(t_0) - M^m(t_0)| < (2K+1)\nu$$

and $|M(t_0) - M^m(t_0)| \to 0$ as $\alpha \to 0$. ■

10.3.3 Higher-order terms and stability

We next develop a perturbation method which enables us to compute the global structure of the perturbed Poincaré map $P_\varepsilon^{t_0}$, and to determine how the sets of subharmonics and homoclinic orbits are related. Our starting-point is Melnikov [18, §7], although we have somewhat modified his transformations.

Since the unperturbed system is Hamiltonian, a symplectic change of coordinates to action angle variables can be found in the *interior* of Γ^0:

$$I = I(u, v)\theta = \theta(u, v). \tag{10.41}$$

Under this change of coordinates (10.18) becomes

$$\begin{aligned} \dot{I} &= \varepsilon\left(\frac{\partial I}{\partial u} g_1 + \frac{\partial I}{\partial v} g_2\right) \stackrel{\text{def}}{=} \varepsilon F(I, \theta, t) \\ \dot{\theta} &= \Omega(I) + \varepsilon\left(\frac{\partial \theta}{\partial u} g_1 + \frac{\partial \theta}{\partial v} g_2\right) \stackrel{\text{def}}{=} \Omega(I) + \varepsilon G(I, \theta, t), \end{aligned} \tag{10.42}$$

where

$$\Omega(I^\alpha)=\frac{\partial H}{\partial I}(I^\alpha)=\frac{2\pi}{T_\alpha}$$

is the angular frequency of the unperturbed orbit $q^\alpha(t-t_0)$ with action $I^\alpha = I(q^\alpha)$. We now consider small perturbations of a resonant orbit $T_\alpha = mT/n$. Letting

$$\begin{aligned} I &= I^\alpha + \sqrt{(\varepsilon)}h \\ \theta &= \Omega(I^\alpha)t+\varphi = \left(\frac{2\pi n}{mT}\right)t+\varphi \overset{\text{def}}{=} \Omega^\alpha t+\varphi, \end{aligned} \tag{10.43}$$

we obtain

$$\begin{aligned} \dot{h} &= \sqrt{(\varepsilon)}F(I^\alpha, \Omega^\alpha t+\varphi, t)+\varepsilon F'(I^\alpha, \Omega^\alpha t+\varphi, t)h, \\ \dot{\varphi} &= \sqrt{(\varepsilon)}\Omega'(I^\alpha)h+\varepsilon[G(I^\alpha, \Omega^\alpha t+\varphi, t)+(\Omega''(I^\alpha)h^2/2)], \end{aligned} \quad +0(\varepsilon^{3/2}), \tag{10.44}$$

where the prime denotes $\partial/\partial I$. Here we have expanded in Taylor series and used the fact that $\Omega' \neq 0$, since $\mathrm{d}T_\alpha/\mathrm{d}h_\alpha \neq 0$. Noting that

$$\frac{\partial I}{\partial u}=\frac{\partial I}{\partial H}\frac{\partial H}{\partial u}=-\frac{1}{\Omega(I)}f_2 \quad \text{and} \quad \frac{\partial I}{\partial v}=\frac{1}{\Omega(I)}f_1,$$

the leading term of (10.44) can be rewritten

$$\begin{aligned} \dot{h} &= \sqrt{(\varepsilon)}\frac{1}{\Omega^\alpha}f[q^\alpha(t-\varphi/\Omega^\alpha)]\wedge g[q^\alpha(t-\varphi/\Omega^\alpha), t] \\ \dot{\varphi} &= \sqrt{(\varepsilon)}\Omega'(I^\alpha)h. \end{aligned} \tag{10.45}$$

Provided that $\Omega'(I^\alpha)$ is bounded, for $\sqrt{(\varepsilon)}$ sufficiently small the averaging theorem (*cf.* Hale [25]) can be applied to (10.45) to yield

$$\dot{\bar{h}} = \sqrt{(\varepsilon)}\frac{1}{\Omega^\alpha}\frac{1}{mT}\int_0^{mT} f[q^\alpha(t-\varphi/\Omega^\alpha)]\wedge g[q^\alpha(t-\varphi/\Omega^\alpha), t]\,\mathrm{d}t$$

or

$$\dot{\bar{h}} = \sqrt{(\varepsilon)}\frac{1}{2\pi n}M^{m/n}\left(\frac{\bar{\varphi}}{\Omega^\alpha}\right), \qquad \dot{\bar{\varphi}} = \sqrt{(\varepsilon)}\Omega'(I^\alpha)\bar{h}. \tag{10.46}$$

Under the averaging theorem, the hyperbolic or elliptic fixed points of (10.46) correspond to small periodic motions of (10.44) and hence to subharmonics of order m/n of (10.18). It is, of course, no coincidence that a necessary and sufficient condition for the existence of such fixed points is that the Melnikov function $M^{m/n}$ have simple zeros and that $\Omega(I^\alpha)\neq 0$ $(\mathrm{d}T_\alpha/\mathrm{d}h_\alpha \neq 0)$.

We note that (10.46) is a structurally unstable Hamiltonian system with Hamiltonian

$$H = \sqrt{(\varepsilon)}\left(\frac{\Omega' h^2}{2} - V(\bar{\varphi})\right), \tag{10.47}$$

where

$$V(\bar{\varphi})=\frac{1}{2\pi n}\int M^{m/n}(\bar{\varphi}/\Omega^{\alpha})\,\mathrm{d}\varphi;$$

thus, to determine the stability and the global behaviour of orbits of the unperturbed system near the resonant orbit q^{α}, we must investigate the terms of $0(\varepsilon)$. Letting

$$f\wedge g=\frac{1}{2\pi n}M^{m/n}\left(\frac{\varphi}{\Omega^{\alpha}}\right)+\tilde{F}(\varphi,t),$$

where $\tilde{F}$ has period T and zero mean, the averaging transformation is

$$h\to\bar{h}+\sqrt{(\varepsilon)}\int\tilde{F}(\bar{\varphi},t)\,\mathrm{d}t,\qquad\varphi\to\bar{\varphi},\tag{10.48}$$

where the antiderivative is defined up to a t-independent term, generally taken to be zero. Using (10.48), (10.44) becomes

$$\begin{aligned}\dot{\bar{h}}&=\sqrt{(\varepsilon)}\frac{1}{2\pi n}M^{m/n}(\varphi/\Omega^{\alpha})+\varepsilon\left[(F'(I^{\alpha},\Omega^{\alpha}t+\bar{\varphi},t)\bar{h}-\frac{\partial}{\partial\varphi}\int\tilde{F}\,\mathrm{d}t\Omega'\bar{h}\right],\\&\qquad\qquad+0(\varepsilon^{3/2}).\qquad(10.49)\\\dot{\bar{\varphi}}&=\sqrt{(\varepsilon)}\Omega'\bar{h}+\varepsilon\left[\frac{\Omega''h^2}{2}+G(I^{\alpha},\Omega^{\alpha}t+\bar{\varphi},t)+\Omega'\int\tilde{F}\,\mathrm{d}t\right],\end{aligned}$$

Since $\tilde{F}$ has zero mean (it is simply a sum of Fourier components), $\int\tilde{F}$ and $(\partial/\partial\varphi)\int\tilde{F}$ also have zero mean and on a second application of averaging to the $0(\varepsilon)$ terms of (10.49) we obtain (dropping the bars)

$$\left.\begin{aligned}\dot{h}&=\sqrt{(\varepsilon)}\frac{1}{2\pi n}M^{m/n}(\varphi/\Omega^{\alpha})+\varepsilon\bar{F}'(\varphi)h,\\\dot{\varphi}&=\sqrt{(\varepsilon)}\Omega'h+\varepsilon\left(\frac{\Omega''h^2}{2}+\bar{G}(\varphi)\right),\end{aligned}\right\}+0(\varepsilon^{3/2}),\tag{10.50}$$

where $\bar{F}'$, $\bar{G}$ are the averages of F' and G. As Morosov [26] notes, this second-order averaging generally suffices to determine the stability of the fixed points and hence of the bifurcating subharmonics, at least for $\Omega'<\infty$ and ε sufficiently small. However, as we shall see in our application in Sections 10.4 and 10.6, one can also obtain global information on the Poincaré map by considering the time T flow maps of the averaged systems (10.50) in the neighbourhood of each resonant and nonresonant periodic orbit. These results on the full Poincaré map P_{ε} follow from application of the averaging theorem [25].

10.4 Global bifurcations of Duffing's equation

In this section we apply the results of Section 10.3 to the Duffing equation with weak sinusoidal forcing and damping. Written as a system,

we have

$$\begin{aligned}\dot{u} &= v \\ \dot{v} &= u - u^3 + \varepsilon(\gamma \cos \omega t - \delta v),\end{aligned} \tag{10.51}$$

where the force amplitude, γ, frequency, ω, and the damping δ are variable parameters and ε is a small scaling parameter. For $\varepsilon = 0$ the system has centres at $(u, v) = (\pm 1, 0)$ and a hyperbolic saddle at $(0, 0)$. The level set,

$$H(u, v) = \frac{v^2}{2} - \frac{u^2}{2} + \frac{u^4}{4} = 0. \tag{10.52}$$

is composed of two homoclinic orbits, Γ_+^0, Γ_-^0 and the point $p_0 = (0, 0)$. The unperturbed homoclinic orbits are given by

$$\begin{aligned}q_+^0(t - t_0) &= [\sqrt{(2)}\,\mathrm{sech}(t - t_0), -\sqrt{(2)}\,\mathrm{sech}(t - t_0)\tanh(t - t_0)], \\ q^0 &= -q_+^0.\end{aligned} \tag{10.53}$$

Within each of the loops $\Gamma_\pm^0$ there is a one-parameter family of periodic orbits which may be written

$$\begin{aligned}q_+^k(t - t_0) &= \left[\sqrt{\left(\frac{2}{2 - k^2}\right)}\,\mathrm{dn}\left(\frac{t - t_0}{\sqrt{(2 - k^2)}}, k\right), \frac{-\sqrt{(2)}k^2}{2 - k^2}\right. \\ &\quad \left.\times\, \mathrm{sn}\left(\frac{t - t_0}{\sqrt{(2 - k^2)}}, k\right)\mathrm{cn}\left(\frac{t - t_0}{\sqrt{(2 - k^2)}}, k\right)\right], \\ q_-^k(t - t_0) &= -q_+^k(t - t_0),\end{aligned} \tag{10.54}$$

where sn, cn and dn are the Jacobi elliptic functions and k is the elliptic modulus. As $k \to 1$, $q_\pm^k \to q_\pm^0 \cup \{0, 0\}$, and as $k \to 0$, $q_\pm^k \to (\pm 1, 0)$. We have selected initial conditions at $t = t_0$:

$$q_\pm^0(0) = (\pm\sqrt{(2)}, 0), \qquad q_\pm^k(0) = \left[\pm\sqrt{\left(\frac{2}{2 - k^2}\right)}, 0\right]. \tag{10.55}$$

The orbits lying outside $\Gamma_+^0 \cup \{(0, 0)\} \cup \Gamma_-^0$ are given by

$$\begin{aligned}\hat{q}^k(t - t_0) &= \left[\sqrt{\left(\frac{2k^2}{2k^2 - 1}\right)}\,\mathrm{cn}\left(\frac{t - t_0}{\sqrt{(2k^2 - 1)}}, k\right),\right. \\ &\quad \left.-\frac{\sqrt{(2)}k}{2k^2 - 1}\,\mathrm{sn}\left(\frac{t - t_0}{\sqrt{(2k^2 - 1)}}, k\right)\mathrm{dn}\left(\frac{t - t_0}{\sqrt{(2k^2 - 1)}}, k\right)\right],\end{aligned} \tag{10.56}$$

where $k \in (1/\sqrt{(2)}, 1)$ and $\hat{q}^k \to q_+^0 \cup \{(0, 0)\} \cup q_-^0$ as $k \to 1$ and $\hat{q}^k$ becomes unbounded as $k \to 1/\sqrt{(2)}$.

We note that the Hamiltonian (10.52) can be rewritten within Γ_+^0 (or Γ_-^0) in terms of the elliptic modulus k.

$$H(q^k) = \frac{k^2 - 1}{(2 - k^2)^2} \stackrel{\text{def}}{=} h_k. \tag{10.57}$$

Moreover, the period of these orbits is given by

$$T_k = 2K(k)\sqrt{(2-k^2)}, \tag{10.58}$$

where $K(k)$ is the complete elliptic integral of the first kind. T increases monotonically in k with $\lim_{k\to 0} T_k = \sqrt{(2\pi)}$, $\lim_{k\to 1} T_k = \infty$ and

$$\frac{dT_k}{dh_k} = \frac{dT_k/dk}{dH/dk} > 0 \tag{10.59a}$$

and

$$\lim_{k\to 1} \frac{dT_k}{dh_k} = \infty. \tag{10.59b}$$

We first compute the Melnikov function for q_+^0 (the computation for q_-^0 is identical):

$$\begin{aligned} M(t_0, \gamma, \delta, \omega) &= \int_{-\infty}^{\infty} v^0(t-t_0)[\gamma \cos \omega t - \delta v^0(t-t_0)]\, dt \\ &= -\sqrt{(2)}\gamma \int_{-\infty}^{\infty} \operatorname{sech}(t-t_0) \tanh(t-t_0) \cos \omega t \, dt \\ &\quad -2\delta \int_{-\infty}^{\infty} \operatorname{sech}^2(t-t_0) \tanh^2(t-t_0)\, dt. \end{aligned} \tag{10.60}$$

The integrals are easily evaluated (the first by the method of residues) to yield

$$M(t_0, \gamma, \delta, \omega) = -\frac{4\delta}{3} - \sqrt{(2)}\gamma\pi\omega \operatorname{sech}\left(\frac{\pi\omega}{2}\right) \sin \omega t_0. \tag{10.61}$$

If we define

$$R^0(\omega) = \frac{4 \cosh(\pi\omega/2)}{3\sqrt{(2)}\pi\omega}, \tag{10.62}$$

then it follows from Theorem 10.1 that if $\gamma/\delta > R^0(\omega)$, $W^s(p_\varepsilon)$ intersects $W^u(p_\varepsilon)$ for ε sufficiently small and if $\gamma/\delta < R^0(\omega)$, $W^s(p_\varepsilon) \cap W^u(p_\varepsilon) = \varnothing$. Moreover, since $M(t_0, \gamma, \delta, \omega)$ has quadratic zeros when $\gamma/\delta = R^0(\omega)$, it follows from Corollary 10.1 that there is a bifurcation curve in the γ, δ plane for each fixed ω tangent to $\gamma = R^0(\omega)\delta$ at $\gamma = \delta = 0$ on which quadratic homoclinic tangencies occur. Some calculated bifurcation curves are shown in Fig. 10.9 on page 192.

We next compute the Melnikov function for the resonant periodic orbits. We will only consider those within Γ_+^0, $q_+^k(t-t_0)$. The resonance condition is, from (10.58),

$$2K(k)\sqrt{(2-k^2)} = \frac{2\pi m}{\omega n} \tag{10.63}$$

and for each choice of m, n with $(2\pi m/\omega n) > \sqrt{(2)}\pi$, (10.63) can be solved to give a unique resonant orbit $q_+^{k(m,n)}$. Computing

$$M^{m/n}(t_0, \gamma, \delta, \omega) = \int_0^{mT} v^{k(m,n)}(t-t_0)[\gamma \cos \omega t - \delta v^{k(m,n)}(t-t_0)]\, dt, \tag{10.64}$$

using the Fourier series for (10.54) and Remark 10.3 following Theorem 10.1, we obtain

$$M^{m/n}(t_0, \gamma, \delta, \omega) = -\delta J_1(m, n) - \gamma J_2(m, n, \omega) \sin \omega t_0, \tag{10.65}$$

where

$$J_1(m, n) = \frac{2}{3} \frac{[2-k^2(m, n)]2E[k(m, n)] - 4k'^2(m, n)K[k(m, n)]}{[2-k^2(m, n)]^{3/2}},$$

and

$$J_2(m, n, \omega) = \begin{cases} 0, & n \neq 1 \\ \sqrt{(2)}\pi\omega \operatorname{sech} \dfrac{(\pi m K'[k(m, 1)]}{K[k(m, 1)]}, & n = 1. \end{cases}$$

Here $E(k)$ is the complete elliptic integral of the second kind and k' is the complementary elliptic modulus $k'^2 = 1 - k^2$. Defining

$$R^m(\omega) = \frac{J_1(m, 1)}{J_2(m, 1, \omega)}, \tag{10.66}$$

we conclude from Theorem 10.2 and Corollary 10.2 that if $\gamma/\delta > R^m(\omega)$ then there is a pair of subharmonics of order m (period $2\pi m/\omega$) which appear on a bifurcation curve tangent to $\gamma = R^m(\omega)\delta$ at $\gamma = \delta = 0$.

Routine computations verify that

$$\lim_{m\to\infty} M^{m/1}(t_0, \gamma, \delta, \omega) = M(t_0, \gamma, \delta, \omega), \tag{10.67}$$

that *the limit is approached from below* and that the rate of convergence is extremely rapid (*cf.* Fig. 10.18, p. 209).

Similar computations for the orbits lying outside $\Gamma_+^0 \cup \{(0, 0)\} \cup \Gamma_-^0$ give the Melnikov function

$$\hat{M}^{m/n}(t_0, \gamma, \delta, \omega) = -\delta \hat{J}_1(m, n) - \gamma \hat{J}_2(m, n, \omega) \sin \omega t_0 \tag{10.68}$$

with

$$\hat{J}_2(m, n) = \frac{2}{3} \frac{[2k^2(m, n) - 1]4E[k(m, n)] + 4k'(m, n)^2 K[k(m, n)]}{[2k^2(m, n) - 1]^{3/2}}$$

and

$$\hat{J}_2(m, n, \omega) = \begin{cases} 0, & n \neq 1, \quad m \text{ even}, \\ 2\sqrt{(2)}\pi\omega \operatorname{sech}\left(\dfrac{m\pi K'(m, 1)}{K(m, 1)}\right), & n = 1, \ m \text{ odd}. \end{cases}$$

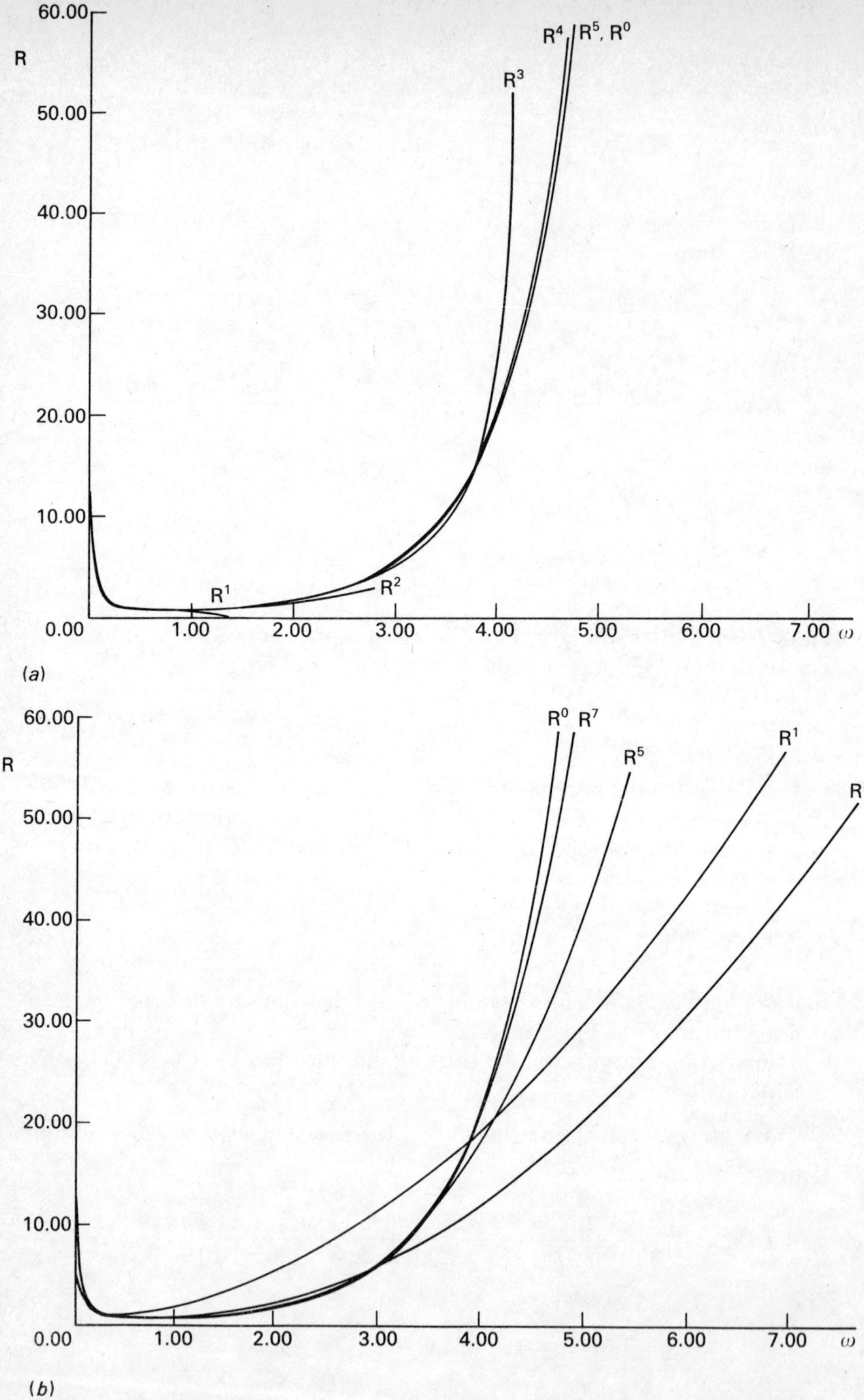

Fig. 10.9 Subharmonic and homoclinic bifurcations (*a*) inside Γ_+^0 and (*b*) outside Γ_+^0

In this case we obtain a sequence of bifurcation curves

$$\gamma = \frac{\hat{J}_1(m, l)\delta}{\hat{J}_2(m, l, \omega)} \stackrel{\text{def}}{=} \hat{R}^m(\omega)\delta$$

accumulating on $\gamma = R^0(\omega)\delta$ from above. Here $k(m, n)$ is the unique solution of the resonance relation

$$\hat{T}_k = 4K(k)\sqrt{(2k^2-1)} = \frac{2\pi m}{\omega n}. \tag{10.69}$$

For more details on the above computations see Greenspan [20].

In Figure 10.9 we show some of the approximate bifurcation curves $\gamma/\delta = R^0(\omega)$, $R^m(\omega)$ for orbits both inside and outside the separatrix. Below, in Fig. 10.18 we show how the subharmonic bifurcation curves accumulate on the homoclinic bifurcation curve as $m \to \infty$ for fixed ω. We leave the discussion of the detailed structure of the Poincaré maps for various parameter values (γ, δ, ω) to Section 10.6.

10.5 Smale horseshoes, Newhouse sinks and chaotic motions

In this section we provide a brief review of some useful results in the abstract theory of dynamical systems, and in particular of two-dimensional diffeomorphisms. For background material see [17, 27–30].

10.5.1 The Smale horseshoe

Here we outline an important example of a planar diffeomorphism due to Smale [27]. We shall adapt the map to our specific application.

Consider a map $F: Q \to \mathbb{R}^2$ defined on the square $Q = [0, 1] \times [0, 1]$ such that F is linear on the two horizontal strips $H_1 = [0, 1] \times [0, \alpha]$. $H_2 = [0, 1] \times [1-\alpha, 1]$, with

$$DF(p) = \begin{bmatrix} \beta & 0 \\ 0 & 1/\alpha \end{bmatrix}, \quad p \in H_1 \quad \text{and} \quad DF(p) = \begin{bmatrix} -\beta & 0 \\ 0 & -1/\alpha \end{bmatrix}, \quad p \in H_2 \tag{10.70}$$

where $0 < \beta < \alpha < \frac{1}{2}$. F is chosen such that the image of the strip $[0, 1] \times (\alpha, 1-\alpha)$ falls outside Q but $F(H_1) \stackrel{\text{def}}{=} V_1$ and $F(H_2) \stackrel{\text{def}}{=} V_2$ are two disjoint vertical strips lying in Q as shown in Fig. 10.10. By our construction, F has a hyperbolic saddle-point $(0, 0)$ and F/Q has constant Jacobian $\beta/\alpha < 1$ and is thus a *dissipative* map [24]. The *nonwandering set* $\Lambda \subset Q$ of F is obtained by intersecting all forward and backward iterates:

$$\Lambda = \Lambda_v \cap \Lambda_h, \qquad \Lambda_v = \bigcap_{n \geq 0} F^n(Q), \qquad \Lambda_h = \bigcap_{n \geq 0} F^{-n}(Q) \tag{10.71}$$

Thus $\Lambda_v^1 = Q \cap F(Q) = V_1 \cup V_2$ is a pair of vertical strips each of width β. It is not hard to see that $\Lambda_v^k = \bigcap_{n=0}^k F^n(Q)$ is 2^k disjoint vertical strips

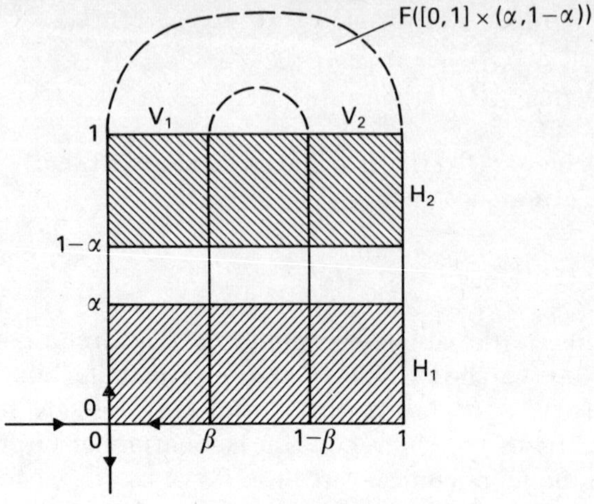

Fig. 10.10 Horseshoe

each of width β^k and thus that $\Lambda_v = C_\alpha \times [0, 1]$ is a Cantor set of vertical intervals. Similarly, $\Lambda_h^k = \bigcap_{n=0}^{k} F^{-n}(Q)$ is 2^k disjoint horizontal strips each of width α^k and $\Lambda_\kappa = [0, 1] \times C_\alpha$ is a Cantor set of horizontal intervals. Thus $\Lambda = C_\beta \times C_\alpha$ is also a Cantor set.

To describe the orbits of $F|_\Lambda$ we assign to each point $x \in \Lambda$ a bi-infinite symbol sequence $\{a_j(x)\} = \{\ldots a_{-2}a_{-1} \cdot a_0a_1a_2 \ldots\}$ chosen such that $a_j(x) = 1$ (resp. 2) if $F^j(x) \in H_1$ (resp. H_2) and $a_{-j}(x) = 1$ (resp. 2) if $F^{-(j-1)}(x) \in V_1$ (resp. V_2). The *action* of F on Λ then corresponds to the *shift* σ on the space Σ of all such symbol sequences, and in fact there is a homeomorphism φ such that the following diagram commutes:

$$\begin{array}{ccc} \Lambda & \xrightarrow{F} & \Lambda \\ \downarrow{\scriptstyle\varphi} & & \downarrow{\scriptstyle\varphi} \\ \Sigma & \xrightarrow{\sigma} & \Sigma \end{array}$$

To every periodic sequence there corresponds a periodic orbit of $F|_\Lambda$; for example the fixed point (0, 0) has a symbol sequence $\{\ldots 111 \cdot 111 \ldots\}$ and there is a second fixed point with sequence $\{\ldots 222 \cdot 222 \ldots\}$. For details see [30]. This method of *symbolic dynamics* enables us to prove the following.

Proposition 10.1. *The map F has a countable set of periodic points, with points of arbitrarily high period, and an uncountable set of nonperiodic recurrent motions. The periodic points are dense in Λ and there is an orbit dense in Λ. All the periodic points are of saddle type.*

The last assertion follows directly from the form of the linearized map *DF*. It is more difficult to prove that $F|_Q$ is structurally stable, but this latter important fact implies that any sufficiently C^1-close map $\hat{F}$ also possesses an invariant Cantor set $\hat{\Lambda}$ homeomorphic to Λ. Thus the complicated dynamics of F cannot be removed by small perturbations.

The 'piecewise linear' map F represents an idealization of a map which is embedded in the dynamics of any system with transverse homoclinic orbits. This result, the Smale–Birkhoff homoclinic theorem (*cf.* [27, 30, 31]) follows from the lambda Lemma and the homoclinic structure of Fig. 10.5, although it can be proved independently. We select a small rectangle S bounded by pieces of the stable and unstable manifolds $W^s(p)$, $W^u(p)$, with a transverse homoclinic point q at one corner. There are then integers l_1, l_2, n_i such that $A = p^{-l_1}(A)$, $B = P^{l_2}(S)$ are disjoint and lie in any neighbourhood $U_\nu(p)$, and the map $P^{n_i}: B \to A$ is well approximated by its linearization $DP^{n_i}(p)$. This enables one to find horizontal strips $H_1, H_2 \subset B$ whose images under $P^{l_1+l_2+n_i}$ are vertical strips V_1, $V_2 \subset B$. The map $P^{(l_1+l_2+n_1)\cdot(l_1+l_2+n_2)}$ then has a horseshoe. A careful study of the linearized map enables one to obtain estimates of hyperbolicity and show that the behaviour of $P^{(l_1+l_2+n_1)(l_1+l_2+n_i)}$ is homeomorphic to that of the 'idealized' map F. Specifically, for our problem we have:

Theorem 10.4. *If the Poincaré map $P_\varepsilon^{t_0}: \Sigma^{t_0} \to \Sigma^{t_0}$ possesses a transverse homoclinic point q_ε to a hyperbolic saddle-point p_ε, then, in a neighbourhood of q_ε, some power $(P_\varepsilon^{t_0})^N$ possess an invariant zero-dimensional Cantor set Λ on which $(P_\varepsilon^{t_0})^N$ is conjugate to a shift on two symbols.*

Corollary 10.3. *$(P_\varepsilon^{t_0})^N/\Lambda$ possess a dense set of periodic points, there are points of arbitrarily high period and there is a nonperiodic orbit dense in Λ. Moreover, Λ is structurally stable in the sense that if $\hat{P}$ is sufficiently C^1 close to $P_\varepsilon^{t_0}$, then $(\hat{P})^N$ has a nonwandering set Λ and there is a homeomorphism h such that $(P_\varepsilon^{t_0})^N \circ h = h \cdot \bar{P}^N$.*

The horseshoe's nonwandering set Λ is extremely complicated. It is of saddle type and has one-dimensional stable and unstable bundles $W^s(\Lambda)$, $W^u(\Lambda)$ which are locally the products of Cantor sets and curves. Letting $W^s(0, 0)$, $W^u(0, 0)$ be the stable and unstable manifolds of the fixed point $\{\ldots 111 \cdot 111 \ldots\}$ of the idealized horseshoe, F, we have $W^s(\Lambda) = \mathrm{Cl}[W^s(0, 0)]$ and $W^u(\Lambda) = \mathrm{Cl}[W^u(0, 0)]$, and since such a Λ can be chosen arbitrarily close to any transverse homoclinic point q_ε and hence to the saddle-point p_ε in our application, it follows that

$$W^s(\Lambda) = \mathrm{Cl}[W^s(p_\varepsilon)]$$
$$W^u(\Lambda) = \mathrm{Cl}[W^u(p_\varepsilon)].$$

Thus the structure of the stable and unstable manifolds $W^s(p_\varepsilon)$, $W^u(p_\varepsilon)$ is doubly important not just for establishing the existence of horseshoes, but also in determining their structure. We shall return to this in Section 10.6.

10.5.2 Homoclinic tangencies and Newhouse sinks

We now turn to the bifurcations in which horseshoes are created. Newhouse [17, 24, 32] has studied the situation and has shown that the lack of hyperbolicity which occurs when stable and unstable manifolds have a quadratic tangency persists under small perturbations. We first require a definition. A hyperbolic (basic) set $\Lambda(F)$ for a two-dimensional diffeomorphism F is called *wild* if there is a C^r-neighbourhood N of F such that for any $G \in N$ there are points $x, y \in \Lambda(G)$ such that $W^u(x)$ and $W^s(y)$ are tangent somewhere.

Theorem 10.5 ([24]). *Let p be a dissipative saddle-point of a two-dimensional C^r diffeomorphism F ($|\det[DF(p)]| < 1$). Suppose $W^u(p)$, $W^s(p)$ are tangent at some point q. Then arbitrarily C^r near F there is a diffeomorphism G having a wild hyperbolic set near the orbit of q.*

Wild hyperbolic sets are important because they imply the existence of countably many stable periodic orbits:

Theorem 10.6. *Suppose F is a two-dimensional C^r diffeomorphism with a wild hyperbolic set Λ containing a dissipative periodic point. Then there is a neighbourhood N of F and a residual subset $B \subset N$ such that if $G \subset B$, then G has countably many periodic sinks.*

For a proof of Theorem 10.6 see [17]. One step in the proof closely resembles a result obtained independently by Gavrilov and Shil'nikov [33, 34], who showed that, for certain two-dimensional diffeomorphisms, the homoclinic tangencies were the limit of an infinite sequence of saddle-node bifurcations in which pairs of periodic sinks and saddles are created. To illustrate their argument, consider a one-parameter family of maps F_μ such that for $\mu < 0$, $W^u(p) \cap W^s(p) = \varnothing$ while for $\mu > 0$ we have transversal intersections and at $\mu = 0$, F_0 has a quadratic tangency as in Theorem 10.5. We pick a coordinate system (x, y) such that, in a neighbourhood $U_\nu(p)$, $W^s(p)$ and $W^u(p)$ are given by $y = 0$ and $x = 0$ respectively. Let $q_2 = (0, \bar{y})$ and $q_1 = (\bar{x}, 0) = F^n(q_2)$ be (nontransverse) homoclinic points and A_1, A_2 be neighbourhoods of q_1, q_2 lying in U_ν as shown in Fig. 10.11. The map $F_0^k : A_1 \to A_2$ is well approximated by

$$[DF_0(p)]^k = \begin{bmatrix} \lambda^k & 0 \\ 0 & \gamma^k \end{bmatrix}, \tag{10.72}$$

where $0 < \lambda < 1 < \gamma$ and $\lambda\gamma < 1$ and $k \to \infty$ as we pick domains in A_1 closer and closer to $W^s(p)$. The form of the map $F_0^n : A_2 \to A_1$ (n is fixed) may also be approximated since, by hypothesis, the vertical line $W^u(p) \cap A_2$ is mapped to the parabola $W^u(p) \cap A_1$ and the parabola $W^s(p) \cap A_2$ to the horizontal line $W^s(p) \cap A_1$. When $\mu \neq 0$ the images also undergo a vertical translation. Using these facts, and normalizing the maps to

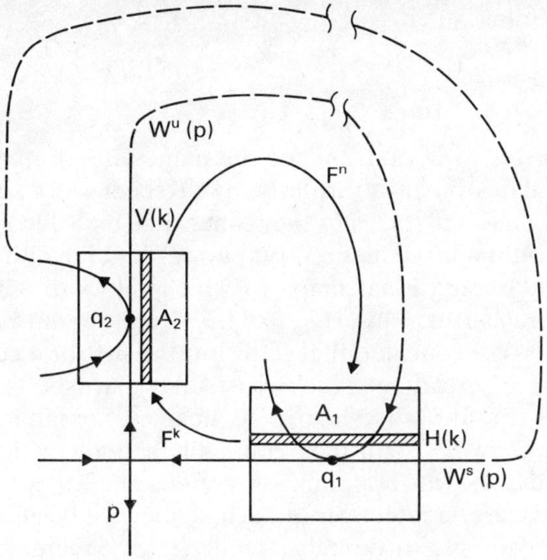

Fig. 10.11 Quadratic homoclinic tangencies

remove redundant parameters, we arrive at an approximate quadratic map

$$F_{\mu}^{n+k} \approx f_{\mu,k} : (x - \bar{x}, y) \to [-(\gamma^k y - \bar{y}), \lambda^k x + a(\gamma^k y - \bar{y})^2 - \mu]. \quad (10.73)$$

(We have selected the orientation such that $a > 0$.) It is interesting to note that this is essentially the same quadratic map studied by Hénon [35] and *conjectured*, on the basis of numerical evidence, to possess a strange attractor.

We fix a horizontal strip $H(k) \subset A_1$ and an integer k such that $F^k[H(k)] \cap A_2$ is a vertical strip and hence $F^{n+k}[H(k)] \cap A_1$ is non-empty. Equation (10.73) then permits us to compute the approximate bifurcation value for which a pair of periodic orbits of period $n+k$ appear in a saddle-node bifurcation

$$\mu_k = \lambda^k \bar{x} - \gamma^{-k} \bar{y} - \frac{1}{4a} (\lambda^k + \gamma^{-k})^2. \quad (10.74)$$

It is easy to check that $\mu_k \to 0^-$ as $k \to \infty$, showing that the homoclinic tangency at $\mu = 0$ is the limit of a countable sequence of saddle-node bifurcations. For sufficiently large k, these subharmonics of order $m \stackrel{\text{def}}{=} n+k$ are precisely those found by Melnikov's method in Sections 10.3 and 4, since the perturbation methods used there yield *all* subharmonics which pass once through U_ν.

The two fixed points of (10.73) existing for $\mu > \mu_k$ are a saddle and a sink and another straightforward analysis shows that the sink bifurcates to a sink of period two at

$$\mu'_k = \lambda^k \bar{x} - \gamma^{-k} \bar{y} + \frac{3}{4a}(\lambda^k + \gamma^{-k})^2, \tag{10.75}$$

leading to an orbit of period $2m$ for the map. The bifurcation sequence μ'_k also accumulates on $\mu = 0$ from below. Recent work of Lanford [36] has shown that quadratic maps of the plane with small Jacobian determinant undergo countable sequences of period-doubling bifurcations precisely as do one-dimensional maps to which they are sufficiently close [37, 38]. The present map has $|Df_{\mu,k}| = (\lambda\gamma)^k$ and therefore falls into this class for k large. We conclude that μ'_k is but the first of a countable set of period-doubling bifurcations which must take place before $\mu = 0$, since for $\mu > 0$ there is a full horseshoe in A_1 near q_1, containing points of all periods for the map F^{n+k}. In fact, countable sets of further bifurcations must also occur, in which points of period $2^j \cdot 3(n+k)$, $2^j \cdot 5(n+k)$, $j = 1, 2, \ldots$, etc., are created. Since such sequences occur for each sufficiently large k we have a countable sequence of countable sequences of bifurcations all accumulating on $\mu = 0$. One can thus find sinks of arbitrarily high period in any neighbourhood of a nontransverse homoclinic point q, and finite sets of these sinks persist when the homoclinic tangency is broken. Newhouse's results [17, 24] give more information and in particular show that, in the case of tangencies, the closure of the set of homoclinic points of a suitably perturbed map is contained in the closure of the set of periodic sinks. This implies that any homoclinic orbit can be approximated arbitrarily closely (for finite, but arbitrarily long times) by a stable periodic motion and thus the horseshoe might become 'physically observable'. On the basis of this work Newhouse [24] conjectured that some of the numerically observed strange attractors (such as that of Hénon [35]) might in fact be orbits lying in the stable manifolds of long period sinks. The question is still open, and we shall return to it in the next section.

Before closing we note that, even without Theorem 10.5, the existence of a single parameter value μ at which a homoclinic tangency occurs immediately implies a countable set of such values, since the stable and unstable manifolds accumulate on one another, by the lambda Lemma. Thus the 'first' tangency is immediately followed by infinitely many others, each with their attendant saddle-node and period doubling bifurcations.

10.6 Global structure of solutions of Duffing's equation

In this final section we make use of the bifurcation and perturbation computations of Sections 10.3 and 4 and the abstract theory of Section 10.5 to obtain a partial understanding of Duffing's equation (10.51) for ε small. We start by considering the phase portraits of the averaged systems

(10.46) and (10.50) near each resonant periodic orbit inside the separatrix Γ_+^0. Similar conclusions apply within Γ_-^0 and outside the separatrix.

10.6.1 Subharmonics and their domains of attraction

The material in this and the following section is deduced directly from the theory of Section 10.3, the results of Section 10.4, and standard results in dynamical systems theory such as that reviewed in Section 10.5. In these two sections we collect various proven facts concerning Duffing's equation, although we do not state precise theorems. In the concluding Section 10.6.3 we go on to make some conjectures which we feel have a firm basis in these facts.

Within Γ_+^0 the action transformation gives

$$I^k = I(k) = \frac{[1/(3\pi)][2(2-k^2)E(k)-4k'^2K(k)]}{(2-k^2)^{3/2}} \tag{10.76}$$

where the elliptic modulus k depends implicitly upon u and v through the relationships

$$H(u, v) = \frac{v^2}{2} - \frac{u^2}{2} + \frac{u^4}{4} = \frac{k^2-1}{(2-k^2)^2}.$$

Using (10.63) we have

$$\begin{aligned}\Omega'(I^k) &= \frac{\mathrm{d}}{\mathrm{d}I}\left(\frac{2\pi}{T(I)}\right)\bigg|_{I=I^k} = -\frac{(2\pi/T_k^2)(\mathrm{d}/\mathrm{d}k)T(k)}{(\mathrm{d}/\mathrm{d}k)I(k)} \\ &= -\frac{\pi^2(2-k^2)[(2-k^2)E(k)-2k'^2K(k)]}{2k^4k'^2K(k)^3} \equiv \Omega'(m),\end{aligned} \tag{10.77}$$

where $k = k(m)$ is defined by the relationship $K(k)\sqrt{(2-k^2)} = \pi m/\omega$ on a resonant orbit of period $2\pi m/\omega$ and $k'^2 = 1-k^2$. Thus (10.46) and its integral (10.47) become

$$\begin{aligned}\dot{\varphi} &= \sqrt{(\varepsilon)}\Omega'(m)h, \\ \dot{h} &= \sqrt{(\varepsilon)}\frac{1}{2\pi}[-\delta J_1(m) - \gamma J_2(m)\sin(m\varphi)],\end{aligned} \tag{10.78}$$

and

$$H = \sqrt{(\varepsilon)}\left[\frac{\Omega'(m)}{2}h^2 + \frac{1}{2\pi}\left(\delta J_1(m)\delta\varphi - \gamma\frac{J_2(m, \omega)}{m}\cos(m\varphi)\right)\right], \tag{10.79}$$

where we have used Eq. (10.65) (with $J_i(m, 1) \stackrel{\text{def}}{=} J_i(m)$) and the fact that $\Omega(I^\alpha) = \omega/m$ at resonance. Noting that $\delta J_1(m)$, $\gamma J_2(m, \omega)$ are positive while $\Omega'(m)$ is negative, we easily sketch the phase portraits of (10.78) for the three cases $\gamma/\delta <, =, > R^m(\omega)$, as in Fig. 10.12(*a*, *b*, *c*). Note that, for $\gamma/\delta > R^m(\omega)$, in each resonant band there are m saddles and m 'centres' corresponding to a pair of subharmonics of order m in the original system.

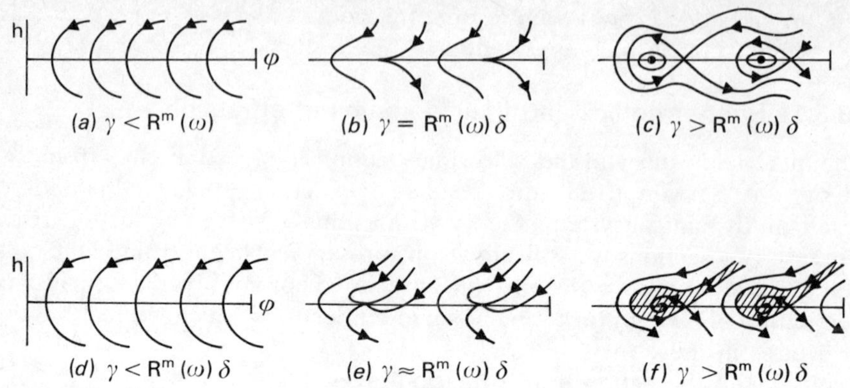

Fig. 10.12 Phase portraits of Eq. (10.78), $m=2$; in (d, e, f), $0(\varepsilon)$ terms have been added
(a) $\gamma<R^m(\omega)\delta$ (b) $\gamma=R^m(\omega)\delta$ (c) $\gamma>R^m(\omega)\delta$
(d) $\gamma<R^m(\omega)\delta$ (e) $\gamma\approx R^m(\omega)\delta$ (f) $\gamma>R^m(\omega)\delta$

To study the actual stability type of the fixed points which appear as centres in the $0(\sqrt{\varepsilon})$ approximation, we compute the averaged higher-order terms $\bar{F}'$, $\bar{G}$ of Eq. (10.50). After tedious but routine computations we obtain

$$\begin{aligned}\overline{F'}(\varphi)&=-\gamma K_2(m,\omega)\sin(m\varphi)-\delta K_1(m)\\ \overline{G}(\varphi)&=-\gamma K_2(m,\omega)\cos(m\varphi)/m\end{aligned}\tag{10.80}$$

where K_1 and K_2 are positive constants found by integrating products of elliptic and trigonometric functions, much as in the first-order terms computed in Section 10.4. We note that we do not need Ω'' explicitly, and that the trace (up to $0(\varepsilon)$) is given by

$$\frac{\partial}{\partial h}(\varepsilon\overline{F'}(\varphi)h]+\frac{\partial}{\partial\varphi}\left[\varepsilon\left(\frac{\Omega''h^2}{2}+\bar{G}(\varphi)\right)\right]=-\varepsilon\delta K_1(m)$$

(*cf.* Eq. (10.50)). These fixed points are thus found to be sinks, and the bifurcations in which they are created simple saddle-nodes, rather than the doubly degenerate Hamiltonian bifurcations occurring in (10.78) when $\gamma/\delta=R^m(\omega)$. We show the phase portraits in Fig. 10.12(d, e, f). The stable manifolds or domains of attraction of the sinks are shaded.

The methods used here are quite standard and include a check on the nonexistence of closed orbits by Bendixon's criterion. Morosov [26, 39] obtained similar results for a related equation.

It is important to realize, however, that unlike the existence Theorem 10.2 these results are *not* uniformly valid since the factor $\Omega'(m)$ in (10.78) becomes unbounded as $m\to\infty$; in fact, using the asymptotic behaviour $K(k)\sim\ln(4/k')$ as $k\to 1(k'\to 0)$, we have

$$\Omega'(m)\sim\frac{-\pi^2}{4k'^2[\ln(4/k')]^3}=-\frac{4\omega^3\exp(2\pi m/\omega)}{\pi m^3},\tag{10.81}$$

as $m \to \infty$. Thus averaging is valid for successively smaller regions $0 < \varepsilon \leq \varepsilon_0(m)$ as m increases, and, while the invariant manifolds and periodic points of the Poincaré map P_ε are qualitatively similar to those of Fig. 10.12(*d, e, f*) in any given resonance band, we cannot carry our stability results uniformly to the limit of homoclinic bifurcations as in Theorem 10.3. Our results are, however, good for subharmonics of all finite orders $m < \infty$ and hold within each resonance band near the unperturbed resonant orbit of period $2\pi m/\omega$.

Noting that $J_1(m, n) > 0$ while $J_2(m, n, \omega) = 0$ for $n \neq 1$ (Eq. (10.65)), we conclude that, between the resonance bands, all orbits decay 'inward' ($h(t)$ decreases). This implies that the unstable manifolds $W^u(s_k)$ of the saddles of period k intersect the stable manifolds $W^s(s_{k-1})$ of those of period $k-1$ and one therefore obtains the global structure of connected resonance bands indicated in Fig. 10.13 (*cf.* Fig. 3 of [40]). This structure is repeated between each pair of resonance bands and, using the lambda Lemma, it is possible to prove that in any neighbourhood U of the stable manifold $W^s(s_k)$ of a saddle of period k there will be pieces of the stable manifolds $W^s(s_j)$, $W^s(a_j)$ of saddles and sinks of all lower periods $j < k$. The resultant violent winding and packing of the alternating domains of attraction of the stable subharmonics leads to a sensitive dependence on initial conditions, especially for the higher-order subharmonic motions, since the basic width of the domains of attraction, calculated from the integral (10.79) of the $0(\sqrt{\varepsilon})$ Hamiltonian approximation, is

$$\Delta I(m) = \left(\frac{\varepsilon}{\pi \Omega'(m)} (\gamma J_2(m) \cos(m\hat{\varphi}) - \delta J_1(m)\hat{\varphi})\right)^{1/2} + 0(\varepsilon) \tag{10.82}$$

where $\hat{\varphi}$ is the root of

$$\sin \varphi = -\frac{\delta J_1(m)}{\gamma J_2(m)} = -\delta R^m(\omega) \tag{10.83}$$

corresponding to the centre. For m large, while $J_1(m)$, $J_2(m, \omega)$ are bounded (Theorem 10.3 and Section 4), $\Omega'(m) \to \infty$ and we find

$$\Delta I(m) \sim \sqrt{\left(\varepsilon \frac{m^3}{4\omega^3}\right)} \exp(-\pi m/\omega) \tag{10.84}$$

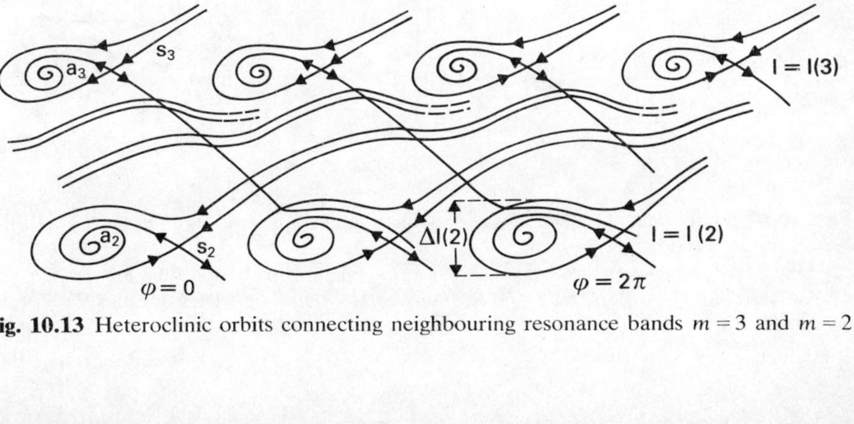

Fig. 10.13 Heteroclinic orbits connecting neighbouring resonance bands $m = 3$ and $m = 2$

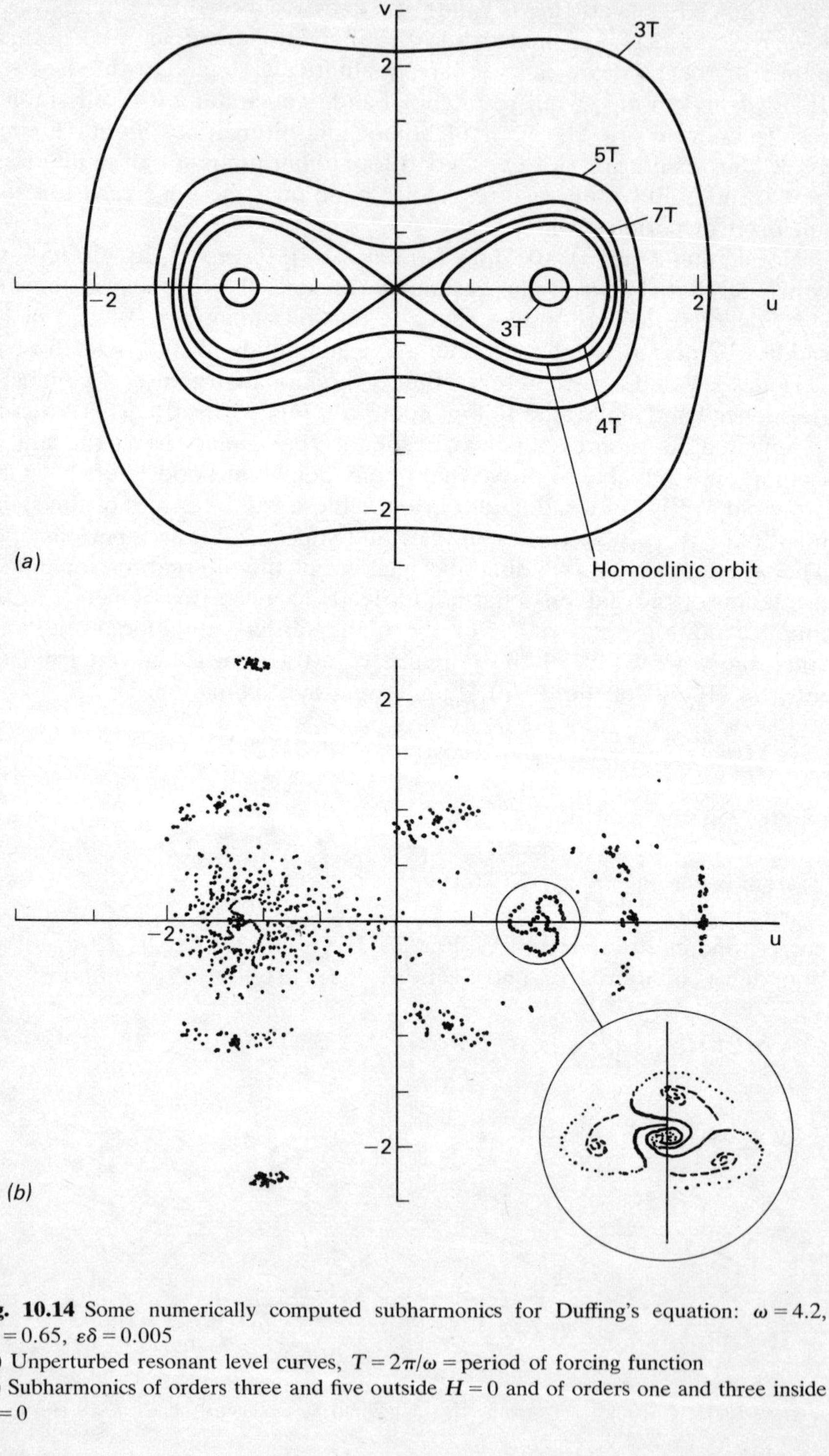

Fig. 10.14 Some numerically computed subharmonics for Duffing's equation: $\omega = 4.2$, $\varepsilon\gamma = 0.65$, $\varepsilon\delta = 0.005$
(a) Unperturbed resonant level curves, $T = 2\pi/\omega$ = period of forcing function
(b) Subharmonics of orders three and five outside $H = 0$ and of orders one and three inside $H = 0$

and thus the domains shrink rapidly in size as m increases. In typical computations all but the lowest order ($m \leq 5$, say) subharmonics lie within a band whose area is less than 1 per cent of the interior of Γ_+^0!

In Fig. 10.14 we illustrate our results with numerical computations of the Poincaré map, showing unperturbed resonant orbits and the perturbed subharmonics. For more detailed and extensive computations on different versions of Duffing's equation, see [40–43]. The interaction of the subharmonics of order $m=2$ and 3 with the fixed points near $(u, v)=(\pm 1, 0)$ has been studied using conventional averaging methods by Holmes and Holmes [44].

10.6.2 Homoclinic orbits and chaotic motions

As we pointed out in Section 10.6.1, the results based on the averaged systems (10.46) and (10.50) are invalid near the homoclinic orbit, since $|\Omega'(m)| \to \infty$ as $m \to \infty$. We can, however, use the abstract results of Section 10.5 to study the global structure of the Poincaré map in this region, and we do know from Theorems 10.2 and 10.3 and Corollary 10.2 that subharmonics of all orders exist here for $\gamma > R^0(\omega)\delta$, although we cannot compute their stability types from Eq. (10.50).

We start by noting that the symmetry of solutions of (10.51) (if (u, v, t) is a solution, then $[-u, -v, t+(\pi/\omega)]$ is also a solution) together with the presence of transverse homoclinic points for $\gamma/\delta > R^0(\omega)$ immediately imply that we can find strips S_+, S_- 'parallel' to $W^s(p)$ and in a neighbourhood $U_\nu(p_\varepsilon)$, whose images $P_\varepsilon^N(S_\pm)$ under some iterate P_ε^N of the Poincaré map are as shown in Fig. 10.15. The orbit of any point $x \in S_+$(resp. S_-) will remain near the unperturbed loop Γ_+^0 (resp. Γ_-^0) until it re-enters U_ν. In this way, using the symbolic dynamics approach of Section 10.5.1, we can find orbits of P which visit neighbourhoods of Γ_+^0 and Γ_-^0 in *any specified order.*

This goes some way toward explaining the chaotic snap-through motions of Fig. 10.3 (*cf.* [5, 6, 8]), but we must recall that all these orbits are of unstable saddle type, their dynamics being dominated by the linearized map $DP_\varepsilon(p_\varepsilon)$. They would thus be expected to give rise only to *transient*

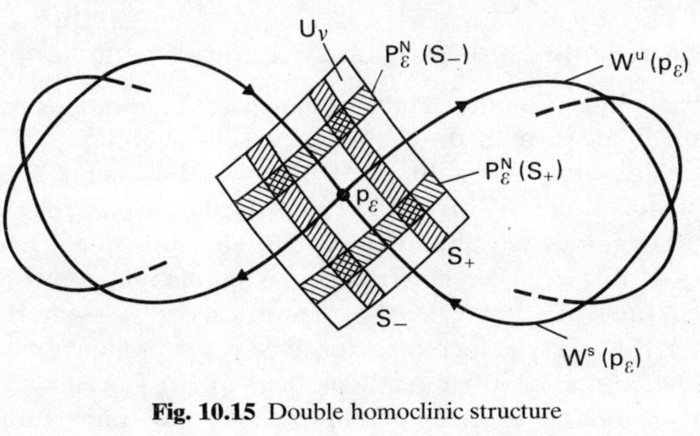

Fig. 10.15 Double homoclinic structure

chaotic motions and this is, indeed, precisely what is observed when numerical integrations are performed in certain parameter ranges; almost all orbits converging to fixed points or subharmonics lying within Γ_+^0 or Γ_-^0 (*cf.* Fig. 10.14). However, in view of Newhouse's results on persistent tangencies and wild hyperbolic sets, it is reasonable to expect that there exist *stable* long-period motions which approximate such saddle-type chaotic orbits arbitrarily closely on finite time intervals. We return to this point in Section 10.6.3.

Note that while such 'chaotic subharmonics' cannot be found by the perturbation procedures of Section 10.3, since they pass through $U_\nu(p_\varepsilon)$ arbitrarily often, their existence can be inferred from our perturbation calculations of $W^s(p_\varepsilon)$ and $W^u(p_\varepsilon)$.

We also note that, by arguments similar to those of Section 10.6.1, we can conclude that pieces of the unstable manifold $W^u(p_\varepsilon)$ accumulate on all the unstable manifolds $W^u(s_k)$, $k=1, 2, \ldots$, of all periodic saddle-type motions within Γ_+^0 and Γ_-^0. It follows that any attracting periodic motions within this region lie in $\mathrm{Cl}[W^u(p_\varepsilon)]$.

In earlier work [7] we showed that all orbits of the damped Duffing equation remain bounded and enter some bounded simply connected set $A\times S^1\subset\mathbb{R}^2\times S^1$ as $t\to\infty$. Letting $A\in\Sigma^{t_0}$ be a section of such a set, if we define the *attracting set* $\mathscr{A}$ as

$$\mathscr{A}=\bigcap_{n\geqslant 0}(P_\varepsilon)^n(A); \tag{10.85}$$

it follows that

$$\mathscr{A}=\mathrm{Cl}[W^u(p_\varepsilon)]. \tag{10.86}$$

Here we assume that any sinks lying inside $\mathscr{A}$ but outside Γ_0^+, Γ_0^- are sufficiently close to the latter so that $W^u(p_\varepsilon)$ intersects their stable manifolds. However, as we show in the next section, it is not likely that $\mathscr{A}$ is a 'nice' *attractor*—an invariant attracting set which is *indecomposable* and *contains a dense orbit.* Certainly for parameter values (γ, δ, ω) for which tangencies occur, $\mathrm{Cl}(W^u(p_\varepsilon)$ contains periodic sinks and hence is decomposable.

10.6.3 The strange attractor: numerical results and conjectures

We start by summarizing the results of numerical integrations and computations of the Poincaré map, drawing in the results of Holmes [7], Shaw [45] and, principally, Ueda [46]. In Fig. 10.16 we show plots of the stable and unstable manifolds $W^s(p_\varepsilon)$, $W^u(p_\varepsilon)$ for fixed $\varepsilon\delta=0.25$, $\omega=1.0$ and $\varepsilon\gamma$ varying. These curves are obtained by iterating short line segments lying near $W^s_{\mathrm{loc}}(p_\varepsilon)$, $W^u_{\mathrm{loc}}(p_\varepsilon)$. We note that the manifolds have just intersected for the first time at $\varepsilon\gamma=0.19$, the theoretical value from Eq. (10.62) being $\varepsilon\gamma=0.188$. As γ is increased for fixed δ the homoclinic bifurcation and resulting formation of horseshoes near $W^u(p_\varepsilon)$ occurs while stable fixed points continue to exist within each of the (unperturbed) loops

Γ_+^0, Γ_-^0. However, with appreciably higher γ and for a wide range of γ, δ values, numerical computations suggest that there are no stable low-period attractors and one observes the characteristic 'strange attractor' motions of Fig. 10.17. Comparing Fig. 10.17(*a*) with Fig. 10.16(*d*), we note that this single orbit appears to fall on $\mathrm{Cl}[W^u(p_\varepsilon)]$, as predicted in Section 10.6.2. For typical power spectra of these motions see [7] or [42, 43]. (The term 'strange attractor' is used only for convenience; we should more correctly say 'apparently nonperiodic attracting motion'. As indicated at the end of Section 10.6.2, we define a strange attractor for the map P as an attracting set $\mathscr{A}$ which is indecomposable, has a dense orbit and is neither a fixed point nor a periodic orbit.)

In Fig. 10.18 we indicate regions in $\varepsilon\gamma$, $\varepsilon\delta$ parameter space for $\omega = 1$ in which the various attractors are observed. We also superpose some bifurcation curves from our theory. In each region we sketch a representative motion projected onto u–v space. These numerical results are due to Ueda [46]. Note how the subharmonic bifurcations all occur very close to the homoclinic bifurcation curve $\gamma = R^0(\omega)\delta$, apart from those of order $m = 1$. Above $\gamma = R^1(\omega)\delta$ subharmonics of all orders exist both inside and outside the unperturbed separatrix $\Gamma_0^+ \cup \{p_0\} \cup \Gamma_0^-$, although only low-period ones are observable. The theoretical value of R^1 agrees well with that observed. For small damping ($\varepsilon\delta < 0.1$), sustained strange attractor motions are not observed and we find only fixed points (T-periodic motions) or subharmonics of relatively low orders ($< m = 5$) coexisting with the transient chaos characteristic of horseshoes, *cf.* Fig. 10.14. On the other hand, for higher damping levels we observe stable T-periodic motions outside $\Gamma_0^+ \cup \{p_0\} \cup \Gamma_-^0$ (*cf.* [7]). One apparently requires both moderate forcing and damping for the strange attractor to exist. We note that in some parameter ranges (with multiple shading in Fig. 10.18) two or more attractors (periodic and/or strange) coexist and that one can observe hysteresis in the jumps from one attractor to another as the parameters (ε, γ, δ, ω) are slowly varied.

We now attempt to describe the various attracting sets in the light of the theory developed in this chapter. For low damping, the stable subharmonic motions and horseshoes predicted in Sections 10.3 and 4 account reasonably well for the numerical observations, especially in view of the small characteristic size of the domains of attraction of high subharmonics (*cf.* (10.84)) and the associated implication that small machine errors will be sufficient to perturb orbits from the stable manifold of one such subharmonic to a neighbouring one, leading to a 'transitionally chaotic orbit' (*cf.* [42]). Thus, although there certainly are parameter values for which such long-period sinks exist, they are not observable in practice. Franks [47] has some interesting results on the addition of random perturbations to a motion originally attracted to a stable period 7 sink of the Hénon map (*cf.* [35]). The addition of a small random vector after each iteration leads to an orbit which appears to fill out the same 'strange attractor' as one observes for nearby parameter values *without intentional perturbations.* This observation is one of the motivations for the conjecture which follows.

The second pertinent observation is also based jointly on analysis and numerical integrations. The study of near resonant excitation of the stable periodic motions near $(u, v) = (\pm 1, 0)$ in [7] and more particularly [44] shows that for forcing frequencies ω near $\sqrt{2}$ or $2\sqrt{2}$, one obtains harmonic and subharmonic bifurcations as γ is increased, so that the stable fixed point of the Poincaré map can become a saddle-point and (in the second case) throw off an orbit of period two. As in the case of subharmonics near the homoclinic orbit (*cf.* Section 10.5.2), there is some reason to think that this period doubling bifurcation is the first in a

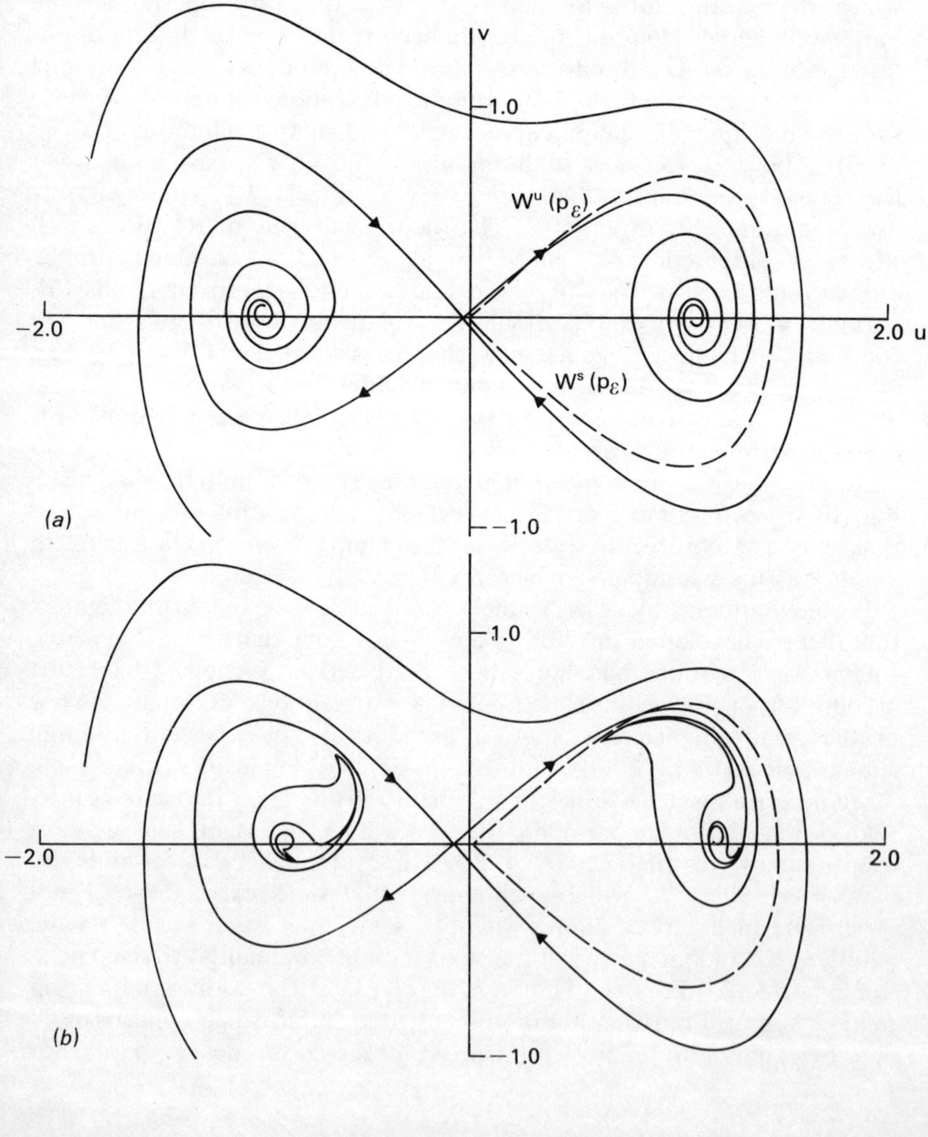

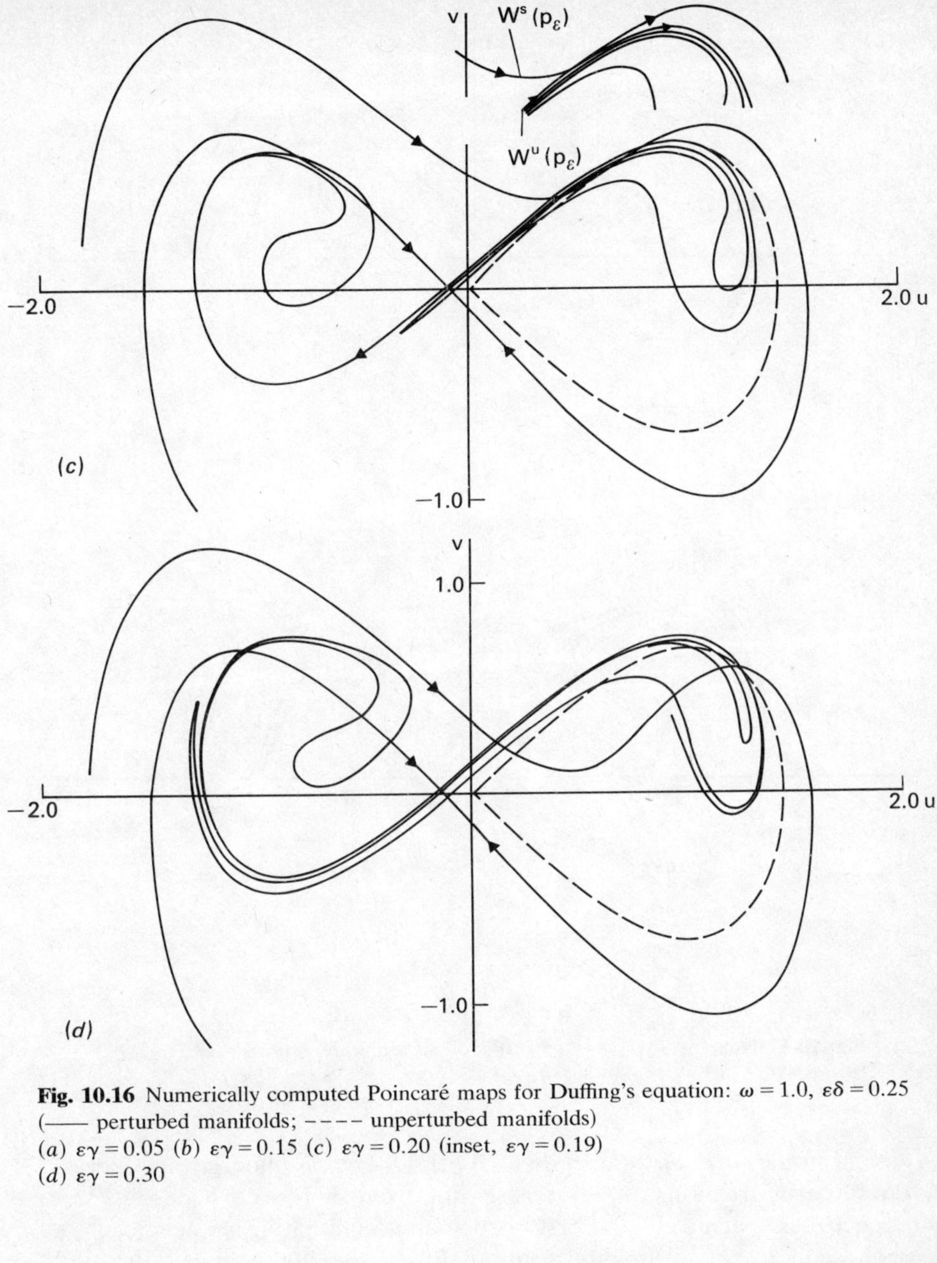

Fig. 10.16 Numerically computed Poincaré maps for Duffing's equation: $\omega = 1.0$, $\varepsilon\delta = 0.25$ (—— perturbed manifolds; ---- unperturbed manifolds)
(*a*) $\varepsilon\gamma = 0.05$ (*b*) $\varepsilon\gamma = 0.15$ (*c*) $\varepsilon\gamma = 0.20$ (inset, $\varepsilon\gamma = 0.19$)
(*d*) $\varepsilon\gamma = 0.30$

countable sequence which has an accumulation point (*cf.* [37, 38]). The studies of Eq. (10.51) and an associated cubic mapping of the plane

$$(u, v) \to (v, -\beta u + \alpha v - v^3) \tag{10.87}$$

strongly support this idea [7]. In fact, in view of the results of Gavrilov and Shil'nikov outlined in Section 10.5.2, we can conclude that *all* the

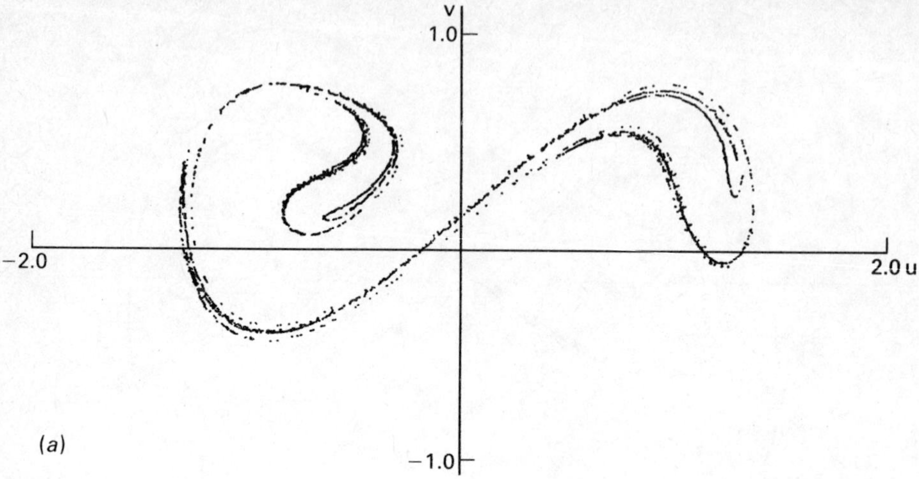

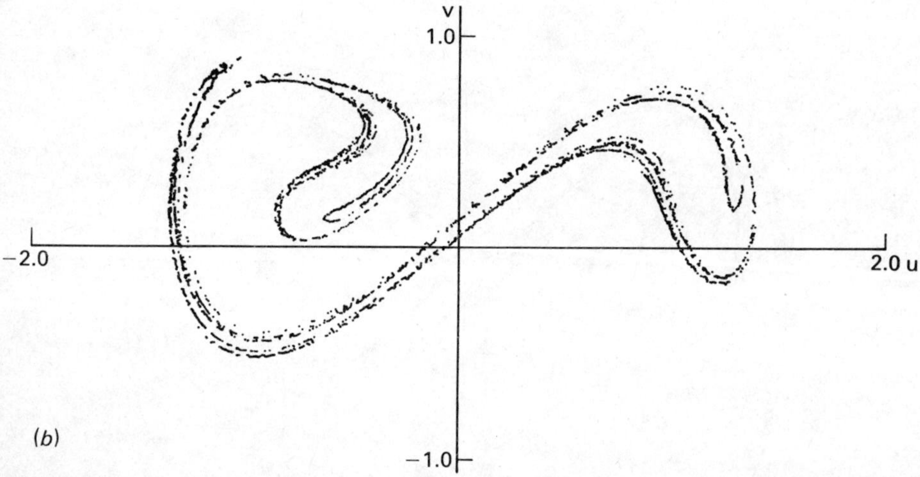

Fig. 10.17 Numerically computed orbits of the Poincaré map: $\omega = 1.0$, $\varepsilon\gamma = 0.30$ (*a*) $\varepsilon\delta = 0.25$ (*b*) $\varepsilon\delta = 0.20$

subharmonic sinks *of sufficiently high period* $2\pi m/\omega$ undergo such period-doubling bifurcations as γ increases for fixed δ. It seems reasonable to expect these each to be the first of such an accumulating sequence, and all such sequences presumably accumulate on the homoclinic bifurcation. This provides a mechanism by which all sinks of periods less than some fixed integer N can become unstable, leading to our conjecture:

Conjecture 10.1. *For any fixed integer* $N<\infty$ *there are open sets of parameter values in* $(\varepsilon, \gamma, \delta, \omega)$ *space for which the attracting set* $\mathcal{A} = \bigcap_{n\geq 0} P^n(A) \equiv \mathrm{Cl}[W^u(p_\varepsilon)]$ *of the Duffing equation* (10.51) *contains a shift*

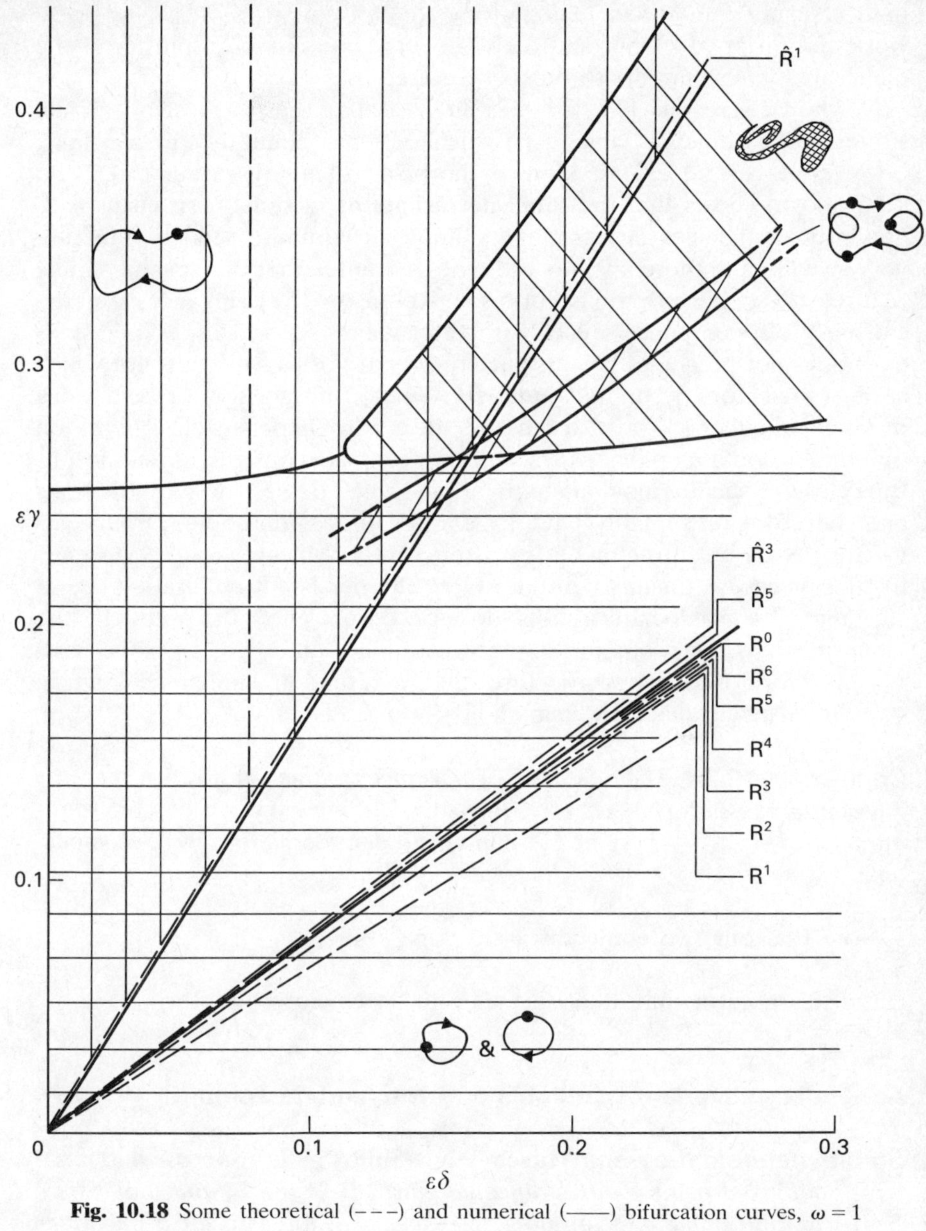

Fig. 10.18 Some theoretical (– – –) and numerical (——) bifurcation curves, $\omega = 1$

on two symbols with a countable set of saddle-type periodic orbits and finite sets of periodic sinks, none of the latter's periods being less than N.

This conjecture implies that $\mathscr{A}$ is not a strange attractor, since it can be decomposed into a finite set of simple periodic attractors. Such an attracting set is structurally stable. We can of course, prove this conjecture for the case $N = 2$ [44]. The conjecture can be proven for arbitrarily high N using a modified Duffing–van der Pol equation [48]. Numerical work of Chirikov [49, 50] and Lieberman [51] on diffeomorphisms of the plane also lends support to our conjecture.

Newhouse's results [17, 24, 32] suggest that there are also residual subsets of parameter values for which one has countably many sinks; however, a careful examination of the proof of his theorems (10.5 and 10.6 above) shows that the small perturbations of diffeomorphisms with quadratic tangencies necessary to obtain such infinite sets of sinks are very special, and there appears to be no guarantee that our specific family realizes these perturbations (but see Robinson [52]: in any case, the existence of open dense sets of parameter values for which periodic sinks exist does not preclude the existence of sets of positive measure for which strange attractors occur [*note added in proof*]). In fact some recent work of Guckenheimer [53] which borrows ideas for Yakobson's [54] work on one-dimensional maps indicates that one can also construct diffeomorphisms close to such homoclinic bifurcations which possess no periodic sinks and therefore presumably have a genuine (indecomposable) strange attractor. (Such an attractor has been proven by Misiurewicz [55] to exist for the piecewise linear Lozi map. Here the persistent quadratic tangencies and their associated sinks do not occur since the folds in the manifolds become *points* in the piecewise linear manifolds.) On the basis of this work one can conjecture that a situation similar to that in one-dimensional quadratic maps holds:

Conjecture 10.2. *There is a set of parameter values of positive measure but containing no open sets for which the Duffing equation has a strange attractor* $\mathcal{A} = \mathrm{Cl}[W^u(p_\varepsilon)]$, i.e., *A contains a dense orbit and is indecomposable.*

Note that our two conjectures are compatible.

On this note of unbridled speculation, let us close our story.

Appendix. Invariant manifolds and the lambda Lemma

Stable manifold theorem (Hirsch–Pugh–Shub [15]). *Suppose that p is a hyperbolic fixed point of a* C^r *diffeomorphism* $P:\mathbb{R}^n \to \mathbb{R}^n$ *and that* $E^s(p)$, $E^u(p)$ *are the stable and unstable eigenspaces of the associated linearized mapping* $DP(p)$. *Then, in a neighbourhood* $U(p)$ *there are* C^r *invariant submanifolds* $W^s_{\mathrm{loc}}(p)$, $W^u_{\mathrm{loc}}(p)$, *tangent to* $E^s(p)$, $E^u(p)$ *at p. The global manifolds* $W^s(p)$, $W^u(p)$ *are injectively immersed copies of* $\mathbb{R}^s$, $\mathbb{R}^u$, *where* $s = \dim E^s(p)$, $u = \dim E^u(p)$ $(s+u=n)$. *Moreover, if P depends smoothly upon parameters* $\mu \in \mathbb{R}^m$, *so do* $W^s_{\mathrm{loc}}(p)$, $W^u_{\mathrm{loc}}(p)$.

λ-Lemma (Palis [16]). *Let P be a* C^r *diffeomorphism with a hyperbolic* *fixed point p and let D^u be a u-disk in $W^u(p)$. Let Δ be a u-disk meeting $W^s(p)$ transversely at some point q. Then $\bigcup_{n>0} P^n(\Delta)$ contains u-disks arbitrarily C^r close to D^u.*

These results also apply to periodic orbits $\{p_i\}$ $i = 1, \ldots, n$, $P^n(p_i) = p_i$ and an analog of the first result applies to flows, such as that of the suspended system (10.12).

Acknowledgement

The research reported in this chapter was supported by NSF Grant ENG 78-02891 and ARO contract DAAG-29C-0086.

References

1. Holmes, P. J. and Marsden, J. E., A partial differential equation with infinitely many periodic orbits: chaotic oscillations of a forced beam, *Archive for Rational Mechanics and Analysis,* **76,** 135–166, 1981.
2. Holmes, P. J. and Marsden, J. E., Horseshoes in perturbations of Hamiltonian systems with two degrees of freedom, *Communications in Mathematical Physics,* **82,** 523–544, 1982.
3. Holmes, P. J. and Marsden, J. E., Melnikov's method and Arnold diffusion for perturbations of integrable Hamiltonian systems, *Journal of Mathematical Physics,* **23,** 669–675, 1982.
4. Holmes, P. J. and Marsden, J. E., Horseshoes and Arnold diffusion for Hamiltonian systems on Lie groups, *Indiana University Mathematics Journal,* **32,** 273–309, 1983.
5. Tseng, W. Y. and Dugundji, J., Nonlinear vibrations of a buckled beam under harmonic excitation, *Journal of Applied Mechanics,* **38,** 467–476, 1971.
6. Moon, F. C. and Holmes, P. J., A magnetoelastic strange attractor, *Journal of Sound and Vibration* **65,** 275–296, 1979.
7. Holmes, P. J., A nonlinear oscillator with a strange attractor, *Philosophical Transactions of the Royal Society A,* **292,** 419–448, 1979.
8. Moon, F. C., Experiments on chaotic motions of a forced nonlinear oscillator: strange attractors, *Transactions of the ASME Journal of Applied Mechanics,* **47,** 638–644, 1980.
9. Chillingworth, D. R. J., *Differentiable Topology with a View to Applications,* London, Pitman, 1976.
10. Iooss, G. and Joseph, D. D., *Elementary Stability and Bifurcation Theory,* New York, Springer-Verlag, 1980.
11. Chow, S. N. and Hale, J. K., *Methods of Bifurcation Theory*: New York, Springer-Verlag, 1982.

12. Chow, S. N., Hale, J. K. and Mallet-Paret, J., An example of bifurcation to homoclinic orbits, *Journal of Differential Equations*, **37,** 351–373, 1980.
13. Andronov, A. A., Vitt, E. A. and Khaiken, S. E., *Theory of Oscillators*: Oxford, Pergamon, 1966 (transl. Immirzi, F.).
14. Hartman, P., *Ordinary Differential Equations*: New York, Wiley, 1964.
15. Hirsch, M. W., Pugh, C. C. and Shub, M., *Invariant Manifolds*, Lecture Notes in Mathematics No. 583, New York, Springer-Verlag, 1977.
16. Palis, J., On Morse Smale dynamical systems, *Topology*, **8,** 385–405, 1969.
17. Newhouse, S. E., Lectures on dynamical systems, in *Dynamical Systems, CIME Lectures Bressanone, Italy, June* 1978, Progress in Mathematics No. 8: Boston, Birkhauser, 1980, 1–114.
18. Melnikov, V. K., On the stability of the center for periodic perturbations, *Transactions of the Moscow Mathematical Society*, **12,** 1–57, 1963.
19. Holmes, P. J., Averaging and chaotic motions in forced oscillations, *SIAM Journal of Applied Mathematics*, **38,** 65–80 (Errata and addendum, *ibid.*, **40,** 167–168), 1980.
20. Greenspan, B. D., *Bifurcations in Periodically Forced Oscillations: Subharmonics and Homoclinic Orbits*, PhD Thesis, Center for Applied Mathematics, Cornell University, 1981.
21. Sanders, J. A., *A Note on the Validity of Melnikov's Method*, Report No. 139: Amsterdam, Wiskundig Seminarium, Vrije Universiteit, 1980.
22. Arnold, V. I., Instability of dynamical systems with several degrees of freedom, *Soviet Mathematics—Doklady*, **5,** 581–585, 1964.
23. Andronov, A. A., Leontovich, E. A., Gordon, I. I. and Maier, A. G., *Theory of Bifurcations of Dynamic Systems on a Plane*: Jerusalem, Israel Program for Scientific Translations, 1971.
24. Newhouse, S. E., The abundance of wild hyperbolic sets and non-smooth stable sets for diffeomorphism, *Publications Mathématiques de l'Institut des Hautes Etudes Scientifiques*, **50,** 101–151, 1979.
25. Hale, J. K., *Ordinary Differential Equations*: New York, Wiley, 1969.
26. Morosov, A. D., Approach to a complete qualitative study of Duffing's equation, *USSR Computational Mathematics and Mathematical Physics*, **13,** 1134–1152, 1973.
27. Smale, S., Diffeomorphisms with many periodic points, in Cairns, S. S. (Ed.), *Differential and Combinatorial Topology*: Princeton University Press, 1963, 63–80.
28. Smale, S., Differentiable dynamical systems, *Bulletin of the American Mathematical Society*, **73,** 747–817, 1967.
29. Nitecki, Z., *Differentiable Dynamics*: MIT Press, 1971.

30. Moser, J., *Stable and Random Motions in Dynamical Systems*: Princeton University Press, 1973.
31. Birkhoff, G. D., *Dynamical Systems*: Providence, R. I., AMS Publications, 1927.
32. Newhouse, S. E., Diffeomorphisms with infinitely many sinks, *Topology*, **13,** 9–18, 1974.
33. Gavrilov, N. K. and Shil'nikov, L. P., On three-dimensional dynamical systems close to systems with a structurally unstable homoclinic curve, I, *Mathematics of the USSR—Sbornik*, **88**(4), 467–485, 1972.
34. Gavrilov, N. K. and Shil'nikov, L. P., On three-dimensional dynamical systems close to systems with a structurally unstable homoclinic curve, II, *Mathematics of the USSR—Sbornik*, **90**(1), 139–156, 1973.
35. Hénon, M., A two-dimensional mapping with a strange attractor, *Communications in Mathematical Physics*, **50,** 69–77, 1976.
36. Lanford, O. E., personal communication, 1981.
37. Feigenbaum, M., Quantitative universality for a class of nonlinear transformation transformations, *Journal of Statistical Physics*, **19,** 25–77, 1978.
38. Collet, P., Eckmann, J. P. and Lanford, O. E., Universal properties of maps on an interval, *Communications in Mathematical Physics*, **76,** 211–254, 1980.
39. Morosov, A. D., A complete qualitative investigation of Duffing's equation, *Differential Equations*, **12,** 164–174, 1976.
40. Hayashi, C., The method of mapping with reference to the doubly asymptotic structure of invariant curves, *International Journal of Nonlinear Mechanics*, **15,** 341–348, 1980.
41. Hayashi, C., *Selected Papers on Nonlinear Oscillations*: Osaka, Nippon Publishing, 1975.
42. Ueda, Y., Steady motions exhibited by Duffing's equation: a picture book of regular and chaotic motions, in Holmes, P. J. (Ed.), *New Approaches to Nonlinear Problems in Dynamics*: Philadelphia, SIAM Publications, 1980.
43. Ueda, Y., Explosion of strange attractors exhibited by Duffing's equation, *Annals of the New York Academy of Sciences*, **357,** 422–434, 1980.
44. Holmes, C. and Holmes, P. J., Second order averaging and bifurcations to subharmonics in Duffing's equation, *Journal of Sound and Vibration*, **78,** 161–174, 1981.
45. Shaw, S., A numerical study of Duffing's equation (unpublished notes), Department of Theoretical and Applied Mechanics, Cornell University, 1980.
46. Ueda, Y., personal communication, 1981.
47. Franks, J., Studies of the Hénon attractor (unpublished notes).
48. Greenspan, B. D. and Holmes, P. J., Repeated resonance and homoclinic bifurcation in a periodically forced family of oscillators, *SIAM Journal on Mathematical Analysis*, 1983 (in press).

49. Chirikov, B. V., A universal instability of many-dimensional oscillator systems, *Physics Reports*, **52,** 263–379, 1979.
50. Chirikov, B. V. and Izrailev, F. M., A degeneration of turbulence in simple systems, Preprint 80–128, Institute of Nuclear Physics, Novosibirsk, 1980.
51. Lieberman, M. A., personal communication, 1981. (See also Lichtenberg, A. J., Lieberman, M. A. and Cohen, R. H., Fermi acceleration revisited (preprint), Department of Electrical Engineering, University of California, Berkeley.)
52. Robinson, C., Bifurcation to infinitely many sinks (preprint), Northwestern University, 1982.
53. Guckenheimer, J., personal communication, 1981.
54. Yakobson, M., Topological and metric properties of one-dimensional endomorphisms, *Doklady Akademii Nauk USSR*, **243,** 866–869, 1978.
55. Misiurewicz, M., Strange attractors for Lozi mappings, in Helleman, R. H. (Ed.), Nonlinear dynamics, *Annals of the New York Academy of Sciences*, **317,** 348–358, 1980.

11 A Review of Interactions of Hopf and Steady-state Bifurcations

W F Langford

11.1 Introduction

The seminal papers of Lorenz [1] and Ruelle and Takens [2] have inspired a flood of papers on simple three-dimensional dynamical systems which display chaotic behaviour analogous to turbulence. However, as yet there has been little success in demonstrating that such examples explain the turbulence that is actually observed in fluid dynamics experiments. On the other hand, the onset of turbulence in experiments is often preceded by transitions which can be identified with generic bifurcations which are mathematically well understood even for infinite-dimensional systems, for example the Hopf bifurcation of a periodic orbit or the simple bifurcation of steady-state solutions. Recent work [3–10] indicates that these generic bifurcations can interact nonlinearly to produce more complicated solution behaviour, including quasiperiodic, aperiodic, homoclinic, intermittent, entrainment or phase-locking, and different types of chaotic behaviour. The formal investigation of such interactions for a given problem involves little more than the computation of a finite number of coefficients analogous to the usual Hopf coefficients. This brings us closer to providing mathematical explanations of at least some physical examples of turbulence, however much work remains to be done. The examples presented here raise challenging questions which have yet to be answered.

This chapter is limited to a review of some of the more interesting phenomena arising from the interactions of a Hopf bifurcation (associated with eigenvalues i, $-i$) and a steady-state bifurcation (associated with eigenvalue 0). These phenomena have come to light only in the past few years, following the fundamental discovery [9] of the existence of invariant tori of solutions arising from such interactions. The Hopf bifurcation is taken to have minimum degeneracy (with one exception), while the steady-state bifurcation is taken to be any one of the four types defined precisely in Section 11.3: transcritical, pitchfork, saddle-node and hysteresis. We proceed to a review of background material in Section 11.2, then to a summary of five elementary bifurcations of steady-state and periodic solutions, and one case of bifurcation of tori, in Section 11.3. The reader familiar with these cases may skip directly to Section 11.4 which presents the four examples of bifurcations which are essentially three-dimensional (that is, displaying behaviour not possible in two dimensions), and includes some phenomena not previously published. In the concluding Section 11.5, we remark on the relationship of these results to the works cited in the opening sentence.

11.2 Preliminaries

Consider an ordinary differential equation (finite-dimensional for convenience):

$$\frac{\mathrm{d}u}{\mathrm{d}t}=F(u), \qquad F:U\subset\mathbb{R}^n\to\mathbb{R}^n, \quad U \text{ open.} \tag{11.1}$$

Assume F lies in the class C of smooth vector fields defined on $\bar{U}$, transverse to the boundary of U, and possibly satisfying additional symmetry conditions to be defined later. Then (11.1) (or equivalently F) is called *structurally stable* if there exists a neighbourhood V of F in C such that for any $G\in V$ there exists a homeomorphism which maps trajectories of F onto trajectories of G, preserving the time orientation. This says that sufficiently small perturbations of F do not qualitatively change the solutions of (11.1). We are concerned here with the converse case in which qualitative changes do occur.

Next consider a smooth one-parameter family of differential equations

$$\frac{\mathrm{d}u}{\mathrm{d}t}=f(u,\lambda), \qquad f:U\times I\subset\mathbb{R}^n\times\mathbb{R}\to\mathbb{R}^n, \quad I \text{ open,} \tag{11.2}$$

such that for each $\lambda\in I$, $f(\cdot,\lambda)$ has the properties assumed for F in (11.1). We say that (11.2) (or equivalently f) has a *bifurcation* at $\lambda_0\in I$ if the following two hypotheses are satisfied:

Hypothesis 1. $f(\cdot,\lambda_0)$ is not structurally stable.

Hypothesis 2. There exists an open interval I_0, $\lambda_0\in I_0\subset I$, such that $\lambda_0\neq\lambda\in I_0$ implies $f(\cdot,\lambda)$ is structurally stable.

Now let D be the class of one-parameter families of vector fields as in (11.2). By definition, f has a *generic* (one-parameter) *bifurcation* at λ_0 if Hypotheses 1 and 2 above are satisfied and in addition:

Hypothesis 3. There exists a neighbourhood W of f in D such that $g\in W$ implies there exists a homeomorphism on $U\times I$ of the form

$(u, \lambda) \mapsto [v(u, \lambda), \mu(\lambda)]$ which maps trajectories of $f(\cdot, \lambda)$ onto trajectories of $g(\cdot, \mu)$, preserving the time orientation.

Otherwise f is said to have a *degenerate bifurcation* at λ_0. This chapter is concerned with *local bifurcations* (which can be either generic or degenerate): we say that (11.2) or f has a (local) bifurcation at (u_0, λ_0) if there is a neighbourhood U_0 of u_0 such that the relevant Hypotheses (1–3) hold for any neighbourhood U satisfying $u_0 \in U \subset U_0$.

The examples of Section 11.4 are all degenerate local bifurcations. Our goal is to introduce additional parameters which unfold the degenerate bifurcation and yield only generic bifurcations, but more than one of them. This idea has been employed fruitfully in [7, 9, 10–12]. We maintain a distinction between the original bifurcation parameter λ and the new unfolding parameters α, and so consider parametrized families of one-parameter families

$$\frac{du}{dt} = f(u, \lambda, \alpha), \qquad \alpha \in \mathbb{R}^m. \tag{11.3}$$

An advantage of this approach is that the canonical examples of generic bifurcations are well understood, and so the sequence of generic bifurcations encountered as λ varies can be conveniently depicted on a bifurcation diagram which is easily interpreted for applications. We then classify these bifurcation diagrams as α varies; this is often simpler than a full classification of phase portraits. A *bifurcation diagram* is a plot of some real functional x of the solution u versus the bifurcation parameter λ. The functional $x(u)$ is usually taken to be a norm or an amplitude or a component of u on an eigenspace.

An alternative approach is to suppress any distinction between λ and α but allow $\lambda \in \mathbb{R}^k$. Often $k = m + 1$. This is the approach employed in elementary catastrophe theory [13, 14] and it has been used also in dynamical systems [6, 15–17].

11.3 Elementary bifurcations

In this section we summarize the features of six classical bifurcations which arise in the examples of the next section. Four of these involve changes only in steady states and one-dimensional dynamics. The Hopf bifurcation gives rise to a periodic orbit with essentially two-dimensional dynamics, and the Naimark–Sacker bifurcation produces an invariant torus in a space of at least three dimensions. For the four steady-state bifurcations we set $du/dt = 0$ in (11.2) and consider

$$f(u, \lambda) = 0, \qquad f: U \times I \to \mathbb{R}^n. \tag{11.4}$$

Suppose that (u_0, λ_0) is a solution of (11.4) and that the Jacobian matrix $D_u f(u_0, \lambda_0)$ has no eigenvalues either zero or pure imaginary. Then

u_0 is called a *hyperbolic* steady state of (11.2) and there is a neighbourhood of u_0 on which $f(\cdot, \lambda_0)$ is structurally stable. The implicit function theorem guarantees the existence of a unique smooth branch of solutions of (11.4) $[u(\lambda), \lambda]$ through (u_0, λ_0), and the eigenvalues of $D_u f[u(\lambda), \lambda]$ remain off the imaginary axis for λ near λ_0. Hence structural stability is preserved for λ in a neighbourhood of λ_0.

A necessary condition for (11.2) to have a local bifurcation at (u_0, λ_0) is that $D_u f(u_0, \lambda_0)$ have zero or pure imaginary eigenvalues. The case of simple pure imaginary eigenvalues leads to the Hopf bifurcation discussed later; we first consider the case of a simple zero eigenvalue, with corresponding eigenvector φ and adjoint (or left) eigenvector ψ. ('Simple' means algebraic multiplicity one.) We assume that

$$f(u_0, \lambda_0) = 0, \qquad D_u f(u_0, \lambda_0)\varphi = 0, \quad |\varphi| = 1. \tag{11.5}$$

Even though the implicit function theorem now fails to guarantee a unique branch $[u(\lambda), \lambda]$ through (u_0, λ_0), one can still calculate (non-unique) branches of solutions of (11.4) by taking into account higher derivatives of f.

11.3.1 Saddle-node bifurcation

Assume (11.5) and

$$a \equiv \psi D_\lambda f(u_0, \lambda_0) \neq 0 \qquad b \equiv \psi D_{uu} f(u_0, \lambda_0)\varphi\varphi \neq 0. \tag{11.6}$$

Then in a neighbourhood of (u_0, λ_0) there exists a unique branch of solutions of (11.4) of the form $[u(z), \lambda(z)]$ for small z in $\mathbb{R}$, given by

$$u(z) = u_0 + z\varphi + 0(z^2) \qquad \lambda(z) = \lambda_0 + z^2 \frac{-b}{2a} + 0(z^3). \tag{11.7}$$

The bifurcation diagram is Fig. 11.1(a). In this case, (μ_0, λ_0) is often called a turning point (or limit point) of solutions of (11.4), and a saddle-node of the differential equation (11.2). The derivation of (11.7) (as for all of the first five cases of this section) is by the Lyapunov–Schmidt method: project (11.4) onto the eigenspace and range space; in our case they are complementary. Solve first in the range space using the fact that $D_u f(u_0, \lambda_0)$ restricted to the range is invertible. Here, as in the following examples, z is made unique by the condition that the $0(z^2)$ terms in $u(z)$ lie in the range space. It remains only to satisfy the eigenspace projection of (11.4), but the assumption $a \neq 0$ in (11.6) guarantees that this is uniquely solvable for λ, and the result is (11.7).

Returning to the differential equation (11.2), we want to know the stabilities of the steady state (11.7). Only cases in which at least one solution is asymptotically stable are of much interest. An easy calculation shows that exactly one eigenvalue changes sign as z passes through 0 [18]. Therefore, if one of the solutions u for a given $\lambda \neq \lambda_0$ in (11.7) is asymptotically stable, and if $n \geqslant 2$, the other one must be a saddle. As

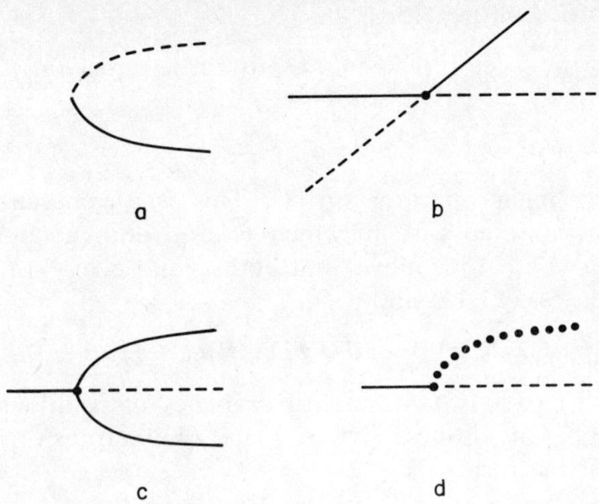

Fig. 11.1 Generic bifurcation diagrams:
(*a*) Saddle-node; (*b*) Transcritical; (*c*) Pitchfork; (*d*) Hopf
In these and the following bifurcation diagrams, solid (broken) lines indicate stable (unstable) steady states, solid (open) circles indicate stable (unstable) periodic orbits

$\lambda \to \lambda_0$ from the appropriate side, an unstable saddle-point and a stable node (or node-focus) coalesce and disappear. Hence the name saddle-node.

This example is the only bifurcation at a zero eigenvalue which is generic for the general definition in Section 11.2. However, if the vector field is restricted to a smaller class, other examples become generic.

11.3.2 Transcritical bifurcation

Suppose (11.4) has a 'trivial solution' $[u(\lambda), \lambda]$ for all $\lambda \in I$, then without loss of generality $u \equiv 0$, i.e.,

$$f(0, \lambda) = 0 \qquad \text{for all} \quad \lambda \in I. \tag{11.8}$$

Now restrict the class D of vector fields in Section 11.2 to those satisfying (11.8). This excludes (11.6) since $D_\lambda f(0, \lambda) = 0$. A classical result is the following [18, 19]. Assume (11.5), (11.8) and

$$c \equiv \psi D_{u\lambda} f(0, \lambda_0)\varphi \neq 0, \qquad b \equiv \psi D_{uu} f(0, \lambda_0)\varphi\varphi \neq 0. \tag{11.9}$$

Then there exists a unique smooth branch of nontrivial solutions $[u(z), \lambda(z)]$ for small z in $\mathbb{R}$ of the form

$$u = z\varphi + 0(z^2), \qquad \lambda = \lambda_0 + z\frac{-b}{2c} + 0(z^2). \tag{11.10}$$

Again (11.10) is derived by the Lyapunov–Schmidt method; the typical bifurcation diagram is shown in Fig. 11.1(*b*).

11.3.3 Pitchfork bifurcation

Now restrict the vector fields $f \in D$ to those satisfying the symmetry condition

$$Sf(u, \lambda) = f(Su, \lambda), \qquad S\varphi = -\varphi, \tag{11.11}$$

where S is a linear operator on $\mathbb{R}^n$. This is a generalization of the frequently encountered case in which f is an odd function of u; then $S = -I$. Under (11.11) it follows that both a and b in (11.6) are zero. If we assume (11.5), (11.11) and

$$c \equiv \psi D_{u\lambda} f(u_0, \lambda_0)\varphi \neq 0, \qquad d \equiv \psi D_{uuu} f(u_0, \lambda_0)\varphi\varphi\varphi \neq 0, \tag{11.12}$$

then there exist exactly two smooth branches of solutions $[u(z), \lambda(z)]$ through (u_0, λ_0). One of these is 'trivial' and can be transformed to (11.8), the other is of the form

$$u(z) = z\varphi + 0(z^2), \qquad \lambda(z) = \lambda_0 + z^2 \frac{-d}{6c} + 0(z^4), \tag{11.13}$$

and $\lambda = \lambda(z)$ is an even function. The bifurcation diagram is typically as shown in Fig. 11.1(c).

11.3.4 Hopf bifurcation

Now suppose $D_u f(u_0, \lambda_0)$ is invertible but has a simple pair of purely imaginary eigenvalues $\pm i$, with corresponding complex eigenvectors φ, $\bar{\varphi}$ (bar denoting complex conjugation) and adjoint eigenvectors ψ, $\bar{\psi}$. The absence of a zero eigenvalue gives, by the implicit function theorem, a unique branch of steady-state solutions $[u(\lambda), \lambda]$, $\lambda \in I$, which without loss of generality we can translate to the trivial solution (11.8). The main result is the following [15, 18, 20–22]. Assume (11.5), (11.8) and

$$c = \psi D_{u\lambda} f(0, \lambda_0)\varphi \neq 0. \tag{11.14}$$

Then in a neighbourhood of $(0, \lambda_0)$, Eq. (11.2) has a smooth family of periodic solutions $[u(t; r), \lambda(r)]$ of period $T = T(r)$, for small r in $\mathbb{R}$, of the form

$$\begin{aligned} u &= r \operatorname{Re}\{\varphi \exp[i(2\pi t/T + \theta)]\} + 0(r^2) \\ \lambda &= \lambda_0 + \lambda_2 r^2 + 0(r^4) \qquad T = 2\pi + T_2 r^2 + 0(r^4). \end{aligned} \tag{11.15}$$

Here r is essentially the amplitude of the orbit, θ is an arbitrary phase, and the functions $\lambda(r)$ and $T(r)$ are even in r. Explicit formulae for λ_2 and T_2 can be found in [18, 21, 22]. Generically $\lambda_2 \neq 0$, and then the amplitude of the oscillation satisfies

$$r \sim (\lambda - \lambda_0)^{1/2}. \tag{11.16}$$

If either $\lambda_2 = 0$ or (11.14) fails, or both, the Hopf bifurcation is degenerate in the sense of Section 11.2. These cases are treated in detail in [21].

11.3.5 Hysteresis bifurcation

Suppose that in the previously defined saddle-node bifurcation the condition $b \neq 0$ (11.6) fails. Then we can still carry out the Lyapunov–Schmidt reduction. The result is a branch of solutions $[u(z), \lambda(z)]$ for small $z \in \mathbb{R}$ of the form

$$u(z) = u_0 + z\varphi + 0(z^2) \qquad \lambda(z) = \lambda_0 + z^3\lambda_3 + 0(z^4) \tag{11.17}$$

as in the bifurcation diagram of Fig. 11.2 with $\alpha = 0$ (if $\lambda_3 \neq 0$). However, this is a degenerate bifurcation according to Section 11.2. Small perturbations can remove the degeneracy and give either two saddle-nodes or no bifurcation, as shown in Fig. 11.2. We therefore introduce a 'splitting' or 'unfolding' parameter α, which has this effect, as follows. Let $f = f(u, \lambda, \alpha)$ as in (11.3), satisfying (11.5) for $\alpha = 0$, and

$$\begin{aligned} \psi D_\lambda f(u_0, \lambda_0, 0) &\neq 0 \\ \psi D_{uu} f(u_0, \lambda_0, 0)\varphi\varphi &= 0 \\ \psi D_{uuu} f(u_0, \lambda_0, 0)\varphi\varphi\varphi &\neq 0 \\ \psi D_{u\alpha} f(u_0, \lambda_0, 0)\varphi &\neq 0. \end{aligned} \tag{11.18}$$

The first three of these conditions just ensure the stated form of the solution for $\alpha = 0$ and the fourth ensures that $\alpha \neq 0$ unfolds the degeneracy. Without loss of generality we can also set

$$D_\alpha f(u_0, \lambda_0, 0) = 0. \tag{11.19}$$

Then (11.3) has steady-state solutions $[u(z, \alpha), \lambda(z, \alpha)]$ of the form

$$\begin{aligned} u(z, \alpha) &= z\varphi + \ldots \\ \lambda(z, \alpha) &= \lambda_0 + c_1\alpha z + c_3 z^3 + \ldots \end{aligned} \tag{11.20}$$

where ... denotes higher order terms in α and z. This follows by the same Lyapunov–Schmidt and implicit function theorem argument as before. The bifurcation diagrams are now as shown in Fig. 11.2. We remark that this 'hysteresis bifurcation' is similar to the 'cusp' elementary catastrophe [13, 14], except for the distinction made here in the roles of λ and α.

Using techniques of singularity theory perfected in [12], one can show for $C^\infty f$ the existence of nonlinear symmetry covariant contact transformations which eliminate the higher-order terms in the above five solu-

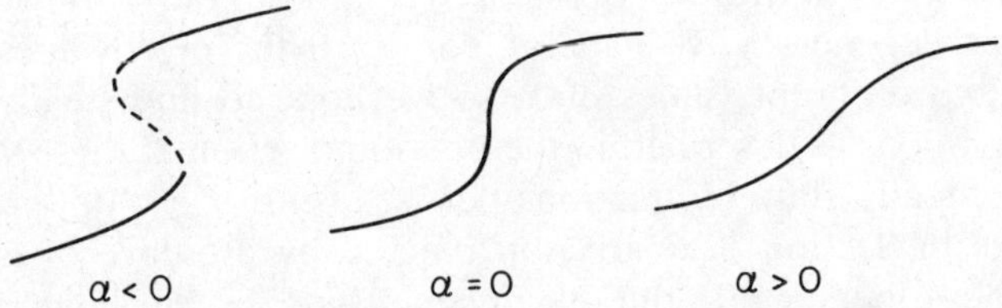

Fig. 11.2 Hysteresis bifurcation diagram, and unfoldings

tions and yield the following normal forms under contact equivalence [12, 21]:

saddle-node $\lambda \pm z^2 = 0$

transcritical $\lambda z \pm z^2 = 0$

pitchfork $\lambda z \pm z^3 = 0$

Hopf $\lambda r \pm r^3 = 0$

hysteresis $\lambda \pm \alpha z \pm z^3 = 0$

It is important to realize that, while these normal forms preserve branches of steady-state and periodic solutions, they do not preserve the full dynamics of the differential equation (11.2).

11.3.6 Naimark–Sacker torus bifurcation [2, 15, 18, 22–24]

Suppose $u = p(t)$ is a T-periodic solution of (11.2). Then p is (asymptotically orbitally) stable if the Floquet multipliers ρ all have modulus $|\rho| < 1$. Equivalently, the ρ are eigenvalues of the derivative of the Poincaré map at $p(0)$. If a simple pair of complex multipliers ρ, $\bar{\rho}$ satisfies

$$\begin{aligned} &|\rho(\lambda_0)| = 1, \qquad \frac{\mathrm{d}}{\mathrm{d}\lambda}|\rho(\lambda_0)| > 0, \\ &\rho(\lambda_0)^k \neq 1, \qquad k = 1, 2, 3, 4, \end{aligned} \tag{11.21}$$

then for generic f an invariant torus of solutions bifurcates from the periodic orbit p. Obviously the phase space must be at least three-dimensional for this to happen. A major difficulty in applying this theorem stems from its nonlocal character: the multipliers ρ can rarely be computed analytically because this requires knowledge of the whole orbit $p(t)$, $0 \leq t \leq T$.

11.4 Interactions of steady-state and Hopf bifurcations

Throughout this section we consider the differential equation as in (11.3)

$$\frac{\mathrm{d}u}{\mathrm{d}t} = f(u, \lambda, \alpha) \tag{11.22}$$

and we assume that $f(0, 0, 0) = 0$, and that $D_u f(0, 0, 0)$ has simple eigenvalues 0, i, $-i$ and no others pure imaginary. Then f has a degenerate bifurcation at $\lambda = 0$ (for $\alpha = 0$) and we choose the α-dependence to remove this degeneracy. In general, for $\alpha \neq 0$ the eigenvalues 0 and $\pm i$ occur for two different values of λ, so we hope to find steady-state and Hopf bifurcations as described in the previous section at these values of λ. In fact the steady-state bifurcations persist even for $\alpha = 0$, because these calculations in Section 11.3 are not affected by presence of eigenvalues $\pm i$. On the other hand, the eigenvalue 0 does invalidate the Hopf bifurcation theorem, leading to new phenomena.

We consider four examples, each of which corresponds to a coalescence of a Hopf bifurcation with one of the four types of steady-state bifurcation in the previous section. In each example, for $\alpha \neq 0$, we recover the expected local steady-state bifurcation and Hopf bifurcation of Section 11.3; these we call the primary bifurcations. In the two cases with the trivial solution, we find additional bifurcations of periodic orbits from nontrivial branches; these we call secondary bifurcations. In all four examples we show the possibility of additional bifurcations giving rise to three-dimensional dynamics (such as invariant tori); these we call tertiary bifurcations.

The main tool of this section is the method of Poincaré–Birkhoff normal forms described, for example, in [6, 10, 15, 16, 18]. This procedure consists of successive near-identity nonlinear transformations of coordinates which systematically eliminate most of the lower-degree terms of the power series representation of f in terms of u. Certain terms cannot be eliminated; they are called resonant, and they are fundamental in determining the dynamics. Unlike the contact equivalence normal forms of Section 11.3, these normal forms preserve the full dynamics in a neighbourhood of the singularity; however, the disadvantage is that the procedure does not converge in general. For the cases considered here, the neighbourhood of validity of the normal form can tend toward zero as the computation proceeds to higher-degree terms; the series is asymptotic rather than convergent. This fact has an important effect on the dynamics to be expected in a finite neighbourhood.

The first step is a linear change of coordinates to bring $D_u(f(0, 0, 0)$ to block-diagonal form, consisting of a 3×3 block with eigenvalues 0, i, $-i$ and a $(n-3) \times (n-3)$ block with all eigenvalues off the imaginary axis (we will assume them in the negative half-plane for reasons of stability). Then, as shown in [10], the last $(n-3)$ components of the solution can be suppressed to arbitrarily high order relative to the first three components. Therefore we proceed from here on as if $n=3$. We use Cartesian coordinates (x, y, z) or cylindrical coordinates (r, θ, z), where z is the coordinate in the eigenspace corresponding to 0 and (x, y) or (r, θ) are coordinates in the two-dimensional eigenspace corresponding to $\pm i$.

11.4.1 Transcritical–Hopf

This was the first example in which tertiary bifurcation was discovered [9]. The Poincaré normal form is

$$\begin{aligned}
\frac{dz}{dt} &= \lambda z + az^2 + br^2 + 0(3) \\
\frac{dr}{dt} &= r(\lambda - \alpha + cz) + 0(3) \qquad (11.23) \\
\frac{d\theta}{dt} &= 1 + 0(2)
\end{aligned}$$

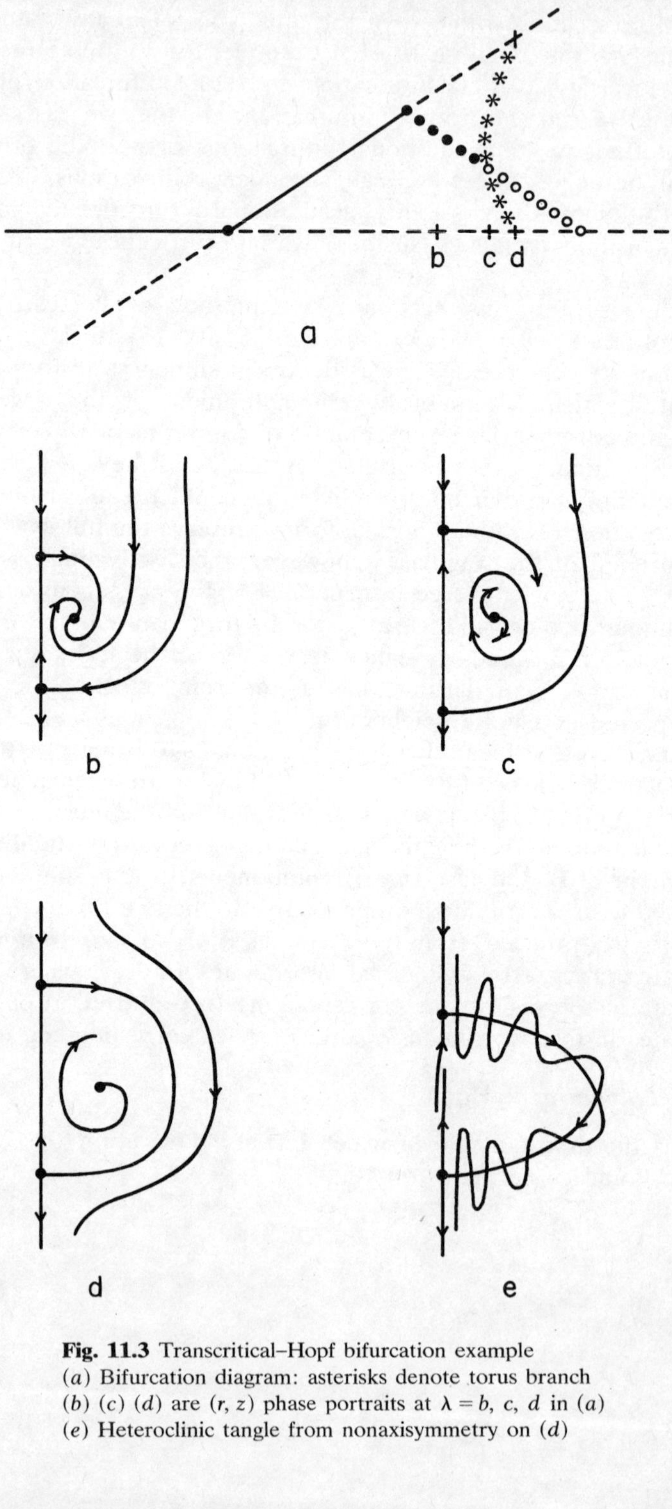

Fig. 11.3 Transcritical–Hopf bifurcation example
(a) Bifurcation diagram: asterisks denote torus branch
(b) (c) (d) are (r, z) phase portraits at $\lambda = b, c, d$ in (a)
(e) Heteroclinic tangle from nonaxisymmetry on (d)

where $0(n)$ denotes terms of order n, the coefficients a, b, c are generically nonzero and depend smoothly on the parameters, and we have allowed an invertible transformation of (λ, α). Clearly (11.23) has the trivial solution $z = r = 0$, and neglecting the $0(n)$ terms we regain the transcritical steady-state bifurcation with $r = 0$ at $\lambda = 0$, and a Hopf bifurcation with $r \neq 0$ at $\lambda = \alpha$. There are six generic bifurcation diagrams depending on a, b, c as documented in [9]. All of these exhibit a secondary Hopf bifurcation, but only one has a tertiary bifurcation (Fig. 11.3(a)). In this case, the branch of periodic solutions itself changes stability, a pair of complex Floquet multipliers crosses the unit circle near 1, and the Naimark–Sacker theorem in Section 11.3 implies that for generic higher-order terms there exists an invariant torus of solutions surrounding the periodic orbit, either supercritical and stable or subcritical and unstable. This result was first demonstrated in the Appendix of [9]. Several authors [3, 5, 7, 8] have noted that the truncated equations (11.23) (i.e., with $0(n)$ term deleted) are exactly integrable at the torus bifurcation point, giving a family of nested tori bounded by a saddle–saddle connecting cycle in (r, z). Although this situation would not be realized generically, it can be of use in calculating the generic torus bifurcation as a perturbation of the exact integral. Figure 11.3 shows typical phase portraits for (11.23), and Fig. 11.4 is a numerical integration of a special case of (11.23) with cubic terms included, which has an attracting torus.

The second interesting feature of this case is the unfolding of the saddle–saddle connection for the (r, z) equations, see Fig. 11.3(d). The θ-variable would rotate this into a two-dimensional manifold of solutions joining the two saddles, for the axisymmetric truncation. However, recall that in general the normal form calculation does not converge, so that the exact equations must remain nonaxisymmetric. This leads, as in Fig. 11.3(e), to transverse intersections of the two-dimensional stable

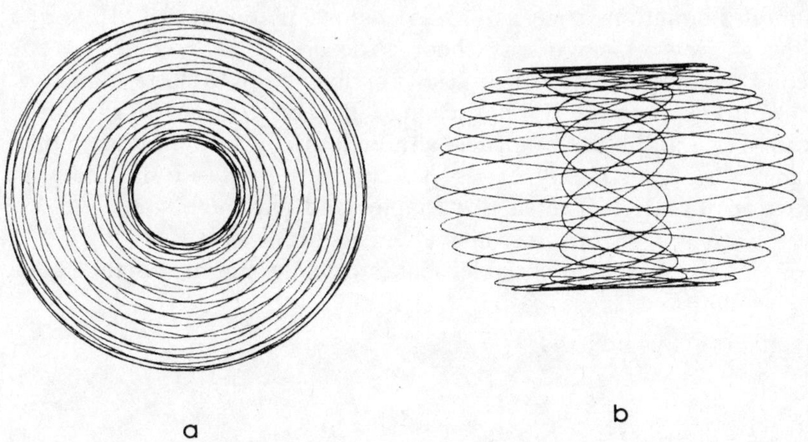

Fig. 11.4 Numerically computed torus corresponding to Fig. 11.3(c): (a) (x, y) plot; (b) (x, z) plot

and unstable manifolds of the two saddle-points and, as observed by Guckenheimer [5, 16] and Holmes [6], to Smale 'horseshoes' and hence to chaotic behaviour. The underlying mechanism is related to work of Shil'nikov, see [16, 25, 26].

11.4.2 Pitchfork–Hopf

This case, studied in [6, 7, 10, 16, 27–29], has the normal form

$$\begin{aligned}\frac{dz}{dt}&=z(\lambda+az^2+br^2)+0(5)\\ \frac{dr}{dt}&=r(e\lambda-\alpha+cz^2+dr^2)+0(5)\\ \frac{d\theta}{dt}&=1+0(2).\end{aligned}\tag{11.24}$$

Clearly this has the trivial solution and, neglecting the $0(n)$ terms, we find also primary bifurcations $r=0$, $\lambda=-az^2$ (pitchfork), and $z=0$, $\lambda=(\alpha-dr^2)/e$ (Hopf). The case with a, d both negative has both primary bifurcations supercritical, and gives secondary but not tertiary bifurcations [27]. For $ad<0$, however, we find tertiary bifurcations to invariant tori [6, 10]. Some of these tori are bounded by saddle-connections as in the previous example, while others are unbounded, at least for the truncated normal equations [6, 7, 16].

The observation that in applications subcritical branches almost always turn back supercritical, motivates a modification of (11.24). Suppose $d>0$, and include an r^5 term with negative coefficient in the r-equation. Then, formally, the Hopf branch starts out subcritical but turns back supercritical. This negative r^5 term also has the effect of stabilizing the torus bifurcation and making it supercritical. If we suppose that all of the other 0(5) terms are such that they do not modify this picture, we are led to model equations such as (11.25) below, taken from [28]. (From the point of view of singularity theory, for $d\to 0$, d becomes in effect a second unfolding parameter. It is possible then to show that the 0(5) terms other than r^5 do not change the bifurcation diagram for the steady-state and Hopf branches, although of course the full dynamics can change. In fact the dynamics do change for $d\to 0$ in these model equations (11.25).) The small variables and parameters, x, y, z, λ, α, d, have been 'blown up', rescaling them to order one. This in turn transforms t to a slow time variable, makes ω large and makes the coefficient of r^5 small.

$$\begin{aligned}\frac{dx}{dt}&=xH-\omega y\\ \frac{dy}{dt}&=\omega x+yH\\ \frac{dz}{dt}&=z(\lambda+x^2+y^2-z^2)+Q\end{aligned}\tag{11.25}$$

where

$$H = 5\lambda - 1 + 2(x^2 + y^2) - 3z^2 - \varepsilon(x^2 + y^2)^2$$

$$Q = 0 \quad \text{if} \quad x < 0 \quad \text{or} \quad y > 0 \quad \text{or} \quad H > 0,$$

$$= \frac{qHx^2y^2(x+y)}{3} \quad \text{otherwise.}$$

The bifurcation diagram for (11.25) is given in Fig. 11.5. Figures 11.6–8 show numerical solutions of (11.25) for $\omega = 17$, $\varepsilon = 0.01$, $\lambda = 2.0$, and two values of q: $q = 0.0$ (axisymmetric) and $q = 0.03$ (nonaxisymmetric). The tori in Fig. 11.6 appear very different from the one in Fig. 11.4. Note that the 'doughnut-hole' has become very small, and that the trajectory spends most of its time in the vicinity of a saddle-point which is just below the bottom of the torus in the (x, z) plots of Fig. 11.6. The direction of the flow is downward through the centre hole, then spiralling out from the bottom and upward around the outside.

Fast Fourier transforms of the $x + z$ components in Fig. 11.6 are shown in Fig. 11.7. In Fig. 11.7(a) we see a typical quasiperiodic spectrum with two basic frequencies, while Fig. 11.7(b) shows a significant decrease in the heights of most of the sharp peaks and a corresponding increase in broadband noise associated with the nonaxisymmetric perturbation.

The interesting feature in Fig. 11.8(b) is that nonaxisymmetry produces an 'S-fold' of the cross-section of the torus analogous to the quadratic mapping of Hénon. The S's get flattened and stretched as they pass down through the saddle region in the lower left of Fig. 11.8(b). These folds become pressed arbitrarily close together as a result of the normal hyperbolicity, yet they remain distinct because the Poincaré map is one-to-one. One conjectures that a transverse cut of Fig. 11.8(b) contains a Cantor set. This is a mechanism for chaos quite distinct from the saddle

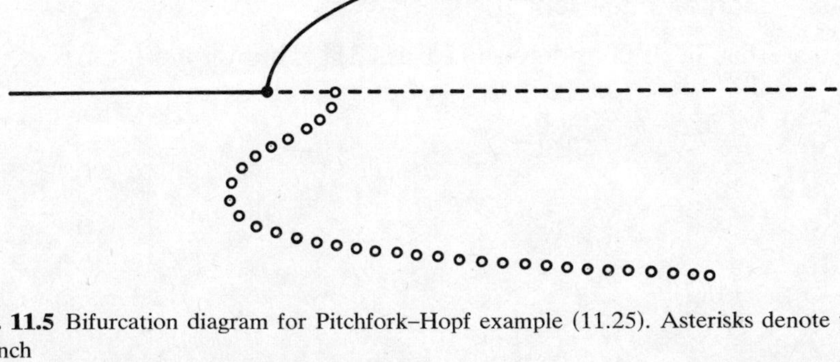

Fig. 11.5 Bifurcation diagram for Pitchfork–Hopf example (11.25). Asterisks denote torus branch

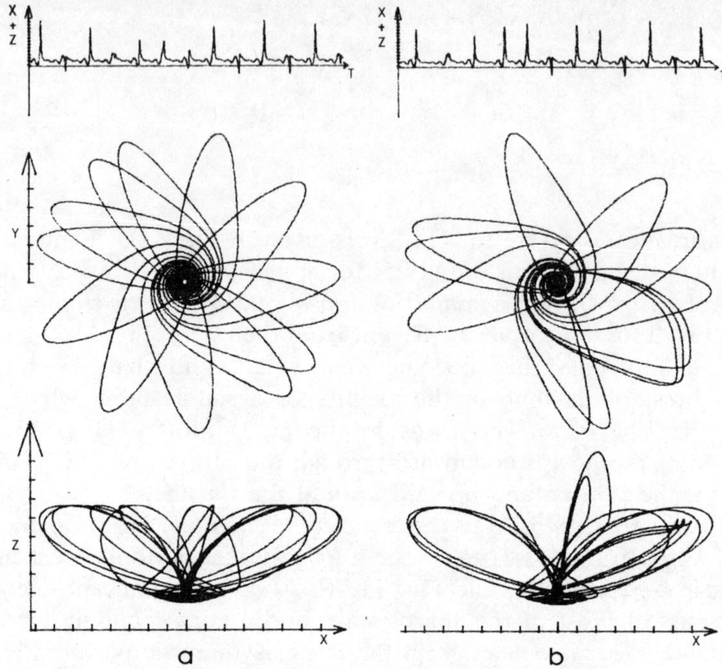

Fig. 11.6 Numerically computed torus for Eq. (11.25): (a) axisymmetric, $q=0$; (b) nonaxisymmetric, $q=0.03$

connection of the previous example. This chaos is robust in that it exists (numerically at least) over a large parameter range, however it is also weak in that much order remains, as seen in Figs 11.6 and 7. This is analogous to 'weak turbulence' in fluid dynamics [28, 30].

It should be remarked that the choice of nonaxisymmetric term Q in (11.25) is not very 'generic'. Our particular choice of Q perturbs the vector field only in a fairly small sector of phase space. This allows the flow to return toward its unperturbed, axisymmetric behaviour (Figs 11.6(a), 11.8(a)) between the perturbing kicks, thus making the effects of the perturbations easier to visualize (Fig. 11.8(b)). More generic choices of Q can give qualitatively similar numerical results.

11.4.3 Saddle-node–Hopf

This case has been considered in [5, 16, 31]. The normal form is

$$
\begin{aligned}
\frac{\mathrm{d}z}{\mathrm{d}t} &= \lambda + az^2 + br^2 + 0(3) \\
\frac{\mathrm{d}r}{\mathrm{d}t} &= r(\alpha + cz) + 0(3) \\
\frac{\mathrm{d}\theta}{\mathrm{d}t} &= 1 + 0(2).
\end{aligned}
\tag{11.26}
$$

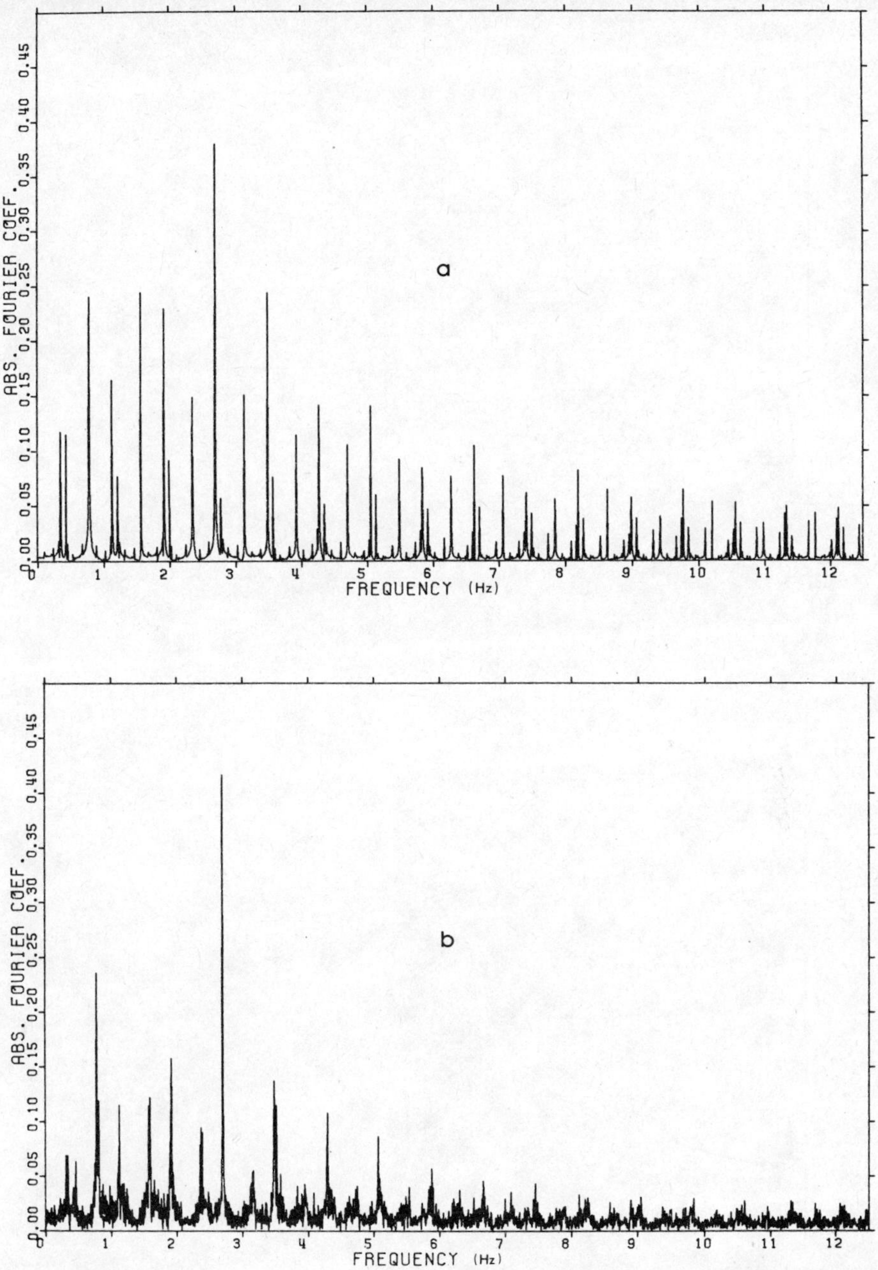

Fig. 11.7 Fast Fourier transform of the $(x+z)$ component of Fig. 11.6: (*a*) axisymmetric; (*b*) nonaxisymmetric.

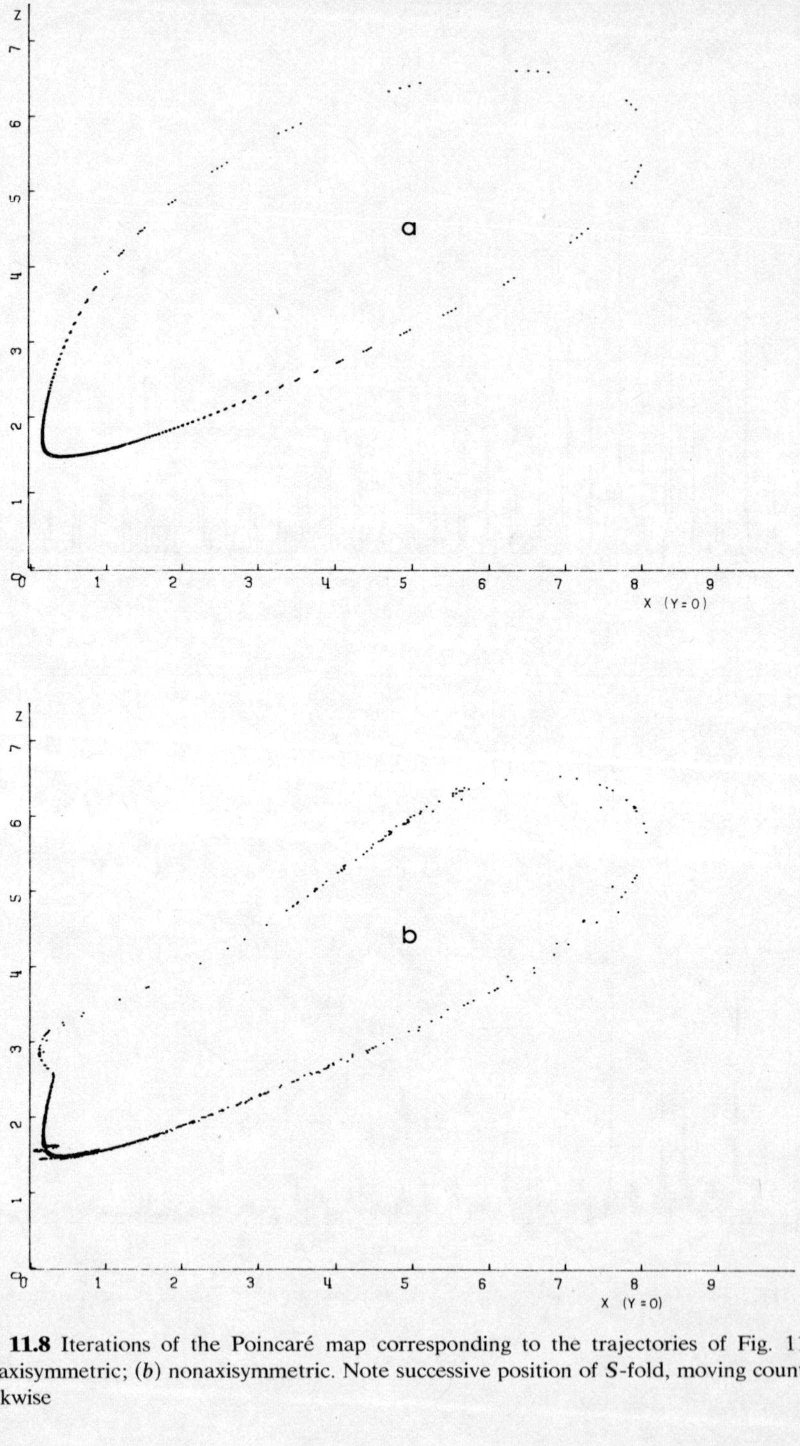

Fig. 11.8 Iterations of the Poincaré map corresponding to the trajectories of Fig. 11.6: (*a*) axisymmetric; (*b*) nonaxisymmetric. Note successive position of *S*-fold, moving counter-clockwise

Truncating, we find the saddle-node bifurcation ($\lambda = -az^2$, $r = 0$) and the Hopf bifurcation ($z = -\alpha/c$, $\lambda = -br^2 - (a\alpha^2/c^2)$) and no secondary bifurcations of new stationary or periodic solutions. The basic steady-state and Hopf branches extend to the same side of $\lambda = 0$ in the case $ab > 0$, and go in opposite directions if $ab < 0$. Naimark–Sacker torus bifurcation is possible only if $bc < 0$, and then there is a saddle-connection only if $ab > 0$. Again the 0(3) terms are required for the three-dimensional bifurcations to be nondegenerate. All the features of this example are contained in the next one. For more details see Guckenheimer [5, 16].

11.4.4 Hysteresis–Hopf

The normal form is

$$\begin{aligned}\frac{dz}{dt} &= \lambda + \alpha z + az^3 + br^2 + 0(3)\\ \frac{dr}{dt} &= r(-\beta + cz) + 0(3) \qquad (11.27)\\ \frac{d\theta}{dt} &= 1 + 0(2).\end{aligned}$$

From the truncated equations we immediately recover the steady-state hysteresis ($\lambda = -\alpha z - az^3$, $r = 0$) and Hopf ($z = \beta/c$, $br^2 = -\lambda - (\alpha\beta/c) - (a\beta^3/c^3)$) bifurcation branches. There are no secondary bifurcations of new stationary or periodic solutions. The classification of generic bifurcation diagrams will be presented elsewhere. Three distinct types of bifurcation involving tori can occur. These are the Naimark–Sacker bifurcation and the saddle-connection bifurcation, which have been seen in earlier examples, and a new type: a 'large' torus can disappear simultaneously with the saddle-node bifurcation of steady states. The following model equations (obtained from (11.27) in the same manner as (11.25)) display all three types.

$$\begin{aligned}\frac{dx}{dt} &= x(z - \beta) - \omega y\\ \frac{dy}{dt} &= \omega x + y(z - \beta) \qquad (11.28)\\ \frac{dz}{dt} &= \lambda + \alpha z - \frac{z^3}{3} - r^2(1 + qx + \varepsilon z).\end{aligned}$$

Let us consider the case $\alpha = 1$, $\beta = 0$, $0 < \varepsilon < 1$, $|q| < 1$. Then a typical bifurcation diagram for (11.28) is as shown in Fig. 11.9; numerically computed attractors in phase space are shown in Figs 11.10 and 11. As in the previous example, $q \neq 0$ inserts nonaxisymmetric perturbations in the normal form. Consider first the case of q near 0. Then there is a nondegenerate Naimark–Sacker torus bifurcation at λ_1 ($\lambda_1 = 1/\varepsilon$ if $q = 0$).

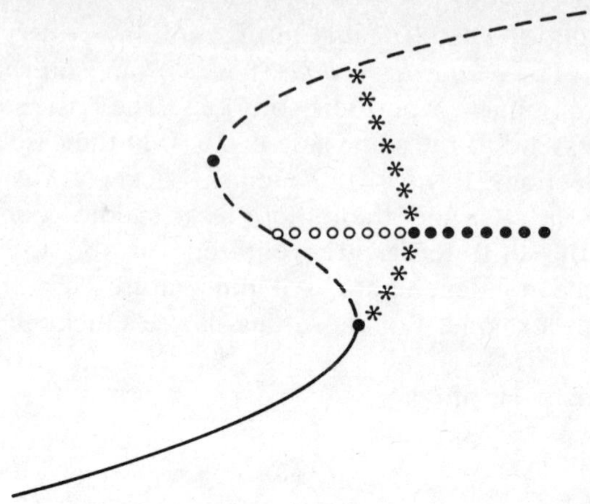

Fig. 11.9 Bifurcation diagram for hysteresis–Hopf example (11.28). Asterisks denote torus branch

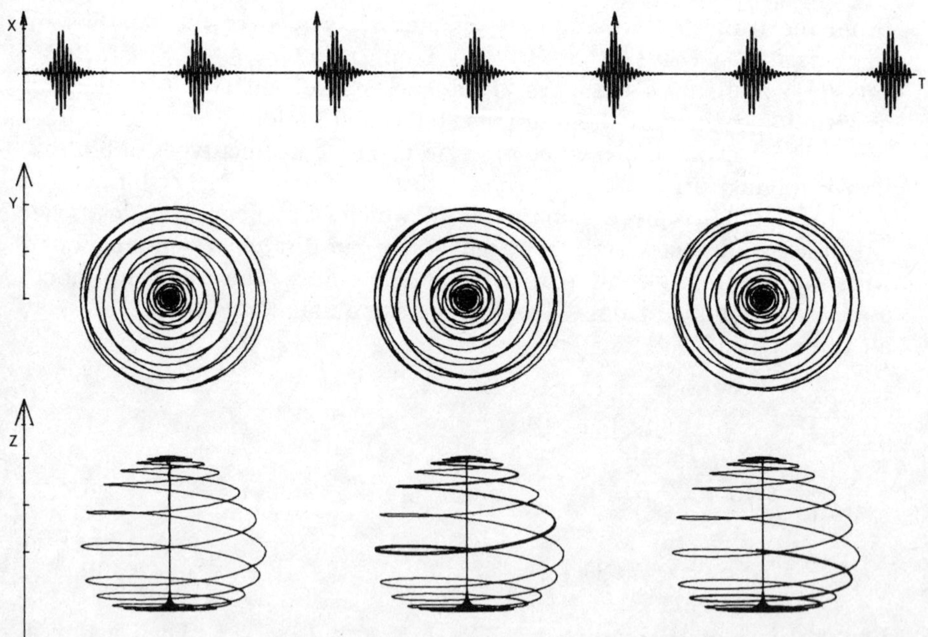

Fig. 11.10 Hysteresis–Hopf example (11.28): intermittent behaviour near saddle-node bifurcation. All plots correspond to the same trajectory on the torus. From left to right, t increases, with the final point of each plot as the initial point of the next

The lower saddle-node steady-state bifurcation occurs at $\lambda=\frac{2}{3}$ for the model equations. The torus exists for $\frac{2}{3}<\lambda<\lambda_1$, and is globally attracting. As λ approaches $\frac{2}{3}$ from above, the torus has the appearance shown in Fig. 11.10 ($\lambda=0.75$, $\omega=15$, $\varepsilon=0.5$, $q=0.7$). The centre hole of the torus shrinks to a thin tube, and the trajectory spends more and more of its time nearly stationary in the region where the saddle-node is about to appear. The upper plot, of x versus t, clearly shows intermittent bursts of oscillations separated by long quiescent intervals. The length of these quiescent intervals becomes infinite as λ approaches $\frac{2}{3}$ from above, while the bursts are relatively unchanged. At $\lambda=\frac{2}{3}$, the torus disappears and is replaced as global attractor by the saddle-node, and then by the stable node for $\lambda<\frac{2}{3}$.

Phase-locking (or frequency entrainment) is easily observed in this model for moderate $q\neq 0$ and sutiable choices of ω. Figure 11.11 ($\lambda=0.8$, $\omega=1.48$, $q=0.2$, $\varepsilon=0.0$) shows an example of 3:1 frequency entrainment. The orbit precesses around the torus at a nonuniform rate for $q\neq 0$, and in this instance asymptotically approaches a stable periodic limit cycle lying in the torus. This is in accord with the general theory of dynamical systems on a torus [15]. It represents another type of bifurcation, as ω or

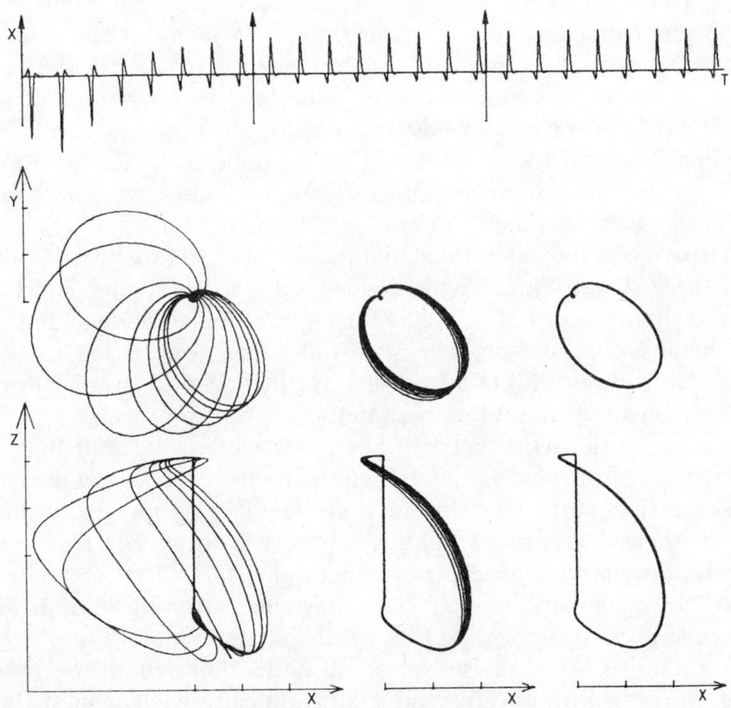

Fig. 11.11 Hysteresis–Hopf example (11.28): phase-locking behaviour. The orbit precesses around the torus, asymptotically approaching a stable limit cycle on the torus. From left to right, the final point of each plot is the initial point for the next

q varies, from an attracting torus of quasiperiodic solutions to an attracting limit cycle, in the phase space of the differential equation (11.28). It can be associated with a saddle-node bifurcation of periodic points of the Poincaré map, much as the Naimark–Sacker bifurcation for tori corresponds to a Hopf bifurcation for maps.

11.5 Conclusion

We have considered four examples of degenerate bifurcations which occur at the coalescence of a Hopf bifurcation point with a point of bifurcation of steady states, for a family of ordinary differential equations. The emphasis has been on finding the interesting dynamical behaviour (e.g., nonstationary, nonperiodic attractors) in the unfoldings of these degenerate bifurcations. The examples indicate the variety of phenomena which can exist, but represent only a very small sampling of all possible cases. Considerable progress toward complete classifications of the unfoldings of these degenerate bifurcations has been achieved very recently [3–10, 16, 27, 29] and still more recent work will appear shortly. In earlier work Takens [17] classified the vector-field singularities associated with eigenvalues 0, i, $-i$, but did not consider their unfoldings. From a physical point of view, such singularities are virtually unobservable (unless there are severe constraints on the system). However, their unfoldings are observable in a generic sense. The first demonstration of generic unfoldings which included nontrivially three-dimensional dynamics (via the Naimark–Sacker theorem) was given for the transcritical–Hopf case in [9]. Other cases then followed quickly. An excellent survey of degenerate bifurcations with 'codimension two' has been written by Guckenheimer [16]. Experience indicates that codimension two bifurcations with nontrivial three-dimensional dynamics, arising through the mechanisms described here, are not globally attracting. In order to find three-dimensional global attractors in such unfoldings, one is led to consider 'codimension three' bifurcations as in this chapter (one bifurcation parameter plus two unfolding parameters, in our approach).

Returning to the work cited in Section 11.1, let us recall that Ruelle and Takens [2] proposed that transition to turbulence could occur via three successive 'Hopf-type' bifurcations, leading from a steady state to a cycle T^1, then to a torus T^2 and finally to a 3-torus T^3. This sequence ends here, because a 3-torus is not structurally stable and can be expected to give way to a 'strange attractor', perhaps satisfying Axiom A. The sequence is usually interpreted as involving the coupling of three nonlinear oscillators in a phase space of dimension $\geqslant 6$. The transitions leading to T^3 are physically and mathematically well-defined, but the strange attractor is not.

The strange attractor of Lorenz [1] involves relatively simple dynamics in $\mathbb{R}^3$ and is now very well understood. However the lack of a physical interpretation for transitions leading up to this strange attractor is a drawback for applications.

The mechanisms described in this chapter resemble the Ruelle–Takens scenario in that they involve well-defined sequences of physically realizable transitions, leading from a steady state eventually to chaotic behaviour. The whole sequence can take place in $\mathbb{R}^3$, as for the Lorenz case, where the dynamics are easy to visualize. Furthermore, there is a standard procedure (the computation of normal forms) for determining whether one of these examples is realized in a given dynamical system.

Acknowledgement

The research reported in this chapter was supported in part by The Natural Sciences and Engineering Research Council of Canada.

References

1. Lorenz, E. N., Deterministic nonperiodic flow, *Journal of the Atmospheric Sciences*, **20,** 130–141, 1963.
2. Ruelle, D. and Takens, F., On the nature of turbulence, *Communications in Mathematical Physics*, **20,** 167–192, 1971.
3. Bissett, E. J., Interaction of steady and periodic bifurcating modes with imperfection effects in reaction-diffusion systems, *SIAM Journal of Applied Mathematics*, **40,** 224–241, 1981.
4. Chow, S.-N. and Hale, J. K., Bifurcation and nonlinear oscillations (preprint), Brown University, 1980.
5. Guckenheimer, J., On a codimension two bifurcation, in Rand, D. A. and Young, L. S. (Eds), *Dynamical Systems and Turbulence, Warwick, 1980*, Lecture Notes in Mathematics No. 898: Berlin and New York, Springer-Verlag, 1981, 99–142.
6. Holmes, P. J., Unfolding a degenerate nonlinear oscillator: a codimension two bifurcation, *Annals of the New York Academy of Sciences*, **357,** *Nonlinear Dynamics*, 473–488, 1980.
7. Iooss, G. and Langford, W. F., Conjectures on the routes to turbulence via bifurcations, *Annals of the New York Academy of Sciences*, **357,** *Nonlinear Dynamics*, 489–505, 1980.
8. Keener, J. P., Infinite period bifurcation and global bifurcation branches, *SIAM Journal of Applied Mathematics*, **41,** 127–144, 1981.
9. Langford, W. F., Periodic and steady-state mode interactions lead to tori, *SIAM Journal of Applied Mathematics*, **37,** 22–48, 1979.
10. Langford, W. F. and Iooss, G., Interactions of Hopf and pitchfork bifurcations, in Mittelmann, H. D. and Weber, J. (Eds), *Bifurcation Problems and their Numerical Solution*, ISNM 54: Basel, Birkhauser, 1980, 103–134.

11. Bauer, L., Keller, H. B. and Reiss, E. L., Multiple eigenvalues lead to secondary bifurcation, *SIAM Review*, **17,** 101–122, 1975.
12. Golubitsky, M. and Schaeffer, D., Imperfect bifurcation in the presence of symmetry, *Communications in Mathematical Physics*, **67,** 205–232, 1979.
13. Poston, T. and Stewart, I. N., *Catastrophe Theory and its Applications*: London, Pitman, 1978.
14. Thom, R., *Structural Stability and Morphogenesis*: Reading, Mass., W. A. Benjamin, 1975.
15. Arnold, V., *Chapitres Supplémentaires de la Théorie des Equations Différentielles Ordinaires*: Moscow, Mir, 1980.
16. Guckenheimer, J., Multiple bifurcation problems of codimension two (preprint) University of California, Santa Cruz, 1981.
17. Takens, F., Singularities of vector fields, *Publications Mathématiques de l'Institut des Hautes Etudes Scientifiques*, **43,** 47–100, 1973.
18. Iooss, G. and Joseph, D. D., *Elementary Stability and Bifurcation Theory*: New York, Springer-Verlag, 1980.
19. Crandall, M. G. and Rabinowitz, P. H., Bifurcation from simple eigenvalues, *Journal of Functional Analysis*, **8,** 321–340, 1971.
20. Andronov, A. A., Leontovich, E. A., Gordon, I. I. and Maier, A. G., *Theory of Bifurcations of Dynamical Systems on a Plane* (English translation): New York, Wiley, 1973.
21. Golubitsky, M. and Langford, W. F., Classification and unfoldings of degenerate Hopf bifurcations, *Journal of Differential Equations*, **41,** 375–415, 1981.
22. Marsden, J. and McCracken, M., *The Hopf Bifurcation and its Applications*: New York, Springer-Verlag, 1976.
23. Naimark, J., On some cases of periodic motions depending on parameters, *Doklady Akademii Nauk USSR*, **129,** 736–739, 1959.
24. Sacker, R. J., *On Invariant Surfaces and Bifurcation of Periodic Solutions of Ordinary Differential Equations*, IMM-NYU 333: New York University, 1964.
25. Arneodo, A., Coullet, P. and Tresser, C., Possible new strange attractors with spiral structure, *Communications in Mathematical Physics*, **79,** 573–579, 1981.
26. Shil'nikov, L. P., A case of the existence of a denumerable set of periodic motions, *Soviet Mathematics—Doklady*, **6,** 163–166, 1965.
27. Keener, J. P., Secondary bifurcation in nonlinear diffusion reaction equations, *Studies in Applied Mathematics*, **55,** 187–211, 1976.
28. Langford, W. F., Modulation instability and weak turbulence, in Bednarek, A. R. and Cesari, L. (Eds), *Dynamical Systems II*: New York, Academic Press, 1982, 589–593.
29. Shearer, M., Coincident bifurcation of equilibrium and periodic solutions of evolution equations (preprint), 1979.
30. Fenstermacher, P. R., Swinney, H. L. and Gollub, J. P., Dynamical

instabilities and the transition to chaotic Taylor vortex flow, *Journal of Fluid Mechanics*, **94,** 103–129, 1979.

31. Lin, J. and Kahn, P. B., Qualitative dynamics of three species predator–prey systems, *Journal of Mathematical Biology*, **5,** 257–268, 1978.
32. Coullet, P., Tresser, C. and Arneodo, A., Transition to turbulence for doubly periodic flows, *Physics Letters*, **77A,** 327–331, 1980. (This paper relates closely to the discussion in Section 11.4 above.)

12 Bifurcations and Chaos in the System of Taylor Vortices—Laboratory and Numerical Experiment

V S L'vov, A A Predtechensky and A I Chernykh

12.1 Introduction

During the past twenty years significant progress has been made in the understanding of one of the most intriguing problems of hydrodynamic turbulence—that of flow transition to turbulence. This problem can be formulated as follows: how can the stochastization of liquid motion at high velocities be reconciled with the dynamical character of Navier–Stokes equations? The first step towards solving this problem was made in 1883 by O. Reynolds, who discovered that laminar fluid flow in a tube becomes unstable when a dimensionless parameter (later named by A. Sommerfeld the Reynolds number) is relatively high. Subsequently various phenomena were studied, such as instability of flow around a cylinder, which produces a Karman vortex street; instability of flow between rotating cylinders (Couette flow) with respect to the formation of Taylor vortices; instability of a plane fluid layer heated from below with respect to the formation of Rayleigh–Bénard convective cells, etc. As the velocity (the Reynolds number) increases, these secondary flows also become unstable and finally the flow appears to be turbulent.

It was recognized much later that this phenomenon can be described by a finite set of ordinary differential equations. This finite set can be obtained, for example, by a reduction of the Navier–Stokes equations using some method of approximating functions (e.g., Galerkin method). This approach considers the onset of turbulence as the appearance of stochastic regimes in finite-dimensional dynamical systems. The study of such systems was stimulated by the discovery and investigation of stochastic attractors by Smale, Lorenz, Anosov, Sinai, Ruelle and Takens *et al.* (see, for instance, the reviews [1–3] and references cited therein). A stochastic attractor (SA) is a limited attracting region in the phase space of a dynamical system in which almost all trajectories are unstable, so that the motion inside it is necessarily complex and tangled. However, the investigation of the problem of correspondence between the onset of turbulence and SA is still in an embryonic state.

First and very interesting steps were made in this field when simple convective flows were studied. The initial stage of the onset of stochasticity in a convection loop was shown [4] to be described by the well-known set of three differential Lorenz equations [5]. In describing thermal convection in a thin plane layer (Hele–Shaw cell), eight equations were used of which the solution showed stochastical behaviour under numerical simulation [3, 6]. The number of equations necessary for describing the more complicated flows [7, 8] is much greater. The reason lies in the arbitrary nature of choice of approximating functions which, generally speaking, are not eigenfunctions even for the simplest regimes arising after the onset of instability. This leads to an unreasonably large number of equations in the system and makes the analysis of its solution very complicated. Truncation of a formal Galerkin series can lead to essential changes in subtle topological characteristics of the original multidimensional attractor, i.e., a stochastic attractor can be transformed into a complicated limit cycle. Further truncation of the model can transmute this cycle into yet another stochastic attractor. Therefore it remains uncertain whether attractors constructed in this way bear any real relation to the problem of onset of flow turbulence. That is why the first question concerns how many and what type of equations should be taken into account for a certain flow at given Reynolds number to be described with reasonable accuracy.

Here we may note that in quite a large number of cases at *Re* values that are transitional from a laminar to a developed turbulent regime, the major part of the fluid energy is concentrated in narrow frequency bands of the power spectrum. Hence the velocity $V(r, t)$ takes the form

$$\mathbf{V}(r, t) = Re\left\{\sum_j \mathbf{A}_j(r, t)\exp(i\omega_j t)\right\} \tag{12.1}$$

where the 'complex envelopes of modes' $\mathbf{A}_j(r, t)$, contrary to $\mathbf{V}(r, t)$, are slow functions of time with characteristic time-scale of the order of inverse line width $(\Delta\omega)^{-1}$. Within a comparatively small *Re* range, changes of the spatial structure of rapid modes ω_i can be neglected and, assuming $\mathbf{A}_j(r, t) = A_j(t)\mathbf{f}_j(r)$, one can write down phenomenological equations for $A_j(t)$, retaining only the resonance terms. Such a phenomenological approach does not merely reflect the lack of experimental information concerning onset of turbulence: it provides a new language in which to present the information obtained and formulate the questions for experimental study. These questions are the following: how can the number of equations to describe onset of turbulence (i.e., the effective space dimensionality) be determined? How can the system trajectory in this space or, in other words, the attractor structure be determined? How does this structure vary with the growth of *Re*? Finally, what is the connection between the attractor structure observed in the experiment and the structure obtained numerically from the phenomenological model?

Here we shall study these questions, taking as an example the transition to turbulence in a Couette–Taylor flow with the inner cylinder rotating. The hydrodynamic cell used ([9], see also Appendix) consisted of two cylinders ($d_1 = 35$ mm, $d_2 = 55$ mm, $h = 300$ mm). Measurements were made by means of Laser–Doppler velocimeter. A computer system for data acquisition and processing [10] enabled us to analyse up to 256 000 samples with necessary sampling time τ.

First of all, in Section 12.2 we consider briefly the process of Taylor vortex formation. In Section 12.3 we discuss the evolution of a power spectrum of azimuthal velocity in a wide range of *Re* numbers (from 10^2 to 10^4) in order to obtain information about the order of appearance and the number of rapid motions. In Section 12.4 more detailed experimental data are presented for the *Re* range 1000 to 1200 in which destruction of azimuthal wave coherence results in stochastic fluid flow. Then in Sections 12.5–7 we propose a set of phenomenological equations (12.7) for amplitudes of azimuthal waves in pairs of Taylor vortices and investigate analytically and numerically the onset of stochasticity in the solutions of these equations. In Section 12.8 we present a statistical description of system behaviour in the domain of stochastic attractors. Using the Wyld diagram technique [11, 12], we obtain a closed set of averaged equations in approximation of direct interaction. We show that its solutions correspond qualitatively to the results of numerical experiments. Some experimental data concerning the evolution of attractor structure in the range $Re > 1200$ are also presented in Section 12.9.

12.2 Formation of Taylor vortices

Couette flow in an infinitely long cylinder is theoretically known to be unstable above some critical Reynolds number Re_c which depends upon radius ratio r_2/r_1. In this case the secondary flow necessarily arises with a periodical structure along the axial direction z (the Taylor vortices). The determination of an actual axial scale of vortices at $Re > Re_c$ is the unsolved problem in such an approach.

However, no problem arises if a cylinder of finite height is considered. If $h \gg (r_2 - r_1)$, the end-plates act as a resonator selecting a set of permissible modes.

At $Re > Re_c$ several permissible modes (the $\tilde{N}$ numbers) exist within the instability region of laminar flow. In a steady state only one of them is realized, which particular one depending on the initial state of the flow. The states with various $\tilde{N}$ are metastable and can persist to rather high *Re* numbers. If *Re* is increased gradually, only one 'optimum' mode $\tilde{N}_0$ enters the instability region at first.

More complete analysis of the boundary conditions leads to the conclusion [13] that this 'optimum' mode appears, not by means of bifurcation, but continuously from the steady state at $Re = 0$. Other permissible modes appear as a result of bifurcations and are separated from $\tilde{N}_0$ by separatrix surfaces. If the relative length of cylinders is increased, then bifurcation points converge and distances between separatrices diminish. Therefore in real flows the optimum mode should arise when the inner cylinder is accelerated adiabatically. This acceleration mainly depends on the length of system N and also on the degree of asymmetry of boundary conditions γ_0 at the end-plates:

$$\frac{1}{Re_c}\frac{dRe}{dt}<\frac{\gamma_0}{N^\alpha}.$$

There is much evidence to suggest that $\alpha \sim 2$. Taking the viscous diffusion time ~ 100 s and $\gamma_0 \sim 1$, we estimate the permissible acceleration $\Delta\, Re/Re < 10^{-5}\, s^{-1}$. $Re = 74 \pm 2$, where

$$Re = \omega_m(r_2 - r_1)\nu \tag{12.2}$$

and ω_m is the angular velocity of inner cylinder rotation. We have observed various numbers of Taylor vortices $\tilde{N}$ (from 22 to 36) depending upon the acceleration regime. Under slow acceleration the 'optimum' number $\tilde{N}_0 = 30$ is realized. Our procedure of slow acceleration is as follows: at $Re = 65$ (for water this corresponds to a rotation period $T \sim 20$ s at $t = 24°C$) the cell was held steady for one hour and then Re was linearly increased to 85 during 5 hours. Spatial structure with thirty Taylor vortices persisted until $Re \sim 10^4$, when the flow was already strongly chaotic in time.

12.3 Power spectrum evolution

In accordance with what has been said above, the number of 'rapid' motions and the order of their appearance should first be determined. This information is given by the evolution of power spectra $J(\omega)$ measured for a set of successively increasing values of Re. The power spectrum

$$J(\omega) = \langle |V_\omega|^2 \rangle \tag{12.3}$$

where V_ω is the Fourier transform of $V(t)$, $\langle\ \rangle$ indicating averaging over successive blocks of velocity data. Spectrum evolution for flows with $\tilde{N} = 28$, 29 and 30 is discussed in [14]. In Fig. 12.1 the spectra for $\tilde{N} = 30$ are presented. One can trace the succession of bifurcations that lead to excitation of new modes of motion. At $Re \sim 1000$ the bends of vortices appear (six bends over the circumference), and a narrow peak is observed at the frequency $\omega_6 = 1.93\omega_m$, with linewidth 10^{-3} at the level 10^{-3} relative to the peak maximum. In the range $Re = 1040$–1140 a number of bifurcations may be observed which leads to the appearance of a narrow ω_6-line. With low-resolution spectra this leads to a slight broadening ($\Delta\omega_6/\omega_6 \sim 10^{-2}$ at $Re = 1040$) of lines. Here a low-frequency motion arises with characteristic times $\sim(\Delta\omega_6)^{-1}$. Subsequently, in the range $Re = 1200$–1300, motions with incommensurable frequencies $\omega_2 = 0.55\omega_m$, $\omega_3 = 0.95\omega_m$ and combination harmonics appear. Later still, the peak widths gradually become comparable with the distances between them. At $Re \sim 1900$, no pronounced maxima are observed in the power spectrum. At still greater *Re* numbers, narrow peaks are observed on the background of continuous spectrum. Later, this system of secondary peaks is destroyed in the same way as the primary one. At very high *Re* numbers (see Fig. 12.1(*f*)) the spectrum becomes continuous and the liquid flow becomes turbulent.

It should be noted that the path to turbulence in Couette–Taylor flow (i.e., the succession of bifurcational Reynolds numbers, the types of flow

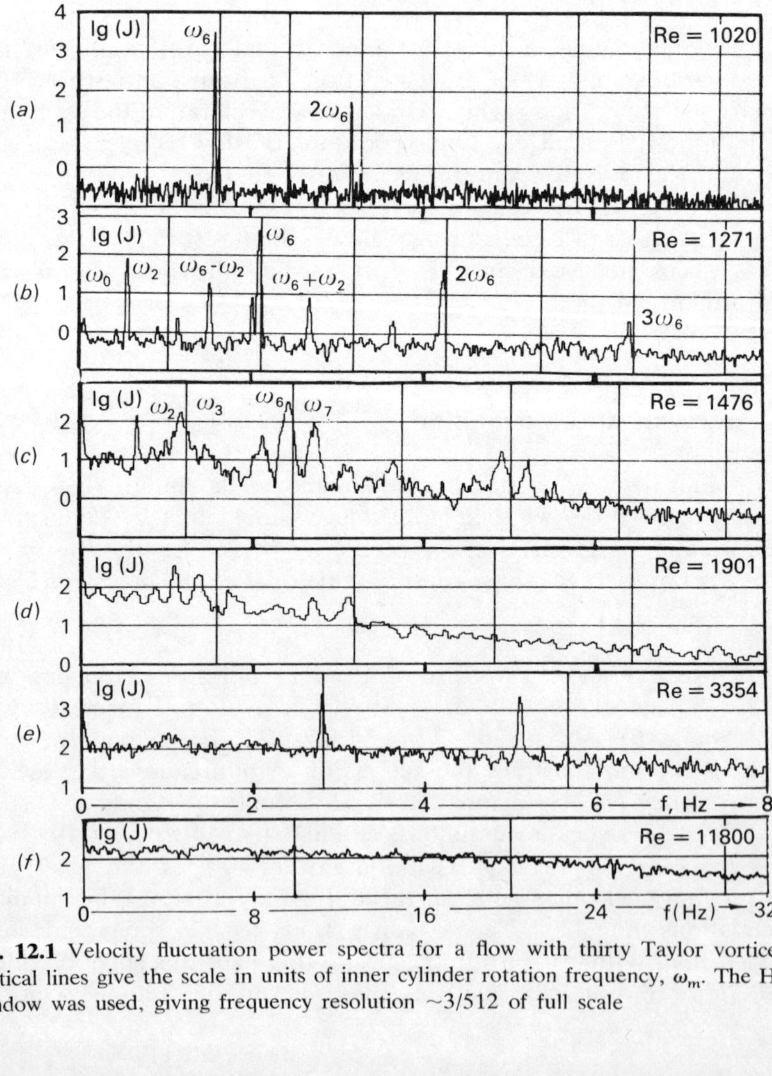

Fig. 12.1 Velocity fluctuation power spectra for a flow with thirty Taylor vortices. The vertical lines give the scale in units of inner cylinder rotation frequency, ω_m. The Hanning window was used, giving frequency resolution $\sim 3/512$ of full scale

Table 12.1 Critical values of parameters for the first bifurcations of Couette flow

Reference	L'vov and Predtechensky [14]		Fenstermacher, Swinney and Gollub [15]
Height/gap ratio	30		20
Gap/inner radius ratio	0.37		0.14
Critical value of *Re* for Taylor vortex	74±2		120
Number of cells	30	28	17
Stability region of Taylor vortices: *Re* =	13 Re_c	6.5 Re_c	1.3 Re_c
Initial number of azimuthal waves	6	1	5?
Stability region of the first azimuthal wave	0.025	0.8	5.4

arising in this case, the character of peak broadening in the power spectrum, etc.) depends not only upon the geometric size of the cylinders but also upon the number of vortices which can arise in the same annular gap (see Table 12.1).

However, the general qualitative aspects of transition pattern coincide [14]. These common aspects include, for instance, the fact that with increasing *Re* bifurcations arise which lead to the appearance of narrow peaks in the spectrum and broadening of the peaks which already exist, and to the appearance of peaks of finite width. The presence of several metastable steady states in the intermediate *Re* range, which is observed so clearly in Couette flow, is also a rather common feature. The above provides definite evidence that Couette flow with a rotating inner cylinder is a convenient model for the study of a number of the common qualitative peculiarities of gradual transition to turbulence.

12.4 Breakdown of azimuthal wave coherence

As has been mentioned, at $Re > 1040$ the ω_6-line of the power spectrum broadens with increasing *Re*. We shall show that this dominant characteristic exemplifies the transition of the flow to stochasticity (i.e., to a flow that is unpredictable in the long term). First, it is natural to consider the

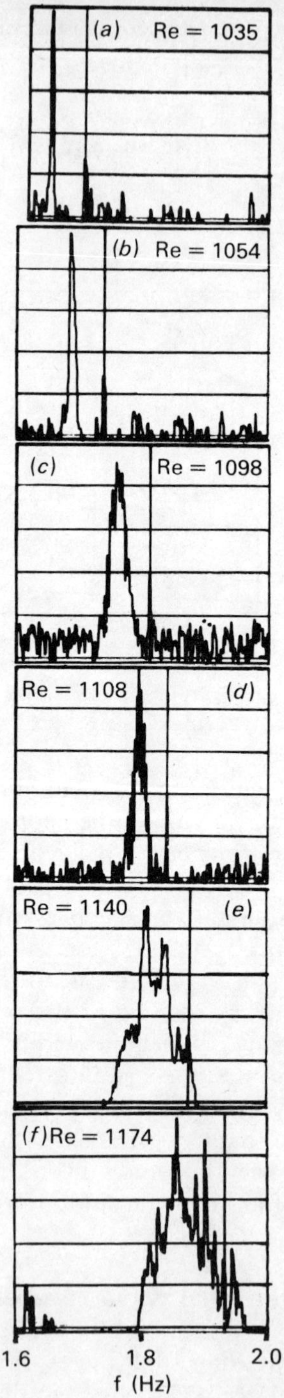

Fig. 12.2 Evolution of the narrow ω_6-peak. The vertical line indicates frequency $2\omega_m$

evolution of the narrow ω_6-line with higher frequency and Reynolds number resolution (Fig. 12.2). These data were obtained from an experimental series with $Re_1 = 1015$.

Numerical evaluation of the time-dependence of the ω_6-motion envelope using one of the narrow-band filtration algorithms is much more informative. These results are presented in Fig. 12.3. The first experiment (Figs 12.2(*a*) and 12.3(*a*) with $Re = 1035$), with supercriticity $(Re - Re_1) \sim 2 \times 10^{-2}\ Re_1$, corresponds to a periodic motion with the frequency ω_6. The image of this motion in the input phase space is a stable limiting cycle.

The next two experiments (Figs 12.2(*b*), 12.3(*c*, *b*)) correspond to $Re = 1043$ and $Re = 1054$ which exceed some bifurcational value Re_1^1, leading to a slow sinusoidal modulation of the basic motion, whose frequency ω_0 grows as supercriticity increases. This modulation is distinctly seen in Fig. 12.3(*b*,*c*) but does not appear in the spectra owing to the limited recording time in this regime. (It should be noted here that in [15], which deals with the experimental study of laminar–turbulent transition in a Couette flow with a narrower gap, a narrow spectral line was also observed. It is interpreted by the authors as a pure periodic regime. However, in the sample of signal shown in [15 (Fig. 5a)] slow modulation is distinctly seen.)

The image of the motion described here is a simple attractor, i.e., a two-dimensional torus. At further increase of *Re* there occurs a series of neighbouring bifurcations, but their observation is hampered by extremely slow transition to the steady state. Therefore the states described

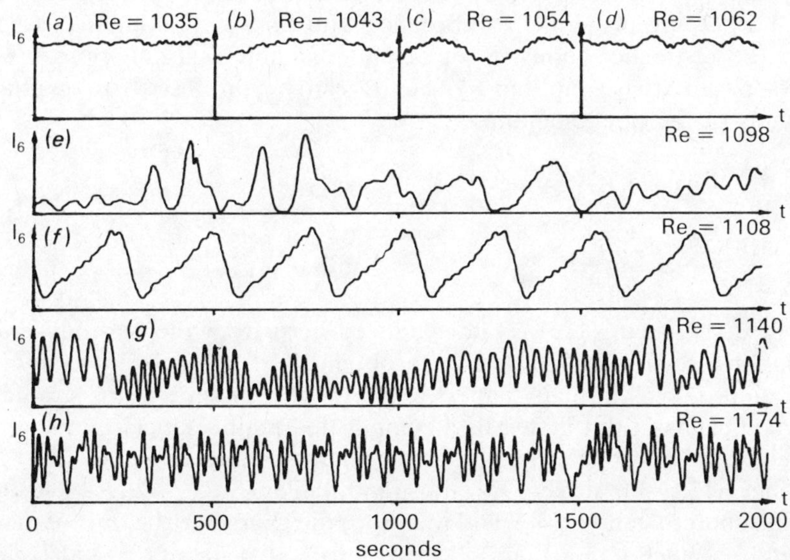

Fig. 12.3 Time-dependence of the ω_6-motion envelope. The frequency window is $\omega_6(1 \pm 0.03)$. A time interval of 2000 s corresponds to ~3500 periods of rapid motion; $Re_1 = 1015$

below do not necessarily immediately follow one another as *Re* is increased. Thus, at $Re = 1062$ the periodic modulation described above does not exist, and a simple limit cycle is observed (Fig. 12.3(*d*)). At $Re = 1098$ a complicated modulation develops which can be considered as chaotic, so far as can be concluded from observation over a relatively short time. The ω_6-peak at this *Re* number (Fig. 12.2(*c*)) is found to be noticeably broadened. It is very interesting that with increasing *Re* the chaotic modulation again changes to a periodic one, this time not sinusoidal but 'sawtoothed' (Figs. 12.2(*d*) and 12.3(*f*) for $Re = 1108$). At $Re = 1140$ (Figs. 12.2(*e*) and 12.3(g)) the character of the modulation becomes complicated again. With some imagination, one can see in Fig. 12.3(g) the alternate fragments of sinusoids of two different frequencies. In agreement with the above, the ω_6-peak in Fig. 12.2(*e*) consists of two broad peaks. It should also be noted that in this case the modulation frequency is significantly higher than that in the previous regime. At $Re = 1174$ (Fig. 12.2(*f*)) on the right-hand slope of the spectral peak one can see a narrow line with two satellites; this indicates the presence of a periodical component in the amplitude modulation which was not observed in the preceding regime. At larger *Re*, ω_2-motion appears.

12.5 Phenomenological model of the system of interacting Taylor vortices

12.5.1 Basic equations

At small supercriticalities above the Re_1-threshold the peaks at frequencies ω_6 and $2\omega_6$ are present. In accordance with this, and taking into account only the resonance terms, a set of phenomenological equations for the envelopes of these motions A and B can be proposed for qualitative analysis of the phenomenon:

$$\begin{aligned} \frac{\mathrm{d}A}{\mathrm{d}t} &= \gamma A + v_1 B A^* + w\,|A|^2\,A, \\ \frac{\mathrm{d}B}{\mathrm{d}t} &= -\tilde{\gamma} B + v_2 A^2. \end{aligned} \tag{12.4}$$

(Note that system (12.4) is not deduced here from the hydrodynamical equations. Such a system has been obtained by a regular procedure for similar problems in many papers, e.g. [16–18]. In general the evaluation of coefficients in this system is a complicated mathematical problem, and here we prefer to consider these coefficients as phenomenological.)

Since $\gamma(Re) = 0$ at $Re = Re_1$, it is reasonable to set $\gamma = \alpha(Re - Re_1)$. At small supercriticality, $\gamma \ll \tilde{\gamma}$ and therefore the characteristic time of motion should be much larger than $\tilde{\gamma}^{-1}$. This means that $\mathrm{d}B/\mathrm{d}t \ll \tilde{\gamma}$ and B is the forced motion:

$$B = v_2 A^2/\gamma. \tag{12.5}$$

The range of *Re* numbers where this relation is valid can be found experimentally. Firstly, it follows from (12.5) that the phases of A and B motions must be correlated rigidly: $2\psi_A - \psi_B = \text{const.}$, and therefore the modulus of the correlator,

$$|K(\omega_6, 2\omega_6)| = |\langle V(\omega_6)V^*(2\omega_6)/|V(\omega_6)V^*(2\omega_6)|\rangle|,$$

must be equal to unity. This value, computed from experimental data, remains equal to 0.98–0.99 up to $Re \sim 1300$, i.e., up to *Re* numbers that are much larger than those discussed. Secondly, the variables $I_6 = |A_6(t)|^2$ and $I_{6.2} = |B(t)|^2$, computed from the velocity data, can be considered to be a parametrical representation of phase trajectory projection onto some $(I_6, I_{6.2})$ plane (Fig. 12.4(*a*)). Since this trajectory has a noisy component, it is reasonable to construct a probability density of sojourn of the system at various points of such a plane and further to analyse the lines of regression—$\langle I_6\rangle$ versus $I_{6.2}$ and $\langle I_{6.2}\rangle$ versus I_6. The results of such processing are shown in Fig. 12.4(*b*), where the vertical and horizontal bars indicate the corresponding dispersions. It can be concluded from Fig. 12.4(*b*) that the regression lines are situated along a parabolic curve corresponding to (12.5): $|B(t)| \sim |A(t)|^2$. This is traced more distinctly if $|B(t)|$ is replaced by $|B(t)|^{1/2}$—see Fig. 12.4(*c*), where the parabola is transformed into a straight line. All this means that the motion at frequency $2\omega_6$ is a forced motion and therefore system (12.4) takes the form

$$\frac{dA}{dt} = \gamma A + (iT - \eta)|A|^2 A \qquad \gamma = \alpha(Re - Re_1), \quad \varepsilon = (Re - Re_1)/Re_1 \tag{12.6}$$

This equation possesses a unique solution giving the Landau law: $|A|^2 = \gamma/\eta = \alpha(Re - Re_1)/\eta$. In our case this law is well justified up to $\varepsilon = 0.02$ when the stable closed orbit exists, as may be seen from Fig. 12.5.

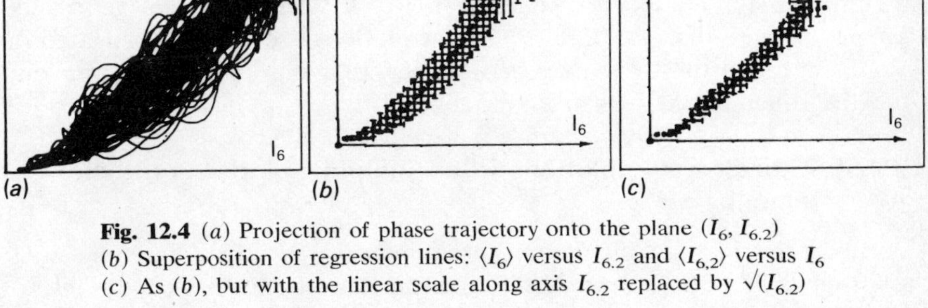

Fig. 12.4 (*a*) Projection of phase trajectory onto the plane $(I_6, I_{6.2})$
(*b*) Superposition of regression lines: $\langle I_6\rangle$ versus $I_{6.2}$ and $\langle I_{6,2}\rangle$ versus I_6
(*c*) As (*b*), but with the linear scale along axis $I_{6.2}$ replaced by $\sqrt{(I_{6.2})}$

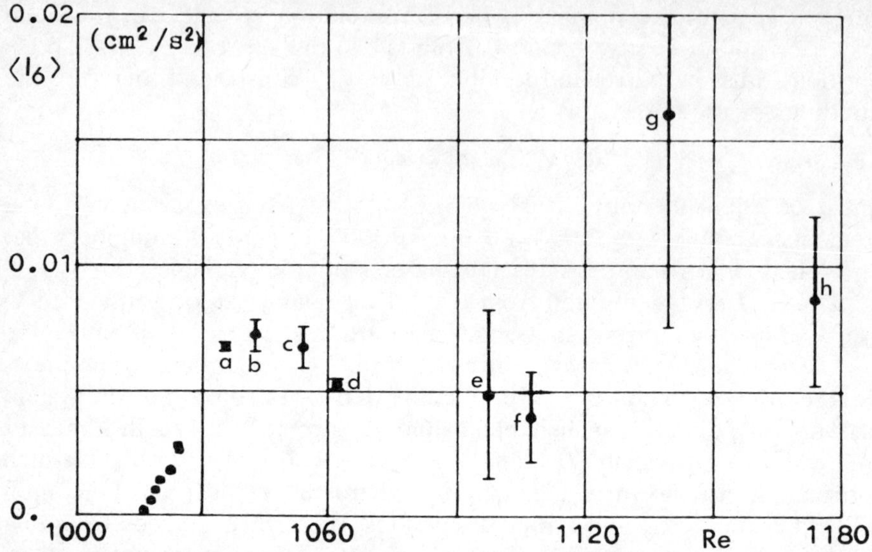

Fig. 12.5 Mean amplitude of ω_6-motion and its dispersion as a function of *Re*

Owing to the absence of nonstationary solutions at $t \to \infty$, Eq. (12.6) cannot explain the appearance of time-dependence of A. While we had assumed the ω_6-flow to be described by a single spatial function, visual observation shows that the boundary between a pair of vortices is almost immovable even when the bending amplitudes are relatively high. This means that oscillations in the pairs of Taylor vortices have some 'freedom'. A separate equation of type (12.6) must therefore be written for the bending amplitude of each vortex pair with serial number n—see the first term in (12.7). In other words, the ω_6-motion possesses a high multiplicity of degeneration related to the high height/gap ratio: $h \gg (r_2 - r_1)$. To complete the construction of a rough phenomenological model we take into account the interaction between neighbouring vortex pairs, but only in linear approximation because this interaction is small compared with the 'internal interaction' within pairs:

$$\frac{\mathrm{d}A_n}{\mathrm{d}t} = \gamma A_n + (\mathrm{i}T - \eta)\,|A_n|^2\, A_n + \tfrac{1}{4}(a + \mathrm{i}b)(A_{n+1} + A_{n-1} - 2A_n). \tag{12.7}$$

Here it may be supposed that $\gamma > 0$, $\eta > 0$. Without loss of generality we may assume that $a > 0$. Since these equations are to be investigated in a narrow range of *Re*, it may be assumed that only γ is *Re*-dependent and the other coefficients are constants.

12.5.2 Steady-state solutions in an infinite cylinder and their stability

Vortex pairs in our cell are rather numerous: $N = 15$. Therefore it is reasonable first to assume $N \gg 1$, ignoring any boundary conditions. It then becomes clear that at $\gamma > 0$ Eq. (12.7) has a nontrivial spatially homogeneous solution which coincides with that of Eq. (12.6):

$$A_n(t) = A^0(t) = A^0 \exp\left(\frac{i\gamma Tt}{\eta}\right), \qquad |A^0|^2 = \gamma/\eta. \tag{12.8}$$

Let us now consider the stability of this solution. We assume that the difference of A_n and A^0 is small,

$$|\delta A_n| = |A_n - A^0| \ll |A^0|,$$

and linearize Eq. (12.7) with respect to δA_n, which we represent in the form

$$\delta A_n = \varepsilon A^0(t) \exp(\nu t + 2i\psi n). \tag{12.9}$$

The set of linearized equations for δA_n is then reduced to one complex equation for ε and ε^*:

$$\varepsilon\left[\gamma + \nu + (a + ib)\sin^2\psi + \frac{i\gamma T}{\eta}\right] = \frac{\varepsilon^*(iT - \eta)\gamma}{\eta}. \tag{12.10}$$

The necessary condition for the existence of nonzero solutions of (12.10) gives the expression for 'modulation instability' increment:

$$\nu_\psi = -(\gamma + a_\psi) \pm \left[\gamma^2 - b_\psi\left(b_\psi - \frac{2\gamma T}{\eta}\right)\right]^{1/2}$$

$$a_\psi = a\sin^2\psi, \quad b_\psi = b\sin^2\psi. \tag{12.11}$$

It is evident that $\nu_0 = 0$. At small ψ,

$$\nu_\psi \simeq (bT - a\eta)\sin^2\psi.$$

Thus, when $bT > a\eta$, $\nu_\psi > 0$ and the spatially homogeneous solution (12.8) is unstable with respect to amplitude modulations (12.9). The boundary of instability region can be found from (12.11) assuming $\nu_{\psi_0} = 0$:

$$\sin^2\psi_0 = 2\gamma\frac{bT - a\eta}{\eta(a^2 + b^2)}. \tag{12.12}$$

From (12.11), (12.12) it follows also that in the opposite case, i.e., at

$$a\eta > bT, \tag{12.13}$$

$Re\{\nu_\psi\} < 0$ at any ψ, and solution (12.8) is stable.

12.5.3 Steady-state solutions in a cylinder of finite height

In the case of a cylinder of finite height it should be taken into account that the equations for boundary vortex pairs (with $n = 1$ and $n = N$) differ from (12.7) because each of them interacts not with two vortex pairs but only with one pair (with $n = 2$, $n = N - 1$ respectively). Formally this can be done by imposing the following 'boundary conditions':

$$A_0 = A_{N+1} = 0. \tag{12.14}$$

When motions are slightly supercritical it is convenient to search for the solution of (12.7) in the form of expansion over harmonics B_n:

$$A_n = \sum_{m=1}^{N} B_m \sin\left(\frac{\pi mn}{(N+1)}\right)$$
$$(N+1)B_m = 2\sum_{n=1}^{N} A_n \sin\left(\frac{\pi mn}{(N+1)}\right) \tag{12.15}$$

which automatically satisfy the conditions (12.14). This representation is convenient also because it diagonalizes the linear term of Eq. (12.7):

$$\frac{dB_n}{dt} = -(\gamma_n + ib_n)B_n + (NL)_n. \tag{12.16}$$

Here the following notations are introduced:

$$\gamma_n = -\gamma + a_n, \qquad a_n = a\sin^2\left(\frac{\pi n}{2(N+1)}\right), \qquad b_n = b\sin^2\left(\frac{\pi n}{2(N+1)}\right). \tag{12.17}$$

The nonlinear term in (12.16) looks rather complicated:

$$(NL)_n = \frac{iD}{2}\sum_{n_1,n_2,n_3=-N}^{N} B_1^* B_2 B_3[\Delta(\tilde{n}) + \Delta(\tilde{n}+2N+2) + \Delta(\tilde{n}-2N-2)], \tag{12.18}$$

where $\tilde{n} = n + n_1 + n_2 + n_3$, $D = (T + i\eta)/2$, $B_i = B_{n_i}$, $B_{-n} = -B_n$ and Δ is defined by the relations $\Delta(0) = 1$, $\Delta(n) = 0$ at $n \neq 0$.

Equation (12.16) allows us to find easily the instability threshold of zero solution with respect to excitation of harmonics with serial number $n : \gamma_n = 0$. Harmonic B_1 is the first to be excited at

$$\gamma > a_1 = a\sin^2\left(\frac{\pi}{2(N+1)}\right). \tag{12.19}$$

As might have been expected, the threshold of excitation of the first harmonic tends to the threshold of excitation of the homogeneous solution (12.8) ($\gamma = 0$) if the height of cylinder is increased ($N \to \infty$). In the region of small supercriticality, i.e., at $\gamma_1 < 0$, γ_2 and $\gamma_3 > 0$, harmonics with $n > 3$ can be neglected in Eq. (12.16). Then (12.16) is reduced to

$$\left.\begin{aligned}
\frac{dB_1}{dt} &= -(\gamma_1 + ib_1)B_1 + \frac{iD}{2}\{B_1[3|B_1|^2 + 4(|B_2|^2 + |B_3|^2) \\
&\quad - B_1B_3^* - 2B_1^*B_3] + 2B_1^*(B_2^2 + B_3^2) + 4B_2\,Re(B_2B_3^*)\}, \\
\frac{dB_2}{dt} &= -(\gamma_2 + ib_2)B_2 + \frac{iD}{2}\{B_2[3|B_2|^2 + 4(|B_1|^2 + |B_2|^2) \\
&\quad + 4\,Re(B_1B_3^*)] + 2B_2^*[B_1^2 + B_1B_3 + B_3^2]\}, \\
\frac{dB_3}{dt} &= -(\gamma_3 + ib_3)B_3 + \frac{iD}{2}\{3|B_3|^2 + 4(|B_1|^2 + |B_2|^2) \\
&\quad + 2B_3^*(B_1^2 + B_2^2) + B_1(2|B_2|^2 - |B_1|^2)\}.
\end{aligned}\right\} \tag{12.20}$$

It is obvious that three types of solution can exist:

(a) $B_2=0$, B_1 and $B_3\neq 0$;
(b) $B_1=B_3=0$, $B_2\neq 0$;
(c) B_1, B_2, $B_3\neq 0$.

The last solution is doubly degenerate because if B_1^0, B_2^0 and B_3^0 is a solution of (12.20), then B_1^0, $-B_2^0$, B_3^0 is also a solution of (12.20). This fact is a manifestation of the symmetry of the initial problem with respect to replacement $n \leftrightarrow N+1-n$ which corresponds to reflection in respect of a middle cross-section of the cylinder. Solution (a) with zero even harmonics is symmetric with respect to such replacement; solution (b) with zero odd harmonics is antisymmetric; two asymmetric solutions of type (c) in the case of such replacement are interchangeable.

If $\gamma_1<0$, $|\gamma_1|<\gamma_3$ and $\gamma_2>0$ then the solution $B_2=0$ is realized, and

$$\begin{aligned} B_1(t)&=|B_1|\exp(i\delta_1 t), \qquad |B_1|^2=4\,|\gamma_1|/3\eta. \\ \delta_1&=-(b_1+\gamma_1 T/\eta), \qquad B_3(t)=|B_3|\exp(i\delta_1 t), \\ |B_3|^2&=\frac{|B_1|^2\gamma_1^2(T^2+\eta^2)}{9\eta^2[\gamma_3^2+(b_1-b_3)^2]}. \end{aligned} \tag{12.21}$$

This solution is obtained from (12.20) under the assumption that $|B_3|\ll|B_1|$, i.e. for slight supercriticality. It can be shown to be stable in this range with respect to excitation of even harmonics.

Now we shall show that with the growth of supercriticality the solution of Eq. (12.7) with zero 'boundary conditions' (12.14) approximates the spatially homogeneous solution (12.8) when it is stable, i.e., when $a\eta> bT$.

We first define how soon the amplitude A_n reaches the asymptotic value (12.8) with increasing n, i.e., as we move away from the boundary. For this purpose, assuming $|\delta A_n|=|A_n(t)-A^0(t)|\ll|A^0|$, we linearize (12.7) with respect to δA_n which we represent in the form

$$\delta A_n=\varepsilon A^0(t)\exp(-2\kappa n). \tag{12.22}$$

Thus we obtain a pair of complex equations for ε and ε^* similar to (12.10), and the condition that its characteristic determinant is zero gives

$$C\equiv sh^2\kappa=\frac{2\gamma(a\eta-bT)}{\eta(a^2+b^2)}. \tag{12.23}$$

Solution of this equation has the form

$$\exp 2\,|\kappa|=2C+1+2\surd(C^2+C) \tag{12.24}$$

From this it follows, for instance, that already at $C=\frac{1}{2}$ the deviation δA_n decreases very rapidly with increasing $n:|\delta A_{n+1}/\delta A_n|\simeq 0.21$. At larger C convergence to the asymptotics (12.8) occurs even faster. For example, at $C=1$, $|\delta A_{n+1}/\delta A_n|\simeq 0.17$.

At $C\geqslant\frac{1}{2}$ it is easy to obtain the bending amplitudes of edge pairs directly from Eq. (12.7), assuming $A_2 = A^0$:

$$|A_1|^2 = |A_N|^2 = |A^0|^2[1 - \tfrac{1}{2}(2C+1)]. \tag{12.25}$$

One can see, for instance, that already at $C = \frac{1}{2}$, $|A_1|^2$ differs only a little from $|A^0|^2$: $|A_1^1|^2 = 0.75\ |A^0|^2$.

Summing up the results of Sections 12.5.1 and 2, it can be said that the steady-state solution of Eq. (12.7) with zero 'boundary conditions' is stable when supercritical if $a\eta > bT$. At small γ/a the profile $|A_n|$ has the form of half-waves according to (12.21). With increasing γ the profile $|A_n|$ is flattened and reaches the value $\sqrt{(\gamma/\eta)}$ in accordance with (12.24) and (12.25). However, if $a\eta < bT$, the stationary solution of (12.7) is stable only when slightly supercritical and when the profile $|A_n|$ takes the form of half-wave (12.21). Numerical experiments described in the next section show that at $a\eta < bT$ the steady-state solution of (12.7) becomes unstable as γ increases and a succession of bifurcations separating the behaviour over time of $A_n(t)$ in different domains is observed.

One may suppose that the 'stable' case $a\eta > bT$ was realized in experiments [15] with a narrow gap as well as in those with a wide gap for a bending mode $m = 1$ in a flow with $N = 14$. For this flow $|A_n|^2 = \text{const.}$ for a wide Re range (see Table 12.1). On the other hand, the complicated dynamical behaviour of $I_6(t)$ described in Section 12.3 should be related to instability of solutions developing at $a\eta < bT$. The onset of stochastic flow at $a\eta > bT$ must be of different character; it can, for example, result from excitation of new rapid motions. It seems that some features of this phenomenon can be described by considering the processes in a single pair of vortices. In [8] these processes were simulated by Galerkin expansion of Navier–Stokes equations retaining thirty-two selected functions. Good agreement with [15] was found concerning the critical number of transition to stochasticity, but the position of the broadband component (B) did not coincide with that observed in experiment.

12.6 Complicated behaviour of the system of interacting Taylor vortices: numerical experiments

12.6.1 Procedure of numerical experiment

We choose the following dimensionless time $\tilde{t}$ and amplitude $\tilde{A}$:

$$\tilde{t} = at, \qquad \tilde{A} = A\sqrt{(\eta/\gamma)}; \tag{12.26}$$

then Eq. (12.7) will take the form

$$\frac{d\tilde{A}_n}{dt} = \Gamma[1 + (i\theta - 1)\,|\tilde{A}_n|^2]\tilde{A}_n + \tfrac{1}{4}(1 + iB)(\tilde{A}_{n+1} + \tilde{A}_{n-1} - 2\tilde{A}_n), \tag{12.27}$$

where $\Gamma = \gamma/a$, $\theta = T/\eta$, $B = b/a$.

When studying these equations we restrict our analysis to the case of

$N=15$, which corresponds to the experimental situation, and to zero 'boundary conditions'.

Simulation of the problem (12.26)–(12.27) consisted in investigation of trajectories $\tilde{A}_n(\tilde{t})$ computed over times much longer than the characteristic period of motion. Therefore a numerical solution required the application of strongly stable difference schemes which would not lead to the accumulation of computation errors. We used the Runge–Kutta method of the fourth order of accuracy. In the case of a stable fixed point this provides an approach to a constant steady-state solution with an accuracy of 10^{-5} at indefinitely long time. In the case of a stable limit cycle the amplitude δA and period τ of oscillations were independent of the time increment with an accuracy of 10^{-5} if $\Delta\tilde{t}<0.02\tau$. In the case of the nonperiodic behaviour of solutions we encountered additional difficulties connected with the rapid divergence of trajectories with neighbouring initial values. For example, two trajectories that started from the same initial condition and differed in time step value only ($\Delta\tilde{t}=0.15$ and $\Delta\tilde{t}=0.3$, or 0.008 and 0.016 of the characteristic period of motion τ) for $\theta=10$, $\Gamma=0.2$ and $B=1.25$, were found to diverge completely (over a distance of the order of attractor size) at time $\tilde{t}\sim 80$, and the 1 per cent loss of accuracy caused by difference approximation error was attained approximately by $\tilde{t}\sim 60$.

Generally speaking, all this makes it impossible to study the details of system behaviour within a time longer than a certain confidence interval. This circumstance is, however, not of significant importance when the statistics of this process is studied, because the dynamic stochasticity caused by hyperbolicity of the system is stronger than the chaos produced by computational errors [1]. Numerical experiments confirm this statement in cases when coefficients of system (12.27) were chosen not too close to bifurcation values. Figure 12.13 (page 262) shows two power spectra of the first harmonic $B_1(\tilde{t})$ calculated for $\theta=10$, $\Gamma=0.2$ and $B=1.25$ with $\Delta\tilde{t}$ values of 0.15 and 0.3. One can see that they coincide within the limits of the confidence interval due to finiteness of the observation time.

12.6.2 Dynamical properties of trajectories

Preliminary numerical experiments consisted in visual observation of trajectories $\tilde{A}_n(\tilde{t})$ at various values of coefficients θ, Γ and B in Eq. (12.27). Characteristic time-dependencies $|\tilde{A}_5(\tilde{t})|$ are given in Fig. 12.6(*a*). One can see that the amplitude $|\tilde{A}_5(\tilde{t})|$ may reach a constant value in a nonperiodic way ($\Gamma=0.024$) or with decaying oscillations ($\Gamma=0.048$ and $\Gamma=0.091$). It also may appear as a periodic function of time or chaotically. A better understanding of the system behaviour can be acquired when projections of phase trajectory onto certain selected planes are studied. Figure 12.7 shows various patterns of reaching the fixed point, (*a*), (*b*), and the limit cycle from almost zero initial

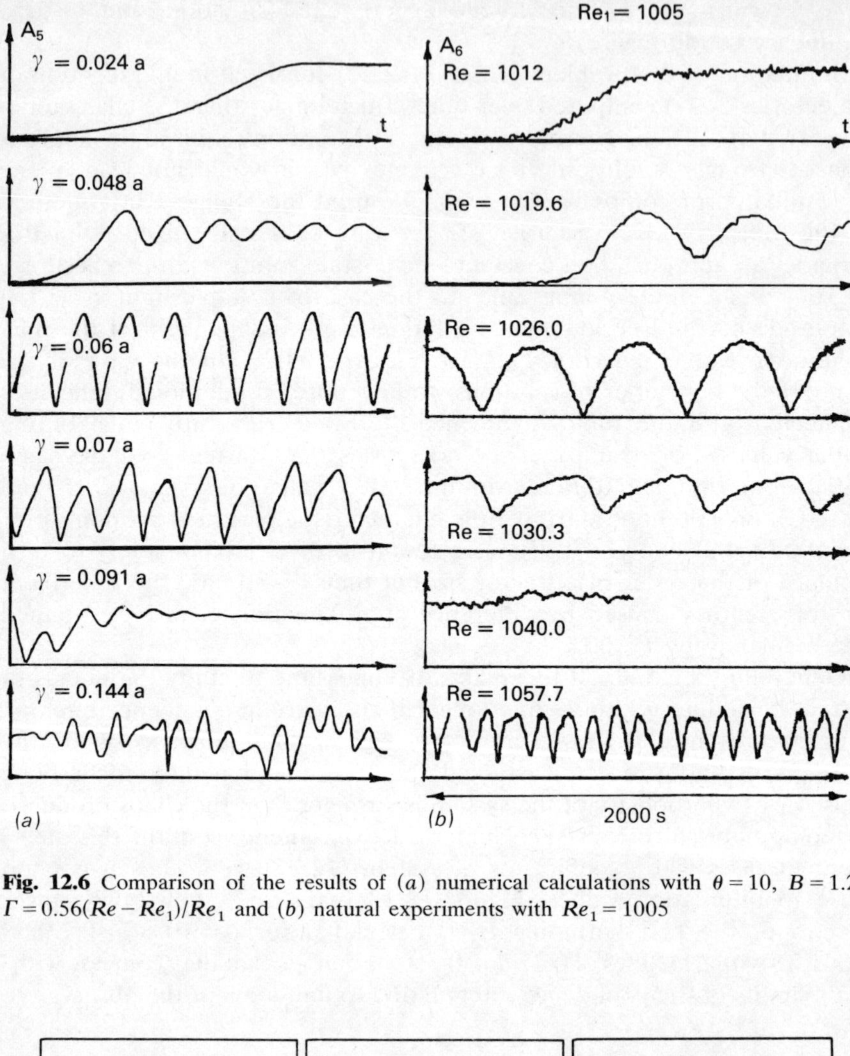

Fig. 12.6 Comparison of the results of (*a*) numerical calculations with $\theta = 10$, $B = 1.25$, $\Gamma = 0.56(Re - Re_1)/Re_1$ and (*b*) natural experiments with $Re_1 = 1005$

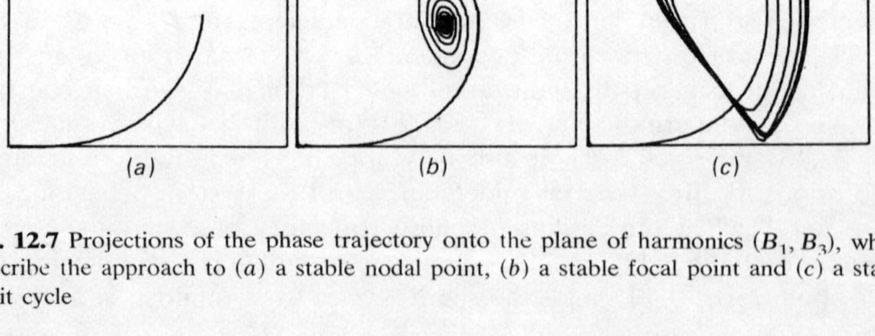

Fig. 12.7 Projections of the phase trajectory onto the plane of harmonics (B_1, B_3), which describe the approach to (*a*) a stable nodal point, (*b*) a stable focal point and (*c*) a stable limit cycle

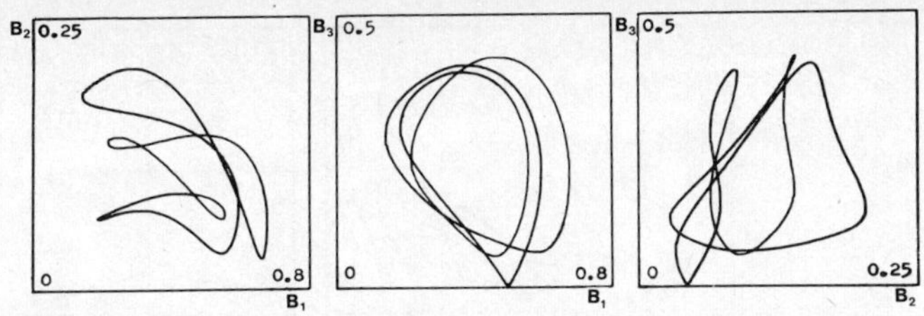

Fig. 12.8 Three different projections of the same threefold limit cycle: $\theta = 10$, $B = 1$, $\Gamma = 0.071$

conditions, (c). In Fig. 12.8 three projections are shown of a threefold limit cycle. And, finally, trajectories in Fig. 12.9 look like stochastic ones.

At this stage we cannot distinguish a fairly complicated limit cycle or a long transitional regime from an actual stochastic attractor. Only relatively simple limit cycles and equilibrium states can be detected with certainty in this way. Results obtained at this level of understanding are shown in Fig. 12.10, where domains of different behaviour of system (12.27) are marked on (B, Γ)-plane at $\theta = 10$ and (θ, Γ)-plane at $B = 1.25$. In both cases the axis of ordinates represents the ratio $\Gamma = \gamma/a$ which is proportional to Reynolds number. By choosing appropriate coefficients B and θ, it is possible to obtain in the numerical experiment the same succession of bifurcations as in the natural experiment with increasing *Re*. For example, at arbitrary $\theta = 10$ this correspondence is observed at $B = 1.20$–1.30. Indeed, when *Re* is increased in a laboratory experiment (see Fig. 12.3) and Γ is increased in a numerical experiment (see Fig. 12.10), a stable fixed point is replaced by a limit cycle. Then a stable point emerges again, which is replaced by a stochastic attractor. In this case a qualitative correspondence is observed between measured and calculated bifurcation Reynolds numbers if we define $\Gamma = 0.56 \times (Re - Re_1)/Re_1$.

Experimental time-dependence of $\tilde{A}(\tilde{t})$ corresponding to stepwise change of Reynolds number is also similar to that obtained numerically (see Fig. 12.6). However, it would be premature to speak about complete correspondence between the laboratory and numerical experiments. In particular, in laboratory experiment we observed a limit cycle following a stochastic attractor (Fig. 12.10) which is absent from numerical experiment. This difference may be connected with the fact that the value of

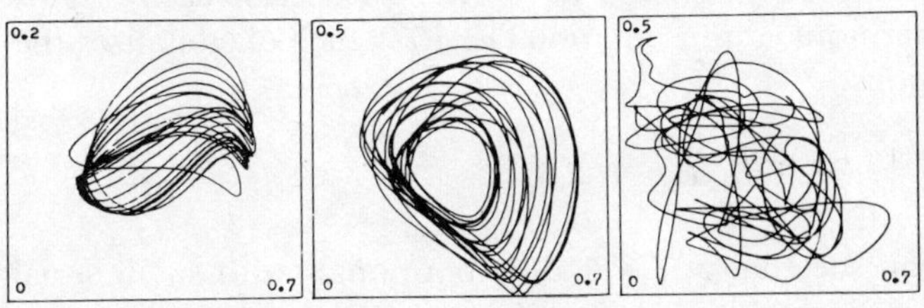

Fig. 12.9 Stochastic trajectories of various degrees of complication

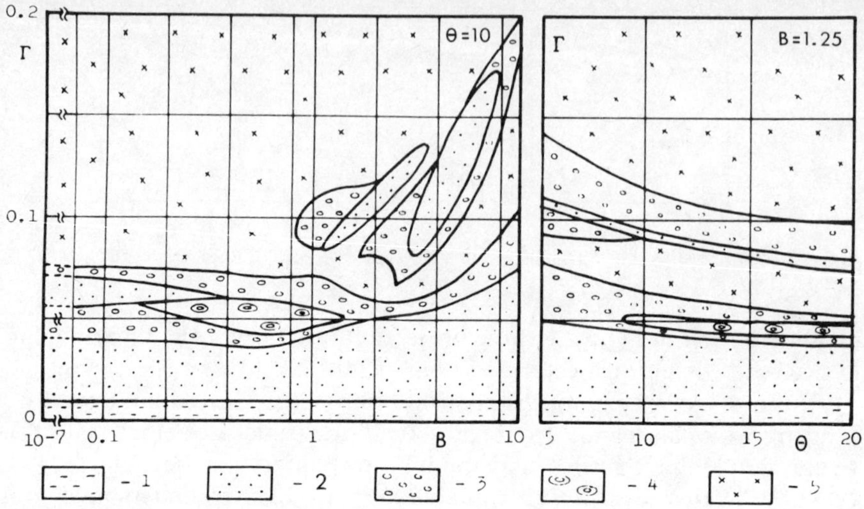

Fig. 12.10 Domains of different behaviour of the solutions of Eqs (12.26), (12.27)
1 Only zero solution is stable
2 Stable fixed point
3 Limit cycle
4 Two-dimensional torus
5 Domain of complicated attractors

ratio θ was chosen arbitrarily. Generally speaking, additional special experiments should be carried out in order to measure directly the coefficients in the phenomenological model (12.26)–(12.27). Nevertheless, we are now persuaded that this describes the initial stage of the onset of Couette–Taylor flow turbulence. A more detailed study of attractor statistical properties in this model has thus become of interest. Some preliminary results of such a study are presented in the next two sections.

12.7 Statistical properties of trajectories in the phenomenological model

12.7.1 Upper estimate of the attractor dimension

Phase space of the set of fifteen complex equations (12.26)–(12.27) has a dimension equal to 30. However, at $\Gamma \ll 1$ a much smaller number of degrees of freedom should be excited. This can easily be proved if we pass to harmonic representation (12.15) which diagonalizes the linear part of equations:

$$\frac{dB_n}{dt} = -(\gamma_n + ib_n)B_n + (NL)_n \tag{12.28}$$

One can see that at $\Gamma \ll 1$ only harmonics with small serial number n fall into the instability region. For example, at $\gamma = 0.05a$ only the first two harmonics are unstable ($\gamma_1 < 0$, $\gamma_2 < 0$, $\gamma_3 > 0$, $\gamma_4 > 0, \ldots$). At $\gamma = 0.2a$,

$\gamma_n < 0$ only for the first four of fifteen harmonics. These are unstable harmonics that should be excited most effectively, and they in turn affect all the others. The strongest response should occur in harmonics with small n which have the least 'reserve' of stability. Harmonics with greater n should be effectively suppressed. Numerical experiments in general confirm this simple theory, see Table 12.2. It can be seen that, except in one case, the first harmonic is the most intensive.

Intensity of excitation decreases monotonously with the growth of serial number in both odd and even harmonic systems. If we assume that weakly excited harmonics (for example, those with relative intensity less than 5 per cent of maximum) do not actually influence the system dynamics, then at $\Gamma \ll 1$ the number of differential equations required is reduced from fifteen to the value shown in Table 12.3. Putting this in terms of geometry, the attractor of the dynamic system under investigation is oriented in phase space almost normal to the axes of reference corresponding to harmonics of high number. However, the representation of harmonics (12.15) is not the optimal linear unitary transformation of variables (choice of orientation for phase space axes of reference) at

Table 12.2 Excitation level of the system at various Γ and $\theta = 10$ (numerical experiment)

The upper part corresponds to $B = 10$, the lower part to $B = 1.25$. The first row of each Γ shows the relative intensities $J_n = \langle |B_n|^2 \rangle$ as a percentage of the intensity of the most excited harmonic. The second row gives the eigenvalue of the matrix $100\ \langle B_n B_m^* \rangle / \max_n \langle |B_n|^2 \rangle$.

Γ \ n	1	2	3	4	5	6	7	8	9	10	11	12	13	14	15
0.08	100	26	28	2.2	6.7	0.2	2.2	0.0	0.9	0.0	0.3	0.0	0.1	0.0	0.0
	111	27	26	2.1	2.0	0.2	0.2	0.0	0.0	0.0	0.0	0.0	0.0	0.0	0.0
0.15	100	61	56	17	20	2.8	5.7	0.8	2.0	0.3	0.7	0.1	0.3	0.0	0.0
	106	61	60	17	16	2.4	2.1	0.7	0.5	0.3	0.2	0.1	0.1	0.0	0.0
0.20	21	100	73	4.1	10	3.6	5.5	0.6	1.6	0.3	0.7	0.1	0.2	0.0	0.0
	8.5	102	99	4.1	1.8	4.4	2.4	0.5	0.3	0.2	0.2	0.1	0.0	0.0	0.0
0.06	100	34	34	10	13	1.5	3.5	0.5	1.3	0.1	0.5	0.0	0.1	0.0	0.0
	109	34	32	11	10	1.3	1.0	0.4	0.2	0.1	0.1	0.0	0.0	0.0	0.0
0.08	100	36	35	12	15	3.0	4.9	0.9	1.8	0.3	0.7	0.1	0.2	0.0	0.0
	109	36	33	12	12	2.9	2.5	0.8	0.5	0.3	0.1	0.1	0.1	0.0	0.0
0.15	100	56	53	32	30	12	12	4.5	5.2	1.9	2.0	0.9	0.9	0.5	0.5
	104	65	60	32	30	12	10	4.5	3.5	1.9	1.4	0.9	0.7	0.5	0.5
0.20	100	87	73	45	39	21	20	9.9	9.5	4.5	4.4	2.5	2.2	1.6	1.5
	101	86	77	45	39	21	18	9.9	7.8	4.5	3.5	2.5	2.0	1.6	1.5

Table 12.3 Number of modes whose excitation level is more than 5 per cent of maximum mode (at $\theta = 10$)

Numbers in parentheses represent eigenvalues of the correlation matrix whose magnitude is >5 per cent of the maximum.

B \ Γ	0.06	0.08	0.15	0.20
10	—	4(3)	6(5)	5(3)
1.25	5(5)	5(5)	8(7)	9(9)

Table 12.4 Normalized correlation matrices $\langle B_n B_m^* \rangle$, $B = 1.25$, $\theta = 10$, $\Gamma = 0.08$ (upper part) and $\Gamma = 0.2$ (lower part)

The matrices are given as percentages of the maximum element. The nondiagonal elements are the coefficients of correlation; its real and imaginary parts are shown above and below the main diagonal, respectively.

	1	2	3	4	5	6	7	8	9	10	11	12	13	14	15
1	100	3.3	32	−0.4	32	−0.7	38	−0.6	44	−0.7	45	−0.1	45	−0.9	30
2	0.3	34	1.5	−5.5	0.3	−7.6	0.8	−10	0.5	−11	1.1	−11	2.1	−6.8	1.0
3	−0.4	0.1	34	1.0	40	−1.3	52	−1.5	59	−0.6	62	−1.8	60	0.0	40
4	0.4	−0.3	0.0	12	1.4	−14	−1.4	−20	−0.1	−19	0.0	−17	0.1	−12	1.6
5	−0.4	−0.2	−1.8	−1.8	15	0.0	55	−2.5	65	−1.0	67	−1.8	64	−1.4	42
6	0.8	1.9	0.4	22	2.2	3.0	1.3	−12	−1.1	−5.7	0.0	−3.4	0.0	−4.6	−2.1
7	0.0	0.6	4.2	0.7	−15	−3.9	4.9	0.7	70	−1.0	72	−1.3	68	−1.6	48
8	0.7	3.9	−2.2	−11	−2.2	9.7	0.0	0.9	0.3	12	−1.13	14	−1.7	15	−0.3
9	0.7	−0.6	−1.7	1.7	−7.6	−1.1	1.8	1.4	1.7	0.4	75	−3.2	74	−0.5	49
10	1.2	−4.6	0.6	−6.8	−1.1	10	1.2	0.4	0.3	0.3	0.3	23	−0.7	16	−0.2
11	−2.8	−1.1	−0.5	1.0	−6.5	−1.7	2.3	−0.4	0.3	−2.9	0.7	−0.3	73	−1.3	49
12	0.1	−10	−0.2	−6.0	−1.6	9.2	0.9	0.8	−0.4	−0.3	0.8	0.1	−0.1	14	−2.2
13	1.4	−0.2	−6.0	1.0	−0.1	−1.2	−0.2	0.3	−1.9	−1.8	−2.6	−0.3	0.2	0.7	46
14	0.3	4.9	0.2	−9.1	−1.4	5.6	1.0	0.5	0.1	−0.5	0.7	−0.8	1.0	0.0	2.4
15	0.0	0.0	0.8	1.1	−7.7	−0.7	0.7	0.0	−0.9	−1.3	−1.1	−0.8	0.5	0.6	0.0
1	100	1.2	1.1	0.3	4.6	−0.1	5.1	0.0	5.8	0.3	7.0	0.1	4.4	0.2	0.0
2	0.4	87	1.0	−0.8	0.0	−0.9	−0.4	−2.1	0.6	−0.1	0.1	1.6	−0.4	−1.2	0.0
3	0.1	−0.6	73	0.6	16	0.3	17.4	1.2	20	0.3	20	0.6	18	0.0	5.8
4	0.5	−0.3	1.0	45	0.2	−4.5	−0.4	−1.7	−1.1	−1.9	−0.5	−0.6	−0.5	2.2	0.0
5	0.0	−0.7	−0.4	0.6	39	0.2	26	0.7	27	0.0	26	0.4	22	−0.2	11
6	0.1	−1.6	0.7	−3.0	−0.7	21	0.4	−2.7	−1.0	−2.1	−0.9	−0.4	0.0	2.2	1.6
7	0.0	−0.6	−3.0	0.8	2.1	1.1	20	1.2	28	0.0	27	0.2	23	1.0	6.8
8	0.7	−8.8	2.2	2.5	−0.8	0.2	−3.0	9.9	0.5	−0.6	0.0	1.6	0.5	−1.1	−0.3
9	−4.7	1.9	7.8	0.4	4.8	2.6	3.2	6.2	9.5	0.3	29	0.0	21	−0.5	8.3
10	4.3	4.7	0.2	0.2	−1.8	−0.9	−4.5	−1.0	−4.5	4.5	1.3	1.5	0.0	2.2	−0.3
11	9.4	0.4	5.8	1.1	4.7	3.1	3.7	5.6	0.9	7.1	4.4	0.0	20	−0.6	8.1
12	−2.6	4.5	−0.6	0.5	−2.7	−0.4	−4.9	−0.7	−5.6	0.3	−5.3	2.5	0.8	2.5	0.1
13	0.7	−1.2	3.9	1.0	2.6	2.8	1.9	4.5	−0.6	5.1	−0.8	4.9	2.2	0.9	5.9
14	−0.2	0.2	−1.1	0.8	−2.3	0.4	−2.7	−0.9	−3.5	0.0	−3.4	0.0	−2.4	1.6	1.1
15	0.4	−0.1	0.7	0.7	1.3	1.7	0.0	1.6	−0.6	1.9	−0.7	1.9	−0.4	1.7	1.5

which the number of intensely excited variables is minimal. To find such variables C_n,

$$C_n = V_{nm}A_n = U_{nm}B_m, \qquad VV^+ = 1, \qquad UU^+ = 1, \tag{12.29}$$

it is necessary to calculate a simultaneous correlation matrix $\langle A_n A_m^* \rangle$ and find a unitary matrix V_{nm} reducing it to diagonal form. This procedure is equivalent to approximation of an attractor by a multidimensional ellipsoid and finding the orientation and values of its principal axis lengths $\langle |C_n|^2 \rangle^{1/2}$. Intensities $\langle |C_n|^2 \rangle$ are the eigenvalues of correlation matrices $\langle A_n A_m^* \rangle$ or $\langle B_n B_m^* \rangle$. In Table 12.4 correlation matrices are presented calculated for $\theta = 10$, $B = 1.25$ and two values of Γ: 0.08 and 0.2.

One can see that odd harmonics are fairly closely correlated and that the correlation coefficient K_{nm} is almost real. It increases with increasing n and m. Correlation between even harmonics is weaker than between odd ones. There is practically no correlation between the motion of even and odd harmonics. With increasing supercriticality (i.e., γ/a ratio) the degree of correlation between harmonics decreases. All these factors (except the fact that K_{nm} is real for odd harmonics) can easily be explained on the basis of Eqs (12.16)–(12.20) for harmonics.

Eigenvalues $\langle |C_n|^2 \rangle$ of the correlation matrix for certain values of parameters are given in Table 12.2. In Table 12.3 the number of eigenvalues $\langle |C_n|^2 \rangle$ whose value is over 5 per cent of the maximum eigenvalue are shown. One can see that the transition to the 'best orientation' of reference axes, C_n, also results in a smaller number of differential equations, as compared with a harmonic representation.

We have thus obtained the upper estimate for the attractor dimension depending upon parameters of the problem, including supercriticity Γ. Of course, in some cases this estimate is excessive. For example, at $\theta = 10$, $B = 1.25$ and $\Gamma = 0.06$, the attractor is a limit cycle, i.e., its dimension equals 1, and if variables are chosen accordingly, the behaviour of the system can be described by a single differential equation. However, the advantage of variables is that they can be directly identified in formula (12.29) with the use of the unitary matrix U which diagonalizes the correlation matrix $\langle B_n B_m^* \rangle$.

12.7.2 Power spectrum

In the preceding section we considered the simultaneous correlation matrix $\langle B_n(t) B_m^*(t) \rangle$. Now we shall discuss the results of the numerical experiment with respect to its diagonal elements at noncoinciding times $J_m(\tau) = \langle B_m(t) B_m^*(t+\tau) \rangle$. Instead of the quantity $J_m(\tau)$ it is convenient to use its Fourier transform

$$J_m(\omega) = \int J_m(\tau) \exp(-i\omega\tau)\, d\tau,$$

i.e., the power spectrum of this harmonic. We calculated $J_m(\omega)$ according to the formula

$$J_m(\omega) = \langle |B_m(\omega)|^2 \rangle,$$

where $B_m(\omega)$ is a Fourier transform of $B_m(t)$, and $\langle \ \rangle$, as in (12.3), indicates averaging over successive blocks of the realization $B_m(t)$ with dimension T. To obtain the Fourier-image of $B_m(t)$ we used the discrete Fourier transform. Time limitation of input data in discrete analysis (the time window T) leads to the well-known 'leakage effect' which limits the output spectra resolution, i.e., results in a broadening of pure spectral lines. Application of the Hamming window enabled us to diminish this effect to three samples in the frequency domain which usually consists of 512 samples within the Nyquist band. This broadening of lines can be seen in Figs 12.11 and 12 which correspond to the simple attractors.

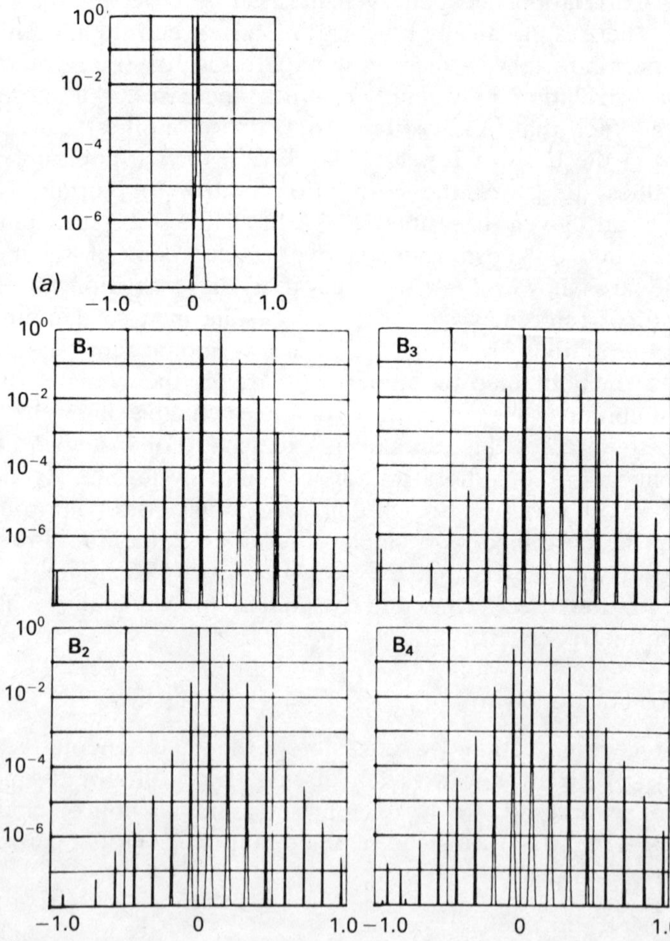

Fig. 12.11 Power spectra at $\theta = 10$, $B = 1.25$, corresponding (a) to the state of equilibrium ($\Gamma = 0.03$) and (b) to the limit cycle ($\Gamma = 0.06$)

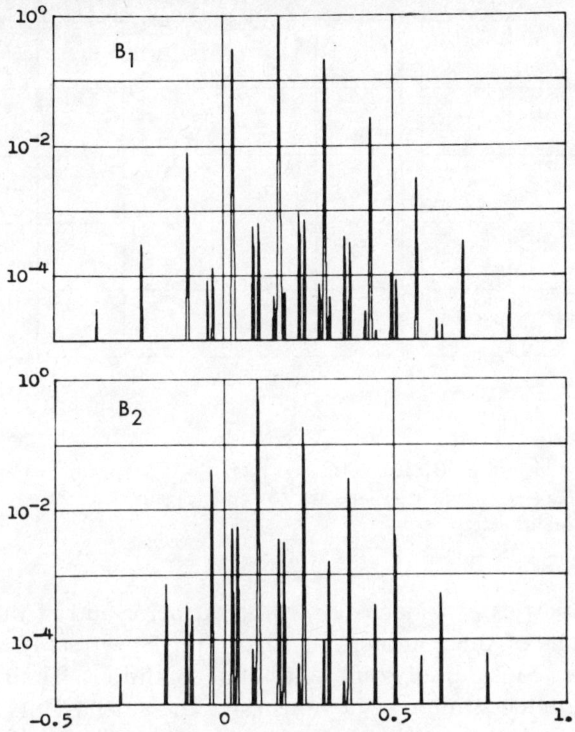

Fig. 12.12 Power spectra corresponding to a twofold limit cycle: $\theta = 10$, $B = 1.25$, $\Gamma = 0.066$

The spectrum of Fig. 12.11(a) corresponds to the simple limit cycle with trivial time-dependence $B_n(t) \sim \exp(i\Omega t)$. In the system of coordinates rotating with the frequency Ω the state of equilibrium corresponds to this situation. Regions with such behaviour are marked on the plane of parameters (Fig. 12.10) by the character 2. The remaining spectra given in Fig. 12.11 were calculated in Region 3 of the plane of parameters where the limit cycles occur (in the rotating system of reference). It is interesting that intensities of satellites diminish exponentially with distance from the central line Ω. In Fig. 12.12 the spectra corresponding to a twofold limit cycle are presented. One can see the splitting of lines which occurs because the periods of the first and the second revolutions do not coincide along a closed orbit.

The spectra shown in Figs 12.13–12.16 were calculated in Region 5 of the plane of parameters where the trajectory of the system appears to be stochastic. To obtain these data we used sufficiently representative blocks of realization which contained up to ten characteristic times γ^{-1} of motion.

The spectra in Fig. 12.13 correspond to the fixed set of coefficients θ, Γ, B. Spectra (a) and (b) differ in choice of time increment, and the fact that they practically coincide gives evidence of the structural stability of the

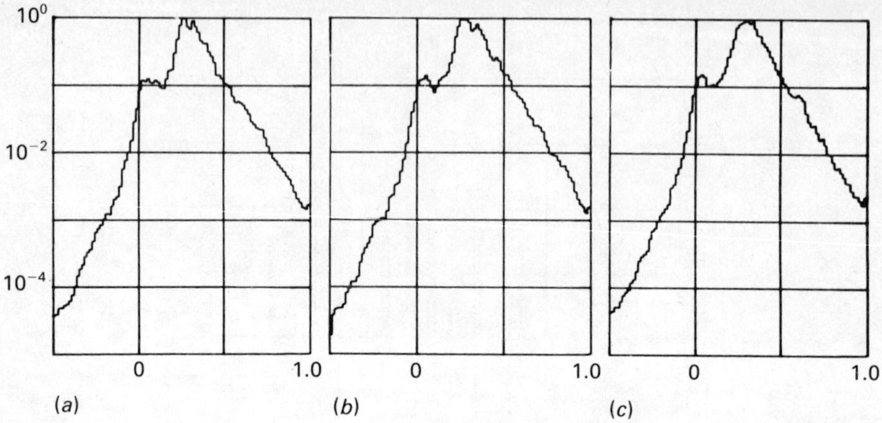

Fig. 12.13 Power spectra of B_1 at $\theta = 10$, $B = 1.25$, $\Gamma = 0.2$. Spectra (*a*) and (*b*) differ in time increment ($\Delta t = 0.15$ and 0.3); spectra (*b*) and (*c*) are determined at considerably different initial conditions

statistical properties of trajectories in the attractor and of their independence of details of the computation algorithm. Spectra (*b*) and (*c*) were obtained under considerably different initial conditions. Their coincidence within the confidence interval demonstrates the stochasticity ('forgetting' of the initial conditions) and stationary character of the system trajectory.

The spectra of harmonics shown in Figs 12.14 and 15 were calculated in the region of stochastic behaviour at $\theta = 10$, $B = 1.25$ and different values of Γ. In the same figures the plots of 'excess' $K_m(\omega)$ are presented, calculated according to the formula

$$K_m(\omega) = [\langle |B_m(\omega)|^4 \rangle / J_m^2(\omega)] - 1. \tag{12.30}$$

The quantities $K_m(\omega)$ will be discussed in the next section. Here we shall consider only the evolution of spectra $J_m(\omega)$ with increasing supercriticality (at fixed $\theta = 10$ and $B = 1.25$).

The value $\Gamma = 0.08$ is quite close to regions of dynamic behaviour, since at $\Gamma = 0.06$ and 0.1 the trajectories are periodical. Therefore at $\Gamma = 0.08$ weakly unstable limit cycles may exist in the attractor. Since the trajectory can remain near these cycles for a rather long time, the power spectrum of odd harmonics with small serial numbers, $m = 1, 3, 5$, contains some definite peaks. It is interesting to note that the asymptotics of the spectra for ω from the centre of Ω line is with good accuracy exponential (a straight line on a logarithmic scale):

$$J_m(\omega) \sim \exp(-|\omega - \Omega|/\Delta_\pm) \tag{12.31}$$

and the exponents $\Delta_\pm$ for left and right slopes are practically independent of the serial number of the harmonics.

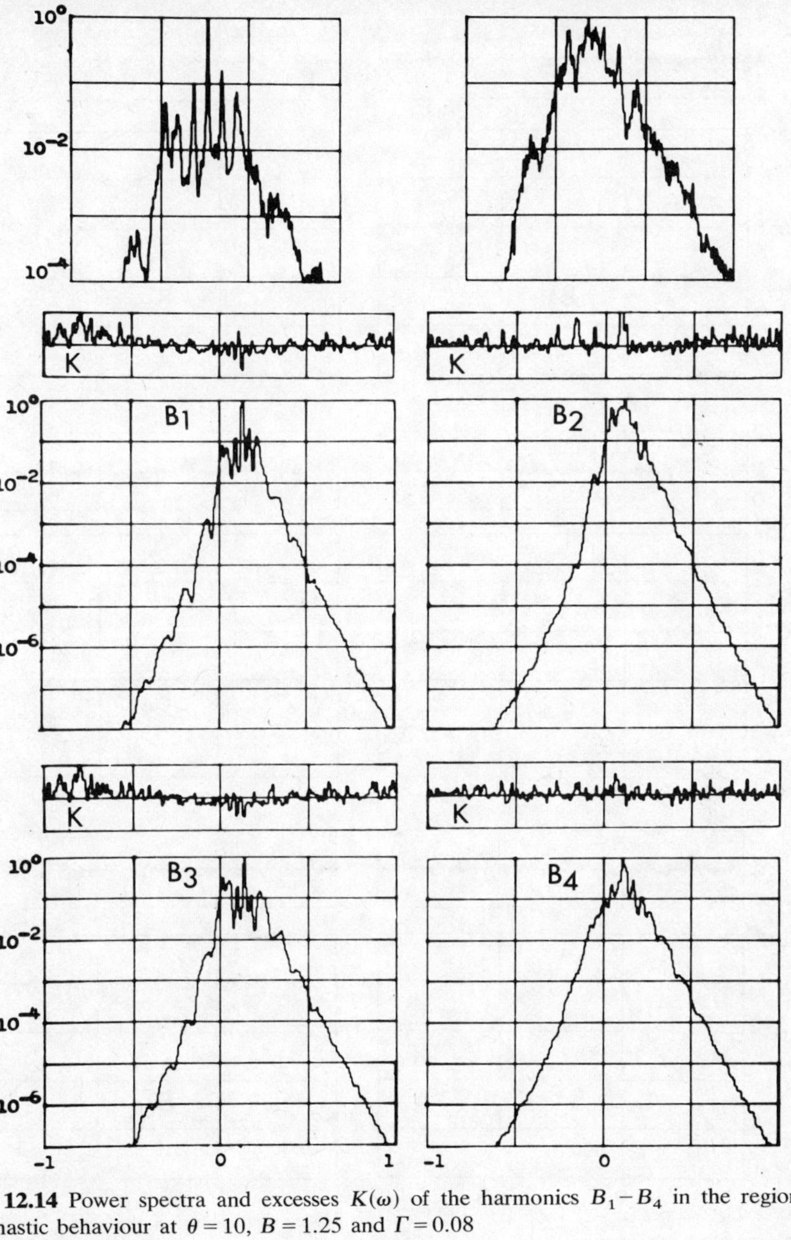

Fig. 12.14 Power spectra and excesses $K(\omega)$ of the harmonics $B_1 - B_4$ in the region of stochastic behaviour at $\theta = 10$, $B = 1.25$ and $\Gamma = 0.08$

For larger supercritical $\Gamma = 0.15$ and 0.2, we move away from the region of stability of limit cycles and they cease to appear in the spectra (Figs 12.15 and 16). It is instructive that the spectra of all harmonics are very close and can be described even better by the exponential dependence (12.31). The exponents increase with increasing Γ—see Table 12.5.

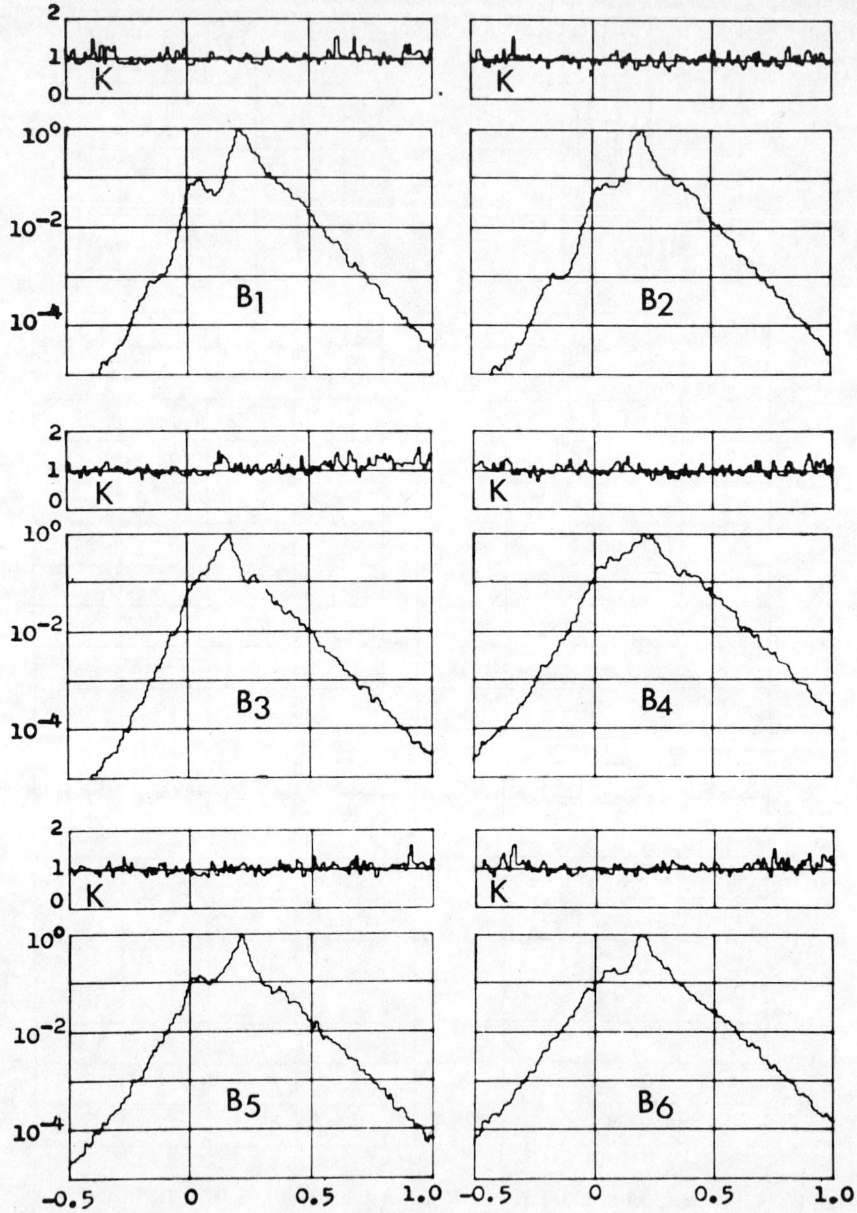

Fig. 12.15 Power spectra and excesses $K(\omega)$ of the harmonics $B_1 - B_6$ at supercriticality $\Gamma = 0.2$, $\theta = 10$, $B = 1.25$

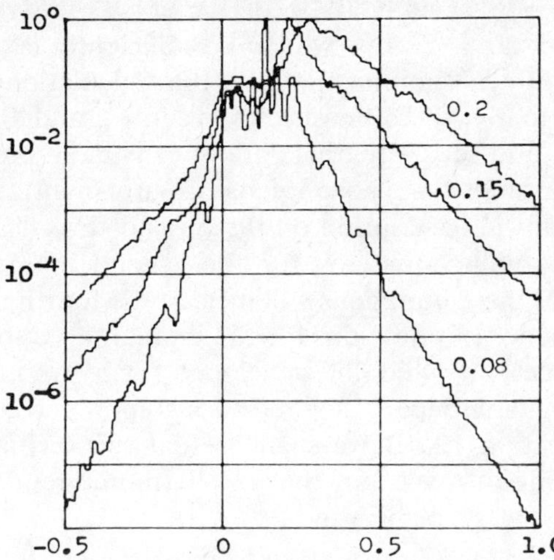

Fig. 12.16 Comparison of the spectra of B_1 at different values of supercriticality Γ

Table 12.5 Exponent $\Delta_\pm$ in (12.31) which governs the decrease of power spectra ($\theta = 10$, $B = 1.25$)

Γ	Δ_+/a	Δ_-/a	Δ_+/γ	Δ_-/γ
0.08	0.051	0.045	0.64	0.57
0.15	0.079	0.063	0.53	0.42
0.20	0.11	0.079	0.54	0.39

12.8 Statistical description of the system of interacting Taylor vortices in the direct interaction approximation

One of the principal aspects of the investigation of hydrodynamic turbulence consists in passing from dynamic description in terms of amplitudes to statistical description in terms of the moments of distribution functions. In our case it is necessary to pass from dynamic equations for amplitudes to the averaged equations for correlation matrices $\langle A_n(t)A_m(t')\rangle$, $\langle A_n(t)A_m(t')A_k(t'')\rangle$, etc. Instead of $A_n(t)$, any linearly independent set of variables $C_n(t)$ can be used:

$$C_n(t) = V_{nm}A_m(t), \qquad VV^+ = 1. \tag{12.32}$$

In this case the correlation matrices transform in a proper way and their properties change. Optimum variables C_n will apparently be those for which the correlation matrix $\langle C_n(t)C^*_m(t)\rangle$ is a diagonal one. In such a case the assumption that this matrix is diagonal at noncoinciding times, which makes the calculation much easier, will be closest to reality.

We shall first use a simpler set of variables—the representation of spatial harmonics B_n (12.15) which is sufficiently close to optimum variables C_n. Besides simplification of the calculations, we have two further reasons: firstly, the difference between C_n and B_n decreases with the growth of Γ, and secondly, this difference is large only for harmonics with high serial number n whose intensities are small.

Making a statistical description on the basis of Eqs (12.16)–(12.18) for $B_n(t)$, one can obtain equations for the second moment $\langle B_n B_m^* \rangle$; the right-hand side of these equations will include the fourth moments, etc. To analyse the sequence of equations for the moments we shall use the Wyld diagram technique [11] which he developed for hydrodynamic equations and upon which many papers devoted to incompressible fluid turbulence are based. In Section 12.8.1 we shall present this technique in terms of Eq. (12.16), taking into account that the harmonic representation differs somewhat from that of plane waves.

12.8.1 Diagram technique

Following Wyld's idea, we introduce into the right-hand side of the input equations of motion (12.16) the random force $f_n(t)$ which represents the random pushes from the 'environment':

$$\frac{\mathrm{d}B_n}{\mathrm{d}t} = -(\gamma_n + \mathrm{i}b_n)B_n + (NL)_n + f_n(t). \qquad (12.33)$$

Assume that $f_n(t) = -f_{-n}(t)$ and its statistical properties are of Gaussian type with the correlator

$$\langle f_n(t) f_m(t') \rangle = \delta(t-t')[\Delta(n-m) - \Delta(n+m)] f_n^2. \qquad (12.34)$$

In ω-representation the Eqs (12.33), (12.18) will take the form

$$B_n(\omega) = G_n^0(\omega)\Bigg\{\frac{D}{2}\sum_{n_1,n_2,n_3=-N}^{N}\int \delta(\omega+\omega_1-\omega_2-\omega_3)\,[\Delta(n+n_1+n_2+n_3) + \Delta(n+n_1+n_2+n_3+2N+2) + \Delta(n+n_1+n_2+n_3-2N-2)] \times B_1^* B_2 B_3 \,\mathrm{d}\omega_1\,\mathrm{d}\omega_2\,\mathrm{d}\omega_3 + f_n(\omega)\Bigg\},\quad B_i = B_{n_i}(\omega_i), \qquad i = 1, 2, 3, \qquad (12.35)$$

where D is given by formula (12.18) and $G_n^0(\omega)$ is the gauge Green function:

$$G_n^0(\omega) = [\omega + b_n - \mathrm{i}\gamma_n]^{-1}. \qquad (12.36)$$

Here b_n and γ_n are defined by formulas (12.17). By iterating these equations, B_n can be represented in the form of a power series of $f_n(\omega)$.

Now let us define the Green function,

$$\left\langle \frac{\delta B_n(\omega)}{\delta f_m(\omega')} \right\rangle = G_n(\omega)\delta(\omega-\omega')[\Delta(n-m) - \Delta(n+m)] \qquad (12.37)$$

and introduce the graphic notations:

$$G_n(\omega) = \rightarrow, \qquad G_n^*(\omega) = \leftarrow.$$

We define the pair correlator as

$$\langle B_n(\omega) B_m^*(\omega') \rangle = J_n(\omega)\delta(\omega - \omega')[\Delta(n-m) - \Delta(n+m)] \tag{12.38}$$

and denote

$$J_n(\omega) = \rightsquigarrow, \qquad J_n^*(\omega) = \leftsquigarrow.$$

Here we have assumed that in agreement with the above the correlation matrix $\langle B_n(t) B_m(t') \rangle$ is diagonal over n, m. By summing up weakly connected diagrams in a conventional way [12], one can get a set of Dyson equations for G_n and J_n:

$$G_n(\omega) = [\omega + b_n - i\gamma_n - \Sigma_n(\omega)]^{-1} \qquad J_n(\omega) = |G_n(\omega)|^2 \Phi_n(\omega). \tag{12.39}$$

Here Σ_n and Φ_n are the sums of compact diagrams. In Σ_n it is possible to pass from input to output along the lines G. All Φ_n can be cut in a unique way into two parts across the lines J.

The first diagrams are:

$\Sigma_n =$ 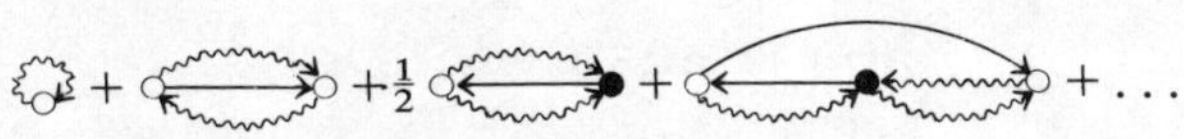 (12.40)

$\Phi_n = \frac{1}{2}$ 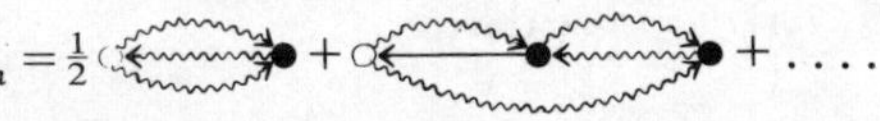 (12.41)

Here the proper number n and index ω should be assigned to each line, and each white vertex corresponds to expression

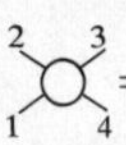

$$= D \sum_{\sigma_1, \sigma_2, \sigma_3, \sigma_4 = \pm 1} \delta(\omega_1 + \omega_2 - \omega_3 - \omega_4) \left[\Delta\left(\sum_{i=1}^{4} \sigma_i n_i \right) + \Delta\left(\sum_{i=1}^{4} \sigma_i n_i + 2N + 2 \right) + \Delta\left(\sum_{i=1}^{4} \sigma_i n_i - 2N - 2 \right) \right].$$

The black vertices are complex conjugates to the white ones. All series should be summed with respect to all internal numbers n_i from 1 to N and integrated with respect to the internal frequencies from $-\infty$ to $+\infty$.

12.8.2 Direct interactions approximation

If we retain only the first diagram in the series (12.40) and set $\Phi_n = 0$, then a comparatively simple set of equations will be obtained for $J_n(\omega)$:

$$\left\{ -i\omega - (\gamma_n + ib_n) + iD \int \left[\sum_{m=-N}^{N} J_m(\omega') + J_n(\omega') \right] d\omega' \right\} J_n(\omega) = 0 \tag{12.42}$$

or, in τ-representation,

$$\frac{dJ_n}{d\tau} = J_n(\tau) \left\{ -(\gamma_n + ib_n) + iD \left[\sum_{m=-N}^{N} J_m(0) + J_n(0) \right] \right\}. \tag{12.43}$$

These equations can be obtained directly from (12.16)–(12.18) by replacing the fourth-order momentum with a product of the pair momenta. From (12.42)–(12.43) one can see that this approximation is equivalent to the selfconsistent field approximation when all the remaining harmonics in the equation for $J_n(\tau)$ are replaced by their time-averaged values. Of course, this approximation is rather far from reality. However, it renders one tendency quite correctly: the region (of serial numbers n) of effectively excited harmonics may prove to be narrower than the region in which the linear increment is positive (i.e., $\gamma > a_n$). In particular, the expression for the threshold of excitation of the second harmonic follows from (12.43):

$$\gamma - a_2 = 2(a_2 - a_1). \tag{12.44}$$

A much more realistic approximation is obtained if diagrams with two vertices are also taken into account in series (12.41) for Σ_n and Φ_n. In this case,

$$\begin{aligned}\Sigma_n(\omega) = D\int \Big[\sum_{n_1=-N}^{N} J_1 + J_n(\omega_1)\Big]\, d\omega_1 + \tfrac{1}{2}\int \delta(\tilde{\omega}) \\ \times \sum_{n_1,n_2,n_3=-N}^{N} \Delta(\tilde{n})[1+\Delta_1+\Delta_2+\Delta_3][|D|^2 G_1^* J_2 J_3 \\ + D^2 J_1(J_2 G_3 + J_3 G_2)]\, d\omega_1\, d\omega_2\, d\omega_3,\end{aligned} \tag{12.45}$$

$$\begin{aligned}\Phi_n(\omega) = \frac{|D|^2}{2}\int \delta(\tilde{\omega}) \sum_{n_1 n_2, n_3=-N}^{N} \Delta(\tilde{n})[1+\Delta_1+\Delta_2+\Delta_3] \\ \times J_1 J_2 J_3\, d\omega_1\, d\omega_2\, d\omega_3.\end{aligned} \tag{12.46}$$

Here

$$\tilde{\omega} = \omega + \omega_1 - \omega_2 - \omega_3, \quad \tilde{n} = n + n_1 + n_2 + n_3,$$
$$\Delta_i = \Delta(n + n_i), \quad J_i = J_{n_i}(\omega), \quad i = 1, 2, 3.$$

A similar approximation in the theory of developed hydrodynamic turbulence proceeding directly from Navier–Stokes equations was made by Kraichnan [19], who called it 'direct interactions approximation'—DIA. This approximation is in fact something more than simply taking into account an interaction of the order of D^2 (i.e., in our case, splitting the sixth-order correlators into pairs). It also includes the procedure of selfconsistency which leads to replacement of initial decay and frequency of harmonics, $(\gamma_n + ib_n)$, by effective values $(\gamma_n + ib_n) + \Sigma_n(\omega)$ that depend on characteristics of motion. To define the limits of applicability of DIA it is necessary to obtain solutions of Eqs (12.39), (12.45), (12.46) and calculate the three-vertex diagrams in the series (12.40), (12.41) for Σ_n and Φ_n which give the first correction to DIA. It may be assumed that DIA will be reasonably far from the bifurcation lines that separate, on the plane of parameters, the domain of the stochastic attractor from the domain of regular behaviour (i.e., those of limit cycles and equilibrium states). Plots of excess $K_n(\omega)$ shown in Figs 12.14 and 15 support this assumption. One can see that K_n is close to unity, as it should be in the case of Gaussian statistics of random processes $B_n(t)$ in accordance with formula (12.30).

12.8.3 Preliminary analysis of DIA equations

The statistical equations (12.39), (12.45), (12.46) represent a rather complicated nonlinear set of $2N$ integral equations. Their detailed analysis, both analytical and numerical, is performed later. In the present section we shall just make first steps in this direction.

12.8.3.1 The first summand in (12.45) for Σ_n does not depend on ω in general. We shall neglect the dependence on ω in the second summand too, assuming in it $\omega=\omega_0$, where ω_0 is the root of the equation

$$\omega=-b_1+Re\{\Sigma_1(\omega)\}. \tag{12.47}$$

At small Γ the dependence of Σ_n on n can also be neglected as compared with the dependence of gauge values a_n and b_n upon n. Then the Green function $G_n(\omega)$ will take the simple form

$$G_n(\omega)=[(\omega+b_n-b_1)-\mathrm{i}(a_n+\tilde{\gamma})]^{-1}. \tag{12.48}$$

Here ω is taken as the distance from the centre of the line $\omega=\omega_0$ and the interaction has the effect of renormalization of damping:

$$\tilde{\gamma}=-\gamma+\mathrm{Im}\{\Sigma_1(\omega_0)\}. \tag{12.49}$$

In this approximation, instead of the second equation (12.39) one can write

$$J_n(\omega)=\frac{\Phi_n(\omega)}{(\omega+b_n-b_1)+(a_n+\tilde{\gamma})^2}. \tag{12.50}$$

12.8.3.2 According to the experiment, the number of excited harmonics is rather small as compared with N. We propose a very strong simplifying assumption that $J_n\neq 0$ only if $n=1$. Then (12.50) and (12.46) reduce to

$$J_1(\omega)=\frac{4.5\,|D|^2}{\omega^2+(\tilde{\gamma}+a_1)^2}\int \delta(\tilde{\omega})J_1(\omega_1)J_1(\omega_2)J_1(\omega_3)\,\mathrm{d}\omega_1\,\mathrm{d}\omega_2\,\mathrm{d}\omega_3. \tag{12.51}$$

It has been shown in [20, 21] that this equation has a single stable solution

$$\sqrt{(4.5)}\,|D|\,J_1=\mathrm{ch}^{-1}[\pi\omega/2(\tilde{\gamma}+a_1)]. \tag{12.52}$$

It is not difficult to see that solution (12.52) is also valid for the case when several more harmonics are excited besides the first one, but this packet remains narrow ($n\leqslant m$), so that

$$a_m-a_1<\tilde{\gamma},\qquad b_m-b_1<\tilde{\gamma}. \tag{12.53}$$

Hence, instead of (12.52) we obtain:

$$\frac{|D|}{2m}\sqrt{(2m^2+6m+1)}\sum_{n=1}^{m} J_n(\omega)\simeq \mathrm{ch}^{-1}(\pi\omega/2\tilde{\gamma}) \tag{12.54}$$

where m is the number of excited harmonics. By integrating this expression with respect to ω, we obtain the estimate for the total intensity of excited harmonics

$$I=\int \mathrm{d}\omega \sum_{n=1}^{m} J_n(\omega)\simeq\frac{2\pi m\tilde{\gamma}}{|D|\sqrt{(2m^2+6m+1)}}\simeq\frac{\tilde{\gamma}}{|D|}. \tag{12.55}$$

It can be shown that, when the wide packet of harmonics is excited and condition (12.53) is violated, the asymptotics of $J_n(\omega)$ at high ω still remains exponential:

$$J_n(\omega)\sim \exp(-|\omega|/\Delta_{\pm}). \tag{12.56}$$

12.8.3.3 To obtain the dependence of J_n on the number n we consider the case with high N when in (12.46) one can replace the summation over n_i by integration. For the sake of simplicity we also expand a_n and b_n:

$$\begin{aligned} a_n &= \alpha n^2, & b_n &= \beta n^2, \\ \alpha &= \pi^2 a/4(N+1)^2, & \beta &= \pi^2 b/4(N+1)^2. \end{aligned} \tag{12.57}$$

Using (12.50) and (12.46), we obtain

$$J_n(\omega)=\frac{|D|^2\int \delta(\tilde{n})\delta(\tilde{\omega})J_1J_2J_3\,\mathrm{d}n_1\,\mathrm{d}n_2\,\mathrm{d}n_3\,\mathrm{d}\omega_1\,\mathrm{d}\omega_2\,\mathrm{d}\omega_3}{2[(\omega+\beta n^2)+(\tilde{\gamma}+\alpha n^2)^2]}. \tag{12.58}$$

The analytical solution of (12.58) is not yet known. However, the asymptotics of $J_n(\omega)$ can be obtained if $\omega\leqslant\tilde{\gamma}$ and $\alpha n^2>\tilde{\gamma}$, $\beta n^2>\omega$:

$$|D|\,J_n(\omega)\simeq n\tilde{\gamma}^{-1}\sqrt{(\alpha^2+\beta^2)}\exp(-\delta\,|n|), \qquad \delta^2\simeq\frac{\sqrt{(\alpha^2+\beta^2)}}{\tilde{\gamma}}. \tag{12.59}$$

According to (12.52) and (12.53) the main contribution to $J_n(\omega)$ is due to the region $\omega\leqslant\tilde{\gamma}$, therefore for the intensities of harmonics $J_n=\int J_n(\omega)\,\mathrm{d}\omega$ one can obtain from (12.59)

$$|D|\,J_n=n\sqrt{(\alpha^2+\beta^2)}\exp(-\delta\,|n|). \tag{12.60}$$

According to expression (12.59) for δ, this yields again estimate (12.55) for the total intensity, I.

12.8.3.4 Now let us briefly discuss the problem of dependence of total intensity I on the supercriticality γ. For this purpose, using (12.45) and (12.49), we estimate the effective damping $\tilde{\gamma}$:

$$\tilde{\gamma}=-\gamma+2\eta I+|D|^2\,G''I^2. \tag{12.61}$$

Roughly evaluating the imaginary part of the Green function G'' as $1/\tilde{\gamma}$ and using estimate (12.55), we obtain

$$|D|\,I=[-\gamma+2\eta I+|D|\,I]d, \quad d\simeq 1$$

or

$$I=\frac{\gamma}{[2\eta+\mu^2\,|D|]} \tag{12.62}$$

where μ is a numerical coefficient of the order of unity.

12.8.3.5 Let us now compare the analytical conclusions concerning the dependencies $J_n(\omega)$ with the results of numerical experiment. First of all it should be noted that both theory (formulae (12.52), (12.54)) and experiment ((12.31) and Table 12.5) give the exponential decrease of $J_n(\omega)$ at high ω. In fact, the exponents $\Delta_\pm$ in numerical experiment are different for the right-hand and left-hand slopes of spectra, while the final expression (12.54) is symmetric with respect to replacement $\omega \to -\omega$. This false symmetry is the result of considerable simplification of equation (12.50).

The exponential decrease of $J_n(\omega)$ with the growth of n which was predicted by formula (12.59) is also well fulfilled in the experiment—see Fig. 12.17 based on the results of Table 12.2. The difference in behaviour of even and odd harmonics observed in the experiment also arises from the theory when we proceed to the limit $N \to \infty$.

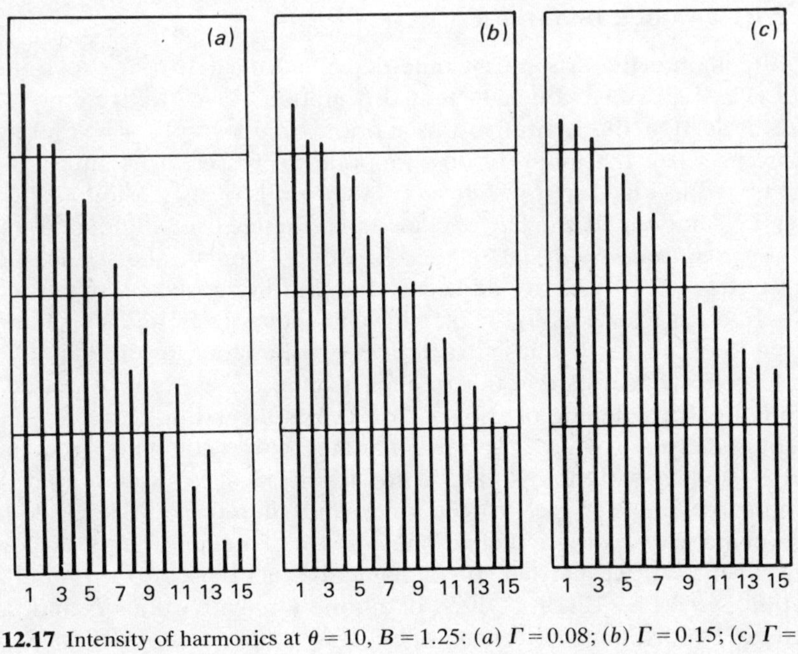

Fig. 12.17 Intensity of harmonics at $\theta=10$, $B=1.25$: (*a*) $\Gamma=0.08$; (*b*) $\Gamma=0.15$; (*c*) $\Gamma=0.2$

The estimate (12.62) shows direct proportionality between the total intensity I and the supercriticality γ. For $N=15$ from (12.62) and (12.18) it follows that

$$\frac{\eta I}{\gamma}=\left\{2+0.5\mu^2\sqrt{\left[\left(\frac{T}{\eta}\right)^2+1\right]}\right\}^{-1}=\text{const.} \tag{12.63}$$

In the numerical experiment (at $\theta=10$ and $B=1.25$) this ratio is practically independent of Γ and is 0.313, 0.314, 0.317 for $\Gamma=0.08$, 0.15, and 0.20 respectively. For μ^2 it gives the value 0.2.

It can be said in conclusion that the statistical description of the system of interacting Taylor vortices in the direct interaction approximation qualitatively corresponds to the results of numerical experiment. The problem of quantitative correspondence can be discussed after a more detailed analysis of the DIA equations.

12.9 Evolution of the attractor structure with increasing Reynolds number

In Sections 12.4–7 the attractor centred at the single rapid motion ω_6 was investigated. With an increase in *Re* the mode ω_2 and other rapid motions are excited, so the question arises of the behaviour of the system in the effective phase space in such a case. We shall restrict ourselves to an account of experimental data.

12.9.1 ω_2-mode behaviour

If we compute the ω_2-motion time-dependence $I_2(t)$ from experimental data (Fig. 12.18) we conclude that this motion in no way resembles the limit cycle. On the contrary, this motion occurs by 'splashes' at those moments when the intensity of ω_6-motion decreases. This alternation is well traced in Fig. 12.18 which shows the results of regression analysis of I_2 and I_6-motions. It can be concluded that there exist two domains in the phase space, or two quasiattractors. In the first domain the system moves with frequencies which are adjacent to ω_6 but in the second one it moves with frequencies near ω_2, while the characteristic frequency of movements between these quasiattractors is considerably lower than ω_2 and ω_6. Experiment shows that this frequency increases as *Re* is increased and results in a broadening of the spectral peaks ω_2 and ω_6.

The spectrum in Fig. 12.1 shows that motions on combined frequencies $\omega_6 \pm \omega_2$ appear with an increase in *Re*. It is natural to suppose that these combinative motions are forced ones and therefore do not add new coordinate axes to the effective phase space. To support this suggestion experimentally, we carried out a regressive analysis of I_6, I_2 and $I_{6\pm2}$ motions—see Fig. 12.19(a). First of all, the forcing motions I_2 and I_6 are seen to be distinctly alternate ones; this property at smaller *Re* has

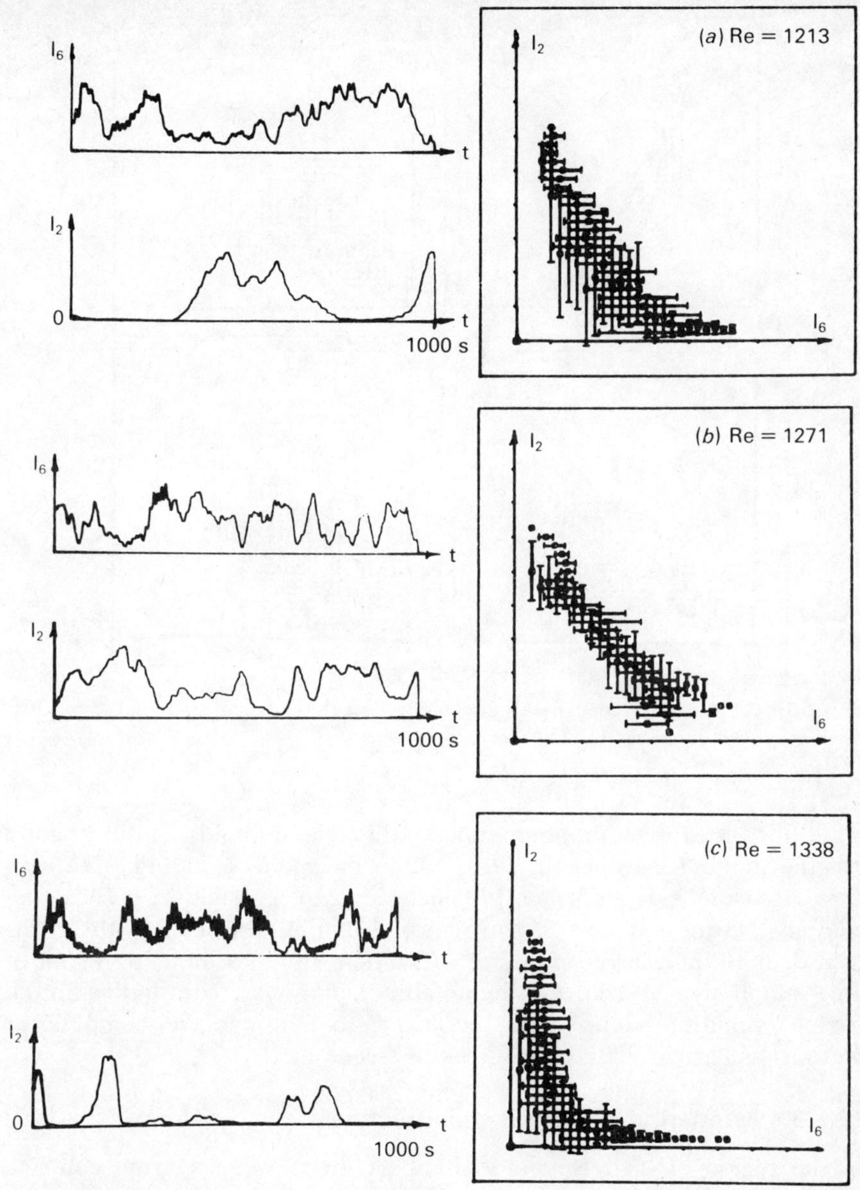

Fig. 12.18 Time-dependence of ω_2 and ω_6-motion envelopes and the results of regression analysis of these modes

already been mentioned. Furthermore, the motions I_{6+2} and I_{6-2} appear to be correlated within the range of small amplitudes only. This correlation practically vanishes at above-average amplitudes. This can be attributed to the fact that at such amplitudes their own nonlinear interactions become essential and violate the simple relation $I_{6\pm2} \sim I_6 \cdot I_2$ which must

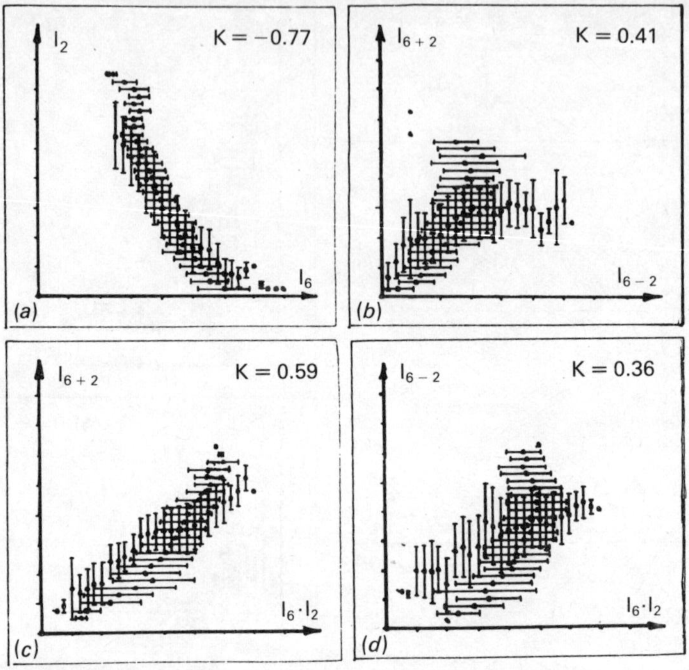

Fig. 12.19 Correlation of combinative motions $I_{6\pm2}$ and forcing motions I_6, I_2: $Re = 1394$, $N = 30$, frequency window $\Omega = 0.05$ Hz

be fulfilled for forced motions. Indeed, this relation holds relatively good for the mode I_{6+2} (Fig. 12.19(*c*)), while the mode I_{6-2} (Fig. 12.19(*d*)) reveals greater independence. In conclusion, it is possible to say that the combinative motions are forced at small amplitudes only. With an increase in *Re* and in the presence of random splashes into the region of great amplitudes at fixed *Re*, the amplitude and phase correlations in the triplets diminish. Hence it is necessary to consider the combinative motions as certain effective degrees of freedom.

12.9.2 Wandering among quasiattractors

In the region of high Reynolds numbers the power spectrum contains quite a number of distinctly pronounced peaks of appreciable widths (see Fig. 12.1(*c*)). Figure 12.20 shows the results of regression analysis of the motions at frequencies ω_2, ω_3, ω_6 and ω_7 for the velocity record which corresponds to the above spectrum. The alternation of the motions ω_2, ω_6 is still clearly traced. Moreover, this property is characteristic for any pair chosen from these four motions. This means that the fluid motion at any of the four frequencies excludes the motions at the three remaining ones. In other words, the velocity record consists always of randomly alternating fragments of sinusoids characterized by the frequencies ω_2, ω_3, ω_6 and

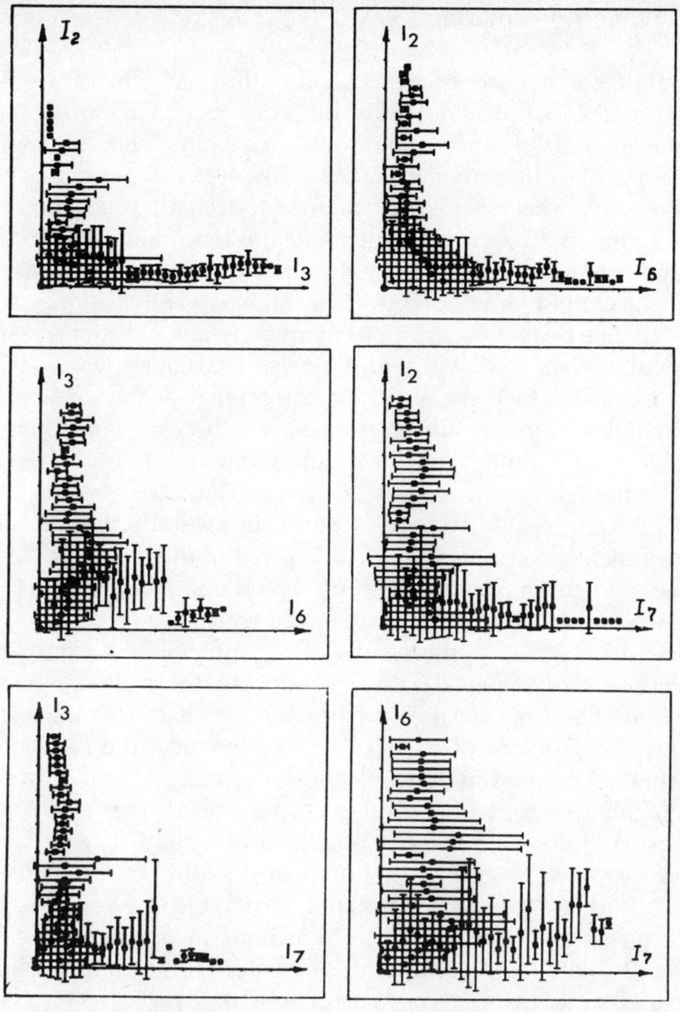

Fig. 12.20 Correlations illustrating the situation when the motion at any one of frequencies ω_2, ω_3, ω_6, ω_7 excludes the motions at the others (see corresponding spectrum in Fig. 12.1(*c*)): frequency window $\Omega = 0.16$ Hz

ω_7. Comparing this situation with those described above, it is possible to claim that there occurs with increasing *Re* an increase in the dimension of the phase space as well as in the number of quasiattractors between which the phase point wanders. At the same time the frequency of wandering also increases, which results in additional broadening of the peaks. When the peak-widths become comparable with the distances between them, the very concept of 'quasiattractor' becomes meaningless and the system performs a tangled motion in a phase space of rather large dimension.

12.10 Concluding remarks

Hydrodynamic turbulence in general and the problem of its onset in particular are the concern not only of 'pure mechanicians' but also of scientists in other fields, such as computer mathematicians and specialists in the theory of differential equations, theoretical physicists and experimenters of various disciplines, engineers designing measuring equipment, etc. Opinions therefore differ as to the main aim of investigations and the importance and validity of the data obtained. Some researchers wish to reduce fluid drag force, while others want to understand the nature of stochasticity in extremely simple dynamic systems which are only remotely connected with fluid flows. Extremists consider that it suffices to carry out high-precision measurements of the velocity field at various Reynolds numbers and everything will be clear, or hope to solve every problem by use of a special supercomputer for the Navier–Stokes equations. Another extreme point of view is that all turbulent flows are described by some simple strange attractor (maybe a universal one) with few degrees of freedom, and the problem is only to discover this.

We adhere to the moderate position which consists in the recognition of the complex character of the problem of transition to turbulence. This dictates the necessity to combine various complementary approaches to the problem. As to the final aim, we hope to show that, by using the traditional circular Couette flow as an example, it is already possible to achieve, for the problem of transition to turbulence, the level of understanding that is accepted in fields of physics (e.g., solid-state physics) that are more highly developed than the 'science of turbulence'. Of course, the results presented do not yet meet this ideal, but they do constitute what we consider to be a promising first step in this direction.

We believe that, starting immediately from the Navier–Stokes equations, it is possible to evaluate the coefficients in our phenomenological equations (12.7) and to determine the limits of their applicability and the character and value of necessary corrections. It is necessary to carry out precise measurements to verify the main hypotheses which form the basis of this model, e.g., the assumption that $\mathbf{A}(\mathbf{r}, t) = A(t)\mathbf{f}(\mathbf{r})$. It is also necessary to investigate the spatial profile of azimuthal waves along the whole cylinder, to obtain the mutual correlations of the oscillations in different vortices and to compare in a more detailed way than we have done previously the results of laboratory and numerical experiments (within the framework of Eq. (12.7)). After this study it will be more reasonable to investigate in detail the structure of a complex attractor in the phase space of Eq. (12.7), e.g., to determine the character of trajectory divergence, to obtain the Poincaré maps, etc.

The questions posed in Section 12.8 are of specific interest. They pertain to the passage from the dynamic description of a system to the statistical description, with the use of approaches developed in the theory of turbulence. In particular, to what extent is the statistics of motion

$A_n(t)$ similar to the Gaussian one? On what factors does this similarity depend and in what domain of θ, B and Γ is the DIA valid? Most of these questions are of interest in themselves but their principal significance appears when they are considered as links in the chain connecting the Navier–Stokes equations with the turbulent motion in Couette flow.

Often the question arises as to whether hydrodynamic flows can be grouped into a small number of classes in which the behaviour is qualitatively similar. If 'qualitatively' is understood in its topological meaning, as accepted in the qualitative theory of differential equations, the answer is as follows. As the number of flow geometries is very large and there is no evidence that the coefficients in the corresponding normal Poincaré forms are fixed in a special manner, there exist as many routes to turbulence as the number of ways in which all the bifurcational diagrams can be traversed from a set of topologically different ones in the vicinity of fixed points, limit cycles, etc. (for example, see [22]). To confirm this, we note that a single Couette flow demonstrates a number of different ways: from the 'soft' stepwise transition with various sequences of bifurcations (when the inner cylinder is rotated) to the 'hard' excitation of turbulence (when the outer cylinder is rotated) which is a usual attribute of transitions in external flows. Nevertheless, it is clear that all the variants of 'soft' transitions are similar in some wider sense than the topological one. Therefore a 'rougher' classification should be developed for such complicated problems as the transition to turbulence.

Appendix

The mechanical design of the experimental apparatus is shown in Fig. 12.21. The main part of the apparatus is the spindle (8) rotating in precision ball bearings (7) which are enclosed in a ball race on the bed plate (11). The massive fly-wheel (4) is placed on this spindle from above; the inner stainless steel cylinder is fixed from below by means of the cone (9). The rotating assembly is supported by air-lubricated thrust bearings (5). The spring hanger (6) is adjusted so that the bearings (7) are free from veritical load, which minimizes friction noise. The fly-wheel is driven by an electric motor (1). Rotation speed is controlled by electronic circuit using a precision tachometer as a feedback probe. These technical measures allow the stability of rotational period to be kept constant to better than $10^{-4}T$ and provide a radial pulse at the bottom of the inner cylinder of less than 5 μm. The outer cylinder (14) in the experiments in question was made from stainless steel with inner diameter 55 ± 0.005 μm. The alignment and coaxiality of the cylinders also ensure that the corresponding pulses provided by the special support (12) are not more than 5 μm. The top and bottom flanges of the annulus have glass windows through which laser beams can penetrate into the water. The apparatus as a whole is contained inside a thermally isolated box in which

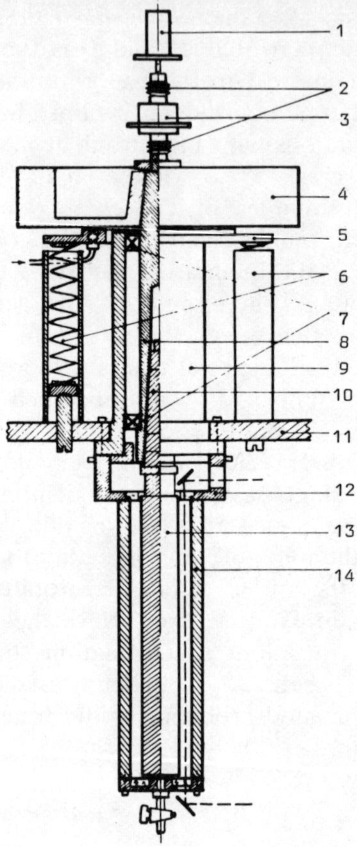

Fig. 12.21 Experimental apparatus

air temperature is controlled to an accuracy of 0.02°C. The temperature of the outer cylinder was measured by Hewlett-Packard 2801A quartz thermometer.

Acknowledgements

The authors would like to express their appreciation to Professor Yu. E. Nesterikhin for his constant interest and encouragement and to Dr F. A. Zhuravel for his great contribution to the design of software for data acquisition and storage system. We are indebted to Z. B. Krugljak and E. N. Utkin for hardware design and for help during the course of experiments. The authors are also grateful to Dr E. A. Kuznetsov and M. D. Spector for valuable discussions and helpful suggestions.

References

1. Sinai, Ya. G., Stochasticity of dynamical systems, in Gaponov-Grekhov, A. V. (Ed.), *Nonlinear Waves*: Moscow, Nauka, 1979 (in Russian).
2. Monin, A. S., On the nature of turbulence, *Uspechi Fizicheskih Nauk*, **125,** 97–122, 1978 (English translation in *Soviet Physics—Uspechi*, **21,** 429–442, 1979).
3. Rabinovich, M. I., Stochastic self-oscillations and turbulence, *Uspechi Fizicheskih Nauk*, **125,** 123–168, 1978 (English translation in *Soviet Physics—Uspechi*, **21,** 443–469, 1979).
4. Creveling, H. F., Paz, J. F., Baladi, J. Y. and Schoenhals, R. J., Stability characteristics of a single phase free convection loop, *Journal of Fluid Mechanics*, **67,** 65–84, 1975.
5. Lorenz, E. N., Deterministic nonperiodic flow, *Journal of the Atmospheric Sciences*, **20,** 130–141, 1963.
6. Lyubimov, D. V., Putin, G. E. and Chernatynsky, V. I., On convective motions in the Hele–Shaw cell, *Doklady Akademii Nauk USSR*, **235,** 554–556, 1977 (in Russian).
7. McLaughlin, J. B. and Martin, P. C., Transition to turbulence in a statically stressed fluid system, *Physical Review A*, **12,** 186–203, 1975.
8. Yahata, H., Temporal development of the Taylor vortices in a rotating fluid, II, *Progress of Theoretical Physics*, **61,** 791–800, 1979.
9. Kuznetsov, E. A., L'vov, V. S., Nesterikhin, Yu. E., Shmojlov, Y. F., Sobolev, V. S., Spector, M. D., Timokhin, S. A., Utkin, E. N. and Vasilenko, Yu. G., On turbulence arising in Couette flow, Preprint 58, Institute of Automation and Electrometry, Novosibirsk, 1977.
10. Zhuravel, F. A., Krygljak, Z. B., Kuznetsov, E. A., L'vov, V. S., Nesterikhin, Yu. E., Predtechensky, A. A., Sobolev, V. S. and Utkin, E. N., Spectrum evolution during the transition to turbulence in Couette flow, Preprint 103, Institute of Automation and Electrometry, Novosibirsk (in Russian).
11. Wyld, H. W., Formulation of the theory of turbulence in an incompressible fluid, *Annals of Physics*, **14,** 142–165, 1961.
12. Zakharov, V. E. and L'vov, V. S., A statistical description of nonlinear wave fields, *Izvestija VUZ Radiofizika* (*USSR*), **18,** 1470–1487, 1975 (English translation in *Radiophysics and Quantum Electronics*).
13. Benjamin, T. B., Bifurcation phenomena in steady flows of a viscous fluid, *Proceedings of the Royal Society, A*, **359,** 1–43, 1978.
14. L'vov, V. S. and Predtechensky, A. A., On the Landau and 'stochastic attractor' pictures in the problem of transition to turbulence, Preprint 111, Institute of Automation and Electrometry, Novosibirsk, 1979.

15. Fenstermacher, P. R., Swinney, H. L. and Gollub, J. P., Dynamical instabilities and transition to chaotic Taylor vortex flow, *Journal of Fluid Mechanics*, **94,** 103–128, 1979.
16. Kuznetsov, E. A. and Spector, M. D., On weak supercritical convection, *Zhurnal Prikladnoi Mekhaniki i Tekhnicheskoi Fiziki*, No. 2, 76–86, 1980 (in Russian).
17. Hall, P., Centrifugal instabilities in finite containers: a periodic model, *Journal of Fluid Mechanics*, **99,** 575–596, 1980.
18. Stuart, J. T. and Di Prima, R. C., On the mathematics of Taylor vortex flows in cylinders of finite length, *Proceedings of the Royal Society, A*, **372,** 357–365, 1980.
19. Kraichnan, R. H., The structure of isotropic turbulence at very high Reynolds number, *Journal of Fluid Mechanics*, **5,** 497–543, 1959.
20. Krutsenko, I. V., L'vov, V. S. and Melkov, G. A., Spectral density of parametrically excited waves, *Zhurnal Eksperimental'noi i Teoreticheskoi Fiziki*, **75,** 1114–1131, 1978 (English translation in *Soviet Physics—JETP*, **42,** 1057, 1976).
21. L'vov, V. S. and Cherepanov, V. B., Collective oscillations and instability of single-frequency state of parametrically excited waves, *Zhurnal Eksperimental'noi i Teoreticheskoi Fiziki*, **75,** 1631–1645, 1978 (English translation in *Soviet Physics—JETP*, **48,** 822–829, 1978).
22. Arnold, V. I., *Additional Chapters of the Ordinary Differential Equations Theory*: Moscow, Nauka, 1978, 234–300 (in Russian).

13 Hyperbolic Attractors of Differentiable Dynamical Systems

R V Plykin

Dedicated to the glowing memory of Vladimir Mikhailovich Alexeev

13.1 Introduction

Opportunities to observe the stochastic behaviour of dynamical systems are frequently associated with the existence of stable attractive limiting sets with a complicated geometrical structure, the so-called *strange attractors.* Two aspects, 'geometrical' and 'applied', can be distinguished, conditionally, of course, in studying these attractors. Difficulties in the 'applied' approach, which is used to handle particular dynamical systems of physical origin, are mainly determined by the fact that in these systems attractors exist along with other limiting sets (structures) whose separation is frequently a difficult task requiring not only extensive use of analytical and geometrical techniques, but also a great deal of computer capacity.

The 'geometrical' approach, which is followed in this chapter, studies manifolds and the related dynamical systems; here a variety of attractor structures manifests itself, so classification problems constitute the main difficulty.

Hyperbolicity assumptions of one kind or another are usually made in studying the properties of strange attractors, so it is natural to formulate a problem of the structure of hyperbolic attractors of diffeomorphisms.

The classification of hyperbolic attractors with codimension 1 given in this chapter shows that such attractors of diffeomorphisms, acting on manifolds with dimension more than 2, are either of the type of Example 13.1 (the orientable case) or of the type of Example 13.2 (the nonorientable case) (see Section 13.2). For diffeomorphisms of surfaces, attractors of the type of Example 13.3 may exist along with attractors of the types of Examples 1 and 2.

The following statement can be made on the basis of the results given below: hyperbolic attractors with codimension 1 for dynamical systems created by diffeomorphisms of manifolds with dimension higher than 2 can exist only on manifolds of fairly complicated topological structure, and this explains the unobservability of phenomena of rigorous hyperbolicity in the analysis of particular systems.

Section 13.2 gives a summary of the main results of the work together with some comments and preliminaries.

13.2 Preliminaries and statement of basic results

Let $f: M \to M$ be a diffeomorphism of a smooth manifold M of class C^m $(m \geq 1)$ with a dimension not less than two. Let d stand for the total Riemannian metric M generated by the Riemannian metric $\|\cdot\|$ on a tangent manifold TM.

An invariant closed set Λ is called an *attractor* (*repeller*) if it consists of nonwandering points and there exists such an open neighbourhood $U \supset \Lambda$ that

$$\bigcap_{n \geq 0} f^n U = \Lambda \left(\bigcap_{n \leq 0} f^n U = \Lambda \right).$$

Let us call an attractor (repeller) *hyperbolic* if for each $x \in \Lambda$ the tangent space $T_x M$ can be represented as the decomposition into a direct sum of its subspaces $T_x M = E_x^u \oplus E_x^s$, which is invariant with respect to the differential operator Df, so that:

(a) $Df(E_x^s) = E_{f(x)}^s$, $Df(E_x^u) = E_{f(x)}^u$;

(b) there are such constants $c > 0$ and $\lambda \in (0, 1)$ that

$$\|Df^n(v)\| \leq c\lambda^n \|v\| \quad \text{if} \quad v \in E_x^s, \ n \geq 0$$
$$\|Df^{-n}(v)\| \leq c\lambda^n \|v\| \quad \text{if} \quad v \in E_x^u, \ n \geq 0.$$

Without loss of generality we may assume $c = 1$.

If Λ is a hyperbolic attractor (repeller) let us define the following sets for any $x \in \Lambda$:

$$W^s(x) = \{y \in M : d(f^n x, f^n y) \to 0, \ n \to +\infty\}$$
$$W_\varepsilon^s(x) = \{y \in M : d(f^n x, f^n y) \leq \varepsilon, \ n \geq 0\}$$
$$W^u(x) = \{y \in M : d(f^{-n} x, f^{-n} y) \to 0, \ n \to +\infty\}$$
$$W_\varepsilon^u(x) = \{y \in M : d(f^{-n} x, f^{-n} y) \leq \varepsilon, \ n \geq 0\}$$

The set $W^s\Lambda = \bigcup_{x \in \Lambda} W^s(x)$ is called the *attraction basin* of the attractor Λ.

According to the theorem on stable manifolds, for sufficiently small $\varepsilon > 0$ and for each $x \in \Lambda$ the following properties are valid:

(a) $W_\varepsilon^s(x)$, $W_\varepsilon^u(x)$ are smooth discs such that

$$T_x W_\varepsilon^s(x) = E_x^s, \qquad T_x W_\varepsilon^u(x) = E_x^u;$$

(Let us recall that C^m-smooth K-disc is a C^m-manifold with an edge, homeomorphic to K-ball.)

(b) $d(f^n x, f^n y) \leq \lambda^n d(x, y)$ for $y \in W^s_\varepsilon(x)$, $n \geq 0$,

$d(f^{-n} x, f^{-n} y) \leq \lambda^n d(x, y)$ for $y \in W^u_\varepsilon(x)$, $n \geq 0$;

(c) $W^s_\varepsilon(x)$, $W^u_\varepsilon(x)$ depend continuously on x;

(d) $W^s(x) = \bigcup_{n \geq 0} f^{-n} W^s_\varepsilon(f^n x)$,

$W^u(x) = \bigcup_{n \geq 0} f^n W^u_\varepsilon(f^{-n} x)$, $x \in \Lambda$.

A hyperbolic attractor (repeller) is called an *expanding attractor* of codimension 1 (a *contractible repeller* of codimension 1) if $W^u(x) \subset \Lambda$ for any $x \in \Lambda$ and its dimension dim $W^u(x)$ is $n-1$ (if $W^s(x) \subset \Lambda$ for $x \in \Lambda$ and dim $W^s(x) = n-1$), $n = \dim M$.

The consideration below will be confined to the case of expanding attractors of codimension 1. According to [1, 2], their local structure is of the form of a product of the Cantor perfect set by an $(n-1)$-disc.

Let $x_1, x_2 \in \Lambda$. Using the hyperbolicity of Λ, we choose $\varepsilon > 0$, $\delta > 0$ such that the neighbourhood

$$U(x_1) = \bigcup_{x \in W^u_\varepsilon(x_1)} W^s_\delta(x)$$

is homeomorphic to R^n.

Assume that $W^s_\tau(x_2) \cap W^u_\varepsilon(x_1) \neq \varnothing$ (this assumption holds if $\tau > 0$ is sufficiently large). Let us choose a direction on $W^s_\tau(x_2)$ and numerate, in accordance with this direction, the points $z_1, z_2, \ldots, z_m$ of successive intersections of $W^s_\tau(x_2)$ with $W^u_\varepsilon(x_1)$. Let us choose next an orientation in $U(x_1)$ generated by a vector tangent to $W^s_\tau(x_2)$ at point z_1 specifying the direction of $W^s_\tau(x_2)$, and by an $(n-1)$th reper of vectors tangent to $W^u_\varepsilon(x_1)$ at z_1. At each point z_i $(i > 1)$ an nth reper arises which is generated by a vector, tangent to $W^s_\tau(x_2)$ at that point, that is determined by the direction of $W^s_\tau(x_2)$ and by the $(n-1)$-reper of vectors tangent to the manifold $W^u_\varepsilon(x_1)$ at the same point, having the same orientation as the $(n-1)$-reper at point z_1. If all the repers which arise in such a way determine the same orientation $U(x_1)$ and this statement does not depend on the arbitrariness in the choice of $x_1, x_2 \in \Lambda$, $\varepsilon > 0$, $\delta > 0$, $\tau > 0$, such an attractor Λ is called orientable. An attractor which is not orientable is called nonorientable.

If G is a connected open domain of manifold M and x_0 belongs to the boundary ∂G, we shall call x_0 an attainable point of the boundary if there exists in G an open arc of which one end coincides with x_0.

A subset $D(G) \subset \partial G$ composed of attainable points will be called an attainable boundary of domain G.

Let Λ be an attractor of diffeomorphism $f: M \to M$ and let $G = M \backslash \Lambda$. The set $C_1, C_2, \ldots, C_m$ of the connectivity components of the attainable boundary $D(M\setminus\Lambda)$ will be called a *connection of power m* or *m-connection* if these components can be positioned in such an order that C_1 and C_m, as well as C_i and $C_{i+1}(1\leq i\leq m-1)$, can be connected by open arcs of one-dimensional stable manifolds entirely contained in the domain $M\setminus\Lambda$. Note that each connectivity component of the set $D(M\setminus\Lambda)$ is homeomorphic to R^{n-1}.

A point $q\in\Lambda$ will be called the *boundary point* of an attractor if one of the connectivity components of the set $W^s(q)\setminus q$ does not intersect Λ. It has been established by Plykin [1, Theorem 1.7] that the set of boundary points is finite and all the boundary points are periodic.

Theorem 13.1 clarifies the geometrical structure of attainable boundaries and distribution of boundary points.

Theorem 13.1. *Let Λ be an expanding attractor of diffeomorphism $f:M\to M$. The attainable boundary $D(M\setminus\Lambda)$ consists of a finite set of connections, each element of any of them containing the boundary point. If* $\dim M=n\geq 3$, *the powers of connections contained in $D(M\setminus\Lambda)$ do not exceed* 2. *In the case of an orientable attractor Λ they are equal exactly to* 2. *The periods of boundary points of each connection are equal to one another. If $n=2$ the boundary points of each connection of the set $D(M\setminus\Lambda)$ have the same period when the orientation of the neighbourhood of each of them is preserved by the diffeomorphism power equal to the period of this point. Otherwise, each connection contains one or two boundary points of an odd period k, as well as an even set of boundary points of period* $2k$ (*this set may be an empty one*).

Let us recall certain further topological results and consider some examples.

Let M and N be manifolds of class $C^m, 0\leq m\leq\infty$, and $f:M\to N$, $g:M\to N$ be maps of smoothness class $C^r, 0\leq r\leq m$. We shall call f and g *C^r-homotopic* (belonging to the same class of C^r-homotopy) if there exists a family of mappings $\varphi_t:M\to N$, $t\in I=[0, 1]$ of smoothness class C^r such that:

(1) $\varphi_0\equiv f$, $\varphi_1\equiv g$;
(2) the mapping $\Phi:M\times I\to N$, where $\Phi(x, t)=\varphi_t(x)$, is continuous;
(3) if $r>0$, then $D\varphi_t:(TM)\times I\to TN$ is continuous.

If in this case f, g and φ_t, $t\in I$, are diffeomorphisms, we shall say that f and g are diffeotopic.

The manifolds M and N have the same homotopic type (are *homotopically equivalent*) if there exist continuous maps $f:M\to N$, $g:N\to M$ such that $gf:M\to M$ and $fg:N\to N$ are C^0-homotopic to identity maps.

The manifold M is called *punctured*, if a point x_0 is distinguished in it. By analogy, we may introduce the concept of C^r-homotopy of mappings of punctured spaces $f, g:(M, x_0)\to(N, y_0)$, $f(x_0)=y_0$, $g(x_0)=y_0$; the family $\{\varphi_t\}$ must preserve the distinguished points, i.e., $\varphi_t(x_0)=y_0$. A set of classes of C^0-homotopic mappings of a punctured circle (S^1, s_0) on a punctured manifold (M, x_0) is denoted as $\pi_1(M, x_0)$. It permits a natural (and unique) group structure, and is called the *fundamental group* M (let us recall that a change of a distinguished point leads to an isomorphic fundamental group). If $f:(M, x_0) \to (N, y_0)$ is a continuous mapping of punctured manifolds, it induces a homomorphism of fundamental groups:

$$f_* : \pi_1(M, x_0) \to \pi_1(N, y_0)$$

These concepts can be deduced by a more geometrical consideration.

Let M be a manifold. A pair consisting of a topological space $\tilde{M}$ and a continuous mapping $\pi : \tilde{M} \to M$ will be called a *covering space* for M or simply a *covering* if it satisfies the condition that for each $x \in M$ there can be found such a neighbourhood U homeomorphic to an n-ball ($n = \dim M$) that each connection component of the set $\pi^{-1}U$ is homeomorphic to U under mapping π. If M is a smooth manifold of class C^m, $\tilde{M}$ can also be provided with a structure of C^m-smooth manifold; if M possesses the Riemannian metric, $\tilde{M}$ can also be provided with a structure of the Riemannian manifold, π being the local isometry.

If the covering manifold $\tilde{M}$ is simply connected, i.e., $\pi_1(\tilde{M}, \tilde{x}_0) = 0$, the covering $\pi : \tilde{M} \to M$ is called *universal* and is unique within a topological transformation. The covering is called k-sheeted if the set $\pi^{-1}x$, $x \in M$, consists of k elements.

The mapping $\sigma : \tilde{M} \to \tilde{M}$ is called the covering transformation if $\pi \cdot \sigma(\tilde{x}) = \pi\tilde{x}$, $\tilde{x} \in \tilde{M}$, and the set of all such mappings forms the covering transformation group $\sigma(\tilde{M})$; each covering transformation different from the identity acts on $\tilde{M}$ without fixed points.

It is known that if the covering $\pi : \tilde{M} \to M$ is universal, the covering transformation group is isomorphic to the fundamental group $\pi_1(M, x_0)$. If the covering transformation group acts transitively on each layer $\pi^{-1}x$, such a covering is called *regular* and is characterized by the fact that the group $\pi_*\pi_1(\tilde{M}, \tilde{x}_0)$ is a normal subgroup of the group $\pi_1(M, \pi\tilde{x}_0)$. For a regular covering the covering transformation group is isomorphic to a factorgroup $\pi_1(M, x_0)/\pi_*\pi_1(\tilde{M}, \tilde{x}_0)$, where $x_0 = \pi\tilde{x}_0$. It follows from this result that each two-sheeted covering is regular and its transformation group is generated by an involution $\theta : \tilde{M} \to \tilde{M}$, i.e., by a homeomorphism having the property $\theta \cdot \theta x = x$.

Let us say that the mapping $\tilde{f} : \tilde{M} \to \tilde{M}$ covers the mapping $f : M \to M$ if the diagram

$$\begin{array}{ccc} \tilde{M} & \xrightarrow{\tilde{f}} & \tilde{M} \\ {\scriptstyle\pi}\downarrow & & \downarrow{\scriptstyle\pi} \\ M & \xrightarrow{f} & M \end{array}$$

is commutative, i.e., the identity $f \cdot \pi(x) = \pi \cdot \tilde{f}(x)$, $x \in \tilde{M}$ holds. It is known that if $f : M \to M$ is smooth, $\tilde{f} : \tilde{M} \to \tilde{M}$ is also smooth, if the covering is universal, and there always exists a mapping $\tilde{f} : \tilde{M} \to \tilde{M}$ covering f which is a diffeomorphism if f is a diffeomorphism. The consideration of covering mapping $\tilde{f}:\tilde{M}\to\tilde{M}$ is often preferred to that of the mapping $f:M\to M$, because the former has a simpler structure and the manifold M can be obtained from M by matching the points belonging to one trajectory under the action of all covering transformations.

Example 13.1. An n-dimensional torus T^n can be obtained from the space R^n via the matching of points occurring in the trajectories under the action of the group of integer shifts $\{f_{\bar{k}}\}$, $\bar{k}=(k_1, k_2, \ldots, k_n)\in Z^n$, where $f_{\bar{k}}(x_1, \ldots, x_n)=(x_1+k_1, \ldots, x_n+k_n)$, and R^n is a universal covering for T^n; the group of integer shifts $\{f_{\bar{k}}\}$ isomorphic to Z^n is a covering transformation group.

Any linear map $A:R^n\to R^n$ specified by an integer matrix preserves the integer lattice $Z^n=\{\bar{k}\}$ and generates the map of torus $A:T^n\to T^n$ which is covered by $\bar{A}$. If the determinant of the matrix $\bar{A}$ is equal, by the magnitude, to one and its spectrum does not contain eigenvalues equal to one in modulus, then the torus mapping A that arises is called a hyperbolic automorphism of torus or Y-diffeomorphism in the Anosov sense. We shall be interested in hyperbolic automorphisms of torus of codimension 1, i.e., in automorphisms for which the determining matrix spectrum has the only eigenvalue less than unity in modulus.

Let $n=2$. Specify the mapping $\bar{A}:R^2\to R^2$ by the matrix $\left(\begin{smallmatrix}2 & 1\\ 1 & 1\end{smallmatrix}\right)$, then it is a simple matter to verify that $\bar{A}$ is a covering mapping for the torus hyperbolic automorphism $A:T^2\to T^2$.

Let us consider an A-invariant set $\mathcal{P}=\{a_0, a_1, a_2, a_3\}$ consisting of a fixed point $a_0=(0, 0)$ and of three points $a_1=(\frac{1}{2}, 0)$, $a_2=(0, \frac{1}{2})$, $a_3=(\frac{1}{2}, \frac{1}{2})$, forming a periodic trajectory of the automorphism A. Following Smale [3], let us modify the automorphism A in small 'rectangular' neighbourhoods of the set $\mathcal{P}$, as shown in Fig. 13.1. After this procedure we obtain a structurally stable DA-diffeomorphism $f^{A,\mathcal{P}}$ whose nonwandering set

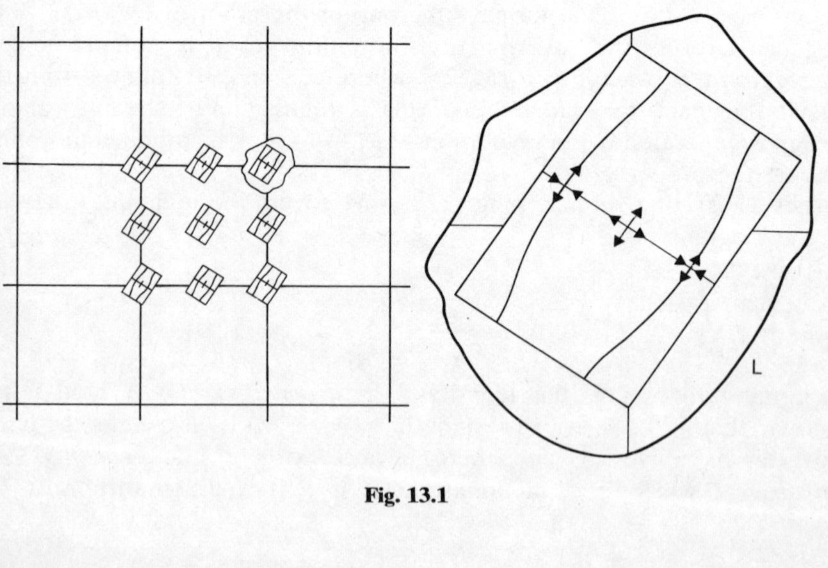

Fig. 13.1

consists of a repelling fixed point a_0, repelling trajectory $\{a_1, a_2, a_3\}$, and one-dimensional attractor Λ (for details of the construction see [3, 4] and Theorem 13.3 below).

Example 13.2. Example 13.1 can be constructed in such a way that the identity $f^{A,\mathcal{P}}(-x) = -f^{A,\mathcal{P}}(x)$ holds. The set $\mathcal{P}$ coincides with the set of fixed points Fix θ of the involution $\theta: T^2 \to T^2$ determined by the equation $\theta x = -x$.

Since the involution θ acts on the manifold $T^2\backslash\mathcal{P}$ without fixed points, one may consider $T^2\backslash\mathcal{P}$ to be a two-sheeted covering of the surface $S^2_{0,4}$ obtained by identifying the pairs $(x, \theta x)$, $x \in T^2\backslash\mathcal{P}$, which, as it is easy to show, is homeomorphic to a two-dimensional sphere with four punctured points. In this case the diffeomorphism $f^{A,\mathcal{P}}$ covers the diffeomorphism $f^{A,\mathcal{P},\theta}: S^2_{0,4} \to S^2_{0,4}$, which possesses a nonorientable hyperbolic attractor of codimension 1.

Let us give an example of a diffeomorphism S^2, possessing an attractor, that cannot be obtained from the DA-mapping.

Example 13.3. Let the diffeomorphism f map into itself a configuration D (Fig. 13.2(a)) homeomorphic to a circle, from which five pairwise nonintersecting discs are cut off in such a way that the transform fD is as shown in Fig. 13.2(b). In this case the operation f_* on a free group $\pi_1(D, x_0)$ with generators x, y, z, t, v, which form a plane union of five similarly oriented closed curves (Fig. 13.3), can be represented as follows:

$$f_*(x) = z^{-1}tzy^{-1}xyz^{-1}t^{-1}z$$
$$f_*(y) = y^{-1}x^{-1}yxy$$
$$f_*(z) = z^{-1}tv^{-1}t^{-1}ztvt^{-1}z$$
$$f_*(t) = y^{-1}x^{-1}y^{-1}xyz^{-1}tzy^{-1}x^{-1}yxy$$
$$f_*(v) = z^{-1}tvt^{-1}z.$$

Following [5, 6], let us construct the mapping in such a way that it is continued to a structurally stable diffeomorphism $F: S^2 \to S^2$, whose nonwandering set is composed of a hyperbolic one-dimensional attractor Λ and six fixed repelling points. The attainable boundary $\mathrm{D}(S^2\backslash\Lambda)$ contains five 1-connections and one 3-connection. Due to the structural stability the example can be realized via the polynomial mapping of a plane into itself.

Let us note that popular examples of polynomial mappings of a plane into itself, such as Henon's example [7], which possess an attracting one-dimensional set located in a simply connected bounded domain whose iteration Lebesgue measures tend to zero, cannot lead to hyperbolic attractors because, according to [5, Theorem 2.2], the complement of a hyperbolic one-dimensional attractor of diffeomorphism of a two-dimensional sphere consists of not less than four domains—this precludes

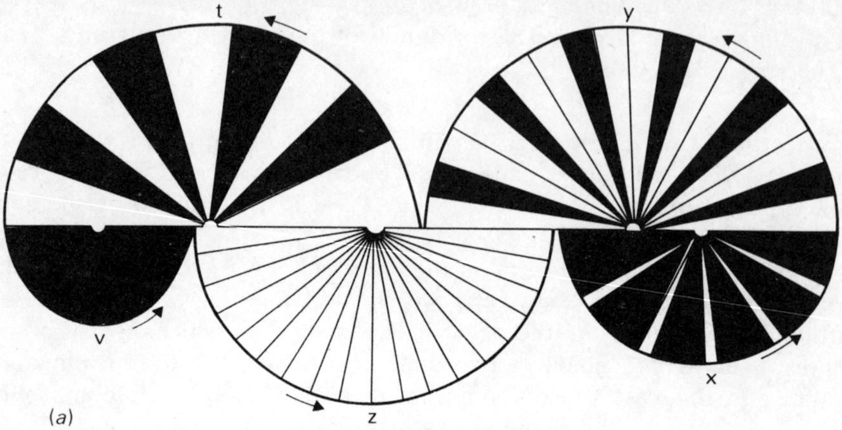

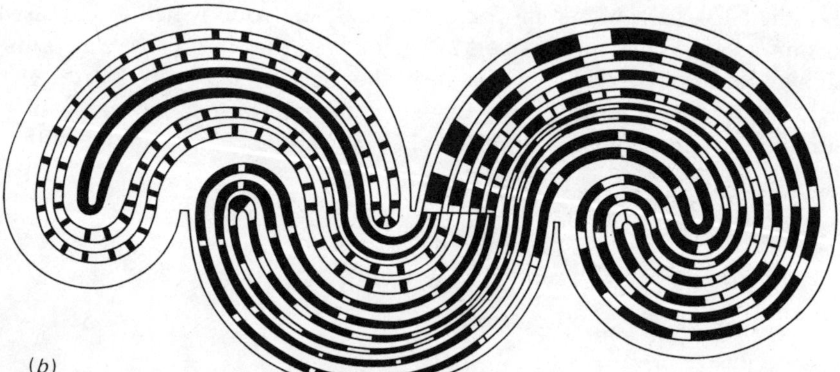

Fig. 13.2

any significant decrease in the iteration Lebesgue measure of a simply connected domain comprising such an attractor.

The next theorem characterizes the orientability property of an attractor Λ as the property of attainable boundary and gives an idea of the topology of a manifold on which a diffeomorphism with an orientable attractor of codimension 1 exists.

Theorem 13.2. *Let dim* $M = n \geqslant 3$ *and* Λ *be an expanding attractor of codimension* 1 *of the diffeomorphism* $f: M \to M$. *The orientability of* Λ *is equivalent to the fact that the set* $\mathrm{D}(M \backslash \Lambda)$ *consists only of* 2*-connections. If* Λ *is orientable and* $\mathrm{D}(M \backslash \Lambda)$ *consists of k connections of power* 2, *and a*

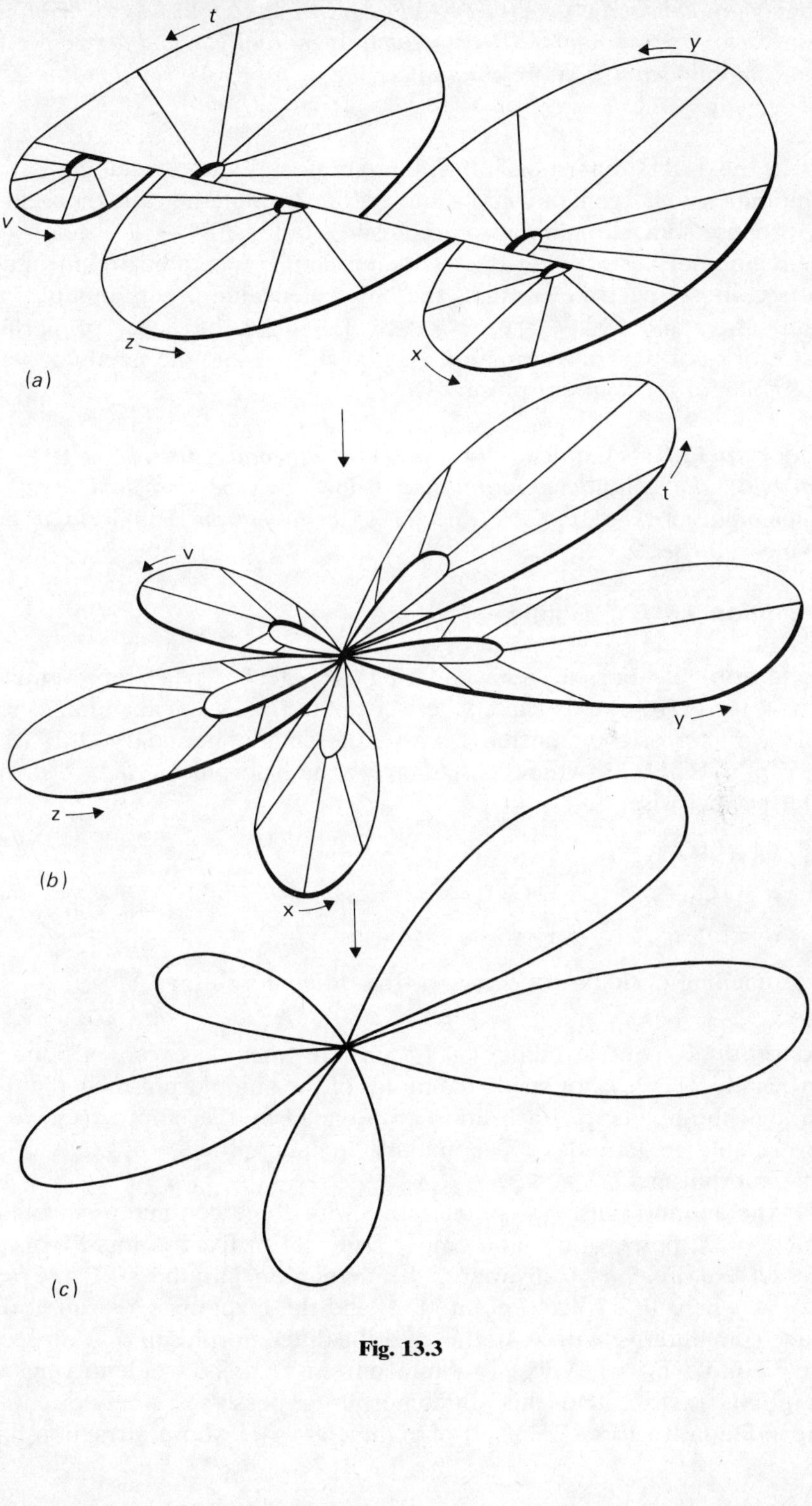

Fig. 13.3

neighbourhood $U(\Lambda)\supset\Lambda$ can be found in the manifold M which has the homotopic type of n-torus, from which k mutually non-intersecting sets homeomorphic to n-ball are eliminated, the group $\pi_1(M, x_0)$ contains Z^n as a subgroup.

The next theorem substantiates Examples 1 and 2 and gives their multidimensional generalization, but before formulating this theorem let us propose some simplifying prerequisites. Let $A: T^n \to T^n$ be a hyperbolic automorphism of torus of codimension 1 generated by an integer matrix whose spectrum contains the only eigenvalue β with modulus less than unity, and let $\mathcal{P}=\{x_1^i, x_2^i, \ldots, x_k^i\}$, $i=\overline{1, l}$, be a set of periodic trajectories of the transformation A. Let $\mathcal{L}$ designate the family of stable manifolds of the automorphism A.

Theorem 13.3. *A structurally stable DA-diffeomorphism $f^{A,\mathcal{P}}: T^n \to T^n$, satisfying the conditions formulated below, can be associated with an automorphism A of torus $T^n, n\geq 2$, of codimension 1 and to a finite A-invariant set $\mathcal{P}$.*

Condition 1. $f^{A,\mathcal{P}}$ is diffeotopic to A.

Condition 2. For a nonwandering set $\Omega(f^{A,\mathcal{P}})$ a representation $\Omega(f^{A,\mathcal{P}})=\Lambda\cup\bigcup_{i=1}^l q_i$ is valid, where $\Lambda=\Lambda(A,\mathcal{P})$ is an expanding orientable attractor of codimension 1 with an attainable boundary $D(T^n\backslash\Lambda)=\bigcup_{i=1}^l\bigcup_{j=1}^{k_i}(C_j^i\cup\tilde{C}_j^i)$, where C_j^i, $\tilde{C}_j^i$ are homeomorphic to R^{n-1}, do not intersect pairwise, and

$$C_{j+1}^i\cup\tilde{C}_{j+1}^i=f^{A,\mathcal{P}}(C_j^i\cup\tilde{C}_j^i),\quad j<k_i$$
$$C_1^i\cup\tilde{C}_1^i=f^{A,\mathcal{P}}(C_{k_i}^i\cup\tilde{C}_{k_i}^i),$$
$$q_i=\{q_1^i, q_2^i, \ldots, q_{k_i}^i\}$$

is a repelling periodic trajectory of the diffeomorphism $f^{A,\mathcal{P}}$.

Condition 3. Stable manifolds $f^{A,\mathcal{P}}/\Lambda$ are lines $\mathcal{L}$, except a finite set $\{L_j^i\}$, $i=\overline{1, l}$; $j=1, k_i$, of stable manifolds of the automorphism A that pass through the points q_j^i. Each line L_j^i is divided by the point q_j^i and forms two stable manifolds of boundary points $p_j^i\in C_j^i$, $\tilde{p}_j^i\in\tilde{C}_j^i$ of the diffeomorphism $f^{A,\mathcal{P}}$.

If the automorphism is commutative with some continuous involution $\theta: T^n\to T^n$, possessing a non-empty finite set of fixed points Fix θ, and the set $\mathcal{P}\supset\text{Fix }\theta$ is θ-invariant, the involution has the structure $\theta x=-x+a$, where a is a fixed point of A and the mapping $f^{A,\mathcal{P}}$ can also be made commutative with θ. In this case the diffeomorphism $f^{A,\mathcal{P},\theta}$ arises on the manifold $T_\theta^n=T^n\backslash\text{Fix }\theta\,|_\theta$ obtained from $T^n\backslash\text{Fix }\theta$ by identifying pairs of points $(x, \theta x)$, and this diffeomorphism possesses a nonorientable expanding attractor $\Lambda(A, \mathcal{P}, \theta)$ of codimension 1, whose attraction basin

$W^s\Lambda(A, \mathcal{P}, \theta)$ coincides with $T^n_\theta \backslash \pi_\theta \bigcup_{i=1}^l q_i$ where $\pi_\theta : T^n \backslash \mathrm{Fix}\, \theta \to T^n_\theta$ is a two-sheeted covering with the covering transformation group generated by the involution θ.

Let us note that this theorem generalizes Williams's theorem [4]. The structure of orientable attractors of codimension 1 is explained by the next theorem whose proof is presented in [8] and which reveals the fundamental character of Example 13.1 suggested by Smale.

Theorem 13.4. *If Λ is an orientable expanding attractor of codimension 1 of the diffeomorphism $f: M \to M$, $\dim M \geqslant 3$, there exists a finite-point compaction $\check{W}^s\Lambda$ of the attraction basin $W^s\Lambda$, homeomorphic to n-torus, and a homeomorphism $\check{f}: \check{W}^s\Lambda \to \check{W}^s\Lambda$ continuing f and topologically similar to some DA-diffeomorphism.*

This theorem reduces the question of classification or orientable attractors, together with their attraction basins, to the problem of topological conjugation of DA-diffeomorphisms. The solution of the latter problem, presented in [8], is given by the following theorem.

Theorem 13.5. *Two DA-diffeomorphisms $f^{A_1 \mathcal{P}_1}$, $f^{A_2 \mathcal{P}_2}$ are topologically conjugated if and only if there exists such a linear mapping $\psi: T^n \to T^n$, being the superposition of the torus automorphism and a shift, that $A_2 = \psi \cdot A_1 \cdot \psi^{-1}$, $\mathcal{P}_2 = \psi \mathcal{P}_1$.*

The next theorem permits one to reduce the study of nonorientable attractors of codimension 1 to orientable ones which arise due to two-sheeted covering of the attraction basin of an attractor.

Theorem 13.6. *Let $\dim M = n \geqslant 2$ and Λ be a nonorientable expanding attractor of codimension 1 of the diffeomorphism $f: M \to M$ such that the attainable boundary $D(M \backslash \Lambda)$ possesses only connections of powers not higher than two. Then there will be a two-sheeted covering $\pi: \tilde{W}^s\Lambda \to W^s\Lambda$ and covering diffeomorphism $\tilde{f}: \tilde{W}^s\Lambda \to \tilde{W}^s\Lambda$ for which $\pi \cdot \tilde{f} = f \cdot \pi$ and $\pi^{-1}\Lambda$ is a hyperbolic orientable attractor of codimension 1.*

Theorem 13.6 forms the basis for a classification of nonorientable attractors along with their attraction basins. This classification is expressed by the theorem formulated below. Let us consider a set Γ of triples $(A, \mathcal{P}, \theta)$, each consisting of a hyperbolic automorphism A of codimension 1 of n-torus and a linear involution θ of the same torus, which is commutative with A and possesses a non-empty set of fixed points $\mathrm{Fix}\, \theta$ and is contained in a finite set $\mathcal{P}$ invariant relative to both θ and A. Let us introduce in Γ the equivalence relation $\sim$, assuming that $(A_1, \mathcal{P}_1, \theta_1) \sim (A_2, \mathcal{P}_2, \theta_2)$ if there exists such a linear mapping ψ of torus, which is a superposition of the automorphism and group shift, that $A_2 = \psi \cdot A_1 \cdot \psi^{-1}$, $\theta_2 = \psi \cdot \theta_1 \cdot \psi^{-1}$, $\mathcal{P}_2 = \psi \mathcal{P}_1$. The equivalence class of the triple $(A, \mathcal{P}, \theta)$

will be called the class of the attractor $\Lambda(A, \mathcal{P}, \theta)$ defined by Theorem 13.3.

Theorem 13.7. *A certain class of the attractor $\Lambda(A, \mathcal{P}, \theta)$ corresponds to each nonorientable expanding attractor Λ of codimension* 1 *of the diffeomorphism $f: M \to M$,* dim $M \geq 2$, *satisfying the conditions of* Theorem 13.6. *Any two diffeomorphisms f_1, f_2 acting on manifolds of the same dimension, possessing nonorientable attractors of codimension* 1, *bounded within attraction basins of their attractors, are topologically conjugated if and only if the classes of their attractors coincide.*

Let us note in conclusion the relationship between the topological structure of a manifold M and the possible number of hyperbolic attractors and repellers of codimension 1.

Theorem 13.8. *If there exists a diffeomorphism $f: M \to M$ of a smooth manifold M,* dim $M \geq 3$, *possessing a hyperbolic attractor or repeller of codimension* 1, *then $\pi_1(M, x_0) \supset Z$.*

It follows, in particular, from Theorem 13.8 that the existence of diffeomorphism of n-sphere, $n \geq 3$, or of n-dimensional space possessing a hyperbolic expanding attractor (or contractible repeller) of codimension 1 is impossible. This result has been formulated in [9].

The properties of attractors of particular dynamical systems with continuous time, which are described by systems of ordinary differential equations in R^n, can often be understood from the behaviour of Poincaré mappings localized on sets homeomorphic to R^{n-1}. One may conclude that such systems do not contain hyperbolic attractors (repellers) of codimension 1.

13.3 Theorem on invariant foliations

Let Λ be an expanding attractor of codimension 1 *of diffeomorphism $f: M \to M$,* dim $M = n \geq 2$, *and let the attainable boundary* $\mathrm{D}(M \backslash \Lambda)$ *consist of k 2-connections. The manifold $W^s\Lambda$ has the homotopic type of n-torus with k punctured points, and there exists a finite-point compactification $\check{W}^s\Lambda \supset W^s\Lambda$ with the homotopic type of n-torus, the limitation of diffeomorphism f on $W^s\Lambda$ being continued into homeomorphism $\check{f}: \check{W}^s\Lambda \to \check{W}^s\Lambda$.*

The manifold $\check{W}^s\Lambda$ admits two $\check{f}$-invariant foliations: one of dimension 1 with leaves homeomorphic to R, which includes stable manifolds of points of Λ as leaves; the other of codimension 1 with leaves homeomorphic to R^{n-1}, which includes unstable manifolds of points of Λ as leaves.

If, moreover, a continuous involution $\varphi: W^s\Lambda \to W^s\Lambda$, without fixed points and commutative with f, is acting on $W^s\Lambda$, it can be extended to the continuous involution $\check{\varphi}: \check{W}^s\Lambda \to \check{W}^s\Lambda$ commutative with f.

If in each φ-invariant domain generated by the φ-invariant connection of the attainable boundary $D(M\backslash\Lambda)$ one can find a φ-invariant manifold transversal to one-dimensional stable manifolds of points of Λ, which, when complemented by a point added during compaction, is homeomorphic to R^{n-1} and invariant with respect to some iteration of the diffeomorphism f, then the above foliations can be made $\check{\varphi}$- and $\check{f}$-invariant, and the foliation of codimension 1 will contain, as leaves, these φ-invariant manifolds complemented by periodic points added during compaction $W^s\Lambda$.

Note on the proofs of the theorems

For outlines of the proofs of Theorems 13.1 and 13.2 and the theorem on invariant foliations, see Plykin [8]. The proof of Theorem 13.8 is based on the Van Kampen theorem (Massey [10]).

Proofs of Theorems 13.3, 13.6 and 13.7 are available direct from the publishers on request. While originally communicated with this chapter, they have been omitted from the book due to lack of space.

Acknowledgements

I am greatly indebted to the late Professor V. M. Alexeev, to Professor D. V. Anošov, Dr L. A. Bunimovich, Professor S. P. Novikov, Professor Ya. G. Sinai and Dr D. B. Fuks for fruitful discussions, and also to my postgraduate students D. A. Kamaev, Yu. I. Ustinov and A. Yu. Zhirov for help in its preparation.

I express my deep gratitude to Professor G. I. Barenblatt whose remarks and advice prompted changes to the final form of the chapter.

References

1. Plykin, R. V., The topology of basic sets of Smale diffeomorphisms (in Russian), *Matematichesky Sbornik* (N.S.), **84**(126), 301–312, 1971 (*Math. USSR-Sb.*, **13,** 297–307).
2. Williams, R. F., Expanding attractors, *Publications de l'Institut des Hautes Etudes Scientifiques*, **43,** 161–203, 1974.
3. Smale, S., Differentiable dynamical systems, *Bulletin of the American Mathematical Society*, **13,** 747–817, 1967.
4. Williams, R. F., The DA-maps of Smale and structural stability, *Proceedings of the Berkeley Symposium on Global Analysis*, **14,** 329–334, 1970.

5. Plykin, R. V., Sources and sinks of A-diffeomorphisms of surfaces (in Russian) *Matematichesky Sbornik* (N.S.), **94**(136), 243–264, 336, 1974.
6. Plykin, R. V., Kamaev, D. A. and Zhirov, A. J., One-dimensional attractors of A-diffeomorphisms on S^2 and diffeomorphisms with infinitely many sinks, *Societé Mathématique de France, Asterisque*, **51,** 355–394, 1978.
7. Henon, M., A two-dimensional mapping with a strange attractor, *Communications in Mathematical Physics*, **50,** 69–77, 1976.
8. Plykin, R. V., On hyperbolic attractors of diffeomorphisms (in Russian), *Uspekhi Matematicheskikh Nauk*, **35**(3–213), 94–104, 1980.
9. Kölmer, H., On hyperbolic attractors of codimension one, *Communications in Mathematical Physics*, **N597,** 330–334, 1976.
10. Massey, W. S., *Algebraic Topology*: *An Introduction*: New York–Chicago–San Francisco–Atlanta, Harcourt, Brace and World, Inc., 1967.

14 The Intermittent Transition to Turbulence

Y Pomeau

14.1 Introduction

Following the work of Ruelle and Takens [1], a number of investigators have tested the idea of a 'strange attractor' for turbulent physical systems, both in theory and experimentally. One of the most impressive successes of this approach has been its ability to describe many details of the onset of turbulence in experiments. In thermoconvection in normal liquid helium, Libchaber and Maurer [2] have observed the cascade of period doubling predicted by Feigenbaum [3] to occur before the onset of turbulence. Another case where theory and experiment fit nicely together is the transition via intermittency [4, 5].

Before explaining what this sort of transition is, I have to make clear the possibility of a sharp 'onset of turbulence'. This requires an unambiguous definition of turbulence or randomness. In turbulent flows one has to consider randomness in space and/or in time. This is intimately associated with the asymptotic decay of correlation functions: memory or 'coherence' is lost in a practical sense after long times or at large distances. Randomness in space is relevant only for flows (or systems) of great extent, in one dimension at least. In what follows I shall be concerned with time-dependent phenomena almost exclusively. In fluid mechanics one may limit oneself to time-dependence for flows in everywhere-bounded geometries, as in Rayleigh–Bénard convection in a brick-shaped box, for instance. In this sort of geometry, fluctuations at different places are so strongly interconnected that only a few degrees of freedom exist. This description also applies to unsteady stirred chemical reactions [5, 6]. If stirring is efficient enough, the concentration of the reagents does not depend on position and the relevant degrees of freedom are the uniform concentrations of the various chemical species in the reaction.

Much progress has recently been made in the study of the turbulent behaviour of deterministic systems with several degrees of freedom, and it is within this framework that intermittent transmission is considered below. Nevertheless, intermittent transition to turbulence is a more widely occurring phenomenon: it was observed by Osborne Reynolds himself in his famous studies on pipe flows [7]. This type of transition has since been the subject of many experimental studies. In my final remarks I give some indication of the theoretical side of this particular problem.

14.2 General description of intermittent transitions

Intermittency at transition is quite common in such dynamical systems as nonlinear differential equations [8] and iterated maps of the real line, the circle or higher-dimensional manifolds. The general picture of this intermittent behaviour is as follows: when a control parameter (say, ε) has a value that is less than critical (say, <0), the behaviour over time is regular (it may be steady or periodic). As ε becomes positive, but while it is still very small, the behaviour is very similar to that observed at $\varepsilon<0$, except that large fluctuations take place from time to time. These large fluctuations finally lead to recovery of the seemingly regular behaviour, and so on. As ε approaches 0_+ there is a sort of continuity [9], in that the mean interval between two bursts (or large fluctuations) increases indefinitely.

It should be noted, however, that this does not imply that the transition at $\varepsilon=0$ is one between ordered and chaotic motion. We can only say that the invariant measure describing the time average [10] is continuous in some weak sense at the intermittency threshold. In particular, the support of this measure may change completely as ε crosses 0. For instance, this support may be a fixed point at $\varepsilon=0_-$ and a strange attractor [11], or a closed curve at $\varepsilon=0_+$.

But other possibilities exist. To make this last point clear a few specific possible intermittent transitions are now considered.

14.2.1 Transition from a stable steady state

Consider a velocity field in $\mathbb{R}^d$, that is, a vector $\mathbf{X}(t)\in\mathbb{R}^d$ and a smooth vector field $\mathbf{f}(\mathbf{X}, \lambda)$ which depends on a deformation parameter λ. The time behaviour of $\mathbf{X}(t)$ is given by

$$\frac{\mathrm{d}\mathbf{X}}{\mathrm{d}t}=\mathbf{f}(\mathbf{X}, \lambda). \tag{14.1}$$

Let $\mathbf{X}_\lambda$ be a fixed point of this flow, that is,

$$\mathbf{f}(\mathbf{X}_\lambda, \lambda)=0.$$

The local stability of this fixed point depends on the tangent flow, i.e., on the eigenvalues of the $d\times d$ matrix

$$F_\lambda\equiv\frac{\partial\mathbf{f}(\mathbf{X}_\lambda, \lambda)}{\partial\mathbf{X}} \tag{14.2}$$

Let $z_1, \ldots, z_d$ be eigenvalues of F_λ with $Re\, z_1 \geqslant Re\, z_2 \geqslant \ldots \geqslant Re\, z_d$. We assume that there is no multiple eigenvalue, two real parts being equal for a pair of complex conjugate eigenvalues only.

The fixed point $\mathbf{X}_\lambda$ loses its stability for a value λ_c of λ such that

$$Re[z_1(\lambda_c)] = 0.$$

This defines a manifold of codimension one in the space of all smooth vector fields in $\mathbb{R}^d$.

Bifurcation theory [12] tells us that, if the first nonlinear terms in the expansion of $f(\mathbf{X}, \lambda)$ around $\mathbf{X}_\lambda$ and $\lambda = \lambda_c$ have the right sign and if $\text{Im}\, z_1 = -(\text{Im}\, z_2) \neq 0$ at $\lambda = \lambda_c$, then a small stable limit cycle bifurcates from $\mathbf{X}_\lambda$ at $\lambda = \lambda_c$. This is the standard case of the so-called normal or supercritical Hopf bifurcation [12]. We shall not consider this any further, but the case where the Hopf bifurcation is inverted or subcritical, and where $\text{Im}\, z_1 \neq 0$ at $\lambda = \lambda_c$.

For λ larger than λ_c, and if the parametrization by λ defines a transverse crossing of the bifurcation manifold, the fixed point becomes an unstable node on the surface defined locally by the eigenvectors of the pair of eigenvalues $(z_1, z_2 = z_1^*)$. Thus any trajectory starting near $\mathbf{X}_\lambda$ in the plane of these eigenvectors diverges along a logarithmic spiral.

This describes the vicinity of the two fixed points $x = y = \pm\sqrt{b(r-1)}$, $z = r - 1$ of the Lorenz model:

$$\frac{dx}{dt} = \sigma(y - x), \tag{14.3a}$$

$$\frac{dy}{dt} = -xz + rx - y, \tag{14.3b}$$

$$\frac{dz}{dt} = +xy - bz. \tag{14.3c}$$

When r becomes larger than $[\sigma(\sigma + b - 1)]/(\sigma - b - 1) \equiv r_T$, the fixed points $z = r - 1$ and $x = y = \pm[b(r-1)]^{1/2}$ become two unstable nodes. It is easy to show that this is a subcritical bifurcation for the values of b and σ chosen by Lorenz ($b = \frac{8}{3}$, $\sigma = 10$).

It is also well known [13] that this does not give rise to an intermittent behaviour for $r \geqslant r_T$. It seems likely that intermittency exists for the Lorenz model when another condition is imposed on the parameters (beyond the condition $r \simeq r_T |_+$) The reason for this is as follows: as shown by Lorenz [8], the time behaviour of his model is well described by a one-dimensional map near $r \simeq r_T$. Consider the map giving the distance to one of the fixed points, say

$$x_0 = y_0 = [b(r-1)]^{1/2} \qquad z_0 = r - 1$$

$$\rho \equiv [(x - x_0)^2 + (y - y_0)^2]^{1/2}$$

at each crossing of the trajectory through the plane $z = r - 1$ with $(dz/dt) > 0$.

For $r \leq r_T$, $\rho = 0$ is a stable fixed point of this map.

For $r \geq r_T$, $\rho = 0$ is still a fixed point, but it is unstable. Hence the map $\rho_{n+1}(\rho_n)$ roughly takes the form shown in Fig. 14.1, and it is clear that $\rho = 0$ is never visited in the iterations if $\varepsilon > 0_+$, since the region $\rho \simeq 0_+$ has no pre-image (no ρ_n exists such that $\rho_{n+1}(\rho_n) \sim 0$ in Fig. 14.1(*b*)). Around the fixed point there is a 'hole' that is not visited during chaotic motion. The radius of this hole is determined at once from the one-dimensional map $\rho_{n+1}(\rho_n)$, see Fig. 14.1(*b*).

Nevertheless an intermittent transition from a stable fixed point to a chaotic behaviour is possible on a manifold of codimension one in space of smooth flows in $\mathbb{R}^d$ with $d \geq 4$. I shall now try to justify this statement. As explained below, this study was motivated by observations made [14] on the unsteady Belousov–Zhabotinski reaction.

Let us consider a situation described by equations of the general form (14.1), and a bifurcation where the first pair of complex eigenvalues of F_λ has a positive real part although the second pair (i.e., z_3 and $z_4 = z_3^*$) still has a negative real part and a nonvanishing imaginary part at λ_c. We assume that the other 'degrees of freedom' (i.e., those corresponding to eigenvalues $z_5, \ldots, z_d$) are irrelevant. In this situation the fixed point is an *unstable* node along the $2d$ eigenspace of (z_1, z_2) and a *stable* node in the $2d$ eigenspace of (z_3, z_4). Consider now what happens in the vicinity of the fixed point. Let (ρ, r) be the (positive) distances measured from the fixed point in the stable and unstable manifolds respectively. Consider furthermore the Poincaré map defined locally by some discrete time map. This yields a pair of iteration laws for (ρ, r) that reads (in dimensionless form)

$$\rho_{n+1} = \mu\rho_n + (\text{H.O.T.}) \tag{14.4a}$$

$$r_{n+1} = (1+\varepsilon)r_n + r_n^3 + (\text{H.O.T.}) \tag{14.4b}$$

where (H.O.T.) stands for 'higher order terms' with respect to the amplitudes of r and ρ, as ρ is measured along the stable manifold, $0 < \mu < 1$. Furthermore, $r = 0$ is a stable (unstable) fixed point for the map (14.4b) if $\varepsilon < 0$ (resp. $\varepsilon \geq 0$). The plus sign of the cubic term in (14.4b) implies that this bifurcation is subcritical.

System (14.4) defines a mapping of the vicinity of the point (0, 0) in the quarter plane $(r > 0, \rho > 0)$. By a convenient rescaling of the coordinates near this fixed point this mapping can be reduced to its linear part in the square $S \equiv [0, 1] \times [0, 1]$. Consider now a standard set of iterates (in short, 'trajectory') in S when $0 < \varepsilon \ll 1$.

Such a standard trajectory starts from $r = r_0$, $0 < r_0 < 1$ and $\rho_0 = 1$. It will tend first in a geometric fashion to $\rho = 0$. If one increases the distance to $\rho = 0$ on a scale η $(0 < \eta \ll 1)$ then the vicinity of $\rho = 0$ will be reached after $(\ln \eta / \ln \mu)$ iterations. Then, because of the very weak instability along the r-direction (recall that $0 < \varepsilon \ll 1$), the number of the iterations $(= n_{\text{esc}})$ needed to escape from S (i.e., to get $r_n > 1$), is such that

$$r_0 \exp(\varepsilon n_{\text{esc}}) = 1$$

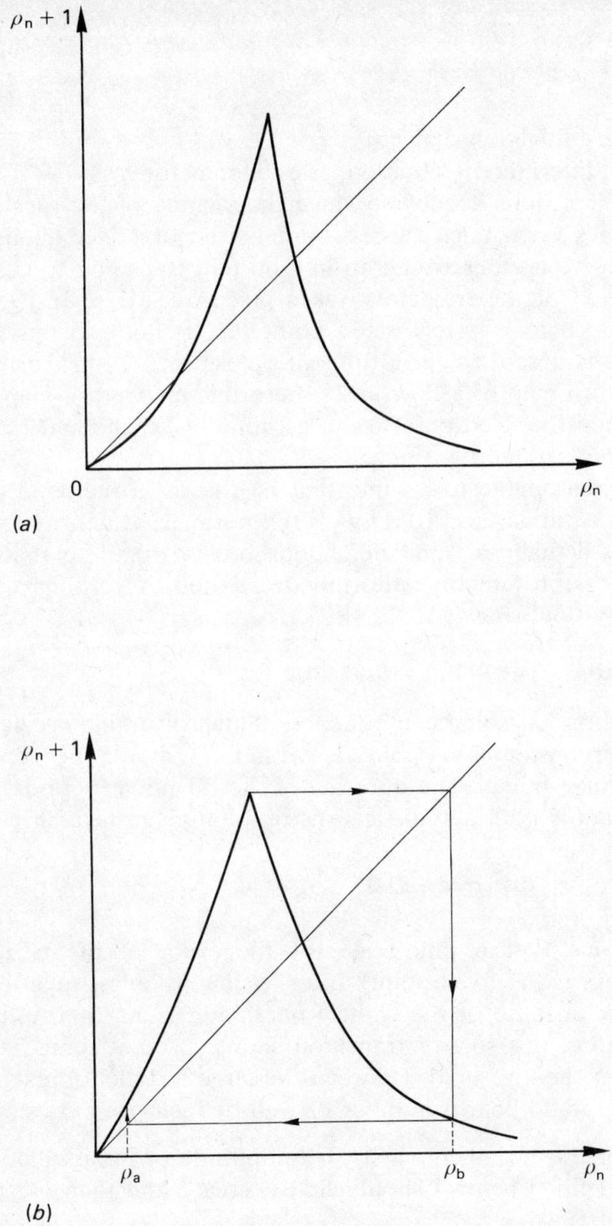

Fig. 14.1 (*a*) Diagrammatic view of the map $\rho_{n+1}(\rho_n)$ in the Lorenz model at $r \leqslant r_T$. ρ is the distance from a trajectory to one of the fixed points when it crosses the plane $z = r - 1$. The curve shown is obtained when this fixed point is linearly stable: the slope of $\rho_{n+1}(\rho_n)$ at $\rho = 0_+$ is less than 1
(*b*) As (*a*), except that $r \geqslant r_T$ so that $\rho = 0$ is an unstable fixed point of the map $\rho_n \to \rho_{n+1}$. The interval $[\rho_a, \rho_b]$ is globally stable under the map. It is bounded by the first two iterates of the maximum of the function $\rho_{n+1}(\rho_n)$

or

$$n_{\text{esc}} \simeq -\frac{1}{\varepsilon}(\ln r_0), \tag{14.5}$$

r_0 being the initial value of r $(0 < r_0 < 1)$.

This is an intermittent situation, as defined before: as $\varepsilon \to 0_+$, the mean number of iterations needed to leave the vicinity of the (unstable) fixed point diverges. As r_0 fluctuates, this number is not a fixed quantity. Let us assume that we are describing an intermittent transition to chaos, thus a randomization of the trajectory takes place outside S. In [9] a strange attractor which may have a fixed point that is both an unstable and a stable node is described for a diffeomorphism in $\mathbb{R}^4$ (this corresponds to the first return map of a flow in $\mathbb{R}^5$; here this first return map should be drawn around the fixed point itself, as done by Ruelle and Lanford [15] for the Lorenz system).

It seems reasonable to assume that r_0 (that is, a radius in a plane) is distributed at random on $[0, 1]$ with the normalized density $2r_0\,\mathrm{d}r_0$. The square S is actually a small neighbourhood of $(0, 0)$ so that the entry points in S are presumably uniformly distributed. Accordingly, the probability distribution for n_{esc} is

$$\mathrm{d}n_{\text{esc}}P(n_{\text{esc}}) = 2\varepsilon\exp(-2\varepsilon n_{\text{esc}})\,\mathrm{d}n_{\text{esc}}. \tag{14.6}$$

This prediction has some consequences. Small diverging oscillations (i.e., where the representative point is within S) should be separated by large-amplitude bursts. The duration of these laminar periods, say τ, is a random quantity with a statistical distribution given near threshold by

$$Q(\tau)\,\mathrm{d}\tau = \frac{2\varepsilon}{T}\exp(-2\varepsilon\tau/T)\,\mathrm{d}\tau,$$

T being some typical time-scale (of the order of the period of the oscillations near the fixed point) and ε a dimensionless quantity proportional to the distance of the control parameter from its threshold value.

Furthermore, this sort of transition should present some well-defined features near the threshold. Between the large turbulent bursts, the small oscillations should bear a number of well-defined properties:

(i) just after the turbulent burst, the amplitude of the oscillations ($\sim$the ρ-variable defined before) should first decrease and then increase up to the next burst;
(ii) the frequencies of the two sets of oscillations (after/before the burst or decreasing/increasing) should be different and independent of each other (in general, $|\mathrm{Im}\, z_1| \neq |\mathrm{Im}\, z_3|$);
(iii) at threshold, the decrement of the decaying oscillations remains finite, although the increment of the amplified oscillations tends to zero;
(iv) the mean duration of a laminar period (between two bursts) increases as ε^{-1} as $\varepsilon \to 0_+$, ε having the same meaning as before. If this transition

is observed in computer experiments, the inverse Lyapunov number increases also as ε^{-1} near threshold (except perhaps for 'renormalization' phenomena, as those observed in type 2 intermittency [16]).

Properties (i), (ii) and (iii) have been seen [14, 17] in the Belousov–Zhabotinski reaction in the chaos–steady state transition at large renewal times.

14.2.2 Intermittent transition from a limit cycle

With Paul Manneville, I have studied [11, 16] the intermittent transition from a periodic to a chaotic state. This transition is continuous in the sense that the Lyapunov number grows continuously from zero (in the periodic case) to a vanishing small value at the onset of turbulence. This sort of transition has been studied already in some detail and only a few points are presented here.

(1) Intermittency appears typically as a qualitative change in time behaviour, but the transition is not necessarily from periodic to chaotic. In the case of diffeomorphisms of the circle consider, for instance, the disappearance of a pair of stable/unstable fixed points. As in [18], and as shown in Fig. 14.2, after the disappearance of the pair of fixed points the iteration will spend a lot of time in a 'channel region'. Nevertheless there is certainly no chaos in this case, since the rotation number still depends continuously on the deformation parameter [19]. The behaviour of a flow should be either periodic or quasiperiodic in the continuous case; the large bursts should occur in a periodic fashion, their periods being either

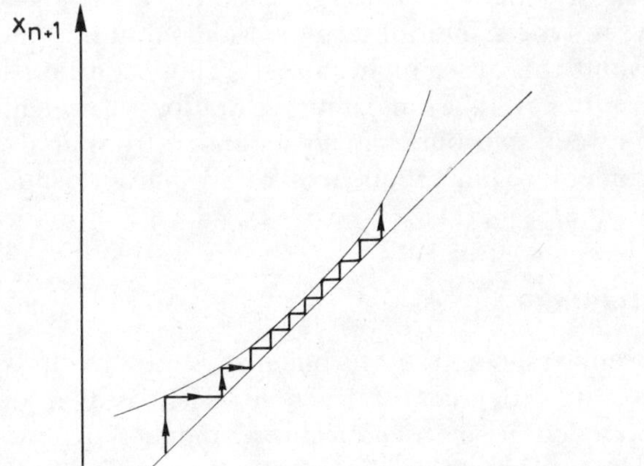

Fig. 14.2 The channel picture of Type 1 intermittency. The curve $x_{n+1}(x_n)$ is slightly above the first bisector and, as iteration proceeds, many points lie in the channel between curve $x_{n+1}(x_n)$ and the first bisector

commensurate (rational rotation number) or incommensurate (irrational rotation number) with the underlying oscillations.
(2) In the much studied quadratic map of $\mathbb{R}$, $x \to 1-ax^2$ (and, more generally in maps, $x \to \lambda f(x)$, where f is a smooth function with a single hump and λ is the deformation parameter) an intermittent behaviour exists just before the first occurrence of a stable period, whenever this stable period does not arise from a period-doubling bifurcation. For instance, for the map of $\mathbb{R}$, $x \to 1-ax^2$, a increasing, period three appears first for $a=\frac{7}{4}$ and period four for $a=\frac{3}{4}(1+4^{1/3})$. When a is just a little smaller than these critical values, an intermittent behaviour is observed: most of the time the iterates (i.e., the successive points x_0 arbitrary in $[-1,+1]$, $x_1=1-ax_0^2, \ldots, x_n=1-ax_{n-1}^2, \ldots$) are concentrated around three (if $a \leqslant \frac{7}{4}$) or four (if $a \leqslant \frac{3}{4}(1+4^{1/3})$) small intervals and drift slowly across a channel when iteration proceeds. Nevertheless, for these values of a the Lyapunov number is still a very erratic function of a, as is well known for this sort of map, and an extremely complicated bifurcation pattern is observed with a structure of internal similarity in the parameter space [20]. This is in sharp contrast to the continuous dependence of the Lyapunov number with respect to the deformation parameter observed in [16]. The difference obviously arises from the expanding or nonexpanding character of the maps outside the channel.
(3) A periodic motion may become unstable in three different ways [12]: the Floquet stability multiplier ($\in \mathbb{C}$) may cross the unit circle either through $+1$, -1 or with a nonzero imaginary part. When it crosses through $+1$, the periodic trajectory is lost and locally one gets the channel picture—this is what we have called 'type 1 intermittency'. Meunier *et al.* [22], by linearizing a homographic map, have been able to prove in a well defined case most statements already made in [11, 16] about this type 1 intermittency. In type 2 intermittency, periodic (but unstable) trajectories still exist beyond the onset of instability. Thus, contrary to the case of type 1 intermittency, the maximum duration of seemingly periodic oscillations between two bursts is not bounded from above, and even in the fully turbulent regime sequences of regular oscillations still occur from time to time.

14.3 Final remarks

The intermittent transition to turbulence is now well understood for systems with a few degrees of freedom, such as that describing fluid motions in bounded geometries or stirred chemical reactions. Nevertheless, the best known example of intermittent transition to turbulence is in parallel flows [7, 22]—the so-called 'transition flows'. The structure of these transition flows is usually attributed to the subcritical character of the instabilities described in their linear version by the Orr–Sommerfeld equations. However, the precise connection between intermittency at the onset and subcriticality of the instability needs to be made more precise.

Intermittency in transition flows probably has little to do with linear stabililty analysis. In pipe flows, for instance, it seems likely that *localized, steady* and *finite-amplitude* fluctuations can exist with a given relative velocity along the flow lines (say along the x-variable) when the Reynolds number reaches a well defined value, dependent on the relative velocity of the fluctuation. If one considers the fluid equations as a dynamical system where the variable x plays the role of time, this sort of localized fluctuation is represented by an exceptional trajectory starting from and returning to the undistributed parallel flow at $x \to +\infty$. The fact that such trajectories exist in a manifold of codimension one (which fixes the value of the Reynolds number for a given relative velocity) in space of all flows is still true for a dynamical system in function space (one must consider function spaces as the local state of the system, i.e., the value of all functions for a given x is an element of an infinite dimension space spanned by all admissible functions of the space variables independent of x). This helps one to understand how a well defined *nonlinear* stability criterion may exist and why, in transition flows, the bursts tend to be widely separated at the threshold.

This picture, if true, needs to be made more precise. It could provide, for instance, a detailed structure for critical fluctuations in concrete examples and the behaviour of the intermittency parameter at the threshold.

References

1. Ruelle, D. and Takens, F., *Communications in Mathematical Physics*, **20,** 167, 1971.
2. Libchaber, A. and Maurer, J., *Journal de Physique*, **41**(C3), 51, 1980.
3. Feigenbaum, M., *Journal of Statistical Physics*, **19,** 25, 1978; *Physics Letters*, **74A,** 375, 1979; *Communications in Mathematical Physics*, **77,** 65, 1980.
4. Bergé, P., Dubois, M., Manneville, P. and Pomeau, Y., *Journal de Physique—Lettres*, **41,** L341, 1980.
5. Pomeau, Y., Roux, J. C., Vidal, C., Bachelart, S. and Rossi, A., *Journal de Physique—Lettres*, **42,** L271, 1981.
6. Ruelle, D., *Transactions of the New York Academy of Sciences*, **35,** 66, 1973.
7. Reynolds, O., *Philosophical Transactions*, **174,** 935, 1883.
8. Lorenz, E. N., *Journal of the Atmospheric Sciences*, **20,** 130, 1963.
9. Pomeau, Y., in Gresillon, D. and Laval, G. (Eds), *Intrinsic Stochasticity in Plasmas*: Orsay, Editions de Physique, 1979.
10. Ruelle, D., in Helleman, R. (Ed.), Nonlinear dynamics, *Annals of the New York Academy of Sciences*, **357,** 1, 1980.
11. Pomeau, Y. and Manneville, P., *Physica*, **1D,** 219, 1980.

12. Iooss, G., *Bifurcation of Maps and Applications*, Mathematical Studies No. 36: Amsterdam and New York, North Holland 1979.
13. Eframovich, V. S., Bikov, V. V. and Shil'nikov, L. P., *Soviet Physics—Doklady*, **22,** 253, 1977; see also Fig. 18 in Rabinovich, M. I., *Soviet Physics—Uspekhi*, **21,** 441, 1978.
14. Roux, J. C., Vidal, C., Bachelart, S. and Rossi, A., *Physica*, **2D,** 395, 1981.
15. Lanford, O. E., III, 1976 *Duke Turbulence Conference*, Duke University Mathematics Series No. 3, 1977.
16. Pomeau, Y. and Manneville, P., *Communications in Mathematical Physics*, **74,** 189, 1980.
17. Vidal, C. and Roux, J. C., private communication.
18. Pomeau, Y. and Manneville, P., *Physics Letters*, **75A,** 1, 1979.
19. The study of diffeomorphisms of the circle and of flows on the torus has a long and glorious history, going back to Poincaré. Many results in this field are summarized in Herman, M. R., *Publications Mathématiques de l'Institut des Hautes Etudes Scientifiques*, **49,** 5, 1979. New and profound results are also proved in his thesis.
20. Derrida, B., Gervois, A., and Pomeau, Y., *Annales de l'Institut Henri Poincaré*, **29,** 305, 1978 and *Journal of Physics*, **A12,** 369, 1979; Collet, P. and Eckmann, J. P., *Iterated Maps of the Interval as Dynamical Systems*, Progress in Physics Vol. 1: Basel and Boston, Birkhäuser, 1980.
21. Meunier, C., Bussac, M. N. and Laval, G., Intermittency at the onset of stochasticity in nonlinear resonant coupling process, *Physica*, **4D,** 236, 1982; Meunier, C., Thesis, University of Paris 6, 1981.
22. Wygnanski, I. J. and Champagne, F. H., *Journal of Fluid Mechanics*, **59,** 281, 1973 (inc. references).

15 Resonance and Bifurcation in Hopf–Landau Dynamical Systems

George R Sell

15.1 Background

Turbulence is a phenomenon that is observed in certain dynamical systems that depend upon one or several real parameters. Couette flow and Bénard flow are fairly simple physical systems in which the onset of turbulence may be seen as the underlying physical parameters change. From a mathematical point of view, it is known that the onset of turbulence is closely connected with a series of bifurcations that occurs as the parameters change. Therefore in order to get a better understanding of the nature of turbulence in a given system, one should study this series of bifurcations.

About forty years ago Hopf [1] and Landau [2] speculated that the series of bifurcations that precedes the onset of turbulence is a series of quasiperiodic motions which bifurcate by developing additional fundamental frequencies (*cf.* [3]). Since the existence of (nonperiodic) quasiperiodic solutions is very unstable (i.e., it can be destroyed with arbitrarily small changes), it is not likely that turbulence will develop in this fashion (*cf.* [4–8]).

We feel that, in their description of the onset of turbulence, Hopf and Landau were restricted by the mathematical syntax of their times. With the advantage of an additional forty years of mathematical insight, we believe that a more appropriate description of the onset of turbulence is given in terms of the existence of invariant manifolds rather than the existence of quasiperiodic solutions. In order to make this more precise, let us recall that if $\varphi(t)$ is a quasiperiodic solution (of an autonomous differential equation) with k fundamental frequencies, then the hull (i.e., orbit closure) $H(\varphi)$ is a k-dimensional torus T^k. The bifurcation that occurs when a quasiperiodic solution develops an additional fundamental frequency (or period) can then be described as a bifurcation of T^k into T^{k+1} [9–11]. We propose that the series of bifurcations envisioned by

Hopf and Landau may be described schematically as

$$T^0 \to T^1 \to T^2 \to T^3 \to T^4 \to T^5 \to \ldots,$$

where T^0 denotes a fixed point and the notation $T^k \to T^{k+1}$ denotes a Hopf–Landau bifurcation. It should be emphasized that our reformation of the Hopf–Landau conjecture about the onset of turbulence is more general than theirs because we do not require that the flow on T^k $(k \geqslant 2)$ be quasiperiodic. Nevertheless, the nature of the flow on T^k at the instant of bifurcation is of considerable interest and we shall say more about that later.

In 1971 Ruelle and Takens [5] offered another speculation about the onset of turbulence. They envisioned a short series of bifurcations which ends with the development of a strange attractor. In schematic form the Ruelle–Takens conjecture about the onset of turbulence can be described by the series

$$T^0 \to T^1 \to T^2 \to \ldots \to T^N \to \text{strange attractor} = \text{turbulence},$$

where N is to be thought of as small.

We showed in [12] that the Ruelle–Takens conjecture and the (extended) Hopf–Landau conjecture are not incompatible. Strange attractors can develop; however, *the occurrence of a strange attractor is, in itself, not an obstacle to a higher-order Hopf–Landau bifurcation* $T^k \to T^{k+1}$ *for any* $k \geqslant 2$. Since a Hopf–Landau bifurcation $T^k \to T^{k+1}$, even in the presence of strange attractors, is possible, this phenomenon should be studied further.

In this chapter we study a class HL of dynamical systems, which we call Hopf–Landau dynamical systems, that depend on a real parameter ε, where ε takes values in an open interval I. We restrict our attention, first, to those Hopf–Landau dynamical systems that are *normally singular.* This means that the system is normally hyperbolic for all ε in I, with the exception of a single (critical) value ε_i which we normalize to be $\varepsilon_i = 0$. Thus the system is normally hyperbolic for all $\varepsilon \neq 0$.

The family of normally singular Hopf–Landau dynamical systems is a rather small collection from the set of all dynamical systems that depend on ε. It is known [9–11] that for a special class BIF of normally singular Hopf–Landau systems there is a Hopf–Landau bifurcation $T^k \to T^{k+1}$ at $\varepsilon = 0$. This fact has some rather surprising repercussions which are of interest in relation to numerical studies or experimental investigations. In order to explain this we need to digress and look at some questions in mathematical modelling.

A differential equation (E) that describes a physical system should be viewed as only an approximation of the actual equations (E_a) of motion. Thus (E_a) lies in some neighbourhood U of (E). Presumably U is 'small'. A theorem of special interest for analysis of physical systems would then have the following form:

If system (E) satisfies Hypothesis H, then there is a neighbourhood V of (E) such that every system ($\hat{E}$) in V has Property P.

Of course, in order to use this theorem to study (E_a) one would also need an approximation theorem which compares the neighbourhoods U and V. (One would need to show that $U \subseteq V$.)

We show below that for every differential system (E) in the class BIF there is a neighbourhood V of (E) with the property that *every* differential system in V has a *nearly instantaneous* Hopf–Landau bifurcation $T^k \to T^{k+1}$. By an (ordinary) Hopf–Landau bifurction $T^k \to T^{k+1}$ one means that a system (E) has a normally hyperbolic invariant T^k for all ε on one side of the critical value ε_c and this bifurcates to a normally hyperbolic T^{k+1} for all ε on the other side of ε_c. The bifurcation is said to be *supercritical* if T^k occurs for $\varepsilon < \varepsilon_c$ and T^{k+1} occurs for $\varepsilon > \varepsilon_c$. Otherwise the bifurcation is said to be *subcritical*. In the case of a nearly instantaneous bifurcation, the point set $\{0\}$ is replaced by a small interval J. Thus, in the supercritical case, an arbitrary system ($\hat{E}$) from the neighbourhood V would have a normally hyperbolic T^k for ε to the left of J and a normally hyperbolic T^{k+1} for ε to the right of J. The behaviour of the system ($\hat{E}$) for ε in J may be very complicated.

The precise statement of this result is given in Theorem 15.2 below. We shall see that the size of the neighbourhood V will depend on the interval J, also that in every such neighbourhood V there will exist dynamical systems that are replete with strange attractors.

Theorem 15.2 is especially useful for applied scientists who are doing numerical studies or experimental investigations of dynamical systems that depend on a real parameter ε. In such a situation a study will be made for the system at discrete values of ε, say for $\varepsilon_1 < \varepsilon_2 < \ldots$. The theory described here helps to indicate what to look for in these experimental or numerical studies.

The theory presented in this chapter is a preliminary report: further investigations, particular in the related problems in approximation theory, are needed. These results will be published elsewhere as they are developed.

15.2 Normal singularity

Consider the differential system

$$X' = f(X, \varepsilon) \tag{15.1}$$

that depends smoothly on X and ε, where X lies in some Euclidean space R^n and ε is a real parameter in an open interval I. We assume that Eq.

(15.1) has a family $\tau(\varepsilon)$ of k-dimensional invariant tori that varies smoothly in ε for ε in I.

The key to studying the behaviour of the flow near $\tau(\varepsilon)$, as ε varies, lies in the study of the linearized flow near $\tau(\varepsilon)$ [4, 6–8, 11, 13–15]. In particular one wants to determine whether or not the flow near $\tau(\varepsilon)$ is normally hyperbolic. In order to do this we need to determine the tangential spectrum $\Sigma_{T,\varepsilon}$ and the normal spectrum $\Sigma_{N,\varepsilon}$ of $\tau(\varepsilon)$ [8, 13]. The manifold $\tau(\varepsilon)$ is normally hyperbolic if there are real numbers $0 \leq a < b$ such that

$$\begin{aligned} |\lambda| &\leq a, \qquad \text{for all } \lambda \in \Sigma_{T,\varepsilon}, \\ |\lambda| &\geq b, \qquad \text{for all } \lambda \in \Sigma_{N,\varepsilon}. \end{aligned} \tag{15.2}$$

(Recall that if $k = \dim \tau(\varepsilon) \geq 1$, then one always has $\lambda = 0$ in $\Sigma_{T,\varepsilon}$.)

If, at $\varepsilon = \varepsilon_0$, $\tau(\varepsilon_0)$ is normally hyperbolic and if $\hat{f}(X)$ is 'close to' $f(X, \varepsilon_0)$ in some C^P-topology, then the perturbed equation $X' = \hat{f}(X)$ has an invariant k-dimension torus $\hat{\tau}$ that is 'close to' $\tau(\varepsilon_0)$, [4, 7, 14]. Furthermore, $\hat{\tau}$ is also normally hyperbolic. The smoothness of $\hat{\tau}$ depends (i) on the degree of smoothness of f and $\hat{f}$, (ii) on the number P and (iii) on the ratio b/a [4, 7, 14].

It follows that, if at $\varepsilon = \varepsilon_0$ the torus $\tau(\varepsilon_0)$ is normally hyperbolic, then for all ε in a neighbourhood J of ε_0 the manifolds $\tau(\varepsilon)$ are normally hyperbolic. Furthermore, for all ε in J there are no Hopf–Landau bifurcations. Also if, for some ε in J, Eq. (15.1) has a strange attractor near $\tau(\varepsilon)$, then the attractor must lie in $\tau(\varepsilon)$ [12].

We shall say that Eq. (15.1) is *normally singular* on the interval I if there is single value ε_c in I with the property that $\tau(\varepsilon)$ is normally hyperbolic for all ε in I with $\varepsilon \neq \varepsilon_c$. Thus the normal hyperbolicity of $\tau(\varepsilon)$ breaks down at $\varepsilon = \varepsilon_c$, which means that Conditions (15.2) are not satisfied at $\varepsilon = \varepsilon_c$. (The value ε_c is called the singular point.) There are several ways in which this breakdown of normal hyperbolicity can occur. A discussion of these scenarios is rather technical and we will defer that until Section 15.4. Since our interest in this chapter is to study the Hopf–Landau bifurcation $T^k \to T^{k+1}$, we focus on the one scenario which gives rise to this type of bifurcation. This leads us to the concept of a Hopf–Landau dynamical system.

15.3 Hopf–Landau dynamical systems

Let us return for the moment to the situation studied in the last section. Here we are given the differential system

$$X' = f(X, \varepsilon) \tag{15.1}$$

with a family $\tau(\varepsilon)$ of k-dimensional invariant tori where $X \in R^n$ and $\varepsilon \in I$. Furthermore, we assume that the tori $\tau(\varepsilon)$ are normally singular on I with $\varepsilon_c = 0$ being the singular point. The concept of a Hopf–Landau dynamical system, which we now define, is simply a way of introducing suitable local coordinates in the vicinity of $\tau(\varepsilon)$. The question of deriving these local coordinates is discussed briefly in Section 15.4 (see also [11, Section V]).

A differential equation on $R^2\times R^l\times T^k$ of the form

$$\begin{aligned} x' &= A_{11}(\varphi, \varepsilon)x + \varepsilon A_{12}(\varphi, \varepsilon)z + F(x, z, \varphi, \varepsilon) \\ z' &= \varepsilon A_{21}(\varphi, \varepsilon)x + [B(\varphi) + \varepsilon A_{22}(\varphi, \varepsilon)]z + H(x, z, \varphi, \varepsilon) \\ \varphi' &= G(x, z, \varphi, \varepsilon) \end{aligned} \tag{15.3}$$

will be called a *Hopf–Landau dynamical system* provided that the following conditions are satisfied:

(i) all the functions in Eq. (15.3) are C^∞-functions in x, z, φ and ε, where $x\in R^2$, $z\in R^l$, $\varphi\in T^k$ and $\varepsilon\in I$;
(ii) the torus T^k is invariant, i.e., $(F, H)=(0, 0)$ when $(x, z)=(0, 0)$ and

$$\varphi' = G(0, 0, \varphi, \varepsilon) \tag{15.4}$$

denotes the flow on $\tau(\varepsilon)$ for ε in I;
(iii) A_{ij} and B are matrices of the appropriate dimensions and $|(F, H)| = 0(|x|^2+|z|^2)$ as $(x, z)\to(0, 0)$;
(iv) the linear equation

$$z' = B(\varphi)z, \tag{15.5}$$

where φ is any solution of Eq. (15.4), admits an exponential dichotomy over $\tau(\varepsilon)$ for all ε in I,
(v) the spectrum of the linear equation

$$x' = A_{11}(\varphi, \varepsilon)x, \tag{15.6}$$

where φ is any solution of Eq. (15.4), lies on different sides of 0 for $\varepsilon<0$ and for $\varepsilon>0$.

We let HL denote the collection of all Hopf–Landau dynamical systems.

The relationship $n=2+l+k$ suggests the connection between Eq. (15.1) and Eq. (15.3). The fact that x is a two-dimensional vector is an assumption which is explained further in Section 15.4.

We let $\mathscr{D}$ denote the collection of all C^∞-dynamical systems defined on $R^2\times R^l\times T^k$ for ε in I. Thus $\mathrm{HL}\subseteq\mathscr{D}$. For various integers $N\geqslant 0$ we consider $\mathscr{D}$ with the appropriate C^N-topology. In particular we are interested in studying all dynamical systems from $\mathscr{D}$ that lie in fixed C^N-neighbourhood of a given Hopf–Landau dynamical system.

We say that a given Hopf–Landau dynamical system is *nonresonant* if the following three conditions are satisfied:

(i) $A_{11}(\varphi, 0)=\begin{pmatrix} 0 & -\omega_0 \\ \omega_0 & 0 \end{pmatrix}$;

(ii) $G(x, z, \varphi, \varepsilon)=\bar{\omega}+\varepsilon L(x, z, \varphi, \varepsilon)$ for some k-dimensional vector $\bar{\omega}$;

(iii) there exist constants $c>0$, $\gamma>0$ such that the $(k+1)$-dimensional vector $\omega=(\omega_0, \bar{\omega})$ satisfies

$$|n\cdot\omega|\geqslant c\,|n|^{-\gamma} \tag{15.7}$$

for all $n\in Z^{k+1}$ with $n\neq 0$.

All other Hopf–Landau dynamical systems are said to be *resonant.* The resonant systems include Hopf–Landau systems with strange attractors on $\tau(\varepsilon)$ at $\varepsilon\neq 0$, provided that $k\geqslant 3$ [16]. The class of all nonresonant Hopf–Landau dynamical systems will be denoted by NR.

Our first result is a consequence of the bifurcation theorem proved in [11].

Theorem 15.1. *For every $N\geqslant 3$ there is an open subset* BIF *of* NR *in the C^N-topology with the following properties*:

(I) *For each system* (E) *in* BIF *there is a Hopf–Landau bifurcation $T^k\to T^{k+1}$ at $\varepsilon=0$. (Let us assume, for simplicity that the bifurcation $T^k\to T^{k+1}$ is supercritical and let us denote the $(k+1)$-dimensional torus by $\hat{\tau}(\varepsilon)$. Also let $\hat{\Sigma}_{T,\varepsilon}$ and $\hat{\Sigma}_{N,\varepsilon}$ denote the tangential and normal spectra of $\hat{\tau}(\varepsilon)$.)*

(II) *For $\varepsilon>0$ the $(k+1)$-dimensional torus $\hat{\tau}(\varepsilon)$ is normally hyperbolic. More precisely the following hold*:

(1) *For every $M\geqslant 1$ there is a K_M such that one has*

$$|\lambda|\leqslant K_M\,|\varepsilon|^M, \qquad \text{for all } \lambda\in\hat{\Sigma}_{T,\varepsilon}. \tag{15.8}$$

(2) *There is a K_0 such that one has*

$$|\lambda|\geqslant K_0\,|\varepsilon|, \qquad \text{for all } \lambda\in\hat{\Sigma}_{N,\varepsilon}. \tag{15.9}$$

The existence of the set BIF with property (I) above is proved in [15]. We can give a rough idea of how BIF is defined, without going into great detail. Let Eq. (15.3) be given and define

$$W=\frac{\partial}{\partial\varepsilon}A_{11}|_{\varepsilon=0}=\begin{pmatrix} w_{11} & w_{12}\\ w_{21} & w_{22}\end{pmatrix}$$

$$\bar{W}=M_\theta[w_{11}\cos^2 2\pi\theta_0+w_{22}\sin^2 2\pi\theta_0]$$

where θ_0 is the angular coordinate in the x-space R^2, M_θ is the mean value operator on T^{k+1} and $\theta=(\theta_0,\varphi)\in T^{k+1}$. Similarly, by looking at appropriate third-order nonlinear terms and by computing the mean value, one defines a constant K (cf. [11, Eq. (4.10)]). BIF is then the set

$$\text{BIF}=\{\text{Hopf–Landau systems: } \bar{W}\neq 0, K\neq 0\}.$$

In order to prove the normal hyperbolicity of $\hat{\tau}(\varepsilon)$, i.e., in order to verify (15.8) and (15.9), one first notes that the equation of motion for the flow on $\hat{\tau}(\varepsilon)$ is given by

$$\theta'=\omega+\varepsilon V(\theta,\varepsilon). \tag{15.10}$$

By using the method of averaging together with (15.7), one can obtain a coordinate system on $\hat{\tau}(\varepsilon)$ so that the flow is given by (15.10) where $|V| = 0(\varepsilon^M)$ as $\varepsilon \to 0$. The verification of (15.8) now follows from standard estimates on the spectrum [8, 13]. The proof of (15.9) is a consequence of the Hopf bifurcation theory as it applies to Eq. (15.3) [11].

Notice that the smoothness of $\hat{\tau}(\varepsilon)$ is bounded by the ratio $(K_0\varepsilon/K_M\varepsilon^M) \to \infty$ as $\varepsilon \to 0$, provided that $M > 1$. Thus we see that the $(k+1)$-dimensional torus is generally smoother for smaller values of ε.

The next result describes the behaviour of all dynamical systems in a full C^N-neighbourhood of a given system in BIF. The proof is a consequence of Theorem 15.1.

Theorem 15.2. *Let $N \geq 3$ be given and let* (E) *be a Hopf–Landau dynamical system in* BIF. *Assume that the Hopf–Landau bifurcation $T^k \to T^{k+1}$ is supercritical. Then there is an $\varepsilon_2 = \varepsilon_2(E) > 0$ with the following property:*

For every ε_1, $0 < \varepsilon_1 < \varepsilon_2$, there is a full C^N-neighbourhood $\mathcal{U}$ of (E) *in $\mathcal{D}$ such that for every dynamical system* (E) *in $\mathcal{U}$ there is a normally hyperbolic T^k for $-\varepsilon_2 < \varepsilon < -\varepsilon_1$ and a normally hyperbolic T^{k+1} for $\varepsilon_1 < \varepsilon < \varepsilon_2$.*

A similar statement holds when the Hopf–Landau bifurcation is subcritical.

In [16] it is shown that in every C^N-neighbourhood of a nonresonant Hopf–Landau dynamical system there is an Axiom A strange attractor. It then follows from Theorem 15.2 that there exist Hopf–Landau dynamical systems with strange attractors near $\varepsilon = 0$ and with the property that there is a nearly instantaneous Hopf–Landau bifurcation $T^k \to T^{k+1}$.

15.4 Open problems

The concept of normal singularity in the vicinity of an invariant manifold has not been studied extensively. The Hopf–Landau bifurcations $T^k \to T^{k+1}$, for $k = 0, 1, 2, \ldots$, illustrate only one aspect of this problem [5, 6, 9–11, 15, 17, 18]. There are other cases which should be analysed.

The inequality (15.2) can fail either with $0 < b$ and $a = b$, or with $b = 0$. In the first case the induced normal flow admits an exponential dichotomy but the growth rates (i.e., Lyapunov exponents) in the tangential and normal directions have equal strength. It would be interesting to find sufficient conditions, in this case, which guarantee that a bifurcation cannot occur near the singularity.

In the second case, where $b = 0$, one must have $\lambda = 0 \in \Sigma_{N,0}$, i.e., the normal flow does not have an exponential dichotomy. For this case, let I_0 denote the spectral interval in $\Sigma_{N,0}$ that contains $\lambda = 0$ and let $\mathcal{V}_0$ denote the corresponding spectral subspace. Also let l denote the dimension of a fibre in $\mathscr{V}_0$, then $l \geqslant 1$. Finally let γ denote the minimum angle between $\mathscr{V}_0$ and TM, the tangent bundle to $\tau(0)$.

The analysis of the Hopf–Landau bifurcation which is given in [9–11] treats the case that $\gamma > 0$ and $l = 2$. For example, a derivation of the Hopf–Landau coordinate system Eq. (15.3) is given in [11] in this case. The x-coordinate in Eq. (15.3) is basically a coordinate in the fibre of $\mathscr{V}_0$.

It would be interesting to study the behaviour near $\tau(0)$ for other values of l, at least under the assumption that $\gamma > 0$. As far as we know, no one has studied this problem when $\dim \tau(0) = k \geqslant 2$.

Another problem arises in the definition and use of normal singularity. We assumed, in our definition, that the singularity ε_c occurs at an interior point of the open interval I. This means that there exists a (normally hyperbolic) invariant manifold $\tau(\varepsilon)$ for ε on both sides of the singular point. A more appropriate assumption would be that the singularity occurs at an end-point of I. In this situation, one could study problems where the existence of a (normally hyperbolic) invariant manifold $\tau(\varepsilon)$ was known for ε only on one side of ε_c, say for $\varepsilon < \varepsilon_c$. For example, in the conclusion of Theorem 15.2 we are given a family of normally hyperbolic $(k+1)$-dimensional tori on the interval $\varepsilon_1 < \varepsilon < \varepsilon_2$. One would like to study these tori as ε increases and approaches an end-point singularity. Does there exist a nearly instantaneous Hopf–Landau bifurcation $T^{k+1} \rightarrow T^{k+2}$ in this case?

One approach to handling the end-point singularity is to find sufficient conditions for the normally hyperbolic manifold $\tau(\varepsilon)$ to persist as ε passes through such a singularity. In [19] Chenciner and Iooss give an informative analysis of this approach.

Note added in proof

An interesting illustration of a nearly instantaneous Hopf–Landau in the presence of resonance is described in [20].

Acknowledgements

Sincere appreciation is expressed to A. Chenciner and M. Herman for some very helpful comments and suggestions.

References

1. Hopf, E., A mathematical example displaying features of turbulence. *Communications in Pure and Applied Mathematics*, **1,** 303ff., 1948.
2. Landau, L. D., On the problem of turbulence, *Doklady Akademy Nauk USSR*, **44,** 311ff., 1944.
3. Joseph, D. D., *Stability of Fluid Motions*, I, II, Springer Tracts in Natural Philosophy Vols 27, 28: New York, Springer-Verlag, 1976.

4. Fenichel, N., Persistence and smoothness of invariant manifolds for flows, *Indiana University Mathematics Journal,* **21,** 193–226, 1971/72.
5. Ruelle, D. and Takens, F., On the nature of turbulence, *Communications in Mathematical Physics,* **20,** 167–192; **23,** 343–344, 1971.
6. Sacker, R. J., *On Invariant Surfaces and Bifurcation of Periodic Solutions of Ordinary Differential Equations,* New York University Technical Report IMM-NYU, 1964.
7. Sacker, R. J., A perturbation theorem for invariant manifolds and Hölder continuity, *Journal of Mathematical Mechanics,* **18,** 705–762, 1969.
8. Sacker, R. J. and Sell, G. R., The spectrum of an invariant submanifold, *Journal of Differential Equations,* **38,** 135–160, 1980.
9. Chenciner, A. and Iooss, G., Bifurcations de tores invariants, *Archive for Rational Mechanics and Analysis,* **69,** 109–198, 1979.
10. Haken, H., Chaos and order in nature, in *Chaos and Order in Nature,* Proceedings of International Symposium on Synergetics: New York, Springer-Verlag, 1981, 2–11.
11. Sell, G. R., Bifurcation of higher dimensional tori, *Archive for Rational Mechanics and Analysis,* **69,** 199–230, 1979.
12. Sell, G. R., Hopf–Landau bifurcation near strange attractors, in *Chaos and Order in Nature,* Proceedings of International Symposium on Synergetics: New York, Springer-Verlag, 1981, 84–91.
13. Sacker, R. J. and Sell, G. R., A spectral theory for linear differential equations, *Journal of Differential Equations,* **27,** 320–358, 1978.
14. Hirsch, M. W., Pugh, C. C. and Shub, M., *Invariant Manifolds,* Lecture Notes in Mathematics No. 583: New York, Springer-Verlag, 1977.
15. Sell, G. R., The structure of a flow in the vicinity of an almost periodic motion, *Journal of Differential Equations,* **27,** 359–393, 1978.
16. Newhouse, S., Ruelle, D. and Takens, F., Occurrence of strange Axiom A attractors near quasiperiodic flows on T^m, $m \geq 3$, *Communications in Mathematical Physics,* **64,** 35–40, 1978.
17. Hopf, E., Abzweigung einer periodischen Lösung eines Differentialsystems, *Mathematisch-physikalische Bericht über die Verhandlungen der Sächsischen Akademie der Wissenschaften zu Leipzig,* **94,** 1–22, 1942 (English translation in Ref. 18 below, 163–193.)
18. Marsden, J. E. and McCracken, M., *The Hopf Bifurcation and its Applications,* Applied Mathematical Sciences No. 19: New York, Springer-Verlag, 1976.
19. Chenciner, A. and Iooss, G., Persistance et bifurcation de tores invariants, *Archive for Rational Mechanics and Analysis,* **71,** 301–306, 1979.
20. Meyer, K., Tori in resonance, *IMA Preprint Series* No. 13, Institute for Mathematics and its Applications, University of Minnesota, 1983.

16 Distinguishing Deterministic and Random Systems

F Takens

16.1 Introduction

16.1.1

In order to illustrate the type of problem we want to attack in this chapter, I first discuss an example. Suppose we have some physical apparatus which, once it is started, gives as experimental output after each unit of time a real number (we assume these real numbers to be uniformly bounded). So the output of running the experiment once is a bounded sequence $\{a_i\}$; the index i takes the values $0, 1, 2, \ldots,$—of course not infinitely many (the experiment has eventually to be stopped), but as many as we please. For the moment we just assume that in such sequences $A=\{a_i\}$ of experimental data, i runs through all natural numbers.

We are interested to know whether, on the basis of such a sequence, anything can be said about whether our apparatus is 'deterministic' or not. Of course, to make this question more precise, we need several definitions. We say that an 'apparatus' as above (i.e., giving a real number for each unit of time) admits a smooth deterministic model if there is a smooth differential equation, or vector field, $\dot{x}=X(x)$ on the phase space $\mathbb{R}^n$, i.e., $x, \dot{x}\in\mathbb{R}^n$ and $X:\mathbb{R}^n\to\mathbb{R}^n$ smooth, and a smooth function $f:\mathbb{R}^n\to\mathbb{R}$ such that

(i) for each observed sequence $A=\{a_i\}_{i=0}^{\infty}$ of experimental data there is a point $x_0\in\mathbb{R}^n$ such that $a_i=f[x_0(i)]$, where $t\mapsto x_0(t)$ is the solution (or integral curve) of the above differential equation (or vector field) starting in x_0, i.e., with $x_0(0)=x_0$;
(ii) for each $x_0\in\mathbb{R}^n$, the solution $x_0(t)$ for $t\geqslant 0$ is bounded.

In this case we call the apparatus *smoothly deterministic.* If only for some given sequence $A=\{a_i\}$, (X, f, x_0) as above can be constructed such that $a_i=f[x_0(i)]$, we say that A admits a *smoothly deterministic explanation.*

Suppose now that our apparatus is not deterministic, or rather, to be more explicit, that in the sequence $\{a_i\}_{i=0}^{\infty}$ of experimental data, each a_i is a random variable, independent of the other a_j's, and with uniform probability distribution on the interval $[0, 1]$. So the set of all possible sequences of experimental data has in this case the form $[0, 1]^{\mathbb{N}}$; the probability measure is the product of Lebesgue measures on $[0, 1]$. The fact that in this case our apparatus behaves differently from any apparatus having a smooth deterministic model is the content of:

Theorem 16.1. *There is a subset $\mathcal{B} \subset [0, 1]^{\mathbb{N}}$ of measure one such that no $B \in \mathcal{B}$ admits a smooth deterministic explanation.*

16.1.2

This type of question was motivated by the problem of understanding the onset of turbulence in the Taylor–Couette experiment. Several explanations were proposed for the transition to chaotic behaviour in this experiment. The careful measurements of Gollub, Swinney and collaborators [1] provide precise experimental data. The usual way to analyse these data is by transforming them into a power spectrum (i.e., the absolute value of the square of the Fourier transform, or the amount of energy per frequency). This power spectrum is useful for the analysis of, for instance, periodicity or quasiperiodicity, but for other aspects of the data it is less satisfactory: for example, there is no way to tell from the power spectrum the difference between a signal generated by a strange attractor (hence admitting a smoothly deterministic explanation) and a random signal. In this chapter the object is to find invariants, in terms of 'experimental data' which will be finite in the deterministic case and infinite in the nondeterministic case. Where possible we try to relate these invariants, for the case of smooth dynamical systems, to known invariants such as entropy, Lyapunov exponents and various dimensions.

16.1.3

Theorem 16.1 is a corollary of more general results which will be derived later. In order to show how a subset $\mathcal{B}$, as in Theorem 16.1, can be constructed, we proceed as follows. Let $A = \{a_i\}_{i=0}^{\infty}$ be some bounded sequence of real numbers. For $\varepsilon > 0$ and $n \in \mathbb{N}$ we define inductively $\mathscr{C}_{n,\varepsilon} \subset \mathbb{N}$ by

$$0 \in \mathscr{C}_{n,\varepsilon};$$

for $i > 0$, $i \in \mathscr{C}_{n,\varepsilon}$ if and only if for all $0 \leqslant j < i$ with $j \in \mathscr{C}_{n,\varepsilon}$, we have

$$\max\{|a_i - a_j|, |a_{i+1} - a_{j+1}|, \ldots, |a_{i+n} - a_{j+n}|\} \geqslant \varepsilon.$$

$C_{n,\varepsilon}$ denotes the number of elements of $\mathscr{C}_{n,\varepsilon}$; if confusion is possible, we write $C_{n,\varepsilon}(A)$ instead of $C_{n,\varepsilon}$. Since we assume A bounded, all $C_{n,\varepsilon}$ are finite.

The set $\mathscr{B}$ in Theorem 16.1 can be chosen as

$$\mathscr{B}=\left\{B\in[0,1]^{\mathbb{N}}\;\middle|\;\frac{\ln C_{n,\varepsilon}(B)}{n-\ln\varepsilon}\quad\text{not bounded for }(n,-\ln\varepsilon)\to\infty\right\}.$$

The fact that $\mathscr{B}$ has measure one is easy to verify: there is even a subset $\mathscr{B}'$ of $[0, 1]^{\mathbb{N}}$ of full measure such that for all $B'\in\mathscr{B}'$, $C_{n,\varepsilon}(B')\simeq(1/2\varepsilon)^n$. $\mathscr{B}'$ consists of all sequences $B'=\{b'_i\}_{i=0}^{\infty}$ such that for each $\varepsilon>0$ and each finite sequence $(\alpha_0,\ldots,\alpha_m)$, there is some i_0 such that $|\alpha_i-b'_{i_0+i}|<\varepsilon$ for all $i=0, 1,\ldots, m$.

The fact that the expression $(\ln C_{n,\varepsilon})/(n-\ln\varepsilon)$ is uniformly bounded as $(n-\ln\varepsilon)\to\infty$ for any sequence admitting a smooth deterministic explanation is proved in Section 16.3.

16.1.4

In the above discussion, some meaning has gradually accrued to 'deterministic' and 'nondeterministic'. These terms are considered here more carefully, as the precursors of other definitions in this chapter.

First of all it seems strange to require, in the definition of 'smoothly deterministic', that each positive integral curve $x_0(t)$, $t\geqslant 0$, of X is bounded. The fact is, however, that without this assumption Theorem 16.1 is wrong, as can be seen from the following example. Let $B=\{b_i\}_{i=0}^{\infty}$ be some (bounded) sequence. Then there is a smooth function $f:\mathbb{R}\to\mathbb{R}$ such that for all $i\in\mathbb{N}$, $f(i)=b_i$. Now, taking as differential equation $\dot{x}=1$ $(x\in\mathbb{R})$ and as initial point $x_0=0$, we see that, without the condition of bounded positive integral curves, any (bounded) sequence would admit a smoothly deterministic explanation.

Because of this boundedness requirement we shall deal in the following section only with compact phase spaces. This restriction is of course stronger than necessary; however, where only positive integral curves or orbits are bounded, or rather have compact closure, we may apply the theory directly to the closure of such a positive orbit.

Next we may ask how much differentiability of X and f we need. In fact we need none, only we need f and 'the flow of X' to be Lipschitz. For f this means that for all $x, y\in\mathbb{R}^n$, $|f(x)-f(y)|\leqslant K\cdot\|x-y\|$ for some constant K. For X this means that there is a continuous map $\Phi:\mathbb{R}^n\times\mathbb{R}\to\mathbb{R}^n$ that satisfies the following conditions:

Condition 16.1. $\Phi(x, 0)=x$.

Condition 16.2. $t\mapsto\Phi(x, t)$ is the unique integral curve of X through x.

Condition 16.3. For some constants L, M and all $x, y\in\mathbb{R}^n$ and $t\in\mathbb{R}$,

$$\|\varphi_t(x)-\varphi_t(y)\|\leqslant L\cdot\exp(M\,|t|)\cdot\|x-y\|,\qquad\text{where }\varphi_t(x)=\Phi(x,t).$$

From this we see that all conditions here can be expressed in terms of distance functions (in $\mathbb{R}^n$ and in $\mathbb{R}$). So, instead of defining smoothly deterministic in terms of a function and vector field on $\mathbb{R}^n$, we may define it in terms of a metric space P (replacing $\mathbb{R}^n$) with a Lipschitz function $f: P \to \mathbb{R}$, i.e., for some K and all $x, y \in P$, we have $|f(x) - f(y)| \leqslant K \cdot \rho(x, y)$, and with a Lipschitz one-parameter group $\{\varphi_t\}_{t \in \mathbb{R}}$ of homeomorphisms of P, i.e., φ_0 is the identity, $\varphi_t(x)$ is continuous in (x, t), $\varphi_t \circ \varphi_s = \varphi_{t+s}$ and:

Condition 16.3A. For some constants L, M and all $x, y \in P$, $t \in \mathbb{R}$,

$$\rho(\varphi_t(x), \varphi_t(y)) \leqslant L \cdot \exp(M|t|) \cdot \rho(x, y).$$

In order to incorporate the condition of 'bounded positive orbits' we required P to be compact. However, in order to be able to apply the above description to the closure of a positive integral curve we must allow $\{\varphi_t\}$ to be only a semi-group, i.e., only to be defined for $t \geqslant 0$, because on (the closure of) a positive integral curve, one cannot in general go backward in time. In some cases we shall even allow the semi-group $\{\varphi_t\}$ only to be defined for $t \in \mathbb{N}$; in all these cases φ_t need *not* be a homeomorphism.

16.1.5

There is one more condition we have to introduce in the $(P, \{\varphi_t\}, f)$ description of 'smooth determinism' before Theorem 16.1 holds in that generality. To see this we take any bounded sequence $\{a_i\} = A$ and construct a compact metric space P_A with Lipschitz function $f: P_A \to \mathbb{R}$, Lipschitz semi-flow $\{\varphi_t\}$ and initial point $\bar{x} \in P_A$ such that

$$\{f[\varphi_i(\bar{x})]\}_{i=0}^{\infty} = \{a_i\}_{i=0}^{\infty} \cdot P_A = \overline{\{(a_i, a_{i+1}, a_{i+2}, \ldots) \in \mathbb{R}^{\infty} \mid i \in \mathbb{N}\}},$$

where the closure is taken with respect to the 'metric'

$$\rho[(x_0, x_1, \ldots), (y_0, y_1, \ldots)] = \sum_{i=0}^{\infty} \frac{1}{2^i} |x_i - y_i|;$$

this is not really a metric on $\mathbb{R}^{\infty}$ but that presents no problem since our sequence A is bounded. P_A is clearly compact. For $t \in \mathbb{N}$, we define φ_t as

$$\varphi_t(x_0, x_1, x_2, \ldots) = (x_t, x_{t+1}, x_{t+2}, \ldots);$$

so φ_t is Lipschitz with Lipschitz constant $\leqslant 2^t$. If one wants to define φ_t for all $\mathbb{R}_+$, this can be done by a suspension construction [2].

Finally we take $\bar{x} = (a_0, a_1, a_2, \ldots)$ and $f(x_0, x_1, \ldots) = x_0$ and we have the required example.

The fact that in this case Theorem 16.1 does not hold is due to the fact that P_A in the above example is in general infinite-dimensional. Hence for the generalization of Theorem 16.1 we shall also require P, the phase space, to be finite-dimensional in the following sense.

Let C_ε, or $C_\varepsilon(P)$, be the minimal cardinality of a subset $S \subset P$ such that for any $p \in P$ there is some $s \in S$ with $\rho(p, s) < \varepsilon$; we define the *dimension* of P as

$$\dim(P) = \limsup_{\varepsilon \to 0} \frac{\ln C_\varepsilon(P)}{-\ln \varepsilon}$$

(note that this notion of dimension is quite different from the usual or topological dimension; this dimension is more related to the Hausdorff dimension; see also the examples in Section 16.4). Any compact subset of $\mathbb{R}^n$ has dimension $\leqslant n$, so the requirement $\dim(P) < \infty$ is not very severe.

16.1.6

One should notice that the dimension, in the above sense, is not determined by the topology only: if (P, ρ) is a compact metric space and we define a new metric ρ' on P by $\rho'(x, y) = [\rho(x, y)]^\alpha$, $0 < \alpha < 1$, then ρ' and ρ induce the same topology but $\dim(P, \rho) = \alpha \cdot \dim(P, \rho')$. On the other hand if we have a space P with two metrics ρ_1 and ρ_2 such that for some constant $K > 0$ and all $x, y \in P$,

$$K^{-1} \cdot \rho_1(x, y) \leqslant \rho_2(x, y) \leqslant K \cdot \rho_1(x, y), \tag{16.1}$$

then it makes no difference whether we compute the dimension of P using ρ_1 or using ρ_2. This means also that if (P_i, ρ_i), $i = 1, 2$, are two compact metric spaces and $h: P_1 \to P_2$ is a homeomorphism such that both h and h^{-1} are Lipschitz, then $\dim(P_1) = \dim(P_2)$. We call such a homeomorphism a *Lipschitz homeomorphism.* Two metrics ρ_1, ρ_2 on the same space P satisfying (16.1) above, are called *metrically equivalent.*

16.1.7

After all the detours we come to the following definition of 'admitting a continuously deterministic explanation' (because the differentiability has disappeared, 'smoothly' is changed to 'continuously'):

Definition 16.1. We say that a bounded sequence $A = \{a_i\}_{i=0}^\infty$ admits a *continuously deterministic explanation* if there is a compact metric space P, with $\dim(P) < \infty$, a Lipschitz function $f: P \to \mathbb{R}$, a Lipschitz semi-group $\{\varphi_t\}$ acting on P (i.e., satisfying Condition 16.3A above), and a point $x_0 \in P$ such that $a_i = f(\varphi_i(x_0))$.

The analog of Theorem 16.1 holds also with this definition; this follows from the somewhat sharper result, which we prove in Section 16.3.

Theorem 16.2. *If A admits a continuously deterministic explanation, then*

$$\frac{\ln C_{n,\varepsilon}(A)}{n - \ln \varepsilon}$$

is uniformly bounded for $(n, -\ln \varepsilon) \to \infty$.

16.1.8

After all the digressions about the meaning of 'deterministic', I have only a few things to say about 'nondeterministic'. This is mainly because of the lack of a simple and satisfactory description of this notion; in Section 16.3 there is an attempt. One difficulty in giving a good description is that although the case where each b_i in $B=\{b_i\}$ is an independent random variable with uniform distribution on [0, 1], as in Section 16.1.1, is certainly nondeterministic, the discrete analog, where b_i can only take a finite set of values, should be excluded. Namely, if $S \subset \mathbb{R}$ is some finite set and $A=\{a_i\}_{i=0}^{\infty}$ is some sequence with all $a_i \in S$, then A admits a continuously deterministic explanation!

16.1.9

In principle our result gives a method of distinguishing between deterministic and nondeterministic systems, but so far only in principle. Some of the complications are the following.

(a) In order to determine $C_{n,\varepsilon}(A)$ one needs the full sequence $A = \{a_0, a_1, \ldots\}$, which is never available. Define $\mathscr{C}_{n,\varepsilon,m}$ as $\mathscr{C}_{n,\varepsilon} \cap \{0, 1, \ldots, m-n\}$ and $C_{n,\varepsilon,m}$ as the number of elements of $\mathscr{C}_{n,\varepsilon,m}$. $C_{n,\varepsilon,m}$ is clearly determined by the first $m+1$ elements of A; for $m \to \infty$ $C_{n,\varepsilon,m}$ monotonously increases to $C_{n,\varepsilon}$. So if one knows only a finite part of the sequence A, say $\{a_0, \ldots, a_{m_0}\}$, then one may compute $C_{n,\varepsilon,m}$ for $m \leqslant m_0$ and decide whether this as a function of m is 'flat enough' for large m for the limit to be estimated with sufficient precision. In this way it is probably possible to find approximations for $C_{n,\varepsilon}$ if $(n, -\ln \varepsilon)$ is not too large.

(b) Next one has to determine whether $(\ln C_{n,\varepsilon})/(n-\ln \varepsilon)$ is bounded. Since we know $C_{n,\varepsilon}$ only for $(n, -\ln \varepsilon)$ sufficiently small, this is in principle impossible. One might however argue that if

$$\max_{\substack{n \leqslant k \\ -\ln \varepsilon \leqslant k}} \frac{\ln C_{n,\varepsilon}}{n-\ln \varepsilon},$$

as a function of k, stops growing well before the limits are reached where $C_{n,\varepsilon}$ can be computed, then the experimental data show evidence that the system is deterministic.

(c) On the whole the situation is more or less like the case where one wants to determine whether the ratio between two measured quantities is rational. Although this too is in principle impossible, it is still possible, for instance in resonance phenomena, to explain certain behaviour on the assumption that a (frequency) ratio is rational and then show (from experimental data) that this ratio is indeed a rational, say p/q where $1/q$ is much bigger than the experimental error.

The above numerical applications of our results to actual experimental data are not yet known to have been carried out with success since they were first proposed in the summer of 1980.

16.1.10

Another aspect of our results is the following. It will be shown that if X is a C^1 vector field on a compact manifold M satisfying certain generic (even 'open and dense') conditions, then there is an open and dense subset $\mathcal{F}_X \subset C^1(M)$ in the C^1-functions on M such that, for any $x_0 \in M$ and $f_1, f_2 \in \mathcal{F}_X$, the asymptotic behaviour of

$$\frac{\ln[C_{n,\varepsilon}(\{f_j[\varphi_i(x_0)]\}_{i=0}^{\infty})]}{n - \ln \varepsilon}$$

is the same for $j=1$ or 2; φ_t denotes the flow of X. To be more precise, for each $\lambda \in [0, \pi/2]$,

$$\lim_{\substack{n\to\infty \\ -\ln \varepsilon \to \infty \\ (-\ln \varepsilon/n)\to \mathrm{tg}\, \lambda}} \begin{matrix}\sup \\ \inf\end{matrix} \left(\frac{\ln[C_{n,\varepsilon}(A_j)]}{n - \ln \varepsilon}\right) = L^{\pm}(\lambda; A_j),$$

with $A_j = \{f_j[\varphi_i(x_0)]\}_{i=0}^{\infty}$ is the same for $j=1$ or 2 (in fact for $\lambda=0$ and $\lambda=\pi/2$ one has to take a slightly different limit, see Section 16.3). This means that, modulo genericity assumptions, we have suppressed the influence of f_j. For this reason we call expressions such as $L^{\pm}(\lambda; A)$ *dynamical invariants*; one should note that all metric invariants of generic (M, X, x_0), which only depend on the flow induced in the closure of the positive integral curve of X through x_0, can be expressed as dynamical invariants, see [3, 4] and Section 16.3. In Section 16.3 we also introduce a generalization of this invariant, denoted by $L^{\pm}(\lambda, \alpha; A)$, with $L^{\pm}(\lambda, 1; A) = L^{\pm}(\lambda; A)$. We show that $L^{+}(0; A) = L^{-}(0; A)$ is the entropy (in the sense of Adler, Konheim and McAndrew, see [5, 6]) of X restricted to the closure of the positive integral curve through x_0 and that $L^{+}(\pi/2; A)$ is the dimension, as defined in Section 16.1.5, of the closure of this positive integral curve. In Section 16.4 we calculate $L^{\pm}(\lambda, \alpha)$ for a number of examples; these examples suggest relations between $L^{\pm}(\lambda, \alpha)$, Lyapunov exponents and various dimensions, at least in the case where the closure of the positive integral curve is an Axiom A attractor.

16.1.11

Finally, some related publications should be mentioned. First there is a theorem of Aeyels [3] on observability of generic dynamical systems which is important when proving that certain expressions are dynamical invariants. Then there is the work of Frederickson, Kaplan and York [7, 8] on the dimensions of attractors; their way of numerically determining the dimension of an attractor is related to the calculation of $L^{+}(\pi/2; A)$ as we propose it. It follows from work of Mallet-Paret and Mañé [9, 10] that our results may be applicable to certain systems described by partial differential equations (and hence with an infinite-dimensional phase space): namely, these authors prove that solutions of such equations, considered as evolution equations on a Hilbert or Banach space, are, under certain conditions, all attracted to some finite-dimensional set (or even manifold); our results should then be applicable to these finite-dimensional sets. Such a reduction to finite dimension has already been carried out by Marsden [11, 12] for the special case of Hopf bifurcations.

16.2 Dynamical systems on compact metric spaces

16.2.1

The dynamical systems that we consider here consist of a phase space P, which is a compact finite-dimensional metric space, and a Lipschitz semi-group $\{\varphi_t\}$ acting on P; i.e. satisfying Condition 16.3A. The domain of t may be $\mathbb{R}_+$ or $\mathbb{N}$; in this last case we also write n instead of t. For such dynamical systems we define the number $C_{t,\varepsilon,\alpha}$. $t \geqslant 0$, $\varepsilon > 0$, $\alpha > 0$, or, if this may be confusing, $C_{t,\varepsilon,\alpha}(P, \{\varphi_t\})$ as follows: $C_{t,\varepsilon,\alpha}$ is the smallest cardinality of a subset $S \subset P$ such that for each $x \in P$, there is some $s \in S$ such that:

in case $\alpha \leqslant 1$,

$$\max_{\tau \in [0, t]} \{\alpha^{-t+\tau} \cdot \rho[\varphi_\tau(x), \varphi_\tau(s)]\} < \varepsilon;$$

in case $\alpha \geqslant 1$,

$$\max_{\tau \in [0, t]} \{\alpha^{\tau} \cdot \rho[\varphi_\tau(x), \varphi_\tau(s)]\} < \varepsilon.$$

In both cases one should of course take the maximum over those $\tau \in [0, t]$ for which φ_τ is defined. From these numbers we obtain

$$L^+(\lambda, \alpha) = \limsup_{\substack{(t, -\ln \varepsilon) \to \infty \\ (-\ln \varepsilon)/t \to \operatorname{tg} \lambda}} \frac{\ln C_{t,\varepsilon,\alpha}}{t - \ln \varepsilon};$$

for $\lambda = \pi/2$ this should be interpreted as $\limsup_{\varepsilon \to 0}$ (as we shall see, this is independent of t), and for $\lambda = 0$ this should be interpreted as $\lim_{\varepsilon \to 0} [\limsup_{n \to 0} (\ \ )]$. $L^-(\lambda, \alpha)$ is similarly defined with lim inf instead of lim sup; we sometimes write $L^{\pm}(\lambda, \alpha; P\{\varphi_t\})$ instead of $L^{\pm}(\lambda, \alpha)$.

16.2.2

We now derive some of the properties of $L^{\pm}(\lambda, \alpha)$. Since $\{\varphi_t\}$ satisfies Condition 16.3A, the metric ρ on P is metrically equivalent to $P_{t,\alpha}$, defined as

$$\left.\begin{aligned}&\rho_{t,\alpha}(x,y)=\max_{\tau\in[0,t]}\{\alpha^{-t+\tau}\cdot\rho[\varphi_\tau(x),\varphi_\tau(y)]\}\qquad\text{for } 0<\alpha\leqslant 1\\&\text{and as}\\&\rho_{t,\alpha}(x,y)=\max_{\tau\in[0,t]}\{\alpha^{\tau}\cdot\rho[\varphi_\tau(x),\varphi_\tau(y)]\}\qquad\text{for } \alpha\geqslant 1.\end{aligned}\right\}\tag{16.2}$$

In fact there are constants L and M (Section 16.1.4) such that

$$\rho_{t,\alpha}\leqslant L\cdot\exp(Mt)\cdot\exp(|\ln\alpha|\,t)\cdot\rho.$$

From this formula we see that

$$C_{t,\varepsilon,\alpha}\leqslant C_{0,\{\exp[(-M-|\ln\alpha|)\cdot t]\cdot\varepsilon\},1};$$

on the other hand, $\dim(P)=\limsup_{\varepsilon\to 0}(C_{0,\varepsilon,1}/(-\ln\varepsilon))$ and this we assumed to be finite. From this it follows that $\ln C_{t,\varepsilon,\alpha}/(t-\ln\varepsilon)$ is uniformly bounded for each fixed α and hence:

Lemma 16.1. *$L^{\pm}(\lambda,\alpha)$ is finite for all $\alpha>0$ and $\lambda\in[0,\pi/2]$.*

Since $\rho_{t,\alpha}$ is equivalent to ρ,

$$\limsup_{\varepsilon\to 0}\frac{\ln C_{t,\varepsilon,\alpha}}{-\ln\varepsilon}=\dim(P).$$

This implies:

Lemma 16.2. *$L^{+}(\pi/2,\alpha)=\dim(P)$ for all $\alpha>0$.*

16.2.3

Lemma 16.3. *$L^{+}(0,1)=L^{-}(0,1)=$ topological entropy of $\{\varphi_t\}$.*

Proof. First we assume that φ_t is only defined for $t\in\mathbb{N}$. Then $L^{+}(0,1)$ is clearly equivalent to Bowen's definition of entropy [13]. We shall, however, relate both $L^{+}(0,1)$ and $L^{-}(0,1)$ directly to the original definition of entropy in terms of open coverings [5].

We use the following notation: if $\mathcal{U}$ is an open cover of P, i.e., a collection $\{U_i\}_{i\in I}$ of open subsets of P such that $\bigcup_{i\in I}U_i=P$, then $\varphi_t^*(\mathcal{U})$ denotes the (open) cover with elements $\varphi_t^{-1}(U_i)$. If $\mathcal{U}$ and $\mathcal{V}$ are open covers with elements $\{U_i\}_{i\in I}$, $\{V_j\}_{j\in J}$ respectively, then $\mathcal{U}\wedge\mathcal{V}$ is the cover with elements $\{U_i\cap V_j\}_{i\in I,\,j\in J}$. $\mathcal{U}=\{U_i\}_{i\in I}$ is said to be finer than $\mathcal{V}=\{V_j\}_{j\in J}$ if for each $i\in I$ there is some $j\in J$ such that $U_i\subset V_j$; in this case we write $\mathcal{U}\ll\mathcal{V}$. If $\mathcal{U}=\{U_i\}_{i\in I}$ and $\mathcal{U}'=\{U_i\}_{i\in I'}$, for some subset $I'\subset I$, are covers of P, we call $\mathcal{U}'$ a subcover of $\mathcal{U}$. For any cover $\mathcal{U}$, $N(\mathcal{U})$ denotes the smallest cardinality of a subcover of $\mathcal{U}$. With all this we can define the entropy $H(\{\varphi_t\}_{t\in\mathbb{N}})$ as

$$\lim_{\mathcal{U},\ll}\left(\lim_{n\to\infty}\frac{\ln[N(\mathcal{U}\wedge\varphi_1^*\mathcal{U}\wedge\ldots\wedge\varphi_n^*\mathcal{U})]}{n}\right).$$

For a proof that $\lim_{n\to\infty}$ converges see [6; §14]; with $\lim_{\mathcal{U},\ll}$ we mean the limit for finer open covers $\mathcal{U}$ of P (which exists because for any two open covers $\mathcal{U}$ and $\mathcal{V}$ there is an open cover, e.g. $\mathcal{U}\wedge\mathcal{V}$, which is finer than both).

We denote by $\mathcal{U}_\varepsilon$ the cover of P consisting of all open ε-neighbourhoods of points of P. Since for each cover $\mathcal{V}$ there is an ε so that $\mathcal{U}_\varepsilon \ll \mathcal{V}$, we may replace in the above definition of entropy $\mathcal{U}$ by $\mathcal{U}_\varepsilon$ and take the limit for $\varepsilon\to 0$. We denote $\mathcal{U}_\varepsilon\wedge\varphi_1^*\mathcal{U}_\varepsilon\wedge\ldots\wedge\varphi_n^*\mathcal{U}_\varepsilon$ by $\mathcal{U}_{n,\varepsilon}$. If p, q are in the same element of $\mathcal{U}_{n,\varepsilon}$ then $\rho[\varphi_i(p), \varphi_i(q)]<2\varepsilon$ for $i=0, 1, \ldots, n$. Hence $C_{n,2\varepsilon,1}\leq N(\mathcal{U}_{n,\varepsilon})$. On the other hand, if $S\subset P$ is a subset as in the above definition of $C_{n,\varepsilon,1}$ then $\{\bigcap_{i=0}^n \varphi_i^{-1}(\mathcal{U}_\varepsilon[\varphi_i(s)])\}_{s\in S}$ is a subcover of $\mathcal{U}_{n,\varepsilon}$ ($U_\varepsilon[\varphi_i(s)]$ denotes the ε-neighbourhood of $\varphi_i(s)$). Hence $C_{n,\varepsilon,1}\geq N(\mathcal{U}_{n,\varepsilon})$. ■

From this the Lemma follows for the case where the domain of t in $\{\varphi_t\}$ is $\mathbb{N}$. If the domain of t is $\mathbb{R}_+$, we define the entropy first restricting to $\mathbb{N}$; this restriction does not change the value of $L^\pm(\lambda, \alpha)$ (although it may change the value of $C_{n,\varepsilon,\alpha}$). Hence our result holds in general.

16.2.4

From the definition of entropy with open covers it is clear that the entropy, and hence $L^+(0, 1)=L^-(0, 1)$, is determined by the topology of P (together with the semi-group $\{\varphi_t\}$) and is independent of the particular choice of the *metric*. For $L^\pm(\lambda, \alpha)$ with $\lambda\neq 0$ or $\alpha\neq 1$ this is in general not the case: take $\varphi_t=$ identity for all t and define a new metric by $\rho'(x, y)=[\rho(x, y)]^\gamma$ with $0<\gamma<1$; see also Section 16.1.6.

16.2.5

In general, if $\lambda\neq 0$ or $\alpha\neq 1$, $L^+(\lambda, \alpha)$ and $L^-(\lambda, \alpha)$ may also differ. In order to give an example where this occurs, we first need a preliminary example.

Let $\Phi:\mathbb{R}\to\mathbb{R}$ be given $\Phi(x)=x-x^k$; $x_0=1$ and $\mathcal{O}^+(x_0)=\overline{\{\Phi^i(x_0)\}_{i\in\mathbb{N}}}$. We take $P=\mathcal{O}^+(x_0)$ with the induced metric, and $\varphi_t=\Phi^t \mid \mathcal{O}^+(x_0)$ for $t\in\mathbb{N}$. We show that in this case $\dim(P)=(k-1)/k$.

Choose $\varepsilon>0$ and construct a subset $S\subset P$ as in the definition of C_ε on page 315. Then S contains at least all points of $S_1=P\cap\{x^k>\varepsilon\}$, or $S_1=P\cap\{x>\varepsilon^{1/k}\}$. Hence the number of elements $|S_1|$ of S_1 is approximately

$$\int_{\varepsilon^{1/k}}^1 \frac{1}{x^k}\,dx=\left.\frac{x^{-k+1}}{-k+1}\right|_{\varepsilon^{1/k}}^1\simeq\frac{\varepsilon^{(1-k)/k}}{-k+1}.$$

On the other hand, for $S_2 = S \cap \{x \leqslant \varepsilon^{1/k}\}$ we have

$$|S_2| \simeq \frac{\varepsilon^{1/k}}{\varepsilon} = \varepsilon^{(1-k)/k}.$$

Hence

$$\ln C_\varepsilon \simeq \frac{1-k}{k} \ln \varepsilon \qquad \text{so} \qquad \dim(P) = \frac{k-1}{k}.$$

In order to arrive at our final example we replace Φ in the last example by $\tilde{\Phi}(x) = x - g(x)$ where g satisfies the following conditions:

(a) for $x \in [0, 1]$, $g(x)$ is positive, increasing and C^1;
(b) $g(0) = 0$;
(c) for some sequence $1 > a_1 > b_1 > c_1 > d_1 > a_2 > b_2 > c_2 > d_2 > a_3 \ldots > 0$ we have

$$g \,|[a_i, b_i] = x^k, \qquad g \,|[c_i, d_i] = x^l$$

with (i) $2 \leqslant k < l$ and (ii) $(a_i/b_i) \to \infty$ and $(c_i/d_i) \to \infty$ for $i \to \infty$.

Note that such a function can even be made to be C^{k-1}. These conditions on g imply that now

$$\limsup_{\varepsilon \to 0} \frac{\ln C_\varepsilon}{-\ln \varepsilon} \geqslant \frac{l-1}{l} \qquad \text{and} \qquad \liminf_{\varepsilon \to 0} \frac{\ln C_\varepsilon}{-\ln \varepsilon} \leqslant \frac{k-1}{k}.$$

Hence in this case $L^+(\pi/2, 1) \neq L^-(\pi/2, 1)$. It is not hard to verify that now also $L^+(\lambda, \alpha) \neq L^{-1}(\lambda, \alpha)$ whenever $\lambda \neq 0$ or $\alpha \neq 1$.

16.2.6

In the special case where we have a dynamical system $(P, \{\varphi_t\}_{t \in \mathbb{N}})$ such that, for some $x_0 \in P$,

$$\overline{\bigcup_{t \in \mathbb{N}} \varphi_t(x_0)} = \mathcal{O}^+(x_0) = P,$$

there is an alternative way to define $C_{t,\varepsilon,\alpha}$ and hence $L^\pm(\lambda, \alpha)$ (this alternative definition of $C_{t,\varepsilon,\alpha}$ may give somewhat different values, but these differences disappear if we pass to the limit to obtain $L^\pm(\lambda, \alpha)$). This definition should be compared with the definition of $C_{n,\varepsilon}$ in Section 16.1.3.

Let $\mathscr{C}_{t,\varepsilon,\alpha} \subset \mathbb{N}$ be the subset of $\mathbb{N}$, inductively defined as: $0 \in \mathscr{C}_{t,\varepsilon,\alpha}$; for $i > 0$, $i \in \mathscr{C}_{t,\varepsilon,\alpha}$ if and only if for all $0 \leqslant j < i$ such that $j \in \mathscr{C}_{t,\varepsilon,\alpha}$, we have $\rho_{t,\alpha}[\alpha_i(x_0), \varphi_j(x_0)] \geqslant \varepsilon$. $\rho_{t,\alpha}$ denotes here the metric defined in formula (16.2). $C_{t,\varepsilon,\alpha}$ is the number of elements of $\mathscr{C}_{t,\varepsilon,\alpha}$. In the next section this will also be applied to the case where we have $\{\varphi\}_{t \in \mathbb{R}_+}$; in that case we first restrict the domain of t to $\mathbb{N}$ and restrict P so that the new (restricted) semi-group has still a dense positive orbit.

16.3 Dynamical systems with an observable

16.3.1

Here we consider dynamical systems as in the previous section, i.e., a compact finite-dimensional space P as phase space with a semi-group $\{\varphi_t\}$ acting on it; here we add a function $f: P \to \mathbb{R}$, which we assume to be Lipschitz. This function will be interpreted as the function which assigns to each $x_0 \in P$ the experimental output $f(x_0)$ obtained if our 'experiment' is in 'state x_0' (see Section 16.1). As already explained, such P, $\{\varphi_t\}$, x_0 can be used to model certain experiments; with the initial condition x_0 we get $\{f[\varphi_i(x_0)]\}_{i\in\mathbb{N}}$ as the experimental result.

We consider the map

$$\Lambda^q : P \to \mathbb{R}^{q+1},$$

defined by $\Lambda^q(x) = [f(x), f(\varphi, (x)), \ldots, f(\varphi_q(x))]$. Because we assumed f to be Lipschitz and because $\{\varphi_t\}$ satisfies Condition 16.3A, Λ^q is Lipschitz for all q (in many cases, Λ^q, for large q, is a Lipschitz homeomorphism onto its image; we return to this at the end of this section). From this fact we shall deduce a basic inequality.

For $x_0 \in P$, let $P' = \overline{\{\varphi_i(x_0)\}_{i\in\mathbb{N}}}$. $C_{n,\varepsilon,\alpha}(x_0)$ is defined as $C_{t,\varepsilon,\alpha}$ in Section 16.2.6 with phase space P' and semi-group $\{\varphi_t\}_{t\in\mathbb{N}}$; we replace t by n.

For a bounded sequence $A = \{a_i\}_{i=0}^{\infty}$ we define (compare Section 16.1.3) $\mathscr{C}_{n,\varepsilon,\alpha}(A) \subset \mathbb{N}$ inductively:

$$0 \in \mathscr{C}_{n,\varepsilon,\alpha}(A).$$

For $i > 0$, $i \in \mathscr{C}_{n,\varepsilon,\alpha}(A)$ if and only if for all $0 \leqslant j < i$ such that $j \in \mathscr{C}_{n,\varepsilon,\alpha}(A)$, we have

$$\max_{0\leqslant n'\leqslant n} \{\alpha^{-n+n'} \cdot |a_{i+m'} - a_{j+n'}|\} \geqslant \varepsilon \qquad \text{where } \alpha \leqslant 1,$$

$$\max_{0\leqslant n'\leqslant n} \{\alpha^{n'} \cdot |a_{i+n'} - a_{j+n'}|\} \geqslant \varepsilon \qquad \text{where } \alpha \geqslant 1.$$

$C_{n,\varepsilon,\alpha}(A)$ is the number of elements of $\mathscr{C}_{n,\varepsilon,\alpha}(A)$. We want to compare $C_{n,\varepsilon,\alpha}(x_0)$ and $C_{n,\varepsilon,\alpha}\{f[\varphi_i(x_0)]\}_{i=0}^{\infty}$. From the fact that Λ^q is Lipschitz it follows that there is a constant $K_{\alpha,q}$, depending on α and q such that

$$C_{n+q,K_{\alpha,q}\cdot\varepsilon,\alpha}(\{f[\varphi_i(x_0)]\}_{i=0}^{\infty}) \leqslant C_{n,\varepsilon,\alpha}(x_0)$$

for all n, ε. When passing to the limit, we find:

Theorem 16.3. *$L^{\pm}(\lambda, \alpha; \{f[\varphi_i(x_0)]\}) \leqslant L^{\pm}(\lambda, \alpha; x_0)$, where*

$$L^{\pm}(\lambda, \alpha; \{f[\varphi_i(x_0)]\} = \lim_{\substack{(n,-\ln\varepsilon)\to\infty \\ (-\ln\varepsilon)/n\to\operatorname{tg}\lambda}} \begin{matrix}\sup\\ \inf\end{matrix} \frac{\ln C_{n,\varepsilon,\alpha}(\{f[\varphi_i(x_0)]\})}{n - \ln\varepsilon},$$

and

$$L^{\pm}(\lambda, \alpha, x_0) = \lim_{\substack{(n,-\ln\varepsilon)\to\infty \\ (-\ln\varepsilon)/n\to\operatorname{tg}\lambda}} \begin{matrix}\sup\\ \inf\end{matrix} \frac{\ln C_{n,\varepsilon,\alpha}(x_0)}{n - \ln\varepsilon};$$

but for $\lambda = 0$, *one has to take*

$$\lim_{\varepsilon\to 0}\left(\lim_{n\to\infty}\ {}^{\sup}_{\inf}(\qquad)\right),$$

and for $\lambda = \pi/2$,

$$\lim_{n\to\infty}\left(\lim_{\varepsilon\to\infty}\ {}^{\sup}_{\inf}(\qquad)\right).$$

Corollary 16.1. *If a bounded sequence* $A = \{a_i\}$ *admits a continuously deterministic explanation then* $L^{\pm}(\lambda, \alpha; A)$ *is finite for all* $\lambda \in [0, \pi/2]$ *and* $\alpha > 0$.

Remark 16.1. It is not hard to show that if, for some q, Λ^q is a Lipschitz homeomorphism onto its image, then

$$L^{\pm}(\lambda, \alpha; \{f[\varphi_i(x_0)]\}) = L^{\pm}(\lambda, \alpha; x_0).$$

16.3.2

Now we want to give a precise meaning to the statement 'for a random sequence A, $L^{\pm}(\lambda, \alpha; A)$ is infinite'. We say that a finite sequence $(\alpha_0, \ldots, \alpha_k)$ is a limiting sequence of $A = \{a_i\}_{i=0}^{\infty}$ if there is some $n_i \to \infty$ such that $(a_{n_i}, a_{n_i+1}, \ldots, a_{n_i+k})$ converges component-wise to $(a_0, \ldots, a_k)$. Since we assume our sequence to be bounded, there are limiting sequences of any length. We say that a bounded sequence A is *random* if for some positive constants $N \in \mathbb{N}$ and ε_0, each limiting sequence $(\alpha_0, \ldots, \alpha_k)$ has two prolongations $(\alpha_0, \ldots, \alpha_k, \alpha_{k+1}, \ldots, \alpha_{k+N})$ and $(\alpha_0, \ldots, \alpha_k, \bar{\alpha}_{k+1}, \ldots, \bar{\alpha}_{k+N})$, which are both limiting sequences and such that for some $1 \leq n \leq N$, $|\alpha_{k+n} - \bar{\alpha}_{k+n}| > \varepsilon_0$. From the observation in Section 16.1.8 that any sequence $\{a_i\}_{i=0}^{\infty}$ which takes only finitely many different values admits a continuously deterministic explanation, it is clear that the above definition of 'random' is still consistent with $L^{\pm}(\lambda, \alpha; A)$ being finite. Therefore we impose the following extra condition, which is based on the idea of continuous dependence in initial condition. We say that A is *continuously random* if it is random and if the extensions $(\alpha_1, \ldots, \alpha_k, \alpha_{k+1}, \ldots, \alpha_{k+N})$ and $(\alpha_1, \ldots, \alpha_k, \bar{\alpha}_{k+1}, \ldots, \bar{\alpha}_{k+N})$ in the definition of random can be chosen so that $(\alpha_{k+1}, \ldots, \alpha_{k+N})$ and $(\bar{\alpha}_{k+1}, \ldots, \bar{\alpha}_{k+N})$ are in the same connected component of $\{(\alpha'_{k+1}, \ldots, \alpha'_{k+N}) \in \mathbb{R}^N \mid (\alpha_1, \ldots, \alpha_k, \alpha'_{k+1}, \ldots, \alpha'_{k+N})$ is a limiting sequence$\}$.

With this it is easy to see that for such a continuously random sequence A,

$$C_{n,\varepsilon,\alpha}(A) \geq \left(\frac{\varepsilon_0}{2\varepsilon}\right)^{n/N},$$

and hence that $L^{\pm}(\lambda, \alpha; A)$ is infinite.

16.3.3

Finally we recall some results from [3] and [4] which show that Λ^q, as defined earlier, is often a Lipschitz homeomorphism onto its image.

Theorem 16.4. *Let M be a compact m-manifold. Then there is an open and dense subset $\mathcal{O} \subset \mathcal{X}^1(M)$ in the C^1 vector fields such that for each $X \in \mathcal{O}$ there is an open and dense subset $\mathcal{F}_X \subset C^1(M)$ of C^1 functions on M such that for each $f \in \mathcal{F}_X$,*

$$\Lambda^{2m} : M \to \mathbb{R}^{2m+1},$$

defined by $\Lambda^{2m}(x) = (f(x), f[\varphi_1(x)], \ldots, f[\varphi_{2m}(x)])$, is a C^1-embedding (φ_t denotes the flow of X).

In [4] the vector field was required to be C^2 but C^1 is enough here. Since M is compact, a C^1-embedding is a Lipschitz homeomorphism onto its image. Hence the inequality derived in Section 16.3.1 is generically an equality.

16.4 Examples

16.4.1

In this section we first calculate the dimension of a number of spaces and then calculate $L^{\pm}(\lambda, \alpha)$ for a class of dynamical systems; finally we discuss what might be the connection between $L(\lambda, \alpha)$ and Lyapunov exponents.

16.4.2 Cantor sets

Let $\{\alpha_i\}_{i=1}^{\infty}$ be a sequence of real numbers satisfying $0 < \alpha_1 < \frac{1}{2}$, $0 < \alpha_k < \frac{1}{2}\alpha_{k-1}$. For each such sequence we construct a corresponding subset of the unit interval $I_0 = [0, 1]$:

$I_1 \subset I_0$ consists of two closed intervals of length α_1,

$I_2 \subset I_1$ consists of four closed intervals of length α_2,

in general and, in more detail, I_k is obtained from I_{k-1} by taking out of each of the 2^{k-1} intervals, of length α_{k-1}, an open interval in the middle of length $\alpha_{k-1} - 2\alpha_k$; so I_k has again 2^k intervals, each of length α_k. The corresponding Cantor set is then defined as $I_\infty = \bigcap_{k=0}^{\infty} I_k$.

From the construction it is clear that $C_{\alpha_k/2}(I_\infty) \geq 2^k$, $C_{\alpha_k}(I_\infty) \leq 2^k$ and

$$\lim_{\varepsilon \to 0} \begin{matrix}\sup\\ \inf\end{matrix} \frac{\ln C_\varepsilon}{-\ln \varepsilon} = \lim_{k \to \infty} \begin{matrix}\sup\\ \inf\end{matrix} \frac{k \cdot \ln 2}{-\ln \alpha_k} = \alpha^{\pm}.$$

Note that for any $0 \leq \alpha^- \leq \alpha^+ \leq 1$ there is a sequence $\{a_k\}_{k=1}^{\infty}$ with $0 < \alpha_k < \frac{1}{2}\alpha_{k-1}$ and $0 < \alpha_1 < \frac{1}{2}$ so that

$$\limsup \frac{k \cdot \ln 2}{-\ln \alpha_k} = \alpha^+ \qquad \text{and} \qquad \liminf \frac{k \cdot \ln 2}{-\ln \alpha_k} = \alpha^-.$$

16.4.3 Product spaces

We assume here that all spaces have the property that

$$\liminf_{\varepsilon\to 0}\frac{\ln C_\varepsilon(\ )}{-\ln\varepsilon}=\limsup_{\varepsilon\to 0}\frac{\ln C_\varepsilon(\ )}{-\ln\varepsilon}.$$

We want to show for such spaces that $\dim(P_1\times P_2)=\dim(P_1)+\dim(P_2)$, where $P_1\times P_2$ has the product metric

$$\rho[(x_1, x_2), (y_1, y_2)]=\max\{\rho_1(x_1, y_1), \rho_2(x_2, y_2)\}.$$

First we observe that if $S_1\subset P_1$ and $S_2\subset P_2$ such that P_1, P_2 is an ε-neighbourhood of S_1, respectively S_2, then $P_1\times P_2$ is an ε-neighbourhood of $S_1\times S_2$, so $\dim(P_1\times P_2)\leqslant\dim(P_1)+\dim(P_2)$.

To prove the inequality in the other direction we take a subset $\tilde{S}_1\subset P_1$ such that (i) any two points of $\tilde{S}_1$ are at least distance ε apart and (ii) $\tilde{S}_1$ has maximal cardinality with property (i). It is clear that $|\tilde{S}_1|\geqslant C_\varepsilon(P_1)$, $|\tilde{S}_1|$ being the number of elements of $\tilde{S}_1$. On the other hand $C_{\varepsilon/2}(P_1)\geqslant|\tilde{S}_1|$ since in a $\frac{1}{2}\varepsilon$-neighbourhood of a point of P_1 there can be at most one point of $\tilde{S}_1$. Next we take a set $S\subset P_1\times P_2$ such that $P_1\times P_2$ is a $\frac{1}{2}\varepsilon$-neighbourhood of S and such that S has minimal cardinality with this property. For each $\tilde{s}_1\in\tilde{S}_1$ we consider $S(\tilde{s}_1)=\{(p_1, p_2)\in S \mid \rho_1(p_1, \tilde{s}_1)<\frac{1}{2}\varepsilon\}$. Then, for $\tilde{s}_1\neq\tilde{s}_1'\in\tilde{S}_1$, $S(\tilde{s}_1)$ is disjoint from $S(\tilde{s}_1')$. Also, P_2 is a $\frac{1}{2}\varepsilon$-neighbourhood of $\pi_2[S(\tilde{s}_1)]$, where π_2 denotes projection on the second factor. So each $S(\tilde{s}_1)$ has at least $C_{\varepsilon/2}(P_2)$ elements, therefore

$$|S|\geqslant|\tilde{S}_1|\cdot C_{\varepsilon/2}(P_2)\qquad\text{or}\qquad C_{\varepsilon/2}(P_1\times P_2)\geqslant C_\varepsilon(P_1)\times C_{\varepsilon/2}(P_2).$$

This implies that $\dim(P_1\times P_2)\geqslant\dim(P_1)+\dim(P_2)$.

16.4.4

Using the above observations we see that the unit cube in $\mathbb{R}^n$ has dimension n: it is easy to verify that $\dim[0, 1]=1$ and the Euclidean metric on $\mathbb{R}^n$ is equivalent to the product metric. Also if M is a smooth compact n-manifold then 'M has dimension n' (although our dimension is only defined for metric spaces it makes sense here too, because the dimension is the same for every metric on M which is induced by a Riemannian metric). Indeed, let ρ be a metric on M induced by a Riemannian metric and let $\Psi:[0, 1]^n\to M$ be a smooth embedding. Then Ψ is a Lipschitz homeomorphism onto its image and so $\dim(\Psi([0, 1]^n))$ is n. M can be covered by a finite number of such 'smooth cubes', so $\dim(M)=n$.

We note in passing that for each compact subset $A\subset M$, M a smooth manifold, one can define $\dim(A)$ in the same way as $\dim(M)$ above: take a metric ρ on M induced by a Riemannian metric on M and define $\dim(A)=\dim(A, \rho\mid A)$; the result is independent of the Riemannian metric. The same works if M is a C^1 Hilbert or Banach-manifold.

So for manifolds our dimension is equal to the topological dimension. For Cantor sets this is not the case: the topological dimension is zero.

Also in Section 16.2.5 we saw spaces with positive dimension but with topological dimension zero. One can make the difference between our dimension and the topological dimension arbitrarily big by taking products.

16.4.5 Horseshoes

We now compute the $L^{\pm}(\lambda, \alpha)$ invariants for a class of dynamical systems which have often a product of Cantor sets as phase space. Let (K_i, ρ_i), $i=1, \ldots, m$, be a collection of compact metric spaces and let $K=K_1 \times \ldots \times K_m$ with the product metric. Let $\varphi: K \to K$ be a Lipschitz continuous map (and hence generator of a semi-group $\{\varphi_t\}_{t\in\mathbb{N}}$ with $\varphi_t=(\varphi)^t$) satisfying the following condition: for some $\varepsilon_0>0$ and fixed $\lambda_1, \ldots, \lambda_m>0$, we have for any pair of points $(x_1, \ldots, x_m)$, $(y_1, \ldots, y_m)$ in K, at distance less than ε_0, that their images $\varphi(x_1, \ldots, x_m)=(x'_1, \ldots, x'_m)$ and $\varphi(y_1, \ldots, y_m)=(y'_1, \ldots, y'_m)$ satisfy $\rho_i(x'_i, y'_i)=\lambda_i \cdot \rho_i(x_i, y_i)$ for all $i=1, \ldots, m$.

Maps of this kind occur in so-called 'horseshoes' of which an explicit example is now given (for literature on this subject see Smale [2]; more recent references on horseshoes and Axiom A basic sets can be found in [14]).

Take $\Phi: \mathbb{R}^2 \to \mathbb{R}^2$, a diffeomorphism such that the unit square I^2 is mapped over itself as indicated in Fig. 16.1. On each component of $\Phi^{-1}(I^2) \cap I^2$, Φ has the form $(x, y) \to (\lambda x + x_0, \mu y + y_0)$, where $|\lambda|, |\mu|$ are fixed while x_0, y_0 and the signs of λ and μ depend on the component of $\Phi^{-1}(I^2) \cap I^2$. Take $K=\bigcap_{n\in\mathbb{Z}} \Phi^n(I^2)$ and $\varphi=\Phi \mid K$. Then K is a product of two Cantor sets and φ satisfies all the above conditions.

We note that in general, if $\varphi: K \to K$, satisfying the above conditions with $\lambda_i \neq 1$, is surjective, then $\lim_{\varepsilon\to 0} [(\ln C_\varepsilon(K_i))/-\ln \varepsilon]$ exists. In the rest of this paragraph we assume the existence of these limits.

In order to calculate $L(\lambda, \alpha)$ for semi-groups generated by $\varphi: K \to K$

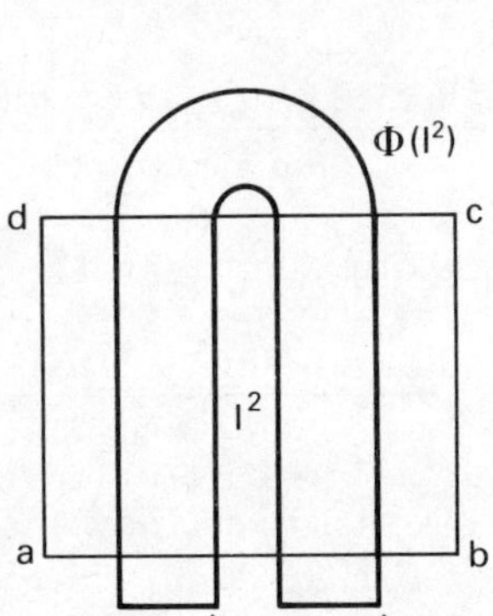

Fig. 16.1

as above, we introduce the metric $\rho_{n,\alpha}$ on K, see also Section 16.2.2:

$$\rho_{n,\alpha}[(x_1,\ldots,x_m),(y_1,\ldots,y_m)] = \begin{cases} \max\limits_{i=1,\ldots,m} \{\max\{\alpha^{-n}\cdot\rho_i(x_i,y_i),\lambda_i^n\cdot\rho_i(x_i,y_i)\}\} & \text{where } \alpha\leqslant 1 \\ \max\limits_{i=1,\ldots,m} \{\max\{\rho_i(x_i,y_i),\alpha^n\cdot\lambda_i^n\cdot\rho_i(x_i,y_i)\}\} & \text{where } \alpha\geqslant 1. \end{cases}$$

From these definitions it follows that $C_{n,\varepsilon,\alpha}(K,\{\varphi_t\})=C_\varepsilon(K,\rho_{n,\alpha})$ for ε sufficiently small. To calculate this last expression, let

$$\delta_i=\lim_{\varepsilon\to 0}\frac{\ln C_\varepsilon(K_{i,\rho_i})}{-\ln\varepsilon},$$

hence $C_\varepsilon(K_{i,\rho_i})\simeq\varepsilon^{-\delta_i}$.

Then, for $\alpha\leqslant 1$,

$$C_\varepsilon(K,\rho_{n,\alpha})\simeq\left[\prod_{i:\alpha\cdot\lambda_i\leqslant 1} C_{\alpha^n\cdot\varepsilon}(K_{i,\rho_i})\right]\cdot\left[\prod_{i:\alpha\cdot\lambda_i>1} C_{\lambda_i^{-n}\cdot\varepsilon}(K_{i,\rho_i})\right].$$

So

$$\ln[C_\varepsilon(K,\rho_{n,\alpha})]\simeq\sum_{i:\alpha\cdot\lambda_i\leqslant 1}-\delta_i\cdot\ln(\alpha^n\cdot\varepsilon)+\sum_{i:\alpha\cdot\lambda_i>1}-\delta_i\cdot\ln(\lambda_i^{-n}\cdot\varepsilon),$$

and

$$L^{\pm}(\lambda,\alpha)=\lim_{\substack{(n,-\ln\varepsilon)\to\infty\\(-\ln\varepsilon)/n\to\operatorname{tg}\lambda}}\frac{\ln(C_{n,\varepsilon,\alpha})}{n-\ln\varepsilon}$$

$$=\frac{(\sum_{i=1}^m\delta_i)\cdot\sin\lambda+[\sum\delta_i\cdot(\max\{\ln\lambda_i,\ln\alpha^{-1}\})]\cdot\cos\lambda}{\sin\lambda+\cos\lambda}$$

Note that, although we have worked with approximate equalities ($\simeq$), the final equality for $L^+=L^-$ is exact.

In the same way we obtain for $\alpha\geqslant 1$ that

$$L^{\pm}(\lambda,\alpha)=\frac{(\sum_{i=1}^m\delta_i)\cdot\sin\lambda+[\sum_{i:\alpha\cdot\lambda_i>1}\delta_i\cdot\ln(\alpha\cdot\lambda_i)]\cdot\cos\lambda}{\sin\lambda+\cos\lambda}$$

So we have for this class of examples that

$$L^+(\lambda,\alpha)=L^-(\lambda,\alpha)=\frac{\dim(K)\cdot\sin\lambda+H(\alpha)\cdot\cos\lambda}{\sin\lambda+\cos\lambda}$$

for some function $H(\alpha)$. In order to see how this function $H(\alpha)$ depends on λ_i, δ_i we first consider Fig. 16.2, in which we assume $\lambda_1>\lambda_2>\lambda_3$. Combining this with a horizontal line at height $\ln\alpha^{-1}$ we get the visualization of $H(\alpha)$ shown by the shaded area in Fig. 16.3(a) for $\alpha\leqslant 1$ and in 16.3(b) for $\alpha\geqslant 1$.

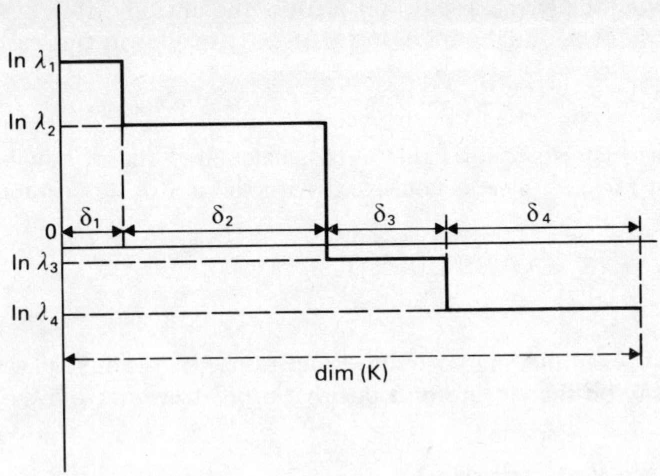

Fig. 16.2

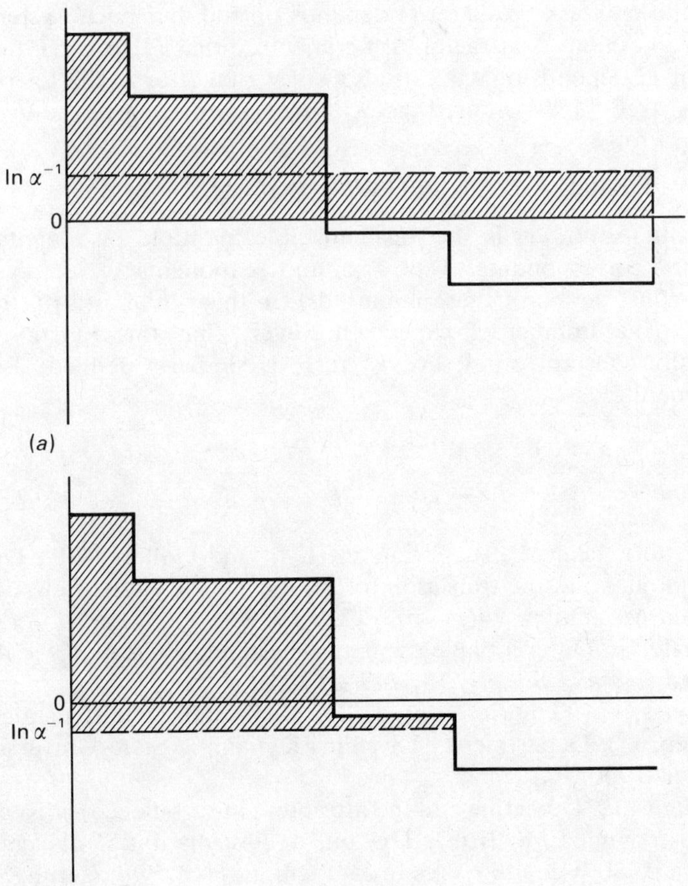

Fig. 16.3

We have already remarked that $H(0)$ is the entropy. If φ is invertible, the entropy of φ equals the entropy of φ^{-1}, so that in this case

$$\sum_{i:\lambda_i\geq 1} \delta_i \cdot \ln \lambda_i = \sum_{i:\lambda_i\leq 1} \delta_i \cdot \ln \lambda_i^{-1}.$$

In this case it also follows that $H(\alpha)$, calculated for φ, equals $H(\alpha^{-1})$, calculated for φ^{-1}. This is consistent with the above description and the fact that

$$L^{\pm}[\lambda, \alpha; (K, \varphi)] = L^{\pm}[\lambda, \alpha^{-1}; (K, \varphi^{-1})].$$

16.4.6

The above example suggests some questions. In the first place, is it in general (say on the closure of a positive orbit) true that $L^{\pm}(\lambda, \alpha)$ has the form

$$\frac{D^{\pm} \cdot \sin \lambda + H^{\pm}(\alpha) \cdot \cos \lambda}{\sin \lambda + \cos \lambda}?$$

If so, one may ask how $H^{\pm}(\alpha)$ depends on the dynamical system. In the case of an Axiom A attractor Λ, I conjecture that $H^{+}(\alpha) = H^{-}(\alpha) = H(\alpha)$ and that H depends on λ_i, δ_i as above if we take for λ_i the Lyapunov exponents (see [15]) ordered as $\lambda_1 > \lambda_2 > \lambda_3 > \ldots > \lambda_u > 0 > \lambda_{u+1} > \ldots > \lambda_{u+s}$, and for δ_i, $i \leq u$,

$$\delta_i = \dim(W^{u,i}_{\text{loc}}(x) \cap \Lambda) - \dim(W^{u,i-1}_{\text{loc}}(x) \cap \Lambda),$$

$x \in \Lambda$, where $W^{u,i}_{\text{loc}}(x)$ is the local unstable manifold of x containing all expansion corresponding to the Lyapunov exponents $\lambda_1, \ldots, \lambda_i$ (see [16] for the construction of these manifolds); in this way δ_i is a *function* on Λ instead of a number. I expect, however, that this function will be essentially constant on Λ. For $i > u$, δ_i is similarly defined, using local stable manifolds.

References

1. Fenstermacher, P. R., Swinney, H. L. and Gollub, J. P., Dynamical instability and the transition to chaotic Taylor vortex flow, *Journal of Fluid Mechanics*, **94**(1), 103–128, 1979.
2. Smale, S., Differentiable dynamical systems, *Bulletin of the American Mathematical Society*, **73,** 747–817, 1967.
3. Aeyels, D., Generic observability of differentiable dynamic systems (preprint), Department of Systems Dynamics, State University Gent, April 1980.
4. Takens, F., Detecting strange attractors in turbulence, in Rand, D. A. and Young, L. S. (Eds), *Dynamical Systems and Turbulence, Warwick* 1980, Lecture Notes in Mathematics No. 898: Berlin, Springer-Verlag, 1980, 366–382.

5. Adler, R. L., Konheim, A. G. and McAndrew, M. H., Topological entropy, *Transactions of the American Mathematical Society*, **114,** 300–319, 1965.
6. Denker, M., Grillenberg, C. and Sigmund, K., *Ergodic Theory on Compact Spaces,* Lecture Notes in Mathematics No. 527: Berlin, Springer-Verlag, 1976.
7. Frederickson, P., Kaplan, J. L. and York, J. A., Dimension of the strange attractor of a class of difference systems (preprint), University of Maryland 1980 (J. A. York, Molecular Physics Building, College Park, Md 20742).
8. Kaplan, J. L. and York, J. A., Chaotic behaviour of multidimensional difference equations, in *Functional Differential Equations and Approximations of Fixed Points, Proceedings, Bonn* 1978, Lecture Notes in Mathematics No. 730: Berlin, Springer-Verlag, 1979.
9. Mallet-Paret, J., Negatively invariant sets of compact maps and extensions of a theorem of Cartwright, *Journal of Differential Equations,* **22,** 204–227, 1976.
10. Mañé, R., *On the Dimension of the Compact Invariant Sets of Certain Non-linear Maps* (manuscript), IMPA, Est. Dona Castorina 110, Rio de Janeiro.
11. Marsden, J. E., The Hopf-bifurcation for non-linear semi-groups, *Bulletin of the American Mathematical Society,* **79,** 537–541, 1973.
12. Marsden, J. E. and McCracken, M., *The Hopf Bifurcation and its Applications,* Applied Mathematical Sciences No. 19: Berlin, Springer-Verlag, 1976.
13. Bowen, R., Entropy of group endomorphisms and homogeneous spaces, *Transactions of the American Mathematical Society,* **153,** 401–414, 1971.
14. Bowen, R., *On Axiom A Diffeomorphisms,* Regional Conference Series in Mathematics No. 35: Providence, American Mathematical Society, 1978.
15. Pessin, Ja. B., Lyapunov characteristic exponents and smooth ergodic theory. *Russian Mathematical Surveys,* **32**(4), 243–264, 1974.
16. Ruelle, D., Ergodic theory of differential dynamical systems, *Publications Mathématiques de l'Institut des Hautes Etudes Scientifiques,* **50,** 27–28, 1979.

17 Energy-conserving Galerkin Approximations for the Bénard Problem

Y M Treve

17.1 Introduction

The method of Galerkin (for a general exposition see, for instance [1]) is particularly well suited to the approximate solution of two-dimensional fluid flow problems since (a) it produces approximations which satisfy the boundary conditions exactly; (b) it is known to converge strongly to the exact solution in the case of the two-dimensional Navier–Stokes equations for a wide class of external forces [2]; and (c) it is computationally efficient when used in conjunction with appropriate fast Fourier transform algorithms [3–6]. But these appealing features are overshadowed by two serious difficulties encountered in its application: first, because of the nonlinear character of the equations it is difficult to estimate how many modes must be carried in order to guarantee that the approximation at least behaves like the exact solution with the same initial flows. However, a new approach has recently been proposed for the determination of such an estimate [7–9] (see also Chapter 8 by C. Foias and R. Témam). The second difficulty, which has already been mentioned in [10], is that arbitrary Galerkin approximations do not in general satisfy the energy balance equations of the problem. Using the Bénard problem as an example, in this chapter we first show how the violation of the energy balance equations arises for arbitrary approximations. We next prove that there is a unique way of constructing Galerkin approximations of any order such that the energy balance equations are exactly satisfied. We then give a simple proof of the boundedness of the approximations so constructed, indicate how bounds can be obtained for quantities of physical interest such as the Nusselt number, and demonstrate the global asymptotic stability of the purely conductive solution for subcritical Rayleigh numbers.

Finally we conclude with some remarks about the possibility of constructing conserving approximations for other fluid flow problems.

17.2 Equations of the Boussinesq model of Bénard convection

We consider the motion of a viscous fluid confined between two horizontal parallel plates separated by a distance H. The fluid layer is heated from below in such a way that the lower plate is maintained at a temperature T_0 while the upper one is at $T_1 = T_0 - \Delta T < T_0$. Both T_0 and T_1 are assumed uniform and constant in time. In the Boussinesq approximation the velocity $\mathbf{v}(t, \mathbf{r})$, pressure $p(t, \mathbf{r})$, and temperature $T(t, \mathbf{r})$ with $\mathbf{r} = (x, 0, z)$, are determined by the equations [11]

$$\rho_0 \frac{\mathrm{D}\mathbf{v}}{\mathrm{D}t} = -\boldsymbol{\nabla} p + \rho_0[1 + \varepsilon(T_0 - T)]\mathbf{g} + \rho_0 \nu \boldsymbol{\nabla}^2 \mathbf{v} \tag{17.1}$$

$$\rho_0 C_v \frac{\mathrm{D}T}{\mathrm{D}t} = \rho_0 C_v \kappa \nabla^2 T \tag{17.2}$$

and some appropriate initial-boundary conditions. In Eqs (17.1) (17.2), $\mathrm{D}/\mathrm{D}t = \partial/\partial t + (\mathbf{v} \cdot \boldsymbol{\nabla})$, $\mathbf{g} = (0, 0, -g)$ where g is the gravitational constant, ρ_0 is the constant mean density, ε is the expansion coefficient of the fluid, C_v is the specific heat at constant volume, and ν and κ are the (constant) coefficients of kinematic viscosity and thermometric conductivity, respectively.

For simplicity we assume 'free' boundary at the bottom and top of the layer:

$$v_z(t, x, 0) = v_z(t, x, H) = 0;$$

with no tangential stresses there,

$$\frac{\partial v_x}{\partial z}(t, x, 0) = \frac{\partial v_x}{\partial z}(t, x, H) = 0.$$

We consider the situation where the motion occurs in the form of rolls of width L parallel to the y-direction. Periodicity in the x-direction requires that

$$v_x(t, 0, z) = v_x(t, L, z) = 0.$$

For the temperature we have $T(t, x, 0) = T_0$, $T(t, x, H) = T_1$ and no heat flux between contiguous rolls, so that

$$\frac{\partial T}{\partial x}(t, 0, z) = \frac{\partial T}{\partial x}(t, L, z) = 0.$$

We find it convenient to pass to dimensionless variables using H/π, $\tau = [H/(g\varepsilon\Delta T)]^{1/2}$, ΔT, $\rho_0 H g \varepsilon \Delta T/\pi^2$, as units of length, time, temperature and pressure, respectively. The dissipative coefficients then transform according to $\nu \to [H/(\pi^2\tau)]\nu$ and $\kappa \to [H/(\pi^2\tau)]\kappa$.

As usual, we treat the problem in terms of the temperature perturbation $\theta(t, x, z)$ of the purely conductive solution, defined by the equation

$$T(t, x, z) = T_0 - \frac{z}{\pi} + \frac{1}{\pi}\,\theta(t, x, z).$$

Our equations are then

$$\frac{\mathrm{D}\mathbf{v}}{\mathrm{D}t}=-\boldsymbol{\nabla}p+\theta\mathbf{g}+\nu\boldsymbol{\nabla}^2\mathbf{v}, \tag{17.3}$$

$$\frac{\mathrm{D}\theta}{\mathrm{D}t}=v_z+\kappa\nabla^2\theta \tag{17.4}$$

where now $\mathbf{g}=(0, 0, -1)$, subject to the boundary conditions

$$v_z(t, x, 0)=v_z(t, x, \pi)=v_x(t, 0, z)=v_x(t, \pi/a, z)=0,$$

$$\frac{\partial v_x}{\partial z}(t, x, 0)=\frac{\partial v_x}{\partial z}(t, x, \pi)=0,$$

$$\theta(t, x, 0)=\theta(t, x, \pi)=\frac{\partial \theta}{\partial x}(t, 0, z)=\frac{\partial \theta}{\partial x}(t, \pi/a, z)=0,$$

where the aspect ratio $a=H/L$.

The energy balance equations of the model are

$$\frac{\mathrm{d}}{\mathrm{d}t}\int_\Omega\left(\frac{v^2}{2}-z\theta\right)=-\nu\int_\Omega|\boldsymbol{\nabla}\mathbf{v}|^2-\kappa\pi\int_0^{\pi/a}\mathrm{d}x\frac{\partial\theta}{\partial z}(t, x, \pi), \tag{17.5}$$

$$\frac{\mathrm{d}}{\mathrm{d}t}\int_\Omega\theta=-\kappa\int_0^{\pi/a}\mathrm{d}x\left[\frac{\partial\theta}{\partial z}(t, x, 0)-\frac{\partial\theta}{\partial z}(t, x, \pi)\right], \tag{17.6}$$

where $\Omega=\{(x, z):0\leqslant x\leqslant\pi/a, 0\leqslant z\leqslant\pi\}$. Equation (17.5) expresses the time rate of change of the sum of the total kinetic and potential energies ($E_{\mathrm{kin}}+E_{\mathrm{pot}}$) in terms of the rate of energy dissipation through friction and of the flux of heat leaving the layer through the top boundary. Similarly, Eq. (17.6) shows that the rate of change of the total internal energy E_{int} associated with the temperature perturbation equals the difference between the fluxes of heat entering the layer from below and leaving it through the top.

Note that ($E_{\mathrm{kin}}+E_{\mathrm{pot}}$) and E_{int} are constants of the motion when $\nu=\kappa=0$.

There is one more balance equation related to entropy [12], namely

$$\frac{1}{2}\frac{\mathrm{d}}{\mathrm{d}t}\int_\Omega(\theta-z)^2=\kappa\int_\Omega(\theta-z)\nabla^2\theta. \tag{17.7}$$

Finally, in terms of the usual stream function $\Psi(t, x, z)$, Eqs (17.3) (17.4) reduce to the system

$$\frac{\mathrm{D}}{\mathrm{D}t}\nabla^2\Psi=\frac{\partial\theta}{\partial x}+\nu\nabla^4\Psi, \tag{17.8}$$

$$\frac{\mathrm{D}\theta}{\mathrm{D}t}=\frac{\partial\Psi}{\partial x}+\kappa\nabla^2\theta, \tag{17.9}$$

where, in $\mathrm{D}/\mathrm{D}t$, $v_x=-\partial\Psi/\partial z$ and $v_z=\partial\Psi/\partial x$.

17.3 Energy-conserving Galerkin approximations

The simplest complete systems of orthogonal functions which satisfy the boundary conditions of our problem are probably $\{\sin max \sin nz\}$, m, $n>0$, for Ψ, and $\{\cos max \sin nz\}$, $m \geqslant 0$, $n>0$, for θ (see, for instance, [13]). We write the corresponding Galerkin approximations in the following convenient form:

$$\Psi^{(k)}(t, x, z) = \sum \Psi^{(k)}_{m,n}(t) \sin max \sin nz, \tag{17.10}$$

$$\theta^{(k)}(t, x, z) = \sum \theta^{(k)}_{m,n}(t) \cos max \sin nz + \sum \theta^{(k)}_{0,n}(t) \sin nz, \tag{17.11}$$

where k will denote the order of the approximation and the precise range of the summations will be specified later in terms of k. For simplicity of notation, the time-dependence of the amplitudes $\Psi_{m,n}$ and $\theta_{m,n}$ will be implied in the following and whenever possible we will omit the superscript (k).

Straightforward application of Galerkin's procedure then yields a system of ordinary differential equations which is given in the Appendix and that we write in vector form as

$$\frac{\mathrm{d}\mathbf{c}}{\mathrm{d}t} = \mathbf{F}(\mathbf{c}) \tag{17.12}$$

where $\mathbf{c}$ denotes the vector whose components are the amplitudes $\Psi_{m,n}$ and $\theta_{m,n}$.

We now show that Galerkin approximations based on an arbitrary set of modes violate the energy and entropy balance equations (17.5)–(17.7), and since the reason for such a violation is the same for the three equations we do that for the simplest, i.e., for Eq. (17.6). This equation, when written in terms of the amplitudes, reads

$$\frac{\mathrm{d}}{\mathrm{d}t} \sum \theta_{0,2l+1} = -\kappa \sum (2l+1)\theta_{0,2l+1}. \tag{17.13}$$

In order for the total internal energy to evolve in time according to Eq. (17.6) it is evidently necessary that Eq. (17.13) be consistent with the differential equations (17.12). However when the left-hand side of (17.13) is evaluated along the orbits of (17.12) we find that

$$\frac{\mathrm{d}}{\mathrm{d}t} \sum \theta_{0,2l+1} = -\kappa \sum (2l+1)\theta_{0,2l+1} + \frac{a}{4}\left[\sum m\Psi_{m,n}\theta_{m,n+2l+1} - \sum m\Psi_{m,n}(\widetilde{2l+1-n})\theta_{m,|2l+1-n|}\right] \tag{17.14}$$

where $(\tilde{x}) = -1$ if $x<0$, $(\tilde{x}) = 0$ if $x=0$, and $(\tilde{x}) = 1$ if $x>0$.

Comparing the right-hand sides of Eqs (17.13) and (17.14) we see that they differ by the presence of quadratic terms and it is easy to verify that in general the sum of these terms does not vanish identically. As a result, the internal energy E_{int} does not evolve as prescribed by Eq. (17.6). A possible remedy to this difficulty is to omit all modes of the form $\sin(2l+1)z$ from the approximation (17.11), but there does not seem to be any mathematical justification for such an omission and physically it is unacceptable since it forces the total internal energy associated with θ to vanish at all times. There is, however, a nontrivial way of selecting the modes so that the undesirable terms appearing in (17.14) cancel out exactly. Because the derivation of the mode selection rule is elementary and somewhat lengthy we shall be content to sketch it.

First we deal with the modes related to the velocity, that is, those used for the representation of the stream function in (17.10); each of these modes is characterized by the pair of integers (m, n), $m, n > 0$. In the theory of the two-dimensional Navier–Stokes equation the Galerkin approximations whose convergence has been demonstrated [2] are based on the eigenfunctions of the problem,

$$\nabla^2\mathbf{w}+\nabla q=-\lambda\mathbf{w}, \qquad \nabla\cdot\mathbf{w}=0, \qquad \mathbf{w}\,|_{\partial\Omega}=0,$$

with positive eigenvalues $\lambda_1 \leqslant \lambda_2 \leqslant \ldots \infty$. The Galerkin approximation of order k is then of the form

$$\mathbf{v}^{(k)}=\sum_{i=1}^{k} c_i\mathbf{w}_i.$$

Thus we define the approximations for the stream function in the following analogous manner: the vectors

$$\begin{aligned}\mathbf{u}_{m,n}&=\nabla\times(0, \sin max \sin nz, 0)\\ &=(-n\sin max\cos nz, 0, am\cos max\sin nz)\end{aligned}$$

are the eigenfunctions of the problem

$$\nabla^2\mathbf{u}+\nabla q=-\rho\mathbf{u}, \qquad \nabla\cdot\mathbf{u}=0, \qquad \mathbf{u}\cdot\mathbf{n}\,|_{\partial\Omega}=0$$

($\mathbf{n}$ = a vector normal to $\partial\Omega$) in a basic cell Ω with eigenvalues

$$a^2+1=\rho_{m_1,n_1}\leqslant\rho_{m_2,n_2}\leqslant\ldots\leqslant\rho_{m_k,n_k}\leqslant\ldots\infty.$$

We therefore take for the approximation of order k of the stream function that which includes all modes $\sin m_i ax \sin n_i z$ with $1 \leqslant i \leqslant k$.

Hence in formula (17.10) the summation is over the set of pairs

$$E(k)\stackrel{\text{def}}{=}\{(m_i, n_i)\}_{i=1}^{i=k}.$$

Then it is not difficult to verify that, in order for the energy balance equations (17.5), (17.6) to be satisfied, the modes of the form $\cos max \sin nz$, with $m, n > 0$ appearing in the representation of $\theta^{(k)}$ must be selected in the same way, that is, the first summation in formula (17.11) must be over the set of pairs $E(k)$.

With the above choice of x-dependent modes it is elementary to show that the undesirable terms appearing in the right-hand side of (17.14) will cancel out exactly if and only if the modes $\sin nz$ with $n=1, 2, \ldots, 2\bar{n}$, where $\bar{n}=max_{1\leq i\leq k}\, n_i$, are included in the representation of $\theta^{(k)}$.

In the following the approximations constructed according to the above prescriptions will be denoted $G^{(k)}$.

It is then easy to verify that all $G^{(k)}$'s also satisfy exactly the balance equations (17.5) and (17.7).

In the next section we indicate some interesting properties of the $G^{(k)}$'s.

17.4 Properties of the $G^{(k)}$'s

First we show that, for any $k\geq 1$, $G^{(k)}$ is bounded for all times whatever the initial conditions.

To this aim we note that after multiplying Eq. (17.6) by any number $\lambda>0$ and adding to Eq. (17.5), the resulting equation can be written as

$$\frac{\mathrm{d}}{\mathrm{d}t}\int_\Omega \left[\frac{v^2}{2}+\frac{\lambda}{2}\left(\theta-\frac{1+\lambda}{\lambda}z\right)^2\right]=-\nu\int_\Omega |\nabla \mathbf{v}|^2+\lambda\kappa\int_\Omega \left(\theta-\frac{1+\lambda}{\lambda}z\right)\nabla^2\theta.$$

When expressed in terms of the amplitudes this equation can be written as

$$\frac{\mathrm{d}P(\mathbf{c})}{\mathrm{d}t}=-Q(\mathbf{c}) \tag{17.15}$$

where

$$\mathbf{c}=(\Psi_{m_1,n_1},\ldots,\Psi_{m_k,n_k},\theta_{m_1,n_1},\ldots,\theta_{m_k,n_k},\theta_{0,1},\ldots,\theta_{0,2\bar{n}})$$

$$P(\mathbf{c})\equiv\tfrac{1}{2}\sum_E \rho_{m,n}\Psi^2_{m,n}+\frac{\lambda}{2}\sum_E \theta^2_{m,n}+\lambda\sum_{n=1}^{2\bar{n}}\left(\theta_{0,n}+2\frac{1+\lambda}{\lambda}\frac{(-)^n}{n}\right)^2$$

$$Q(\mathbf{c})\equiv\nu\sum_E \rho^2_{m,n}\Psi^2_{m,n}+\lambda\kappa\sum_E \rho_{m,n}\theta^2_{m,n}$$
$$+2\lambda\kappa\sum_{n=1}^{2\bar{n}}\left(n\theta_{0,n}+\frac{1+\lambda}{\lambda}(-)^n\right)^2-4\frac{(1+\lambda)^2}{\lambda}\kappa\bar{n}.$$

Now consider the orbit of Eq. (17.12) which starts from an arbitrary point $\mathbf{c}_0$ at time $t=0$, i.e., $\mathbf{c}(0,\mathbf{c}_0)=\mathbf{c}_0$. For R large enough the ellipsoid $E_R=\{\mathbf{c}: P(\mathbf{c})<R^2\}$ contains both $\mathbf{c}_0$ and the ellipsoid $E=\{\mathbf{c}: Q(\mathbf{c})\leq 0\}$ so that, at any point $\mathbf{c}_1$ of the boundary $\partial E_R=\{\mathbf{c}: P(\mathbf{c})=R^2\}$, we have $\mathrm{d}P/\mathrm{d}t<0$. Hence the orbit $\mathbf{c}(t,\mathbf{c}_1)$ enters E_R as t increases from zero. It follows that the orbit $\mathbf{c}(t,\mathbf{c}_0)$ cannot leave the bounded set E_R for $t>0$ and therefore is bounded.

Equation (17.15) can also be used to find upper bounds for the amplitudes according to the following procedure: for a given order k determine the largest value of R, say $\bar{R}$, for which the ellipsoid E_R is tangent to, and contains the ellipsoid E in its interior. It is clear from the foregoing that, for t large enough, $\mathbf{c}(t, \mathbf{c}_0) \in E_{\bar{R}}$ for any $\mathbf{c}_0$. Hence the bounds

$$|\Psi_{m,n}| \leq \left(\frac{2}{\rho_{m,n}}\right)^{1/2} \bar{R}, \qquad |\theta_{m,n}| \leq \left(\frac{2}{\lambda}\right)^{1/2} \bar{R},$$
$$-\frac{\bar{R}}{\lambda^{1/2}} - 2\frac{1+\lambda(-)^n}{\lambda}\frac{}{n} \leq \theta_{0,n} \leq \frac{\bar{R}}{\lambda^{1/2}} - 2\frac{1+\lambda(-)^n}{\lambda n}. \tag{17.16}$$

Inequalities (17.16) can be used, for instance, to bound the Nusselt number

$$\text{Nu} = \frac{\text{heat transferred by conduction} + \text{convection}}{\text{heat transferred by conduction}} = 1 - \sum_{n=1}^{2\bar{n}} (-)^n n\theta_{0,n},$$

giving

$$1 + \left(4\frac{1+\lambda}{\lambda} - \bar{R}\right)\bar{n} \leq \text{Nu} \leq 1 + \left(4\frac{1+\lambda}{\lambda} + \bar{R}\right)\bar{n}.$$

These bounds could then be optimized by an appropriate choice of the arbitrary parameter λ.

Finally we show that the origin $\mathbf{c} = 0$ is globally asymptotically stable for all values of ν and κ such that

$$\frac{1}{\kappa\nu} < \min_{(m,n)\in E} \frac{(a^2m^2+n^2)^3}{a^2m^2}. \tag{17.17}$$

Our approximations $G^{(k)}$ satisfy exactly the equation

$$\frac{d}{dt}\left[\frac{1}{2}\sum_E (\rho_{m,n}\Psi_{m,n}^2 + \theta_{m,n}^2) + \sum_{n=1}^{2\bar{n}} \theta_{0,n}^2\right]$$
$$= -\sum_E (\nu\rho_{m,n}^2\Psi_{m,n}^2 - 2am\Psi_{m,n}\theta_{m,n} + \kappa\rho_{m,n}\theta_{m,n}^2) - 2\kappa\sum_{n=1}^{2\bar{n}} n^2\theta_{0,n}^2.$$

It is easy to verify that the quadratic form in the right-hand side is negative definite for all values of the amplitudes when inequality (17.17) is satisfied. The global asymptotic stability of the origin then follows from the application of Lyapunov's theorem [14].

The stability condition (17.17) can be expressed in terms of the Rayleigh number

$$\text{Ra} = \frac{g\varepsilon\Delta TH^3}{\kappa\nu}$$

as

$$\text{Ra} < \text{Ra}_c \equiv \min_{(m,n)\in E} \pi^4 \frac{(a^2m^2+n^2)^3}{a^2m^2},$$

and, since $\mathbf{c}=0$ corresponds to the purely conductive solution $v^{(k)}\equiv 0$, $T^{(k)}=T_0-(z/\pi)$, we see that this solution is globally asymptotically stable for $\mathrm{Ra}<\mathrm{Ra}_c$.

17.5 Concluding remarks

As illustrated in the preceding section, important properties of the Galerkin approximations can be established with great ease when they satisfy the energy balance equations. This is, in our opinion, one more reason for looking for such approximations in fluid flow problems and we believe that they can be constructed in many cases. In fact, the conserving approximations derived in Section 17.3 have already been extended to an MHD version of the Bénard problem where the electrically conducting fluid is uniformly heated by an externally imposed current [15].

Appendix

Application of Galerkin's procedure to Eqs (17.8) (17.9) produces the following set of ordinary differential equations for the amplitudes ($\dot{}=\mathrm{d}/\mathrm{d}t$)

$$\begin{aligned}\rho_{p,q}\dot{\Psi}_{p,q} &= -\frac{a}{4}\sum \rho_{m,n}\Psi_{m,n}\{(pn+qm)\\ &\quad\times[(p-n)\Psi_{|p-m|,q+n}-(\widetilde{q-n})\Psi_{p+m,|q-n|}]\\ &\quad+(pn-qm)[(\widetilde{p-m})(\widetilde{q-n})\Psi_{|p-m|,|q-n|}-\Psi_{p+m,q+n}]\}\\ &\quad+ap\theta_{p,q}-\nu\rho_{p,q}^2\Psi_{p,q},\end{aligned}$$

$$\begin{aligned}\dot{\theta}_{p,q} &= -\frac{a}{4}\sum \Psi_{m,n}\{(pn+qm)[(\widetilde{q-n})\theta_{p+m,|q-n|}-\theta_{|p-m|,q+n}]\\ &\quad+(pn-qm)[\theta_{p+m,q+n}-(\widetilde{q-n})\theta_{|p-m|,|q-n|}]\}\\ &\quad-\frac{a}{2}p\sum \Psi_{p,n}[|q-n|\,\theta_{0,|q-n|}-(q+n)\theta_{0,q+n}]+ap\Psi_{p,q}-\kappa\rho_{p,q}\theta_{p,q},\end{aligned}$$

$$\dot{\theta}_{0,q}=\frac{a}{4}q\sum m\Psi_{m,n}[\theta_{m,q+n}-(\widetilde{q-n})\theta_{m,|q-n|}]-\kappa q^2\theta_{0,q}$$

where $(\tilde{x})=-1$ if $x<0$, $(\tilde{x})=0$ if $x=0$, and $(\tilde{x})=1$ if $x>0$, and $\rho_{p,q}=a^2p^2+q^2$.

Acknowledgements

We acknowledge many useful conversations with O. P. Manley and thank O. E. Lanford whose comments and suggestions helped us derive the energy conserving approximations.

This work was supported by the US Department of Energy under Contract DE-AS03-78ET53072.

References

1. Kantorovich, L. V. and Krylov, V. I., *Approximate Methods of Higher Analysis*: New York, Interscience, 1958.
2. Ladyzhenskaya, O. A., *Annual Review of Fluid Mechanics*, **7,** 249–272, 1975.
3. Orszag, S. A., *Studies in Applied Mathematics*, **1,** 293–327, 1975.
4. Orszag, S. A., *Physics of Fluids*, Supplement II, **12,** II-250–257, 1969.
5. Orszag, S. A., *Journal of Fluid Mechanics*, **49,** 75–112, 1971.
6. Orszag, S. A., *Physical Review Letters*, **26,** 1100–1103, 1971.
7. Treve, Y. M., *Journal of Computational Physics*, **41,** 217–222, 1981.
8. Manley, O. P. and Treve, Y. M., *Physics Letters*, **82A,** 88–90, 1981.
9. Foias, C. and Treve, Y. M., *Physics Letters*, **85A,** 35–37, 1981.
10. Treve, Y. M. and Manley, O. P., in Laval, G. and Grésillon, D. (Eds), *Intrinsic Stochasticity in Plasmas, International Workshop, Cargèse* 1979: Orsay, Editions de Physique, 1979, 393–402.
11. Chandrasekhar, S., *Hydrodynamic and Hydromagnetic Stability*: Oxford, Clarendon Press, 1961.
12. Howard, L. N., *Journal of Fluid Mechanics*, **17,** 405–482, 1963.
13. Veronis, G., *Journal of Fluid Mechanics*, **26,** 49–68, 1966.
14. Hirsch, M. W. and Smale, S., *Differential Equations, Dynamical Systems and Linear Algebra*: New York, Academic Press, 1974.
15. Treve, Y. M. and Manley, O. P., *Physica*, **4D,** 319–342, 1982.

18 Stochasticity of the Dynamical Systems and the Distribution of Eigenvalues

G M Zaslavsky

18.1 Introduction

We shall consider the boundary problem

$$\Delta\varphi+[\lambda+v(x, y, z)]\varphi=0, \qquad \varphi|_{\mathcal{G}}=0 \tag{18.1}$$

where λ is the eigenvalue of the problem and $\mathcal{G}$ is the boundary of some region. In physics, (18.1) is written as a stationary Schrödinger equation for the wavefunction ψ:

$$\Delta\psi+\frac{2m}{\hbar^2}[E-V(x, y, z)]\psi=0, \qquad \psi|_{\mathcal{G}}=0 \tag{18.2}$$

where the eigenvalue E is the energy of a particle. In this chapter we shall always refer to system (18.2). What can be said about the distribution of eigenvalues in the generic case? Consider a sequence of eigenvalues E_k of increasing size:

$$E_1<E_2<\ldots<E_n<\ldots \tag{18.3}$$

Asymptotic formulae are well known (see, for example, [1]) for the number $N(E)$ of eigenvalues with energy $E_k\leqslant E$:

$$N(E)\sim\frac{\Gamma(E)}{(2\pi\hbar)^d}, \qquad E\to\infty \tag{18.4}$$

where $\Gamma(E)$ is the volume of that part of phase space of the physical system with Hamiltonian

$$H=\frac{p^2}{2m}+V(q), \qquad p\equiv(p_1, p_2, \ldots), \quad q\equiv(q_1, q_2, \ldots) \tag{18.5}$$

and boundary $\mathcal{G}$, which is bounded by hypersurface $H=E$, and d is the dimension of phase space. The reason for the universality of expression (18.4) lies in its independence of the details of the potential V in (18.2) as well as of the details of the form of the region boundary $\mathcal{G}$.

A more subtle characteristic of sequence (18.3) is the distribution of spacings between the pairs of the neighbouring eigenvalues. A nonrigorous definition of this concept is the following. Let us form with the help of (18.3) the sequence

$$\Delta E_1, \Delta E_2, \ldots, \Delta E_k, \ldots, \qquad \Delta E_k = E_{k+1} - E_k. \tag{18.6}$$

Let the sequence (18.6) be sufficiently long. We introduce the function $N(\Delta E)$, which is equal to the number of events in sequence (18.6), where ΔE_k is such that the value ΔE is within a certain given accuracy. In a more general case the function $N(\Delta E)$ depends also on the energy value E in the vicinity of which the values ΔE_k are considered. We shall express this circumstance by writing the dependence in the form $N(E; \Delta E)$. Normalizing $N(E; \Delta E)$, we get the probability $P(E; \Delta E)$ of a given spacing ΔE between the neighbouring levels in the vicinity of the energy level E.

A natural question arises: does there exist a universal form of level-spacing distribution $P(E; \Delta E)$? It turns out that in the generic case such a distribution exists and a rough structure of this distribution may be indicated. The structure is directly related to the generation of 'turbulent' motion of a phase liquid in the phase space of system (18.5) corresponding uniquely to the initial boundary problem (18.2). Before demonstrating this we have, however, to consider some auxiliary topics.

18.2 Some definitions and assumptions

We are interested in the asymptotics of the distribution of large eigenvalues. For Eq. (18.2) this asymptotics is usually called quasiclassical. Equations determining asymptotic values of E_k are called quantization rules. The quasiclassical method is based on establishing a correspondence between the classical dynamical system (18.5) and Eq. (18.2) and between the boundary $\mathcal{G}$ and a perfectly reflecting wall. All asymptotic properties of eigenvalues E_k and eigenfunctions ψ_k may be expressed through the dynamical properties of system (18.5). This allows us to classify the character of energy spectra for various boundary problems of type (18.2) depending on the peculiarities of equivalent dynamical systems.

18.3 Historical observations

Let a dynamical system having M degrees of freedom perform a finite motion and be described by a Hamiltonian:

$$H = H(p_1, q_1; p_2, q_2; \ldots; p_M, q_M) \tag{18.7}$$

where q, p are generalized coordinates and momenta respectively. If all the variables are separable, then the quantization rules have in this case the form (Bohr–Sommerfeld rules)

$$\frac{1}{2\pi}\oint p_i\, \mathrm{d}q_i = n_i \hbar, \qquad i = 1, 2, \ldots, M \tag{18.8}$$

where $n_i > 0$ are arbitrary integers (quantum numbers).

Now suppose that the variables are not separable, but the dynamical system (18.7) has exactly M single-valued and independent integrals of motion. Quantization rules for this case proposed by Einstein [2] take the form

$$S_k \equiv \frac{1}{2\pi}\int_{C_k} \sum_{i=1}^{M} p_i\, \mathrm{d}q_i = n_k \hbar, \qquad k = 1, 2, \ldots, M \tag{18.9}$$

where the contours C_k are defined as follows. If a dynamical system has M degrees of freedom and the same number of single-valued and independent integrals of motion, then its motion in the phase space takes place along the surface of an M-dimensional torus. On this torus one can define exactly M closed contours which cannot be reduced to each other by a continuous deformation. These contours are called irreducible and may be chosen as C_k contours in formula (18.9). The value S_k is a Poincaré integral invariant, and the expression $\sum_{i=1}^{M} p_i\, \mathrm{d}q_i$ is a total differential. Hence the values S_k coincide at different reducible contours. Thus quantization rules for the energy E are determined by the M invariant independent quantum numbers $n_k > 0$, $k = 1, 2, \ldots, M$. These rules have been proved in [3, 4].

What happens to quantization rules when the number of single-valued and independent integrals of the corresponding dynamical system is less than M? This question was posed by Einstein [2]. A possible answer appeared much later, when it became clear that the loss of integrals of motion coincides with the appearance of 'chaos' in the dynamical system; the trajectories of 'chaotic' motion are stochastic [5, 6] or, in other words, a mixing motion arises in the phase space.

The next remark is connected with another problem. The solution of this problem was developed independently but is related to the problem raised by Einstein. From experimental data and quantitative considerations it was clear that it is senseless to describe a system of energy levels of highly excited nuclei or molecules by any system of quantum numbers and corresponding quantization rules. Furthermore, the sequence of such levels resembles a random sequence rather than a regular one. This led to the suggestion of a statistical description of the distribution of energy levels of such complicated objects as, for instance, an excited nucleus (the number of excited levels may reach values of the order of 10^5). The property of repulsion of neighbouring levels would be the basic characteristic of such a distribution, i.e.,

$$P(E; \Delta E) \to 0, \qquad \Delta E \to 0. \tag{18.10}$$

Reasons for this proposed independently by Wigner [7–10] and by Landau and Smorodinsky [11] proceeded from the fact that the very complicated character of the interaction within the excited nuclei should result in the destruction of quantum numbers. The formal theory has been developed by Porter and Rozenzweig [12, 13] and in most elegant form by Dyson [14]. It is based on the hypothesis that the energy levels are distributed in the same way as the eigenvalues of a certain ensemble of random matrices (for example, orthogonal ones). As a result it was found that

$$P(E;\Delta E)=A\,|\Delta E|^{\alpha}\exp(-B\,|\Delta E|^{2}) \tag{18.11}$$

where A and B are certain constants dependent only on E, and parameter α takes the values

$$\alpha=1,2,4 \tag{18.12}$$

depending on the choice of the type of ensemble of matrices.

It can be seen from (18.11) that at small ΔE the spacing distribution has the form

$$P(E;\Delta E)\sim|\Delta E|^{\alpha},\qquad \Delta E\to 0,\quad \alpha=1,2,4 \tag{18.13}$$

which approaches zero with characteristic exponent α. This is the law of repulsion between neighbouring levels.

The basic shortcoming of such an approach was the introduction of a hypothesis which did not result from fundamental principles, i.e., it was not a consequence of the initial statement of the problem. Moreover, as we shall see later, this hypothesis was not confirmed.

18.4 Statement of the problem

Let us formulate more accurately the problem under consideration. We have already noted that the dynamical system (18.5) may be compared to Eq. (18.2) and all results in quasiclassical approximation may be expressed through properties of the trajectories of this dynamical system. Now it is well known that nonlinear dynamical systems under various conditions may perform quasiperiodical as well as stochastic motions. In the first case the system has exactly M single-valued and independent integrals of motion, and in the second case the number of such integrals is less than M. In the 'most stochastic' case all integrals of motion are destroyed except the trivial one—that of bulk energy E (for simplicity we shall consider the case when the bulk momentum and its moments are equal to zero). An example of such a system is given by the Hénon–Heiles model [15]:

$$H=\tfrac{1}{2}(\dot{x}^2+\dot{y}^2)+\tfrac{1}{2}(x^2+y^2)+x^2y-\tfrac{1}{3}y^3. \tag{18.14}$$

At $E<\frac{1}{12}$ the motion described by Hamiltonian (18.14) is quasiperiodic and, in addition to the energy integral $H=E$, one more integral of motion exists. At $E>\frac{1}{12}$ an area of stochasticity appears in the motion of the system and at $E>0.1667$ practically all the phase space is covered by the area of stochastic motion. In this case only one integral of motion E remains.

Transition from the regular to the stochastic motion is possible also as a result of variation of some other parameter of the system.

In the regular case we can use the quantization rules of Einstein (18.9). But what quantization rules and what distribution law of eigenvalues would relate to the case when the motion of an equivalent dynamical system is stochastic? The solution of this problem is given below. Before solving it, we need some information about the stochastic properties of nonlinear dynamical systems.

18.5 Billiard-type stochastic systems

The main feature of stochastic systems is the property of mixing of their trajectories in the phase space. Considered locally, this property may be described as follows. Consider a trajectory from a point 1 with coordinate vector (q_1, p_1), and a trajectory from a point 2 with coordinate vector (q_2, p_2). Assume that the initial distance between trajectories,

$$D(0)=(|q_1-q_2|^2+|p_1-p_2|^2)^{1/2},$$

is very small. If the trajectories are mixing,

$$D(t)=\exp(h_0 t)D(0), \quad h_0>0. \tag{18.15}$$

If t is not too large, the property (18.15) is valid nearly everywhere in that part of the phase space where the flow is stochastic. Exponential instability (18.15) results in fast mixing of a phase drop (Fig. 18.1).

The formal property of exponential mixing is expressed roughly as follows: there exists at least one nontrivial dynamic variable for which the

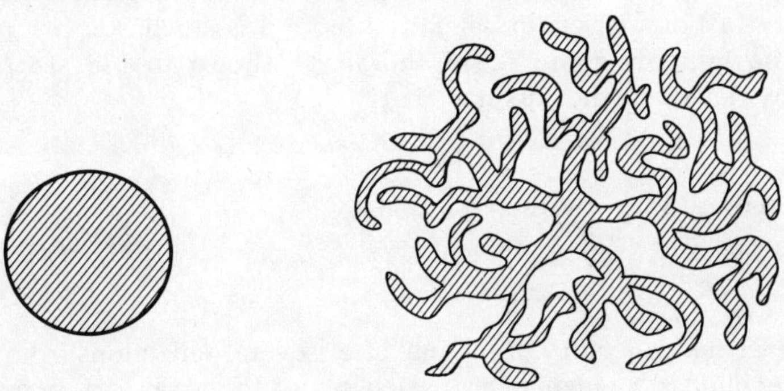

Fig. 18.1

property of exponential correlation decay is valid:

$$\mathrm{R}(t)=\int \mathrm{d}\zeta(\tau)\exp\{\mathrm{i}[\zeta(t+\tau)-\zeta(\tau)]\}=\mathrm{R}_0\exp(-h_1 t). \tag{18.16}$$

It may be shown [5, 6, 16] that the following relation holds:

$$h_0 \sim h_1 \sim \bar{h}_0 \equiv h \tag{18.17}$$

where $\bar{h}_0$ is the mean value of h_0 over the phase space. The quantity h is called Kolmogorov entropy [17, 18] and we shall discuss its properties later.

As an illustration of the stochastic system which is the most convenient for our purpose we shall consider systems of billiard type. Let the billiards game be arranged as in Fig. 18.2(a) with the reflective circle inside. Sinai has shown [19] that ball movement in such a system possesses the property of mixing, which is connected with the existence of a reflective surface of negative curvature. Figures 18.2(b) and (c) show other examples of Sinai's billiards.

According to (18.2) the following boundary value problem corresponds to billiard-type systems:

$$\Delta\psi+\frac{2m}{\hbar^2}E\psi=0, \qquad \psi|_{\mathscr{G}}=0 \tag{18.18}$$

where billiard boundaries give the area $\mathscr{G}$.

Thus, the exact formulation of the problem stated at the end of Section 18.4 is: what is the asymptotic distribution of the eigenfrequencies of a membrane of the form shown in Fig. 18.2(b) and having, for instance, rigidly fixed boundaries?

Consider in more detail another type of billiards game whose boundary consists of some parts of positive curvature and other rectilinear parts. Billiards games of 'stadium' type, consisting of two rectilinear segments and two semicircles, will serve us as an example (Fig. 18.3(a)). Bunimovich [20] proved that the trajectories of the balls in this game are stochastic. He also considered quite a large class of billiards with positive curvature of boundaries, in which the motion is stochastic.

Let the billiards game be of the shape shown in Fig. 18.3(b). For simplicity, let us further assume that

$$a \ll l. \tag{18.19}$$

We set the arc form by the equation

$$\chi=\chi(\xi), \qquad 0\leqslant\xi\leqslant a \tag{18.20}$$

(see Fig. 18.3(b)).

Consider the trajectory of the ball after several reflections from the side of the rectilinear segment. Draw each ray of the trajectory between two reflections on a separate sheet. These sheets may be glued along the

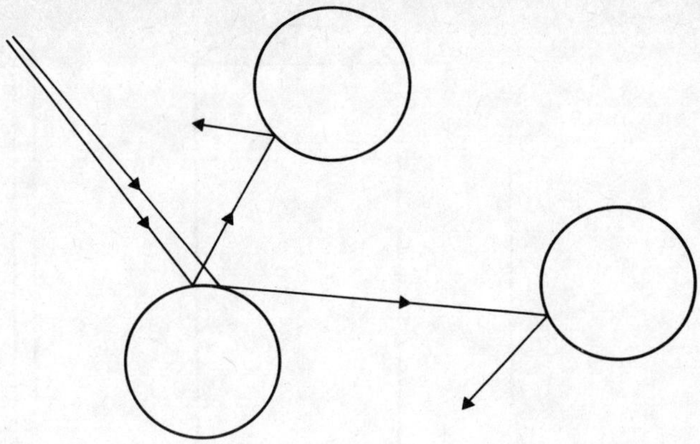

(a)

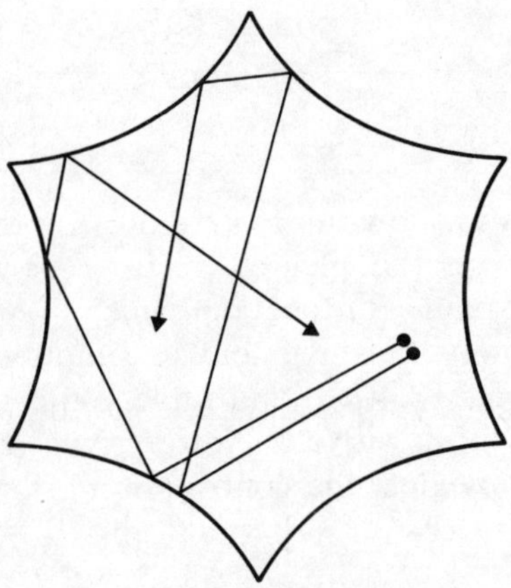

(b)

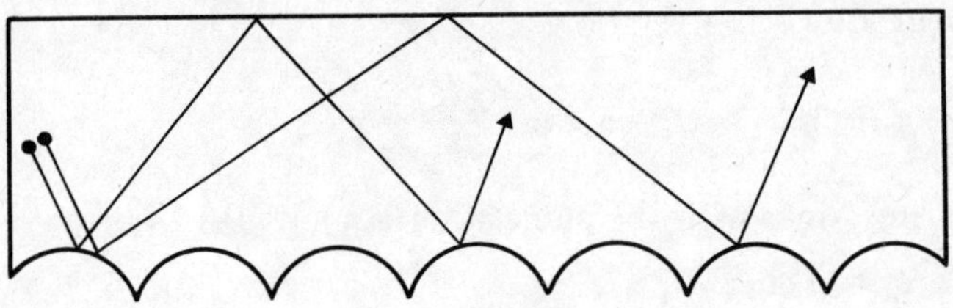

(c)

Fig. 18.2

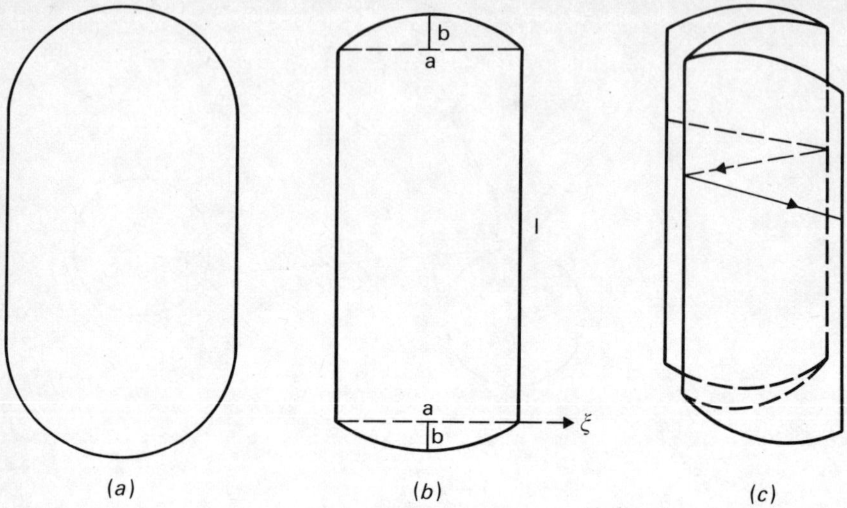

Fig. 18.3

rectilinear segments as shown in Fig. 18.3(*c*), so that the trajectory of the ball is represented on the pleated surface. Now this surface may be stretched into the plane surface on which we may finally study the trajectory. It is, however, convenient to simplify further, using the billiard's symmetry (Fig. 18.4(*a*)), and replace this game by the equivalent one of 'caterpillar with positive curvature' type (Fig. 18.4(*b*)). Inequality (18.19) allows us to neglect the contribution of multiple collisions within one arc.

Assume that

$$\xi = \{x/a\}$$

where the braces {. . .} denote the fractional part of the argument. Then with the help of Figs 18.3(*b*) and 18.4(*b*), and using (18.19), (18.20), it is easy to write down the relations

$$\varphi_{n+1} = \varphi_n - 2\beta_n \qquad \beta = \operatorname{arc\,tg} \frac{d\chi(\xi)}{d\xi} \sim \chi'(\xi). \tag{18.21}$$

We can write down with the same accuracy (18.19):

$$x_{n+1} = x_n + 2l \operatorname{ctg} \varphi_{n+1}.$$

Hence, after division by a, we obtain together with (18.21) the following mapping:

$$\begin{aligned} \varphi_{n+1} &= \varphi_n - 2\chi'(\xi_n) \\ \xi_{n+1} &= \left\{ \xi_n + \frac{2l}{a} \operatorname{ctg} \varphi_{n+1} \right\}. \end{aligned} \tag{18.22}$$

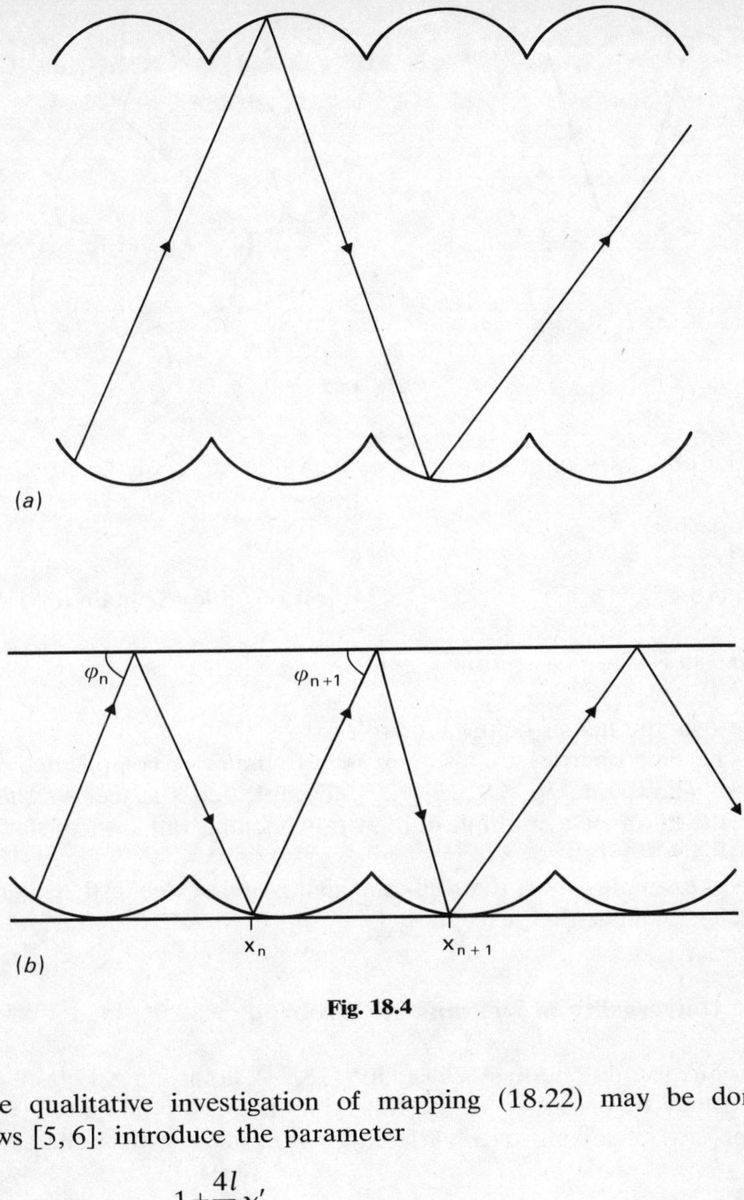

Fig. 18.4

The qualitative investigation of mapping (18.22) may be done as follows [5, 6]: introduce the parameter

$$\tilde{K} = \left|\frac{d\xi_{n+1}}{d\xi_n}\right| = \frac{1+\dfrac{4l}{a}\chi'}{\sin^2\varphi_{n+1}}. \tag{18.23}$$

It is clear from (18.22) and (18.23) that φ is a slow variable ($|\chi'| \ll 1$) and when K is large, ξ may be a fast variable. Stochasticity arises under the condition

$$K = \frac{4l}{a}|\chi'| \gg 1 \tag{18.24}$$

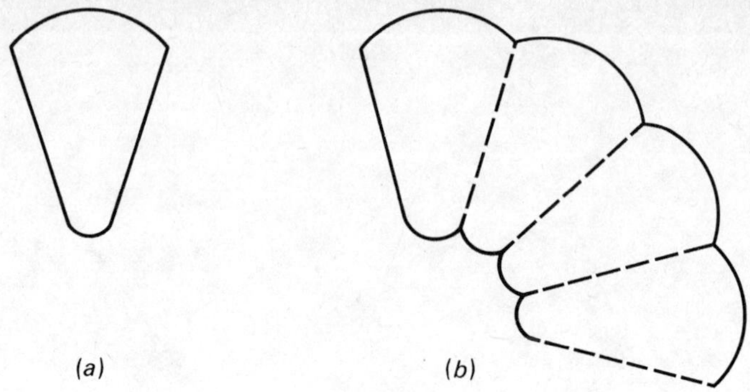

Fig. 18.5

and the boundary of stochasticity is the value of $K \sim 1$, i.e.,

$$\frac{4l}{a}|\chi'| \sim 1. \tag{18.25}$$

From (18.15), (18.17), (18.22)–(18.24) it is possible to obtain the relation

$$h \sim \ln K = \ln \frac{4l}{a}|\chi'| \sim \ln\left(\frac{4lb}{a^2}\right) \tag{18.26}$$

where b is the height of the billiard's arc.

Stochastic properties are also pertinent to the more complicated case of billiards shown in Fig. 18.5(a) [20]. The pleated surface for this case takes the form of a multiple-thread fan opening out into a 'flat' helix (Fig. 18.5(b)).

These examples show that billiards giving rise to stochastic trajectories are generic rather than exceptional.

18.6 Universality of systems with mixing

When the condition of stochasticity (18.24) holds, the motion of the system may be described in the following qualitative manner. At first a fast process of mixing over variable ξ takes place with a characteristic time

$$\tau_c = 1/h. \tag{18.27}$$

There follows a slow diffusion over variable φ, which leads to the establishment of some stationary state. This pattern is typical of many physical situations [5, 6].

The basic process—the process of mixing—is relatively uniform because of the exponential character of local instability (18.15). Departure from uniformity is related to the dependence of h on the region of phase space. This dependence is usually weak and may be neglected in the first approximation.

Thus, physical similarity is found for different systems with the same value of Kolmogorov's entropy h. In other words, systems with the same h have topologically equivalent dynamics (naturally, we are speaking about the process of mixing only). This reveals the property of universality of systems with mixing; in particular, it follows that characteristics of the system that depend only on its mixing properties will be determined by a single parameter and will be the same for systems with the same values of h.

18.7 Basic results

Consider now the boundary value problem of type (18.2). Let the equivalent dynamical system (18.5) have an area of stochasticity in a certain range of values of energy E. Let this area be not too small so that we can consider the distribution of eigenvalues rather far from the stochasticity boundary (this assumption means that mixing of trajectories in phase space takes place relatively uniformly). Then

$$P(E;\Delta E) \sim |\Delta E|^{\alpha}, \qquad \Delta E \to 0, \quad \alpha = \text{const.}/h(E) \tag{18.28}$$

where h is Kolmogorov's entropy. This result was established for a special case in [21] and for a general case in [22] (see also, for a review, [23]). It needs some comments.

The critical index α in (18.28) differs essentially from Dyson's result (18.13) where α takes some definite values ($\alpha = 1, 2, 4$) independent of the dynamic properties of the system. In particular, with increasing h the index

$$\beta = \alpha - 1 = \frac{\text{const.}}{h} - 1$$

changes its sign and the derivative

$$\frac{\mathrm{d}P(E;\Delta E)}{\mathrm{d}\,|\Delta E|} \sim |\Delta E|^{\beta}$$

tends to infinity at $\Delta E \to 0$.

A substantial part of the spacing distribution $P(E;\Delta E)$ is determined only by the entropy h. This demonstrates the universality of the obtained result. We should add that dynamical systems that possess an area of stochastic motion in the phase space are the systems of generic type, and this is why the distribution (18.28) should be characteristic for multidimensional boundary problems with sufficiently complicated functions $V(q)$ or with boundary $\mathscr{G}$ of sufficiently complicated form.

18.8 Proof of the basic result

The first problem to be solved is how to replace the quantization rules (18.9) in the case when some of the integrals of motion are stochastically destroyed. Let us assume that all integrals except the energy integral are destroyed. Then, in a very approximate and rough form, the quantization condition has the form [19, 20]:

$$\exp\left[\frac{i}{\hbar} S_{\Delta\Gamma}(E)\right] = 1. \qquad (18.29)$$

Here $\Delta\Gamma$ is a certain very small area ('area of roughness') in phase space; $S_{\Delta\Gamma}(E)$ is an action over the random trajectory of the particle, the movement of which is described by the equivalent Hamiltonian. The trajectory of the particle should possess the following properties. It should be similar to a closed trajectory, i.e., begin and end in the area $\Delta\Gamma$. Thus, $S_{\Delta\Gamma}(E)$ is an action on an approximate cycle with energy E. Also, the trajectory should be close to a generic one, which means that it is one of those trajectories in the vicinity of which the functional of the distribution over trajectories has a sharp maximum. The existence of the maximum is the result of the large numbers law.

The explanations given above show that the choice of area $\Delta\Gamma$ or trajectory is not essential for the quantization rules (18.29) (by virtue of the uniformity of the process of mixing in phase space).

It follows from (18.29) that if any fixed value E is an eigenvalue, then for the corresponding 'generic' trajectory

$$S_{\Delta\Gamma}(E) = 2\pi\hbar m_0$$

where $m_0 > 0$ is an integer. Now assume $E + \Delta E$ to be the energy that corresponds to the level closest to the level with energy E. Then for a 'generic' trajectory with energy $E + \Delta E$ we have, from (18.29),

$$S_{\Delta\Gamma}(E + \Delta E) = 2\pi\hbar m_1$$

where $m_1 > 0$ is an integer. Hence it follows that

$$|S_{\Delta\Gamma}(E + \Delta E) - S_{\Delta\Gamma}(E)| = 2\pi\hbar\,|m_1 - m_0| \geqslant 2\pi\hbar. \qquad (18.30)$$

This inequality (18.30) governs the computation of $P(E; \Delta E)$.

The form of a trajectory depends on E as a parameter. Let us fix the 'generic' trajectory with energy E. Corresponding to it is a certain typical cycle duration $\tau_{\Delta\Gamma}$. Now let us perturb the parameter E by a small value ΔE and consider asymptotics $\Delta E \to 0$. If $E + \Delta E$ is an eigenvalue, then we have to answer the following question: how does one obtain, at arbitrary small perturbations of the trajectory parameter by ΔE in the finite time interval $\tau_{\Delta\Gamma}$, the finite variation of action $\geqslant 2\pi\hbar$? For the stable case this is impossible. It is clear that the presence of stochastic instability decreases the degree of repulsion between levels.

Note that, if the trajectories do not become statistically independent, action over a trajectory with energy $E + \Delta E$ cannot differ by a finite value $\geqslant 2\pi\hbar$ from the action over the trajectory with energy E. In fact, for times less than the time of stochastic instability development, the variation of action due to perturbations $\Delta E \to 0$ of the energy is small and the inequality (18.30) may not be satisfied at $\Delta E \to 0$. Actions on statistically independent trajectories may, however, differ by an arbitrary value with probability different from zero.

Let us find the condition under which the trajectories with energy E and $E+\Delta E$ become statistically independent. For simplicity, consider a billiard-type problem. From (18.22) it follows that variation of E by ΔE results in a perturbation of the initial condition of the trajectory by the value

$$\delta\xi_0 \sim \Delta E. \tag{18.31}$$

Under the condition of stochasticity (18.24) it follows from (18.31) that

$$\delta\xi_n \sim K^n \delta\xi_0 \sim K^n \, \Delta E. \tag{18.32}$$

Let us find the number N_0 (dimensionless time) of collisions, in which perturbation ΔE will result in $\delta\xi_{N_0} \sim 1$. It is evident that time N_0 is the time after which the perturbation (18.31) makes trajectories with energies E and $E+\Delta E$ statistically independent. From (18.32) we have

$$N_0 \sim \frac{\ln(1/|\Delta E|)}{\ln K}, \qquad \Delta E \to 0. \tag{18.33}$$

On the other hand, as was mentioned above, there exists a typical cycle time, $\tau_{\Delta\Gamma}$, for a 'generic' trajectory with energy E, with a corresponding typical number $\bar{N}$ of collisions. At $\Delta E \to 0$ the value of N_0 grows without changing $\bar{N}$. This is why, at sufficiently small ΔE, we have the inequality

$$N_0 \gg \bar{N}. \tag{18.34}$$

Note that the distribution function $P(N)$ for the number of collisions on different cyclic trajectories would have a sharp maximum at $N = \bar{N}$. The probability $P(N_0)$ is therefore, from (18.34), the probability of a very rare fluctuation and it is possible to write

$$P(N_0) \sim \exp(-\text{const.}\ N_0). \tag{18.35}$$

Thus, the inequality (18.30) may hold only under condition (18.34), and hence

$$P(E;\Delta E) \sim P(N_0) \sim \exp(-\text{const.}\ N_0). \tag{18.36}$$

Substitution of (18.33) into (18.36), taking into account (18.26), gives

$$P(E;\Delta E) \sim |\Delta E|^{\text{const}/\ln K} = |\Delta E|^{\text{const.}/h} \tag{18.37}$$

This is the result (18.28).

We emphasize that the special model of a billiard type has been used only for the sake of simplicity. Formula (18.37) describes the generic situation for all boundary problems of type (18.1), (18.2), in which the equivalent Hamiltonian corresponds to the dynamic systems with mixing.

Acknowledgements

The author expresses his profound gratitude to G. I. Barenblatt, L. A. Bunimovich and Ya. G. Sinai for discussion of the work and helpful comments.

References

1. Courant, R. and Hilbert, D., *Methoden der Mathematischen Physik*, vol. 1, 3rd edn: Berlin, Springer, 1968.
2. Einstein, A., *Deutsche Physikalische Gesellschaft, Verhandlungen*, **19,** 82, 1917.
3. Keller, J. B., *Annals of Physics*, **8,** 180, 1958.
4. Maslov, V. P. and Fedoriuk, M. V., *Quasiclassical Approximations for Quantum Mechanical Equations*: Moscow, Nauka, 1976.
5. Zaslavsky, G. M., *Statistical Irreversibility in Nonlinear Systems*: Moscow, Nauka, 1970 (in Russian).
6. Zaslavsky, G. M. and Chirikov, B. V., *Uspekhi Fizicheskikh Nauk*, **105,** 3, 1971.
7. Wigner, E. P., *Annals of Mathematics*, **53,** 36, 1951.
8. Wigner, E. P., *Annals of Mathematics*, **62,** 548, 1955.
9. Wigner, E. P., *Annals of Mathematics*, **65,** 203, 1957.
10. Wigner, E. P., *Annals of Mathematics*, **67,** 325, 1958.
11. Landau, L. D. and Smorodinsky, Ya. A., *Lectures on Nuclear Theory*: Moscow–Leningrad, GITTL, 1956.
12. Porter, C. E. and Rozenzweig, N., *Suomalaisen Tiedeakatemian Toimituksia*, **AV1,** 44, 1960.
13. Rozenzweig, N. and Porter, C. E., *Physical Review*, **120,** 1698, 1968.
14. Dyson, F. J., *Journal of Mathematical Physics*, **3,** 140/157/166, 1962.
15. Hénon, M. and Heiles, C., *Astronomical Journal*, **69,** 73, 1964.
16. Chirikov, B. V., *Physics Reports*, **52,** 264, 1979.
17. Kolmogorov, A. N., *Doklady Akademy Nauk USSR*, **119,** 861, 1958.
18. Kolmogorov, A. N., *Doklady Akademy Nauk USSR*, **124,** 754, 1959.
19. Sinai, Ya. G., *Izvestiya Akademy Nauk, Seriya Matematika*, **30,** 15, 1966.
20. Bunimovich, L. A., *Communications in Mathematical Physics*, **65,** 295, 1979.
21. Zaslavsky, G. M. and Filonenko, N. N., *Zhurnal Eksperimental'noi i Teoreticheskoi Fiziki*, **65,** 643, 1973.
22. Zaslavsky, G. M., *Zhurnal Eksperimental'noi i Teoreticheskoi Fiziki*, **73,** 2089, 1977.
23. Zaslavsky, G. M., *Uspekhi Fizicheskikh Nauk*, **129,** 211, 1979.